# Digital Sport for Performance Enhancement and Competitive Evolution:
## Intelligent Gaming Technologies

Nigel K. Ll. Pope
*Griffith University, Australia*

Kerri-Ann L. Kuhn
*Queensland University of Technology, Australia*

John J. H. Forster
*Griffith University, Australia*

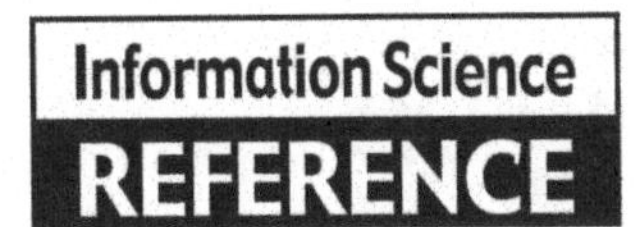


**INFORMATION SCIENCE REFERENCE**
Hershey · New York

Director of Editorial Content: Kristin Klinger
Senior Managing Editor: Jamie Snavely
Managing Editor: Jeff Ash
Assistant Managing Editor: Carole Coulson
Typesetter: Chris Hrobak
Cover Design: Lisa Tosheff
Printed at: Yurchak Printing Inc.

Published in the United States of America by
Information Science Reference (an imprint of IGI Global)
701 E. Chocolate Avenue,
Hershey PA 17033
Tel: 717-533-8845
Fax: 717-533-8661
E-mail: cust@igi-global.com
Web site: http://www.igi-global.com/reference

and in the United Kingdom by
Information Science Reference (an imprint of IGI Global)
3 Henrietta Street
Covent Garden
London WC2E 8LU
Tel: 44 20 7240 0856
Fax: 44 20 7379 0609
Web site: http://www.eurospanbookstore.com



Library of Congress Cataloging-in-Publication Data

Digital sport for performance enhancement and competitive evolution :
intelligent gaming technologies / Nigel K. Ll Pope, Kerri-Ann L. Kuhn and John J. Forster.
p. cm.
Includes bibliographical references and index.
Summary: "This book provides an overview of the increasing level of digitization in sport including areas of gaming and athlete training"--Provided by publisher. ISBN 978-1-60566-406-4 (hardcover) -- ISBN 978-1-60566-407-1 (ebook) 1. Sports--Technological innovations.
2. Performance technology. 3. Sports sciences. 4. Video games. 5. Computer games.. I. Pope, Nigel. II. Kuhn,
Kerri-Ann L. III. Forster, John, 1946- GV745.D55 2009
688.7'6--dc22

2008052440

British Cataloguing in Publication Data
A Cataloguing in Publication record for this book is available from the British Library.

# Editorial Advisory Board

# Table of Contents

# Detailed Table of Contents



This chapter presents an overview of the digital revolution and the role that digital technologies have played in home entertainment generally, and sport specifically. Forster examines how digital technology is transforming the mass consumption, distribution, production and organization of sport. He argues that the impact of digitization has been evolutionary, rather than revolutionary.

**Section I**
**Training and Participation Applications**

*Digital technologies are being used in myriad ways for sports training and particpation applications. This section examines a number of examples. From the use of robotics in sport, to inertial sensors for performance enhancement and electronic gaming to facilitate participation, the chapters in this section illustrate that digital technology is the impetus to improve the experience of elite athletes, as well as casual sports participants.*

*The section begins with a general-purpose taxonomy of computer-augmented sports systems in Chapter II, followed by a discussion of the use of robotics to improve results and diagnostics in sports in Chapter III. Player tracking systems are the focus of Chapter IV, while a video-based marker-less motion capture method for capturing human motion in sport is introduced in Chapter V. The technology for performance enhancement of a tennis player is reviewed in Chapter VI, with the discussion of inertial sensors for athlete assessment extended in Chapter VII, where golf, rowing, cricket, and ski jumping are discussed. The focus shifts from training to sports participation, beginning with Chapter VIII, which presents Computer Supported Collaborative Sports. Two prototypes are presented, which demonstrate how computer games and computer augmented sports can be integrated to affect the social experience of players. A further prototype for social jogging is discussed in Chapter IX. Chapter X contributes to the discussion concerning gaming technologies, with a look at how video games are being used by elite athletes. Chapter XI provides an overview of fantasy sports and their consumption by sports fans. Finally, the section concludes with Chapter XII: computational and robotic pool.*



Sport and digitization are becoming more multi-disciplinary. Chapter II presents a general-purpose taxonomy of computer-augmented sports systems, and in doing so, successfully unravels the increasingly complex domain of technology-augmented sports. The authors present several examples to illustrate their taxonomy, including a system to aid judges in scoring taekwondo matches; a computer vision system that alerts lifeguards to drowning swimmers; a system that uses visual cues to determine an athlete's movements; and the Hawkeye® system, which tracks ball movement. Their taxonomy allows for any computer-augmented sports system to be classified, in terms of its key properties and characteristics.



This chapter relates biomechanics to robotics and presents mathematical models of human dynamics, which can be applied in sports. These models have been derived to cover the kinematics and dynamics of virtually any motion of a human or a humanoid robot, thereby allowing for the calculation of characteristics that could not previously be measured. The authors discuss the use of robotics to improve results and diagnostics in sports, including enhanced human safety, performance and development.



Chapter IV presents an overview of player tracking systems. These systems facilitate the monitoring of human player activity within competition. Compared to manual systems, player tracking technologies benefit players and coaches, offering the key advantages of reduced time and expense, as well as improved sporting performance. Brendan Burkett also highlights that these systems satisfy the demand from media, referees and spectators to know what is happening in a game.

The authors of this chapter present a video-based marker-less motion capture method for capturing human motion in sport. While many technologies exist that capture the human motion, the motion capture technique has to yield accurate and reliable quantitative information when it comes to human biomechanical applications, particularly in real sports environments. The authors review the challenges associated with current technologies and methods, and explain how these are addressed with their system.



The authors look at how technology can be used to improve the performance of a tennis player. They discuss the biomechanics of the various strokes, with particular attention paid to the serve. The chapter introduces the major tools to monitor and analyze the tennis swing, including high speed digital cameras, marker-based optical systems, and inertial sensors.



Traditionally, clinical testing for performance analysis and enhancement of athletes has been conducted in laboratory settings, but with advancements in microelectronics and other micro technologies, testing and monitoring in natural environments is now possible. Chapter VII discusses the use and application of inertial sensors for athlete assessment in the training and competition environments. The authors discuss the emergence of these sensors as a viable tool across a broad range of sporting activities at the elite level. Examples include golf, rowing, cricket and ski jumping.

Computing technology is increasingly being used to augment existing sports experiences. Chapter VIII looks at how computer games and computer augmented sports can be integrated to offer new, shared social experiences. The authors present Computer Supported Collaborative Sports (CSCS), which offer users the opportunity to actually participate in sports, even if they are physically apart. Two prototypes, which illustrate the sportive and collaborative aspects of CSCS, are presented.



Chapter IX also discusses how digital gaming can be merged with physical sports activities, though the focus of this chapter is on social jogging. The author presents a design prototype called "Jogging over a Distance", which offers social joggers the opportunity to run together, although being in two different geographic locations. The system offers users both health and social benefits. Mueller argues that there is potential for computer gaming technology to offer new experiences and support a wide range of sports participants with different objectives.



This chapter presents an overview of commercial sport video games and their use by elite athletes for training purposes. Silberman presents personal insights, results from observational research and anecdotes to support her argument that video games offer educational benefits. By playing as an avatar that simulates the elite athlete, members of his/her team or an opponent, Silberman argues that physical performance can be enhanced from virtual play.



Chapter XI introduces the reader to fantasy sports. Donald P. Roy and Benjamin D. Goss present a conceptual framework of the consumption of fantasy sports by postmodern sports fans. They propose that consumption is impacted by the interplay of psychological characteristics internal to consumers, social interactions, and external influences controlled by fantasy sports marketers. They conclude that fantasy sports offer fans increased opportunities for participation and empowerment.



The authors present computational and robotic pool. They describe the challenges of computationally simulating the game of pool and creating a robot capable of playing cue-sports. The chapter reviews recent work on artificial intelligence methods for strategic play, as well as work in robotic pool. The ultimate challenge is to create a robotic pool system that can compete directly against proficient humans, which the authors argue, will happen one day in the near future.

**Section II**
**Business Applications**

*Digital technology is being used in various ways to support business in sport. Indeed, commercial sport is a business, which is growing in strength. Digital technology has done much to increase fan bases, attendance figures and revenue for professional sport. At the same time, it supports these businesses in their activities, enabling them to offer more to their customers. One could consider the many ways sport is now broadcast, the rise in consumer-generated media such as Web logs, social network sites, fantasy sports leagues, electronic games and so on, all of which allow fans to experience sport in different ways, ultimately as they choose. With technology, power has shifted back to consumers. This section examines digital and other forms of mass sports consumption. The emphasis is not on the use of technology for the athlete or sports participant, as in Section I. Rather, many of the chapters in this section consider the consumer perspective and how digital technologies can be used to enhance their experience. Many also consider how digital technologies can be used for enhanced business success and brand performance.*

*The section begins with Chapter XIII, which looks at how Internet applications are being adopted within sporting associations, while Chapter XIV discusses how the Internet can be used to engage with distant sports fans. Chapter XV highlights that distance is also a barrier when it comes to people visiting the traditional physical Olympic museum: something that the authors address with the use of virtual reality in their presentation of the Virtual Digital Olympic Museum. Chapter XVI serves as an introduction to electronic games, while Chapter XVII provides an overview of sport video game sponsorships and in-game advertising. These chapters are followed by two studies, which explore the effectiveness of promotional messages in games. Chapter XVIII examines attitudes towards in game advertising and effects on brand awareness. Similarly, Chapter XIX focuses on brand awareness outcomes, though the focus is on the effect of arousal on adolescents' short-term memory of placements in advergames specifically. The section concludes with Chapter XX, which presents a technique for analyzing consumer schemas associated with athletes and endorsers. This is a technique that can be used by marketers, sporting management and sports organizations to monitor and evaluate marketing communications, sponsorship efforts and uncontrolled media sources. It therefore represents a tool that could potentially provide feedback on all the business activities discussed in this section.*



With a focus on organized sport at a local level, Chapter XIII looks at how Internet applications are being adopted within associations and clubs, what they are being used for, and what effects they have on the associations and their volunteers. Scott Bingley and Stephen Burgess describe the development of a framework that traces the adoption of an Internet application. They use the adoption of an online statistics program in a local sporting association as an example, to demonstrate how the framework may be applied in a practical situation. The authors suggest that the framework may identify lessons that can help to inform improved decision making in local sporting bodies.



With globalization and advances in communications technology, the potential marketplace for professional sports teams has grown to include foreign sports consumers, or what Anthony K. Kerr calls, "satellite supporters". Chapter XIV presents a series of studies designed to understand the team identification of satellite supporters, which is important for developing marketing strategy. It also highlights how online research methods can contribute to sports fan research, especially when seeking to engage with distant participants. Kerr highlights how mixed methods can be successfully employed to explore sports fandom.



Chapter XV presents the Virtual Digital Olympic Museum (VDOM), which uses virtual reality to extend the main functionalities of the traditional, physical Olympic museum. As such, the authors are able to capture the dynamic elements necessary to present the Olympic Games and create an improved experience for users. They propose solutions for digital-museum oriented data storage and retrieval, modeling and rendering of the digital museum, the virtual demonstration of sports and virtual humans, as well as virtual reality based sports simulation.

This chapter presents an overview of electronic games in all their various forms, with discussion of the industry structure, game characteristics and player motivations. It also explores the relationships games share with sport, in terms of their ability to replicate sports, facilitate sports participation and be played as a sport. Kuhn presents ideas for future research in this important area.



Chapter XVII reviews an increasingly popular form of promotion being used by marketers: sport video game sponsorships and in-game advertising. Beth A. Cianfrone and James J. Zhang discuss the growing trend and uniqueness of sport video games, along with reasons behind the growth of sponsorships and in-game advertising. They identify the advantages associated with this form of promotion and measurement issues associated with studying effectiveness.



The authors of this chapter present a research study, which examined attitudes, recall and recognition of in-game advertising in a sport video game. The authors report that experienced players recalled and recognized significantly more in-game advertisements compared to novice participants, but their attitude toward in-game advertising was unexpectedly low. The authors suggest that marketers must consider many different factors in assessing the individual value obtained from a specific in-game advertisement.



This chapter reports a study that examined factors affecting Mexican adolescents' memory of brand placements contained in advergames. The results indicate that high arousal advergames lead to better recognition and more accurate short term memory than moderate arousal advergames. Hernandez and Chapa recommend that advertisers create stimulating advergames, relying on fast pace or competitive game genres, in order to increase effectiveness.

# Foreword

The development of sport–since the beginning of the 20th century in particular–has been integrally linked to the development of technology. The first radio broadcast of a baseball game from Clarke's Field (Forster & Pope, 2004) created a bond between broadcast media and sport. Television owes a significant part of its commercial viability to its association with sport, something that is reflected in the dominance of pay per view in current rights negotiations.

With that early association between sport and broadcast media, we also witness the growth of sport professionalism and the concomitant development of professional sport as a business. It was not too long before sport itself initiated the next phase of its association with technology: the enhancement of athletic performance. This has, of course, had positive and negative impacts on sport as a spectacle and athletic performance. Training is now safer, more targeted and better controlled. Part of that control, unfortunately, relates to monitoring and detecting the abuse of technology in order to prevent cheating in sport.

As digital technology has developed, so has the quality and impact of sports engineering been enhanced. One of the more intriguing aspects of this is the development of simulations. These can be used to both improve the performance of athletes and to allow sport fans to enhance their own enjoyment of the game. This is technology making a holistic contribution to the sport experience.

Sport still remains a business as well as an activity and it is important that the business implications of using technology in sport are clearly understood. For this reason, it is particularly refreshing to see that this book, *Digital Sport for Performance Enhancement and Competitive Evolution: Intelligent Gaming Technologies*, is edited by two academics from the Griffith Business School.

This cannot have been an easy task, as sections of the book deal with different aspects of the sport and technology interface. The descriptions and discussion of different ways to exploit digital technology in performance development are particularly interesting and present information in an accessible and informative manner, even for the non-scientific reader. The section on simulations and intelligent gaming is equally insightful and gives detailed explanations of the technologies involved. Lastly, the business applications section discusses several studies into how to manage these technologies and gain greater understanding of the engagement of sport consumers.

I commend this book to both the general and technical sportsperson and enthusiast. It is the first of its type to bring together training, participation and business perspectives in an examination of sport and technology. My congratulations go to the editors and the chapter authors.

*Graham Cuskelly*
*Editor*
*Sport Management Review*

***Graham Cuskelly*** *is the editor of Sport Management Review and has published extensively on sport management in journals such as European Sport Management Quarterly and Event Management: An International Journal. In addition, he has authored two highly acclaimed books on sport governance and volunteerism, as well as consulted to national governments on sport policy. He is currently dean of research at Griffith Business School in Australia.*

## REFERENCE

Forster, J., & Pope, N.K. Ll. (2004). *The Political Economy of Global Sports Organisations*. London: Routledge.

# Preface

It was towards the end of the 19th century that the major forms of professional sport, including leagues and tournaments, were formed. As a consequence, so too were the institutional and administrative frameworks that made them work. It is extraordinarily remarkable that this happened in different types of sports and in different countries and locations, all at the same time. We do not know precisely why. The common denominator of location is that the societies appear to have been highly urbanized, and the richest economies in the world at the time. Most notably, two nations were involved: the United States and the United Kingdom. Key sports included soccer, cricket, tennis, and rugby in the United Kingdom; baseball, basketball, and American football in the United States; and golf and horse racing in both. Many organizations in these countries have become the major financial beneficiaries of the electronic recording and distribution of sports. That it was in these two nations that the computer age was initiated, the English language has been cemented as the *lingua franca* of digital and sport worlds, ensuring the ease of their interaction: a relationship examined by Forster.

As an introduction to this volume, Forster presents *Digital Technologies and the Intensification of Economic and Organizational Mechanisms in Commercial Sport.* He presents a model of the digitisation of sport, paying particular attention to the role of the price mechanism as a guiding force in the adoption and impacts of digital technologies. He also considers the nature of home entertainment technologies. Here there has been a revolution. It was a revolution that took place over the long term—a century or perhaps a little less—but it is one that produced enormous changes in sport. That revolution transformed the household from a solitary producer and consumer of its own entertainment, into a networked purchasing and consumption center of entertainment. It is into this well established framework that the "digital revolution" and professional sports, as well as other forms of entertainment, are channeled. Forster argues that digital technologies did not create this revolution, but were able to take advantage of it and intensify it. The impacts of these events and trends upon sport are immeasurable. They have also provided an impetus to the digital revolution.

These technologies have cemented the arrival of the superstar, a phenomenon little seen before the advent of music recording technologies. The recording and replaying of sports events, leagues, teams and tournaments has given a select few superstar status. The economic nature of superstardom is described by Forster, especially the theories that attempt to explain why a few superstars garner the rewards while others of only slightly less talent, or even equal talent, receive very little. The digital phenomenon has apparently cemented these relationships, even though recording and replaying of sport has become dramatically cheaper, and of much more uniform quality.

The phenomenon of superstars means, in part, that much more attention is paid to these individuals, teams and tournaments than others. John J.H. Forster therefore examines the nature of attention. The argument is that, where information (digital content) has become cheap and even costless through ease of recording and reproduction, then it is attention by consumers of information that has become scarce; the attention of consumers therefore has to be bought. Forster reviews tournaments and argues against the major theory of competitors' payments. The standard theory suggests that the payment structure of sports tournaments, with rewards very heavily biased towards the winners, is an incentive structure. That incentive structure is to induce players to perform at their very best. It therefore explains players' rewards on the production side. Forster presents a diametrically opposed argument, which operates on the demand side. It is argued that players are paid according to the attention they garner during the tournament. He uses this analysis to explain the same bias of winners taking a very high proportion of the pay to players. A major part of the discussion is the presentation of a new model of sports tournament incentives to athletes.

The section on training and participation begins with a chapter by Sean Reilly, Peter Barron, Vinny Cahill, Kieran Moran, and Mads Haahr, entitled *A General-Purpose Taxonomy of Computer-Augmented Sports Systems.* This chapter accepts that sport and digitization are becoming more multi-disciplinary. By presenting a common vocabularym as well as a classification of systems by both form and function, they allow for the easier interchange of knowledge between different practitioners and researchers in the area. In Chapter III, Velijko Potkonjak, Miomir Vukobratović, Kalman Babković, and Branislav Borovac extend this concept by presenting the math behind robotic design (*Dynamics and Simulation of General Human and Humanoid Motion in Sports*) along with illustrations to further explain their point to the non-scientific user. Again, this contributes to our multi-disciplinary approach. Importantly, it also discusses the use of robotics to improve human safety and development.

Following this discussion of safety and athletic development, Brendan Burkett in Chapter IV, *Technologies for Monitoring Human Player Activity Within a Competition*, presents an overview of the available technologies for athletic development in order to reduce time and expense, while at the same time improving sporting performance. This is followed in Chapter V, *Video-Based Motion Capture for Measuring Human Movement*, in which Chee Kwang Quah, Michael Koh, Alex Ong, Hock Soon Seah, and Andre Gagalowicz conjoin the biomechanical aspects discussed earlier by Potkonjak et al., with the concept of video monitoring through digital technology. Most impressive about this ongoing research is that it offers a marker-less system of observing human biomechanical movement.

The use of technology training applications for the purpose of enhancing the performance of sports participants is further explored in Chapter VI. One specific example is tennis, where technology is being used to monitor the player's action in order to achieve enhanced stroke play and game improvement. Amin Ahmadi, David D. Rowlands, and Daniel A. James present this in *Technology to Monitor and Enhance the Performance of a Tennis Player*. Their chapter explains some of the various strokes in tennis and the importance of each during the tennis match, though they pay particular attention to the serve, examining the biomechanics of this stroke. Further, the authors detail the major technologies to monitor and analyze the tennis swing, including high speed digital cameras, marker-based optical systems, and inertial sensors. Each of these technologies offers advantages and disadvantages in monitoring the tennis player. Amin Ahmadi, David R. Rowlands, and Daniel A. James present examples of how the technologies can be applied.

Extending the previous chapter, Daniel A. James, Andrew Busch, and Yuji Ohgi present a detailed examination of inertial sensors in Chapter VII, *Quantitative Assessment of Physical Activity Using Inertial Sensors*. They review this technology, its function and implementation, with particular emphasis on the use of accelerometers for the biomechanical quantification of sporting activity. Traditionally, athletic and clinical testing for performance analysis and enhancement has been performed in a laboratory, using simulators and other instrumentation that cannot be easily used in the training and competition environments. Recognizing this, Daniel A. James, Andrew Busch, and Yuji Ohgi argue that inertial sensors are ideal for the portable environment to perform athlete assessment. The authors discuss a number of emerging sporting applications, including the use of inertial sensors in golf, rowing, cricket, and ski jumping.

With advances in computing technology, the market is seeing new commercial products that merge digital gaming with physical sports activities, in order to offer users both health and social benefits. In Chapter VIII, *Computer Supported Collaborative Sports: An Emerging Paradigm,* Volker Wulf, Florian 'Floyd' Mueller, Eckehard F. Moritz, Gunnar Stevens, and Martin R. Gibbs look at the use of technology to affect the social aspects of sport. In particular, they examine the way that computer gaming technology can be applied to increasing communal exercise participation. In *Digital Sport: Merging Gaming with Sports to Enhance Physical Activities such as Jogging* (Chapter IX), Florian 'Floyd' Mueller explores such technologies. He demonstrates the potential for this relationship, with the presentation of a design prototype called "Jogging over a Distance". This system offers social joggers the opportunity to run together, even though they may be in two different locations. Further, it not only supports conversation, but also uses audio to communicate pace, which can serve to motivate joggers and provide a truly shared sportive experience. Unlike many other chapters in this book, Mueller's focus is on casual users seeking to enhance their jogging experience.

Computer gaming technologies may also be used by elite players for training purposes. In Chapter X, *Double Play: How Video Games Mediate Physical Performance for Elite Athletes*, Lauren Silberman describes how video game simulations offer learning opportunities that may benefit elite players. She presents personal insights, results from observational research, and anecdotes with soccer players and other athletes to describe how real players are using game technologies, not just for entertainment purposes, but also education.

Donald P. Roy and Benkamin D. Goss present Chapter XI, *A League of Our Own: Empowerment of Sport Consumers Through Fantasy Sports Participation*. They give an overview of fantasy sports, but also a conceptual framework of the influences on consumption by postmodern sports fans. This is of interest, because there has been little inquiry into the forces that influence one to become a fantasy sports player. Roy and Goss propose that fantasy sports consumption is impacted by the interplay of psychological characteristics internal to consumers, social interactions, and external influences controlled by fantasy sports marketers. This framework serves as a useful tool for marketers to utilize in their examinations of fantasy sports participants.

To end this section, Jean-Pierre Dussault, Michael Greenspan, Jean-François Landry, Will Leckie, Marc Godard, and Joseph Lam offer Chapter XII, *Computational and Robotic Pool.* In this, they examine the game of 8-ball and discuss the problems of simulating the game to an extent that one could create a robot that would be able to compete with a human on a real table. Such a robot would be able to assist with the training of a human player, as well as be a source of great pleasure. The editors wonder if they would cheat!

Digital technology is also being used to support business in sport, so our second section deals specifically with this aspect of sport digitization. For example, Internet technologies are being used in various ways to support activities in business, though little has been written about the adoption and use of such technologies by sporting clubs. In Chapter XIII, *A Framework for the Adoption of the Internet in Local Sporting Bodies: A Local Sporting Association Example*, Scott Bingley and Stephen Burgess explore Internet adoption in local sporting bodies, and the impact on clubs and volunteers. They present a framework that traces the adoption of an Internet application from initial knowledge of the application, through the decision to adopt, and eventual confirmation of the usefulness of the application by continuance/discontinuance of its use. They apply the framework to the adoption of an Internet application by a local cricket association in Australia, finding that the framework provides a useful means to classify the events that lead to the eventual adoption/non-adoption of a particular innovation.

Not only is the Internet being used for operational activities in sporting clubs, but in the domain of sport, it is increasingly being used to engage with distant fans. Globalization and advances in communications technology have expanded the potential marketplace for professional teams, creating fan bases of millions of people who can indirectly consume sport via television and/or the Internet. Kerr refers to these foreign consumers as "satellite supporters". In Chapter XIV, *Online Questionnaires and Interviews as a Successful Tool to Explore Foreign Sports Fandom*, Anthony K. Kerr explores the team identification of satellite supporters, and demonstrates how online research methods can contribute to sports fan research, particularly with regards to distant participants. He adopts a case study approach using mixed methods (including questionnaires and semi-structured interviews in an online environment) to study supporters of the Australian Football League from the Australian Football Association of North America; supporters of Ajax F.C. from Ajax USA; and supporters of Liverpool F.C. from Liverpool F.C.'s Association of International Branches. Kerr concludes that mixed methods can be successfully employed online to explore fandom.

Technology is also being used in the arts. Physical museums possess several limitations including the constraints of time, space and interaction channels. To address these constraints, digital museums are being developed with the aid of computer technologies and other advanced information technologies. However, even here there are limitations, including the over-reliance on static demonstrations, poor virtual simulation, low rendering quality, and an overall lack of interaction with users. These weaknesses are particularly problematic when it comes to presenting a digital museum for something as complex as the Olympic Games. Indeed, all of the aforementioned limitations are evident in the current digital Olympic museum. Using virtual reality, Gaoqi He, Zhigeng Pan, Weimin Pan, and Jianfeng Liu take this a step further to create a *Virtual Digital Olympic Museum* (VDOM) (Chapter XV). In this chapter, the authors discuss the design of the VDOM, its relationship to virtual reality, the use of digital networks and related technologies, as well as the improved experience created for users. What is special about their application is that the VDOM captures the dynamic elements necessary to present the Olympics. It extends the main functionalities of the traditional physical Olympic museum by combining sports, humans, entertainment and education. He, Pan, Pan, and Liu propose solutions for digital-museum oriented data storage and retrieval, modeling and rendering of the digital museum, the virtual demonstration of sports and virtual humans, as well as virtual reality based sports simulation.

There have been a number of changes in the consumption of home entertainment that have impacted digital and other forms of mass sports consumption. One such home entertainment technology that has impacted sport is electronic games. Kerri-Ann L. Kuhn presents a review of this industry in Chapter XVI, *The Market Structure and Characteristics of Electronic Games*. She presents a detailed analysis and classification of all the various forms of games; explores key motivators for game play; examines the game medium and its characteristics; and reviews academic research concerning the effects of play. Of particular significance is Kuhn's discussion of the emerging relationship games share with sport. She suggests that games can replicate sports, facilitate sports participation and be played as a sport. These are complex relationships that have not yet been comprehensively studied.

With the growth of electronic games, the business relationship between sports and advertising has taken on new forms. Beth A. Cianfrone and James J. Zhang present a review of a rising form of promotion in Chapter XVII, *Sport Video Game Sponsorships and In-Game Advertising*. They introduce various types of sponsorships and in-game advertising, and review current literature concerning marketing effectiveness. In examining the stakeholder relationships, they identify that corporations are advertising within sport video games in order to reach sport fans, while game publishers benefit from the enhanced realism the advertisements provide. However, while research has begun in this area and found that sport video game advertising can influence gamers' brand awareness, little is known about the impact on brand attitudes or behavior. Beth A. Cianfrone and James J. Zhang forecast future trends and call for timely examinations in line with technology advancements, in order for this segment of the sport industry to remain successful.

Answering Beth A. Cianfrone and James J. Zhang's call for research, Mark Lee, Rajendra Mulye, and Constantino Stavros explore *In-Game Advertising: Effectiveness and Consumer Attitudes* in Chapter XVIII. Contrary to other studies, these researchers found an unusually low favorability toward in-game advertising, particularly among experienced gamers. They also found that memorability was affected by the game environment, something well within the control of designers.

Monica D. Hernandez and Sindy Chapa also examine in-game advertising (or brand placement), but in online advergames. Specifically, in Chapter XIX, *The Effect of Arousal on Adolescent's Short-Term Memory of Brand Placements in Sports Advergames,* they find more accurate short-term memory (brand recognition) when subjects are exposed to a high arousal advergame than to a moderate arousal advergame. These results resolve contradictory findings in the literature addressing the effect of arousal on memory of brand placements in online settings. The authors recommend that advertisers wishing to target adolescents could strengthen the recognition of their products and brands by relying on fast paced or competitive game genres, which are likely to stimulate stronger emotional responses.

In our twentieth and final chapter (*Schemas of Disrepute: Digital Damage to the Code*), Ellen L. Bloxsome and Nigel J. Ll. Pope offer a worked demonstration of the use of social network analysis to examine consumer schemas, as they relate to perceptions of sponsors and athletes. They show that it is possible to quantify the attitudes of sport fans through analysis of social networking sites and identify any possible damage that an athlete's behavior may do to a sponsor's image.

# Acknowledgment

It is typical of a work of this nature, involving as it does the co-ordination of a large number of people across several continents, that many unseen hands have assisted us in the preparation of this volume. At the top of the list must go our Development Editor with IGI Global, Julia Mosemann, who has been long-suffering, patient and gentle throughout the process. We also sincerely appreciate the hard work and support of our always friendly and helpful Editorial Assistant, Beth Ardner. Unknown by us–and this is written in advance of their work–a thank you to the copy editors whose labour is so often forgotten.

Our thanks go also to the authors of these chapters. Their work is only rewarded by appearing in print and the task, as we know–is arduous. The main bearers of the burden of our work as editors have been our families. They have had to cope with early starts, late nights and absent weekends. It is appreciated.

The last comment is that the good qualities apparent in this volume are the work of those people we identified above. Any shortcomings belong to us.

*Nigel K. Ll. Pope*
*Griffith University, Australia*

*Kerri-Ann L. Kuhn*
*Queensland University of Technology, Australia*

*John J. H. Forster*
*Griffith University, Australia*

# Chapter I
# Digital Technologies and the Intensification of Economic and Organisational Mechanisms in Commercial Sport

**John J. H. Forster**
*Griffith University, Australia*

## ABSTRACT

*One of the major forces shaping modern sport is the application of digital technology. This is transforming the mass consumption, distribution, production, and organisation of sport. A set of frameworks, theories and models for understanding these forces and transformations is presented. Nonetheless, digitisation did not create modern sport. Instead it follows a series of forces and adoptions that have already occurred. Consequently, its impacts can be regarded as evolutionary rather than revolutionary. Three associated aspects of modern sport are examined in this chapter. These are: the phenomenon of superstars; the consumption of sport entertainment in the household; and the competition for the attention of consumers. The events that preceded digitisation are described and analysed, especially in the area of home entertainment, but also in the structures of sports competition, such as leagues and tournaments.*

## INTRODUCTION AND SCOPE OF THE CHAPTER

This chapter considers the role of computer technologies in the distribution and consumption of commercial sport. It considers the economic forces that have promoted and guided these changes. Equally importantly from both methodological and applied viewpoints, it considers the economic forces that these changes have, in turn, unleashed. The more indirect relationships of the technology with some major organisational forms of competition are also placed in context. A primary interest of this chapter lies in determining whether or not the impact of digitisation has been revolutionary or evolutionary.

The concern of many of the chapters in this volume is with the impact of digitisation on the enhanced production of sports, but how does sport come to be digitised and how does that digitisation operate? The schematic shown in Figure 1 is adopted in the present analysis. It shows the central concerns and something of the methodology, which is not biased towards economic and technological mechanisms, but considers social and cultural issues.

In the first instance, the creation of computer technologies and electronic networks is not dependent upon sport. Thus in Figure 1, the block "Digital Technologies and Frameworks", recognises that the development of computer technologies is dependent not only upon purely technological and scientific frameworks, but that these operate within, and are crucially influenced by, societal and economic forces. Within this block are an enormous number of forces and relationships. Sport may be regarded as part of these, but on its own has little influence. This was especially the case in the earlier days of computer technology where the interaction with sport was close to zero. Given these considerations, digitisation is shown in Figure 1 as if it was a *deus ex machina.*

The middle block of this framework shows the areas of present interest. The examination concentrates on the most easily recognised relationship: that of computers and especially computer networks to the distribution of sport. This emphasis allows a very important evolutionary point to be

*Figure 1. The major mechanisms in the rewards to sport digitisation*

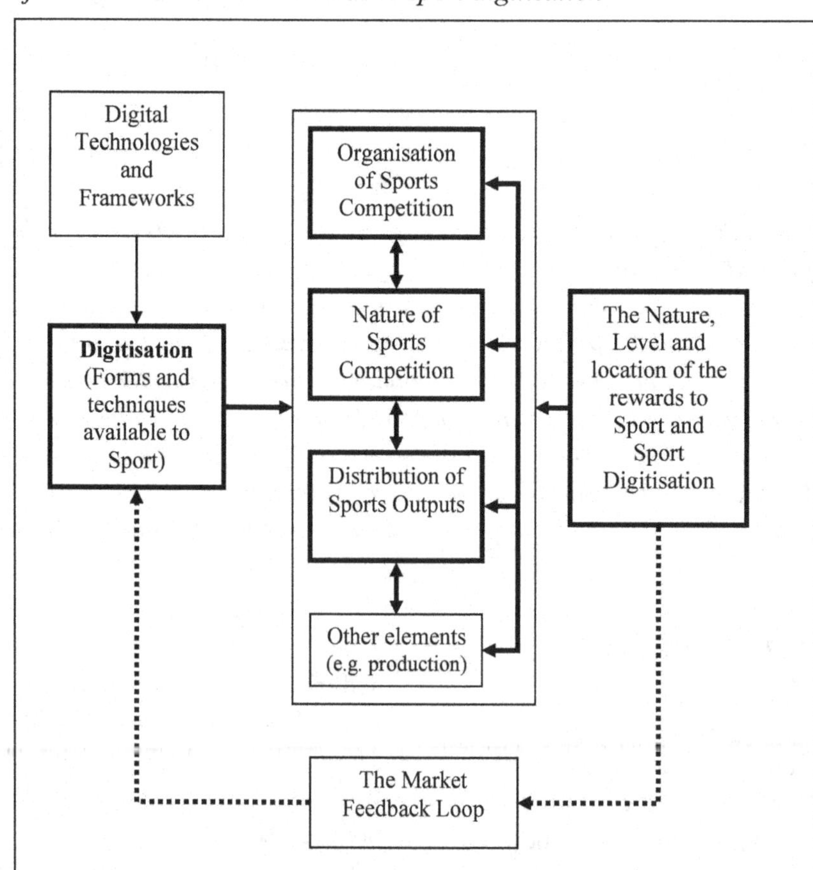

made. The establishment of these networks is not the fruit of digital technologies, but is overlain on models developed in the analogue TV period, and before that, in the days of radio broadcasting of sport. This is in line with the argument that digital technologies have been evolutionary in their impacts rather than revolutionary. They followed in the forms and tracks laid by analogue technologies, especially in sight.

The last and absolutely crucial part of the framework in Figure 1 is the feedback mechanism, that is, market feedback. This works mainly through the price mechanism, as successes (and failures) in the application of digital technologies and related business models are rewarded. This feedback creates incentives for the application of generic digital technologies to the sports industry. It also provides incentives for the development of more specific digital technologies to sport. These operate as both long-term and short-term.

## THE INTERACTION OF HOUSEHOLD TECHNOLOGIES AND SPORTS CONSUMPTION

Rather than immediately and directly consider contemporary sport in the context of digitisation, it is necessary to take an historical viewpoint, and first consider how and in what ways households produce and consume entertainment. In this way, it is possible to more clearly show how sport has not only changed in relation to digitisation, but to indicate all of the other elements that preceded, and have been associated with, that change. Essentially we can then see how sport came to be distributed and consumed in households.

Home entertainment over the last three centuries has undergone a remarkable set of changes and that sport has become part of those changes. The implication is that digitisation has helped a massive change in the viewing of sport, from an external (to the home) and even active participant in the drama, to an internal and largely passive one.

It starts with a European middle-class household in 1600. We see no real change in the forms of home entertainment over the next 200 years. Indeed, little had changed over the previous 1000 years or more. It was also, as per William Manchester's book title, *A world lit only by fire* (Manchester, 1993). Gas and electric lights in the household were to be transformative. So the choice of a European household in 1600 and, even more restrictedly, English households, is not arbitrary. With the benefit of hindsight we know this is where the nascent forms of global forces and structures were being nurtured. Indeed, the Treaty of Westphalia (1648) and the birth of the modern nation state, which has had a formative influence on "inter-national" sport, was of this period. With or without servants, entertainment would be what the family made or purchased itself. Of course, there were commercial forms of entertainment that lay outside the home, but these largely excluded sports. The most studied example is almost certainly the Elizabethan (not to say Shakespearean) theatre. The restrictions placed upon sport and its development by the long hours of darkness of Northern European winters have been little remarked in this context, and deserve more analysis.

During the period 1600 to 1800, books were increasingly in evidence in such households if the inhabitants were literate, which increasingly they were. Over the period, books would also become more entertainment oriented as their absolute and relative costs fell compared to other entertainments. In addition, in non–Puritan households there might be playing cards of various types and wooden board games, including chess. There would be musical instruments in some homes, as well as toys. Oral traditions, such as story telling, would also be strong, depending upon the area. Dedicated sports equipment would be virtually non-existent. What might now be sports equipment (such as archery impedimenta, horse tack and harness) would be matters of work, rather than sport.

Some of these entertainments were purchased items, but many or most were made in the home. The production of entertainment was much more than now undertaken by the family itself. Wealthy families employed household musicians and even players, but these would be few and far between. Consequently, the production and consumption of household entertainment took place simultaneously, and was conducted by the family. The production and consumption of entertainment also took place in isolation in the sense that the house and its entertainment were not physically networked to the world, except culturally and socially. The simultaneity of production and consumption also meant that that production was part of the entertainment and was non-professional.

During the next one hundred years things were to change enormously. By 1900 the household was far more likely than ever before to be urban, and the family to have been urban long enough that it had lost its rural roots. This, it is suggested, has profound implications for the types of entertainment sought and which the household was capable of creating. Home entertainment items would not only include books and playing cards as before, but a wider variety of board games, toys and musical instruments, such as pianos. Dramatically different were the presence of mechanical-acoustic phonographs and gramophones; electric and gas lighting in many households; and magazines, journals and newspapers. Organised sport meant that many middle class and working class households had personal sports equipment. Urbanisation was beginning to create the mass markets that digitisation would later support and help exploit. Wolmar (2007), from a steam technology point of view, also makes the point:

*Every racecourse around Britain soon acquired a station - often equivalent in size to those serving the major towns - since on race days special services would bring in thousands of spectators as well as the racehorses themselves. The very development of professional football and the creation of a national Football League in 1888, as well as a county cricket championship, owe much to the railways. Without them the players and, crucially, the paying fans would not have been able to travel to watch their teams play.* (p. 142)

Thus, steam technology networked sports events with their potential customers in the urban areas in a deliberate manner. This anticipated digital technology in many ways, but was in one sense operating in reverse. Here the network brought the viewer to the event, while digital home technologies brought the events to the consumers.

Important for the adoption of digitisation, while not for entertainment and initially only for the well off, the telephone had arrived. Its significance, as with the electric light, was that households were becoming part of both electrical and electronic networks. Most home entertainment items were now purchased, though some were still home-made. But after hundreds of years not only was change speeding up, it was intensifying the role of market forces. The production and consumption of entertainment still took place simultaneously, but more and more entertainment was purchased in various forms and imported into the home. Most notably, the gramophone diminished the role of music making in the house, and created the first superstars (mostly singers) that were imported into the home.

The 1920s/30s saw further radical changes in home entertainment. The traditional items remained and, in some cases, had increased in importance. Books, playing cards and board games were perhaps more in evidence due to increased incomes, but some items were less used. Musical instruments almost certainly fell into this category. Board games were now commercially purchased. Toys were purchased from an internationalised, if not globalised, toys industry and were of increasing sophistication, sometimes being driven by electrical, clockwork or even steam power. Increasing electrification and the electric light made entertainment move into the hours of darkness.

At first sight, the household consumption of toys has little to do with household consumption of sport, but there are parallels in the way that they have evolved towards professional production. However, the most important reason for mention at this stage is that digitisation has led to a convergence of the consumption of toys and sports. Later in digital form this has become especially true of computer games. Convergence also took the form of sports team clothing, specifically miniaturised for children. Not all of this convergence is electronic.

It was during the inter-war period that the household became an electric one – a clearly necessary pre-cursor to the electrification and then digitisation of the household's consumption. Thus gramophones were gradually changing from acoustic/mechanical to electric/analogue. Meanwhile, telephones carried on their diffusion through well-off households in the middle class. At the same time, universal education and the consequent move to almost universal literacy saw magazines, journals, periodicals and newspapers invade the household. These were crucial to the mass reporting of sport and created a mass market, even though consumption still took place largely outside the realms of the home. These periodicals helped create fashions and superstars, along with increasing returns for advertising and success.

Radio developed before The First World War in a practical form that then rapidly came in to the home in the 1920s. The first commercial radio station began operation in Chicago in 1920. The BBC began to broadcast in England in 1922, and the first sports event was broadcast in the USA in 1921. Many of these dates and the events that took precedence are, of course, obscure and disputed, but the importance of this period is not. This was the start of the expansion of sport into the electric media, although this was clearly nothing compared to visual presentations. TV was not far behind, with broadcasts beginning in the early-mid 1930s in the USA and the UK. All of this was ended temporarily however by The Second World War. These developments were, of course, analogue in form and not digital, but they paved the way. When digitisation arrived several decades afterwards, the channels were already open and sport was a household consumption item.

So again, we see change speeding up and intensifying production and consumption. In addition, the consumption of entertainment took place less and less in isolation – the house and its entertainment were being networked to the world with an installed base of electric lights, magazines, gramophones, records and the telephone. The increasingly sophisticated gramophone/record player, radio, and recorded music industry diminished music making in the house. At the same time, it created increasing numbers of superstars – singers, bands and so on – all of whom were being imported into the house. The creation of a recorded popular music industry rapidly took place and sheet music sales declined dramatically. The moves in the inter World Wars period brought both visual (temporarily) and acoustic elements into home entertainment. One important element that was missing, due to the inflexibility of the analogue system, was the ability to integrate these separate systems via a central control unit. This was not to be achieved for decades. The temporary nature of TV in this period and its expense meant that visual sports entertainment (other than live attendance) was the cinema. This situation lasted into the 1950s.

Following the Second World War the situation reverted, at least superficially, to what had been available before the war. The similarity was superficial in the sense that changes in technology brought about by the conflict meant analogue systems now had far greater capabilities, and these rapidly became available to households. Notwithstanding the presence of radios in households, notable in these advances for wireless technology was radar, and specifically its miniaturisation for use in aircraft. If it fitted in a plane, it easily fitted in a house. At the same time, these technological advances made such systems cheaper in both relative and absolute terms and, in addition, real incomes rose. During this immediate post-war period, the spectre of mass unemployment receded and spending on entertainment items could increase.

Traditional items such as playing cards, chess and upright pianos remained, but by the 1980s many now had electronic versions. The electrification of games became even more prevalent as digital computing rapidly took over from analogue computing. This has affinities to sport, because these games involved competition. Examples included electronic backgammon and chess from dedicated machines. The move to digital implies smaller machines that were more suitable for households, more versatile, dramatically cheaper, and of increasing quality in the level of play.

At this point, a listing of those areas where both electrification and the digital revolution affected the operations of households is appropriate, as it contains clues regarding the importation of sports consumption into the home. Key areas include:

- Toys (a global industry creates and sells toys, and many are electronic and intelligently interactive to a limited degree. Many of these electronic toys, especially for children of pre-literate or early-literacy ages, are promoted as having a teaching/learning component)
- Record players, CDs, DVDs
- Electric light
- Electric and electronic (programmable) equipment in the house, such as vacuum cleaners, washing machines, dish washers, cookers, microwaves (house-work and its management are going electric, electronic, digital and intelligent)
- Telephone (universal and mobile – increasingly multi-functional and intelligent)
- Magazines, newspapers (often support digital equipment and fashions)
- Television
- Radio.

Several comments are in order in relation to this list. In bringing sports distribution and consumption to the household, it is clear that television was, and is, the dominant electronic machine, having both the visual and sound elements necessary for desirable sport entertainment. The visual element speaks for itself, but sound is important in helping provide atmosphere and excitement, as well as commentary. The ability to reproduce sound and sight faithfully and cost-effectively has been taken far beyond what was previously possible due to digital technology.

In summary, the argument is that digital technology has become a part of the movement of the mass of sport entertainment being consumed in the home, rather than at the external venues where sport is produced and consumed live. But this has taken place in an evolutionary manner. It has also been argued that the present manner of sports consumption in the household is the outcome of several centuries of development in household entertainment technologies. The patterns of consumption were established before the digital revolution.

## THE THEORY OF SUPERSTARDOM

Within the entertainment industry, one of the results of being able to record performances and repeat them on demand is that it helps create superstars. The digitisation of sports recording has played its role in the superstar making process in commercial sports. But as with home entertainment systems in general, it is not the creator of this process, nor is digital recording the only means by which superstars can be created. So, if we are to understand the forces working for the increasing digitisation of sport, the mechanisms of superstar creation need to be understood. This section outlines what superstardom is, how it is created and its relationship to the digitisation of sport. This is especially so in the context of sports consumption at leisure in the home. In this sense, the following is an expository section.

The term superstar is very much over-used, but is easily understood as referring to an entity, usually a person, that stands far above the others in its field. Thus Microsoft® Word is the superstar of word-processing software, Manchester United is a superstar of soccer teams, and Michael Phelps is the superstar of swimming. There need not be just one in any sport or profession – both Roger Federer and Rafael Nadal are superstars of the male professional tennis circuit. This further implies that superstardom does not mean that the superstar is the "best" in the field, or even that best can be defined in some unambiguous manner.

In the economic theory of superstars, which is followed here, a superstar is defined as having an income far in excess of others practising in the same field (such as sports and other entertainments) (Rosen, 1981). The founding work of Rosen (1981), Adler (1985) and MacDonald (1988) follows from an empirical observation – that earnings in some professions are highly skewed, with a few having massive incomes and others often struggling (Figure 2).

The percentage figure at any point on the horizontal axis indicates the percentages of the labour force that have incomes above (to the right) or below that point (to the left).

Rosen (1981) suggested that earnings in some professions are highly skewed with respect to talent (Figure 3). That is, small differences in talent at the top end of earnings have a huge impact on incomes. A small difference in talent between

*Figure 2. The skewed earnings of practitioners in a given profession*

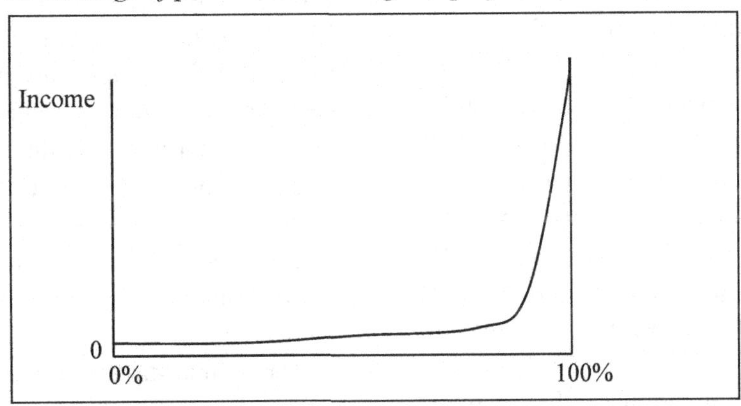


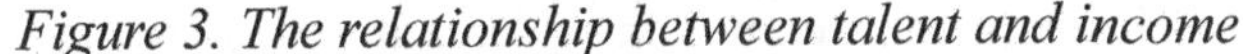
*Figure 3. The relationship between talent and income*

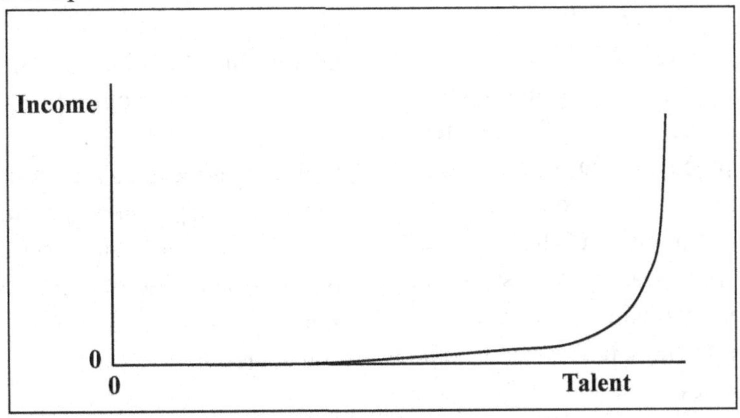

*Figure 4. A small difference in talent leading to a large difference in income*

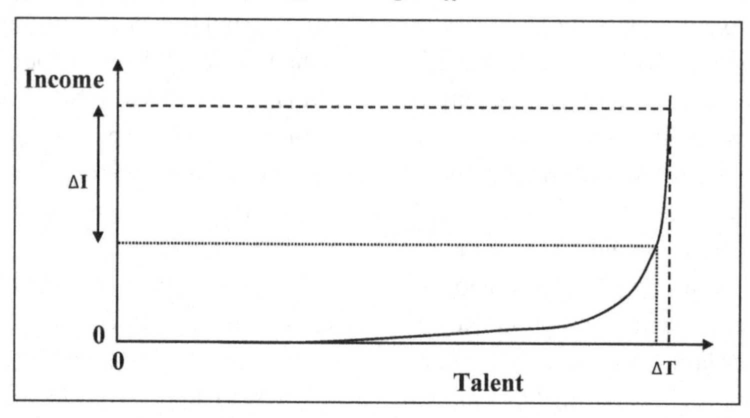


two practitioners may leave one as a superstar (i.e. with an enormous income), while the other has a far smaller income.

This is seen in the diagram (Figure 4), where the small change in talent (ΔT) has an enormous impact on income (ΔI). The same could be true of the differences in performance of a piece of software and its sales revenue.

One of the most important elements in the theory is the explanation of how superstars are able to maintain their position, even when their competitors are willing to charge extraordinarily low prices. Part of this reasoning suggests that technology (as in the digital world) plays an enormous role in the emergence of superstars and their economics.

Rosen assumes for simplicity that there are innate differences in talent (or quality), and that these talent differences are known to society at large. MacDonald (1988) extends Rosen by demonstrating a learning mechanism, by which talents become known through consistent high quality early career performances. Thus in Rosen, there is no theory of the process of stardom creation, but merely an assumed recognition of relative levels of talent. One can contend that this can give rise to circularity of argument, as stars and superstars are often regarded as talented on the basis of their earnings. The same is true of commodities and companies.

It is also suggested that certain types of commodities, especially those in the digital economy, are highly skewed with respect to quality (Figure 5). Thus, small differences in quality have a huge impact on market share (quantity sold compared to other substitute commodities). Analytically, this is virtually identical to individuals (i.e. commodities) as superstars and their personal earnings (i.e. market revenues). The main difference is that the individual is an input to the production process, whereas the team is the firm or company that employs individuals. We therefore have the possibility of superstar teams employing superstar individuals in what is a complementary relationship.

The winner-takes-all idea in the digitally dominated sports world flows from the observation that a few companies, products and individuals achieve dominance in their commercial/professional fields, with massive revenues and market share. Others are often very small by comparison. Economic theory suggests a variety of alternative explanations of the winner takes all phenomenon, including increasing returns and standards. Both of these are of potential relevance to the household consumption of digital sports commodities, as possible outcomes of increased network production and distribution. However, superstardom is considered here.

*Figure 5. The relationship between market revenues and quality of product (such as a professional sports team)*

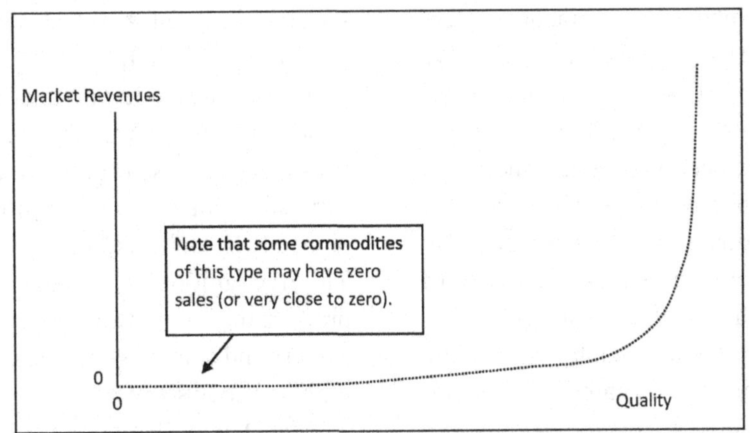


As a matter of methodology, one of the most important questions for Rosen's theory is, can we measure talent independently of income? For some professions, perhaps we can. Rosen in the Phi Beta Kappa journal, *The American Scholar*, in 1983, pointed out that,

*The top five money winners on the pro golf tour have annual stroke averages that are less than 5 percent lower than the fiftieth or sixtieth ranking players, yet they earn four or five times as much money.* (p. 454)

Again, economics can suggest alternative reasons, notably in the form of tournament theory (Lazear & Rosen, 1981; Rosen, 1986). This alternative explanation for golf players' earnings and other players specifically in tournaments is examined later in this chapter. But, it is evident we can measure the performance of some individuals or commodities against others. This is especially so in sports where there are quantitative measures of ability, in terms of stroke averages and so on. It is also worth noting that Rosen (1983) suggests comedians appearing on TV (reproducibility) are disproportionately rewarded, compared to those that do not appear. This, of course, has direct relevance to the previous argument concerning household distribution and consumption technologies.

It is possible to use a simple sports example to show how a superstar in earnings is based upon quality. Consider five football strikers (A, B, C, D and E) regarded as the top five of their profession, playing in the English Premier League. The rest of the strikers in the Premier League are marginally less brilliant players, but are disregarded in this analysis. The aforementioned five strikers are the only ones in the Premier League who can score as a matter of course in top games. Each can play 100 games per year, given the structure of the league and other games (such as Internationals and European championships). The real number is closer to half of this, but it simplifies the example. Striker A has a 90% success rate of scoring in any given game. Again, this is far in excess of reality, but the analysis is simplified by the assumption. Players B, C and D have a 70% success rate. Striker E has a 40% success rate (this would be phenomenally high). From the demand curve shown in Figure 6, the best player, A, takes the 100 games from a team with the greatest ability to pay, so he receives $P_A$. As B, C and D are equal in talent, the competition for the next 300 games (i.e. the next three teams) means that $P_B = P_C = P_D$. Competition forces the price down that each player receives. The lowest ranked, striker E, receives $P_E$. So those teams that cannot pay $P_E$ have to buy someone of less goal-scoring talent than the aforementioned five players. The reason that the teams pay the higher amounts is because these players give them not only greater playing success, but also increased access to the mass market provided by household electronic consumption. Thus there is a cumulative causation, with successful teams able to afford the best players. This creates greater access to households and helps create superstars. Part of the reason for this is that only one game at a time can be observed by consumers. This forces out lesser teams, as the better teams will be more frequently viewed. As a result, in Figure 6 the top striker, A, is a "superstar" with earnings several times the next best, striker B. It is also a property of this example, from the way in which the demand curve is constructed, that striker A will command the same price even if his scoring percentage is only marginally greater than that of B, C and D (i.e. if A has a scoring rate greater than 70%). In reality, such a stark result would be unlikely, particularly given the increased uncertainty about the superiority of A, the closer his strike rate to strikers B, C, D and E.

In Rosen's model, if superstars depend upon relative talent, there is less need for a theory of the superstar-making process. This is because the superstars emerge as their innate talent is "recognised" by informed consumers in the market. This recognition is because these best talents produce the best performances. Does this fit with players and teams as specific commodities in the digitised sports world?

MacDonald (1988) argues for talent differentials appearing as the probabilities of putting in good and bad performances on the sports field. Note that this fits with the exemplar model of super-strikers given here. This means that new sports performers are not necessarily recognised as best talents straightaway, but are required to establish a track record. In this situation, competition forces out the poorer talents and the stronger, more consistent ones emerge – the very best as

*Figure 6. The relationship between relative ability of strikers and their earnings in a team sport in a mass consumption market*

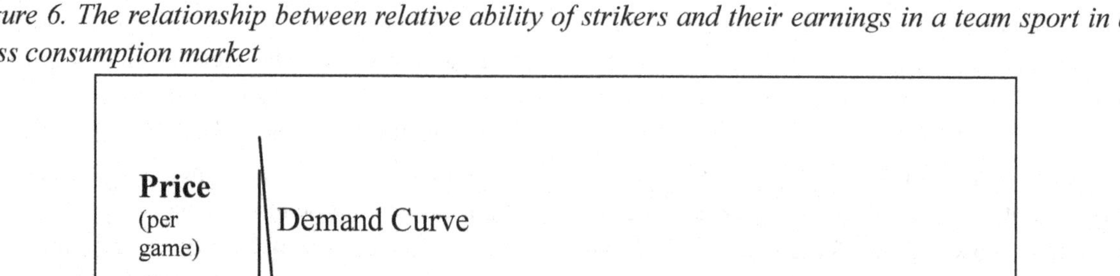

superstars. Against these, Adler (1985) argues against talent differentials being necessary at all. Adler argues that, "…a hierarchy in income could exist without a hierarchy in talent" (p. 208). Performers are chosen at random in their early careers. In this situation, the lucky ones force out the other equally talented, as more and more consumers switch to the successful ones. In other words, early success breeds later success, with chance selecting even further among performers who becomes a superstar. Here, marketing can play a role. Superstars can be manufactured (along with having some luck). Adler's model appears to fit less well with sports performers, but more likely to be consistent with other performers entering the household in digital form. However, evidence from German soccer suggests that an Adler style mechanism can work. Neusch (2007) and Franck and Neusch (2008) suggest that soccer players can increase their earnings, not only by investing in training and having natural talent, but also by cultivating their popularity. At first sight this is seemingly strange where the playing contributions of players are available for masses to see. However, several years of casual observation of soccer fan websites indicates distinct and continuing disagreements about the talents of given players, and being a team sport, soccer can hide some weaknesses of talent, especially in a highly complex and uncertain playing environment. It is also possible that some talented players have advantages in selection and prominence as they add to the markets of a club. This can explain why Asian players are increasingly valuable to European soccer clubs, given the burgeoning of soccer interest in Asia. This fits with Adler's view of superstardom that, "…the phenomenon of stars exists where consumption requires knowledge" (1985, p. 208).

There are several ways in which the modern sports economy and superstars are related. First and foremost, the reproducibility of sports performances as digital products implies the ability to reach a mass audience in the home. There is an ability to reproduce the work faithfully for transmission and home reproduction (although not the same as a live performance), and to reproduce the sports product at a very low per unit cost. The digital sports economy structures are natural monopolies. These help create superstars within areas such as telecommunications, with cable networks able to restrict viewing to create pay-per-view, or some other form of pricing in the household (Hazlett, 1986). Again, this does not rely on digital technology per se, but digital technology inherits these market forms and enhances them. The possibility of superstardom, and the associated but not identical phenomena of "winner-takes-all", is the spur for a great deal of both entrepreneurial and technologically innovative activity. This was true in the early stages of the digital economy and elements of it, such as Internet marketing and computer games. Now these are recognised, they are eagerly exploited, with households increasingly self-trained and acculturated from childhood in consuming from these technologies.

## THE ECONOMICS OF ATTENTION

Attention is now considered to explain how superstars establish themselves and maintain their status. It has been argued here that, in the case of sports, the locus of discovery for the (average) fan has shifted from the physical arena to the household. But the household is now bombarded with information. When there is a mass of information, knowledge or data available, the key to its usefulness (other than its content, of course) is the amount of attention it receives compared to all of the other material available. This is true both in absolute and relative terms (i.e. the amount of attention received and how much is received relative to other available material). Considered in this way, it is no longer information that is scarce, but attention.

*What counts most is what is most scarce now, namely attention. The attention economy brings its own kind of wealth, its own class divisions – stars vs. fans – and its own forms of property, all of which make it incompatible with the industrial-money-market based economy it bids fair to replace.* (Goldhaber, 1997)

Goldhaber makes vital points but misunderstands standard economic principles, which are consistent with much of his argument, despite his claims to the contrary. And certainly these principles are not overthrown by recent developments. His prophecy of the attention economy replacing other economic systems is patently not coming true. Indeed, the attention economy did not arrive with digitisation. As a colleague points out, as far back as the 1920s and 1930s, the problem of scarcity of attention was recognised and considered (M. Kremmer, personal communication, 2005). But the battle for attention has certainly intensified since then.

In Goldhaber's sense and that used here, the technologies that allow easy and cheap importation of digitised sport into households are intensifying this competition for attention. Digital sport is not only competing with other forms of presentation of sport, such as live performances, but also with more traditional forms of household entertainment, such as books and playing cards. This is also true for the impacts of the more active forms of digitised sport, such as computer games, which are particularly "attention-hungry" entertainment technologies. These in turn are changing the way in which educators are having to think about, and deal with, the acquisition of literacy skills (de Castell & Jenson, 2004). Some educational theorists suggest that we now have to deal with multi-media literacies, rather than just those of the printed page (Sanford & Madill, 2007). While the impact on education and society is not the focus of this section, those impacts of the changes in household entertainment already described include an increased reliance upon, and an ability to use the flexibilities of, digitised sport entertainment. In other words, there is a learning process, which reinforces the use of digitised sports entertainment in succeeding generations. It makes the process of attention-giving less reversible than before. Nevertheless, this does not change the intellectual passivity and physical inactivity that attention requires.

In simple terms, the attention received by a digital sport entertainment (among many other competing at-home entertainment activities) can be resolved into three components:

1. Attention gained (before the first visit to a specific digital sport site)
2. Attention retained (during any visit to a specific digital sport site)
3. Attention repeated (subsequent visits to a specific digital sport site).

All of these digital sport sites are assumed to be via the household's digital entertainment platforms. The second two are determined by the content of the information site and its resulting value to the visitor, and are not discussed here. However, the content of the site and its value to a visitor can only be determined by the initial visit. As information goods they are experience goods – you only know if they are what you wanted after you have tried them (Benz, 2007). Consequently, gaining the initial visit is critical to any entertainment site. This implies ensuring that potential visitors know the location of the site where the information is stored. Given there are billions of potential sites, this requires that the visit is not left to chance, as the probability of gaining initial attention will be extraordinarily low. Here it is assumed that much of the competition between sites takes place within the household, but this need not be the case. Often it pays for that initial attention to be gained from sites that are more "visible" and traditional, such as journals.

Goldhaber's contribution was to suggest the idea of a market for attention. In a variety of different forms, this has also been suggested by others. Goldhaber reversed the demand and supply situation, suggesting that information sites are the demanders of attention and it is the potential consumers of the information who are the suppliers of attention. In terms of getting the initial attention then, it is the sites/locations that have to seek out the suppliers of attention, because the competition between the sites/locations is so intense. It is intense, first because there are so many of them; and second because they generally contain a commodity (information) which is duplicated by other sites, such as news, weather and sports results, as well as gaming (i.e. they have no monopoly element). This explains why so much "information" is given away free via the Internet. Thus, unless there are other factors, there is close to perfect competition in the market for attention. The locations of sites therefore have to be advertised beyond the Internet, in order to inform potential users. Most digital sports entertainment is known by reputation (of a club or a tournament) long before the digital intervention. In the case of reputed sports events, the channel can be restricted to subscribers, rather than being "free-to-air".

But what do the sites do with the attention that is supplied to them? They use it to advertise and market commodities that the users of the information might buy. They can often do this better than normal advertising, because they can:

- Change information so it is continually fresh (news, weather, sports results, and so on)
- Be dynamic in their interactions with the user of the site/location in terms of entertaining them
- Provide fresh entertainment, which is most important of all in sports.

This is an implicit market, rather than one where money plays only an implicit role. There are parallels elsewhere, such as in the academic world, for example scientific publishing. For academic authors there is no commercial market as such. Klamer and van Dalen (2001) consider attention paid to scientific publications and state, "The objective is to anticipate the institutions that will help scientists cope with the excess on the web" (p. 3). In thinking about the private market, we can also try to anticipate how the market will cope with the massive redundancy of web sites. It is highly likely that specialised entertainment search engines will eventually become such institutions, but in the household. Klamer and van Dalen also mention "the skewed distribution of attention over all these publications" (p. 3), arguing that this is a kind of superstardom (of the Rosen type). The skewed distribution of attention they consider applies beyond science and academia (meaning that just a few websites get the great mass of attention), and into the household.

## ATTENTION AND TOURNAMENT PLAYERS' PAYMENTS

This section considers forms of competition available in competitive sports. Special emphasis is given to the nature of tournaments, with interpretation of their payments to players as a function of attention received. This is used as a means of interpreting the economics of tournaments in a digital environment.

While the number of forms of competition might be considered infinite, including minority ones such as ladders (often used in club squash and challenge trophies), there are just a few main types that appear in professional sport. These include races, leagues and tournaments. Each has a specificity that determines different ways in which digital and other forms of technology are used.

Racing as a competition structure is enormously flexible. Races can be arranged in heats and against the clock, as opposed to first past the post. But racing has limitations, not always allowing one-to-one competition for beating an opponent. Leagues, on the other hand, require a structure to continue to exist between matches. If that structure is unavailable, the league can easily collapse. A primary example of an inability to build a long lasting structure is the American Negro leagues in baseball, despite having what was, and still is, considered to be some of the finest sports talent. Tournaments, on the other hand, are a form that is especially appropriate to one-on-one competitions, as in tennis and amateur boxing, but can also be used in races. Tournaments enable an overall winner to be established. They can be used in compressed time, as in the four tennis Grand Slam competitions; and over extended time, as in soccer tournaments such as the English F.A. Cup. They can be modified to fit with league structures, as in baseball, football and ice hockey in North America; or with international events, such as Soccer's World Cup and the Olympics. Tournament structures can, therefore, be used with teams, one-on-one sports, or many competitor sports such as races.

While there is a current equilibrium between tournaments, in analysing the relationship one can see tournaments complement, and compete against, each other. The first question is: how do tournaments (within the same sport) complement with each other? Tournaments differentiate themselves by different locales, calendar dates, playing surfaces, course layouts, slight differences in rules, and so on. The ability to do this and to exploit these differences is important in a digital world. So then, how do tournaments compete against other tournaments? In economists' terms, do tournaments compete with each other oligopolistically, perfectly competitively, or monopolistically competitively? For the initial analysis, as others have done, it is convenient to treat tournaments as if they were monopolies. However, recent analyses have begun to treat them as competing for star players, and this has the potential to change the results enormously. It may appear strange that this would affect digitisation, but it changes the incentive structures offered both to tournaments and the players/teams taking part in them. This in turn alters the incentives for the adoption of digital technologies.

The current standard theory of tournaments is that of Lazear and Rosen (1981) and Rosen (1986). They argue in a more general context than just sport, that financial compensation to competitors is an incentive structure. The most common non-sport example used is executive compensation in organisational hierarchies (Lee, Lev & Yeo, 2008; Tsoulouhas, Knoeber & Agrawal, 2007). This means in the first instance that progressing to higher rounds of a tournament (higher managerial levels) implies higher remuneration. This is hardly surprising. However, the theory also seeks to explain why earnings increase more than proportionately as the competitor progresses through the rounds of the tournament, or levels in the executive hierarchy. Earnings can be regarded in many different ways (most importantly in the present context as either reward or incentive, or both). Thus we observe in a grand final that the winner may earn considerably more than the runner-up, even though the contest may be very even and the talent differentials even closer. The argument of tournament theory is that the differentials are not designed just to induce effort in a given game, but throughout the tournament. Thus, the purpose of the grand prize is not just to reward the winner for effort in the final, but also to provide very high incentives to progress further and further in the tournament. Thus the reward has to be increased more than proportionately. Consequently, the winner's purse must be seen as including the winnings from the other rounds in order for the true value to be calculated. This may mean that the analysis of tournaments and their prizes is best conducted in terms of differentials, rather than the actual amounts.

The implication of adopting this theoretical analysis is that sports players have enormous incentives to play hard in early rounds to ensure getting through, rather than attempting to conserve energy. It also implies that players will invest heavily in training and in their equipment in order to gain advantages. This fits with a view that digital technology has had its greatest impact in sports in the production of sport events and the honing of athletic skills (vital when timing technology can time races to within thousandths of a second). However, this does not explain appearance money. Nor does it explain how the winnings (prizes) are related to higher revenues for the tournament via the increased intensity of play from the players. Why this occurs is clear, but is an answer that lies outside the theory itself.

To try to resolve these problems, one way in which tournaments can be analysed in terms of attention theory is the coverage they receive in the media. Over the life of a tournament it can be covered in the newspaper, as well as electronic and other press. One important point is that appearance fees are not explained by tournament theory, as they run counter to any incentive structure. However, they are explained by an attention theory of tournaments. Attracting big name players is a means of attracting attention to the tournament.

A full attention theory of tournaments is not presented here, but a very simple numerical example illustrates how it would operate in determining player compensation. Suppose we have a tournament with 64 players, such as tennis, where they play against each other one-on-one. This is different from golf where the players do not directly play against each other, but against the whole field via their scores. If tennis is played in a standard form as in tournament tennis, this means 32 games in the first round, 16 games in the second round, followed by eight in the third, four in the fourth, then two semi-final games, and a final. This means 63 games in all in order to decide a winner. Given this is an important tournament, each of the games will gather some attention by the media, but not to the same degree.

Let us assume that each round takes the same time to play and receives the same total attention, regardless of the number of games it contains. This means that if the games in a round are played simultaneously, each game in the first round's 32 games is diluted in terms of the potential attention to just 1/32 on average. In other words, if two games are played at the same time, the potential audience is split in two, and if four games are being played at the same time, that audience is split in four. Of course, it is not just dilution at work, as interest in a final game is far greater than any of the individual games in the first round; but for the moment, let us deal just with the dilution element.

This simplified model of audience attention dilution means that we can begin to relate the payment of athletes to the attention their games generate (i.e. to the importance of their games within the tournament). This is a crucial difference with tournament theory. In tournament theory, payment is for effort and prize money differentials are designed to produce more effort. Here the differentials are a reward for generating attention for sponsors and advertisers. The winners are paid according to the attention each game gets. So if dilution operates alone, the winners of the first round might receive $10,000, while the winners of the second round would receive double that amount. If only the winners of each round get paid, this means that winnings are cumulative as a player progresses. The losers of the first round win nothing at all, while the winners of the first round who lose in the second round receive just the $10,000. The winners of the second round receive an additional $20,000 so that they receive $30,000 = $10,000 + 20,000. The winners of the third round receive an additional $40,000 to make $70,000 overall. And ultimately the winner of the final receives cumulative winnings of $10,000 + 20,000 + 40,000 + 80,000 + 160,000 + 320,000 = $630,000. The loser of the final receives $10,000 + 20,000 + 40,000 + 80,000 + 160,000 = $310,000. This means the losing finalist gets less than half the prize money ($310,000) that the winner receives ($630,000). This model of the tournament prize money can explain large differentials, but does so on the basis of attention (and hence the potential for advertising and media revenues to the tournament), rather than effort.

But even first round losers may need compensation. Now let us assume that the loser gets paid some proportion of the prize money for each round. To provide an incentive to win, the loser's proportion of the total prize money for a given round has to be less than 50%. If we make the loser's proportion 20%, then each finalist has gained 0.80 (10,000 + 20,000 + 40,000 + 80,000 + 160,000) = $248,000, which means the losing finalist now gets $248,000 + $64,000 = $312,000. The winner receives $248,000 + $256,000 = $604,000, so the losing finalist gets an amount slightly more than half of the winner. In other words, the prizes are closer when they are shared in some proportion, rather than winner takes all in each round. In what is a simple and natural interpretation, we have the winner receiving much more than the loser, even though there may be only a slight difference in their talent and effort. Indeed, the winner may be the lesser player, but lucky on the day.

The dilution factor gives rise to differences in prize money, assuming that each round receives equal attention. It is natural however, to expect the later, more important rounds to receive a greater intensity of attention from the fans and the media. This is also what is strived for by the tournaments with the use of devices, such as seeding players. Even if attention intensity increases only slightly for each round, then the prizes become less equal with progression further into the rounds.

With standard tournament theory, we can explain those upper level disparities. One element that does not fit with tournament theory is appearance money for the stars. Indeed tournament theory is diametrically opposed to such payments, indicating that they are disincentives to effort on the part of the stars. Now consider the early rounds with these stars playing. These games, assuming players are seeded and do not appear against each other, do not affect the dilution of attention, but rather increase attention during the earlier rounds. This is where the stars earn some of their monies. They make the earlier rounds more interesting by their presence. However, if they play against each other in earlier rounds, the cumulative intensification effect is reduced, especially as one of the stars must lose and their attention gathering services are lost to the tournament.

The attention model also suggests another reason for the disproportionately high winner's purse compared to the second-place getter. The differential and the size of the winner's purse provide not just an incentive for the players in the final, but also extra tension and excitement for the fans. Both the size of the purse and the purse differentials between winner and runner-up send a signal to fans and the media that the game is worthy of attention.

It is noted that in this discussion, as with standard tournament theory, there is no denial of incentive effects, however they are incidental, not central. But what has been achieved is a model of player remuneration that is much more attuned to the way in which tournaments are presented to the public in a digital era.

## CONCLUSION

The major thrust of this chapter has been towards drawing a picture of the impact of digital technologies on the mass distribution of sports entertainment. This was partly to provide a counterbalance to the more emphasised impacts of these technologies on the production side.

Other emphases were on the question of whether or not this constituted a revolution in the delivery of sport. The conclusion was that it did not. Instead, it was concluded that it was part of a long evolution of the delivery of consumer products into the household, stretching over several centuries. This conclusion was based upon an examination of the history of the household consumption in Western societies. It was found that the electrification and electrification of households, which began over one hundred years ago, provided grounds for the easier acceptance of all digital forms of entertainment, including sports. It was argued, albeit briefly, that increasingly digital forms of entertainment will be accepted as the norm by children acculturated to these forms, and that this will help form the basis of strategies to enter households successfully. These strategies and the nature of digital forms has lead to the predominance of individuals, teams, tournaments and certain sports becoming the superstars of their arenas of entertainment. Though, there is competition for attention over all of the other entertainments entering the household. As an example of this analysis, it is suggested that a new theory of tournaments is required that places attention at its centre in this digital age.

## REFERENCES

Adler, M. (1985). Stardom and talent. *American Economic Review, 75*(1), 208-212.

Benz, M-A. (2007). *Strategies in markets for experience and credence goods.* Wiesbaden: DUV Gabler Edition.

de Castell, S., & Jenson, J. (2004). Paying attention to attention: New economies for learning. *Educational Theory, 54*(4), 381-398.

Franck, E., & Neusch, S. (2008). Mechanisms of superstar formation in German soccer:

Empirical evidence. *European Sport Management Quarterly, 8*(2), 145-164.

Goldhaber, M. (1997). The attention economy and the Net. *First Monday.* Retrieved 15 July, 2008, from www.firstmonday.org/issues/issue2_4/goldhaber.

Hazlett, T. (1986). Private monopoly and public interest: An economic analysis of the cable television franchise. *University of Pennsylvania Law Review, 134*(6), 1335-1409.

Klamer, A., & van Dalen, H. (2001). *Attention and the art of scientific publishing.* Tinbergen

Institute Discussion Paper No. TI 2001-022/1, Rotterdam.

Lazear, E., & Rosen, S. (1981). Rank-order tournaments as optimum labor contracts. *Journal of Political Economy, 89*(3), 841-864.

Lee, K., Lev, B., & Yeo, G. (2008). Executive pay dispersion, corporate governance, and firm performance. *Review of Quantitative Finance and Accounting, 30*(3), 315-328.

MacDonald, G. (1988). The economics of rising stars. *American Economic Review, 78*(1), 155-167.

Manchester, W. (1993). *A world lit only by fire: The medieval mind and the renaissance: Portrait of an age*, USA: Back Bay Books.

Nuesch, S. (2007). *The economics of superstars and celebrities.* Wiesbaden: DUV Gabler Edition.

Rosen, S. (1981). The economics of superstars. *American Economic Review, 71*(1), 845-848.

Rosen, S. (1983). The economics of superstars. *American scholar, 52*(4), 449-460.

Rosen, S. (1986). Prizes and incentives in elimination tournaments. *American Economic Review, 76*(4), 701-716.

Sanford, K., & Madill, L. (2007). Understanding the power of new literacies through videogame play and design. *Canadian Journal of Education, 30*(2), 432-455.

Tsoulouhas, T., Knoeber, C., & Agrawal, A. (2007). Contests to become CEO: Incentives, selection and handicaps. *Economic Theory, 30*(2), 195-221.

Wolmar, C. (2007). *Fire and steam: How the railways transformed Britain.* London: Academic Books.

# Section I
# Training and Participation Applications

# Chapter II
# A General-Purpose Taxonomy of Computer-Augmented Sports Systems

**Sean Reilly**
*Trinity College, Ireland*

**Peter Barron**
*Trinity College, Ireland*

**Vinny Cahill**
*Trinity College, Ireland*

**Kieran Moran**
*Dublin City University, Ireland*

**Mads Haahr**
*Trinity College, Ireland*

## ABSTRACT

*The area of computer-augmented sports is large and complex and spans several disciplines. This chapter presents a general-purpose taxonomy of computer-augmented sports systems, which is intended to assist researchers and designers working in this domain. Allowing systems to be classified with regard to form as well as function, the taxonomy is intended to have several uses, including serving as a clear map to aid in the understanding of the domain and as a tool to help researchers analyse the state-of-the-art by characteristics of systems. The taxonomy also offers a common vocabulary to the multidisciplinary teams that work in computer-augmented sports and can be used to identify sparsely populated regions of the domain as promising areas for future research. The authors present and demonstrate the use of the taxonomy using four example systems selected from an extensive review.*

## INTRODUCTION

The use of technology in sport has a range of applications, including training, refereeing and injury prevention. While the tradition of using technology for sport is long-standing (e.g., electric detection systems for fencing), recent advances in mobile and sensor technologies have given rise to a considerable range of sports systems that use the new technologies in interesting ways. The domain of technology-augmented sport systems has therefore increased considerably in complexity, not only with regards to the form of the solutions but also in respect to their functions and scope. While the new developments are exciting for developers and researchers working in the field, it takes a considerable amount of work to understand the domain as its complexity grows.

Taxonomies are used to organise and classify objects in complex domains. A taxonomy familiar to most people is the Linnaean Taxonomy devised by 18th century botanist Carl Linnaeus, which is used to classify living things. The taxonomy as a scientific endeavour has since been used outside biology, for example in the educational field. In the field of computer science, taxonomies are used for a variety of purposes, including aiding in the understanding of complex domains (Meier, 2005) and highlighting design and engineering differences of state-of-the-art systems within a particular domain (Yu, 2005). A hierarchical taxonomy, such as the one presented in this chapter, is a tree-structure of classifications for a given set of objects. At the top of the structure is a single classification, usually called the root, which is applicable to all objects. Nodes below the root represent more specific classifications, which apply to subsets of the total set of objects.

In general, the development of a taxonomy is often a useful technique to help deal with a complex domain. By offering a framework and vocabulary to reason about the domain, a good taxonomy can help reduce the complexity of a large domain with many interacting concerns into a number of well-defined concerns that can be dealt with more easily. The creation of a taxonomy for a given domain is often connected with a maturation of that domain.

This chapter presents a general-purpose taxonomy of computer-augmented sports systems. The taxonomy is based on the findings from an extensive review of the domain of sensor-augmented sports systems in which twenty systems were identified and analysed in detail, incorporating both commercial applications and research projects. To present the taxonomy we employ a number of example systems from the survey to illustrate categories and give an indication of the number of systems that belong to the different categories. As for any taxonomy, the overall purpose of ours is of course to aid in understanding a complex domain. Specifically, our taxonomy is intended to have the following purposes:

- The taxonomy can be considered a map of the complex domain of computer-augmented sports systems. Its clear presentation and separation of concerns can help researchers, designers and developers approach the field in a systematic fashion. They can analyse concerns across many applications and obtain a good level of domain understanding without undertaking expensive and time-consuming analysis of many individual systems.
- The taxonomy allows any computer-augmented sports system to be classified in term of its key characteristics with regard to form as well as function. Any solution can be broken down into a set of constituent components and defining characteristics, which allows it to be placed it in the greater context of the field of computer-augmented sports systems. In this way, the taxonomy can be considered a tool to help analyse the state of the art in this rapidly developing field.
- The field of computer-augmented sport systems is by its very definition multi-dis-

ciplinary. There are many actors within the field, for example sports scientists, coaches, athletes, user interface designers as well as hardware and software engineers – groups that tend to use different vocabularies. The taxonomy can be useful in providing all of these multidisciplinary teams with a framework and common vocabulary with which to discuss applications in this field.

- The taxonomy can aid in the design and implementation of computer-augmented sport systems by allowing users to classify any prospective system they intend to build and compare it to applications that have already been constructed. A system designer can use the taxonomy to analyse the domain of computer-augmented sports systems and place their intended system in the Function side of the taxonomy. They can then use systems which closely resemble the Function classification of their system as a base to begin designing their system. This allows system designers to leverage existing work in the field.
- Because the taxonomy includes an index of a good range of systems currently available on the market and under development in research labs, it allows the reader to identify sparsely populated areas of the domain in a systematic fashion. These are areas which may be particularly promising as topics for future research.

When creating a taxonomy, decisions are made about what the categories of interest are. To a certain extent, this results in taxonomies being subjective in the sense that they are informed by the view taken by their authors. For this reason, many different taxonomies are possible for any one domain. In creating the present taxonomy, analysis of the ease of computer-augmentation and type of augmentation required were the single most important factors when deciding which categories to partition the applications into. For example, when deciding to include Environment as a category in the Sports Domain section of the taxonomy, we concluded that the size of the area to be augmented was a significant factor in any system, one which significantly divides the systems, because a system designed to operate in a 10x10m area is likely to be implemented significantly differently to one designed for an unbounded outdoor environment.

Our taxonomy deals with two essential aspects of the domain: the function of individual solutions (what their intended purposes are) and the form of the solutions (the way they are implemented). The taxonomy's scope is restricted to solutions within a certain set of well-defined functional domains (discussed below) but covers solutions of any form within those functional domains. For each of the two aspects, we adopt the paradigm of ubiquitous computing (Weiser, 1991) as a natural starting point for understanding the computational augmentation of the sporting world. In the following, we describe our approach to function and form in turn and make a case for why ubiquitous computing constitutes a suitable way to approach our chosen domain.

## THE TAXONOMY

This section introduces the taxonomy. We first describe the notation used and outline the method for presentation, then introduce the four example systems that are employed to illustrate the categories. Finally, we present the root node, which is the top category in the taxonomy.

### Organisation

The taxonomy is organised as a hierarchical directory structure with each node representing a property of computer-augmented sports systems. Each node in the directory consists of a textual description of that node and at least one line connecting the node to other nodes. To use the taxonomy to categorise an application, the user moves through the taxonomy selecting the paths through the directory, which are appropriate for the application. After this process, the user is left with a list of nodes with no child nodes, called leaf nodes, which describe the properties of the application. We use 0..n style notation to specify that a particular pathway may have many copies of a sub-directory, e.g., an application can consist of multiple sensor-augmented sports artefacts. The leaf nodes of the directory structure are mutually exclusive, except where specified with 0..n style notation.

The full taxonomy is too large to view comfortably on a single page. For this reason, we present it as a series of diagrams of sub-directories and show the higher level directories in a truncated fashion.

We present the taxonomy by performing a depth-first traversal of the taxonomy directory structure. For each node we motivate the reasons for the inclusion of the node and describe the node itself before proceeding to describe the child nodes in the same manner. For leaf nodes, we give examples using systems from our survey.

## Example Sports Systems

We have selected four systems from our survey, which will be used as examples in order to illustrate the taxonomy as we present it. Each computer-augmented sports system has one or more application scopes, which partially describe what function it is intended to perform. We have selected the example systems based on their application scope, choosing one from each of refereeing, safety, training and sports entertainment.

### SensorHogu®

Chi et al. (2004) have developed SensorHogu, a system to aid judges in scoring taekwondo matches. Taekwondo is an extreme full-contact sport between two opponents. Opponents score when they deliver a powerful and accurate punch or a kick to one of the legal scoring areas on their opponent's body. The SensorHogu system uses piezoelectric sensors embedded in body protectors worn by the competitors to indicate when and where the player has been hit and at what force their opponent has hit them. The data is transmitted wirelessly to a computer that adjusts the scores accordingly and displays the point. The point must also be verified by two of the three judges located around the ring. The system provides each of the judges with a handset, and on observing a hit the judge presses a button which relays the decision wirelessly to a central computer. All the data transmitted by the competitors and judges is received by this computer, which records the score. The system performs under soft real-time constraints. It also encrypts all the wirelessly transmitted data to ensure that competitors cannot tamper with the data. SensorHogu falls within the refereeing application scope.

### Poseidon®

Poseidon (VisionIQ, 2008) is a computer vision system that alerts lifeguards to drowning swimmers. It uses both underwater and overhead cameras to detect motionless swimmers and alerts lifeguards using a pager and via a control desk. This gives the lifeguard the position of the swimmer such that they can rescue them. The system is installed and in active use at a large number of sites across America and Northern Europe and has been credited with saving multiple lives. Poseidon uses cameras with redundancy, a central computer, waterproof alarm pagers to give audible and vibration alerts to lifeguards and a supervisor workstation to monitor the system, zoom in on suspect areas and replay any dangerous situations. The Poseidon application is categorised as having a safety application scope.

## Autonomous Sports Training from Visual Cues

Smith et al. (2003) have developed a system that uses visual cues to determine an athlete's movements. The information obtained allows the system to analyse the movements of the athlete in respect to a technically correct motion that the system has learned. Feedback is provided to the athlete based on the differences between the two. To demonstrate their system, Smith et al. have used the sport of golf. They chose this sport for its well-defined technically correct swing, and for the limited movement of the golfer, which allows the entire golf swing to be seen by a camera without moving. The system uses a single camera facing the golfer in combination with a golf swing analyser mat. Smith et al. use a modified version of the Golftek® (2008) mat. The mat provides the system with extra capabilities for detecting small changes in the club head that can not be picked up through the use of vision techniques. The vision system that is used by Smith et al. captures the posture of the golfer and translates it into a model of the golfer that can be analysed by the system. Once the analysis of the golfer's swing has been completed, the system is able to provide feedback to the golfer. The system highlights the differences in the golfer's swing as compared to a technically correct swing that the system has learned from professional golfers. A video replay is shown with the correct posture superimposed onto the video. Advice on how to improve swing technique is also provided to the golfer. The system is intended for beginners and improving golfers but not to be used as an analysis tool for top golfers. Autonomous Sports Training from Visual Cues is in the training application scope.

## Hawkeye®

Hawkeye (2008) is a vision-based system that uses several cameras positioned around a field of play to create a 3D representation of the ball movement and create replays and attempt to predict future ball location for ball-based sports games. It is used in cricket to attempt to predict 'leg before wicket' (LBW) decisions for television audiences and to give virtual replays from various angles of the action around the crease. In cricket, Hawkeye uses six cameras positioned around the playing ground to track the ball from when it leaves the bowler's hand to when it stops moving. This information is then used to create a virtual representation of the world. The ball's path can then be investigated from any angle and predictions made about where the ball might travel to after it hits the batsman's leg for LBW decisions. Hawkeye can also be used to show statistics about the sport in question, for example the areas to which specific bowlers have been bowling. This information can be supplied to television audiences but would also be of interest to players and coaches of the sport. Because Hawkeye's role is to provide additional information for television audiences, it falls into the sports entertainment application scope.

# The Root

As shown in Figure 1 we categorise every computer-augmented sports system at the highest level as having both function and form. This is a classic division used in many disciplines and we use it here in a similar manner. The function of a system describes the capabilities of a system and at a high level describes the problem that the system was designed to solve. Function is relevant so that we can compare different systems with similar functional requirements and analyse the different forms that were used to implement them. In comparison, form is the implementation of the functional requirements of the system. Form is relevant so that we can better analyse the different technologies and design decisions used. Whilst function and form are inherently linked, there is no one-to-one relationship between the two, and many different forms can be used to realise

*Figure 1. Function and form in augmented sports systems*

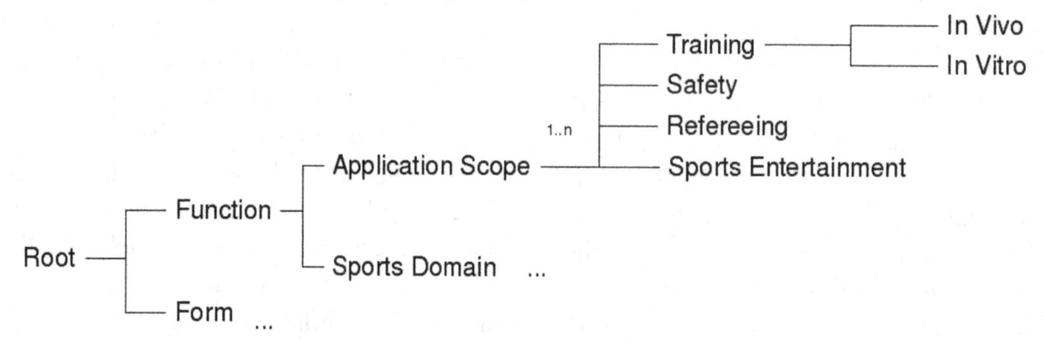


identical functional requirements. The next two sections, which form the bulk of the chapter, discuss function and form in turn.

## FUNCTION

In order to compare different computer-augmented sports systems, we must have some way of categorising the problems that they are intended to solve. The function category of the taxonomy allows us to do just this by categorising the application scope and sports domain of the computer augmented sports system.

### Application Scope

The application scope category captures the intended aims of the system from a sports engineering point of view. Our survey identified four main application scopes; sports entertainment, training, refereeing and ensuring the safety of the participants. These application scopes are displayed in figure 1. Combinations of application scopes are possible, although few solutions of those surveyed had multiple scopes. Of these few that did, combinations of refereeing and sports entertainment were the most common.

A considerable number of sports augmentations are intended to improve spectator experiences. Typically, a sport is augmented with sensors that record statistics and other information and present this data to interested spectators. The Hawkeye system for cricket is one such system. As mentioned in section 2, it provides information such as bowling patterns and analysis of LBW decisions for the television spectators' entertainment.

The largest number of computer-augmented sports technologies surveyed fall under the category of training applications. Training applications cover both professional athletes fine tuning their performance and interested amateurs wishing to advance their game. These training applications typically monitor the user performing activities related to the sport and provide the user with some means to improve these activities. There are two further categories which significantly partition the field of training applications.

Training applications, which are designed to be used while the participant actually takes part in a sport, are described as in vivo. Although use of the training equipment may or may not be permitted in the rules of the sport, what is captured here is the ability of the participant to engage in the full range of movements and interactions in the environment in which the sport is played. In vivo training applications are usually desirable, because they allow the participant to be trained while participating in the sport. This allows the system to analyse the athlete's performance in their normal sporting context and also has the potential for combining the system feedback with the athlete's subjective experience. While none of our four detailed example systems are categorised as in vivo applications, there are examples such as general-purpose heart rate monitors and the popular Nike+iPod sports kit (Nike, 2008). While relatively few applications fall into this category, we consider this a promising type of application for ubiquitous computing research, because it encompasses many of the challenges of the field, including inherent mobility, unobtrusiveness, integration of the computational elements and communicating with the participants while at the same time demanding the minimum of attention focus possible.

When a training application is designed to be used outside of the environment in which the sport is played we describe it as in vitro. These applications tend to focus on one or more elements of an athlete's game and allow them to improve these elements in isolation from the rest. From a technical standpoint, the advantage of this approach is that it is generally easier to augment a restricted lab environment with sensors and computational ability than it is to augment an entire sporting environment. The disadvantages are that the subjects are analysed out of the context of the sport and may develop skills which are not properly incorporated into their game. The research project Autonomous Sports Training from Visual Cues (Smith, 2003) is an example of a sports system which falls into this category.

Our survey found that there is a steady growth in the amount of technology being deployed to aid refereeing of sports. Increased economic value of sports and exposure of refereeing errors–due in part to pervasive television coverage–has led to a larger push for more objective refereeing and hence to computer systems to this end. Some sports have very complicated or precise scoring rules which traditionally have proved error prone and computers and sensor-augmentation has helped overcome this. The SensorHogu system is a good example of a system within the refereeing category because it can be used to help score taekwondo bouts by detecting blows that otherwise might be difficult for referees to observe.

Physical safety of athletes is a major concern in many sports. Most sports involve vigorous physical activity, sometimes in hostile environments. Example sports include mountain climbing, sailing and swimming. A relatively small number of computer systems have been developed to aid athlete's in these situations. We have identified two sub-categories of safety: prevention of chronic injuries and detection of emergency situations.

Many athletes suffer from chronic injuries related to their particular discipline. A large number of these injuries are preventable and can, for example, relate to bad posture, which can be detected and prevented. There is a large amount of ongoing research in the field of sports injury prevention, and for this reason, we consider it an application scope with excellent potential for future research.

A category of safety systems are designed to raise alarms when athletes are potentially in danger. These systems have the potential to make some sports less dangerous and more accessible. The Poseidon Drowning Detection System is an example of such as system.

## Sports Domain

The sports domain category describes the sport to be augmented as opposed to the intended aims of the augmentation captured in application scope. The properties of the sport play an important role in determining the form of the solution.

The first category under which we analyse the sports domain is environment, which is shown in Figure 2. We broadly classify the environment as either indoor or outdoor. Outdoor environments are generally larger than indoor environments which results in them being harder to augment. Also environmental interference (e.g., rain interferes with some electromagnetic waves) and a general lack of infrastructure (e.g., electricity, physical structures on which to attach sensors) make augmenting outdoor environments more difficult.

*Figure 2. Sports domains and application scope*

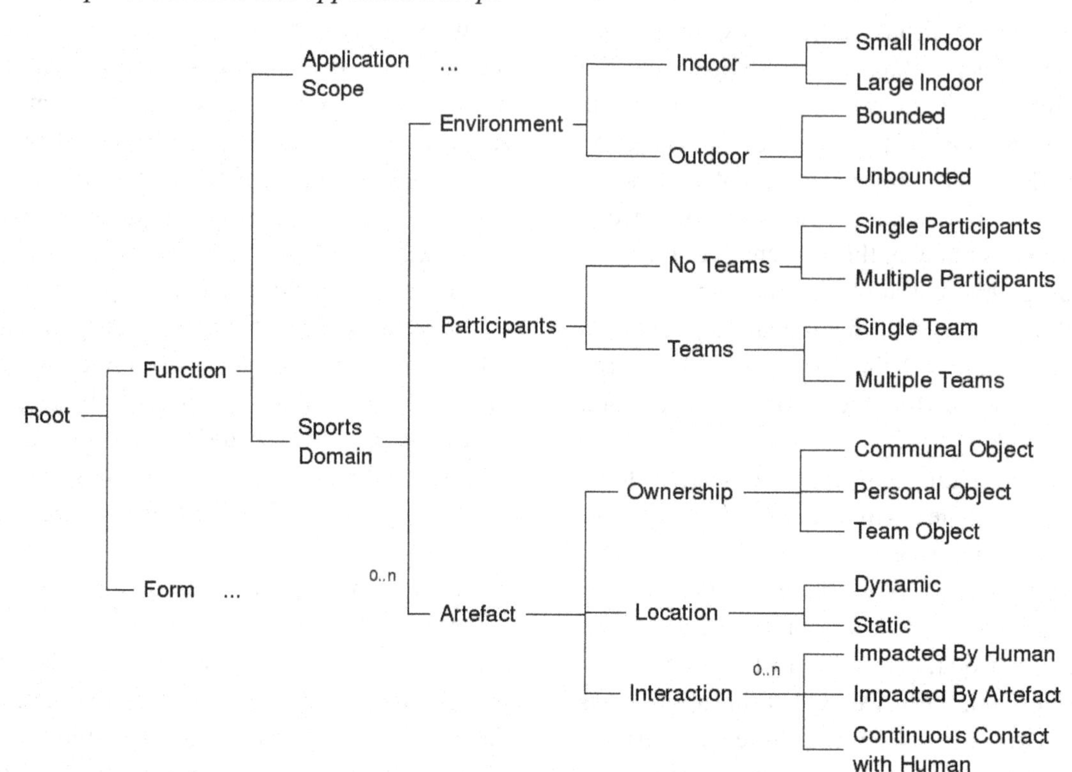


We divide the indoor environment into the categories of small indoors and large indoors. By small we mean environments in the order of hundreds of squared meters. This division is rather loose but is flexible enough to suitably categorise all indoors environments. For example, table tennis or badminton, squash and other racket-based games are examples of sports played in a small indoor environment. Ice hockey, soccer in an indoor stadium, swimming and basketball are on the other hand examples of indoor environments which we would categorise as large. As the environment increases in size, the complexity of augmenting it increases. For example, wireless communications become an issue when moving from small environments to larger ones, because of limitations in wireless transmission range. Bluetooth® and ZigBee® communication typically have ranges of less than ten metres while Wi-Fi ranges from 30 metres to 100 metres depending on environmental conditions.

We divide the outdoor environment into bounded and unbounded environments. A bounded environment is a predefined area in which the entire environment is known prior to the sport taking place and where participants are able to engage in the sport in this area but not outside of it. An unbounded environment on the other hand is one which is not known prior to the start of the sport. Hill walking, rock climbing and cross-country skiing are all examples sports which are typically performed in unbounded outdoor environments.

Because taekwondo competitions occur in a 10 metre squared area, SensorHogu is classified as having a small indoor environment. Cricket, on the other hand, takes place in a large field with marked boundaries, and Hawkeye therefore operates in a bounded outdoor environment.

We classify the environment in which the sport is performed so carefully, because it strongly affects how the environment can be augmented.

Different technologies are suited to different environments and various different considerations must be taken into account for varying environments. For small indoor environments, it is relatively easy to augment the entire environment. Augmentation becomes progressively more difficult when we augment large indoor spaces and even more so when we attempt bounded outdoor spaces. The hardest, and therefore the most challenging, environment is undoubtedly the unbounded outdoor environment.

Another factor is the number of participants and artefacts (e.g., sports requisites) in the sport. As the number of participants increases, the complexity of the system generally increases too, and this is especially true of systems with many artefacts which must maintain some form of global state. Therefore, knowledge of the number of participants and their relationships to each other from a sporting point of view is also of interest.

The number of athletes involved in a sport and their relationship to each other has a significant impact on the computer system that must be created to augment it. Single participant sports are obviously the easiest to augment, all else being equal. Augmentation becomes more difficult as more participants and teams are added. The swimming domain which Poseidon augments is categorised as having many participants and no teams. Hawkeye in contrast in the cricket domain has multiple participants and multiple teams.

We define a sporting artefact as any physical object used whilst taking part in the sport. This includes artefacts explicitly recognised in the sport, e.g., footballs in soccer or tennis rackets in tennis, but also includes physical objects which are permissible in the sport but may not be central to the sport, e.g., items of clothing on the athlete. It is important to note that many sports have more than one artefact.

Ownership is a means of describing the relationship of the artefact to the participating athletes in the sport. We classify an artefact as either a personal object or a communal object. A personal object is one which is associated wholly with one participant. This category includes examples such as soccer players' boots and a cyclist's bicycle. A communal object is one which is shared between one of more participants, e.g., a soccer ball or a table tennis table. Artefacts which are specific to a particular participant allow us to infer additional information from their activities and so are different from shared artefacts. This category allows us to catch this additional information.

The ability of the artefact to move is of importance to us from a computer augmentation point of view, because mobility in general makes artefacts harder to augment. Therefore we have categorised artefacts with respect to location as either static or dynamic. In general, dynamic artefacts are significantly harder to augment than static artefacts. Examples of dynamic artefacts are soccer balls and tennis rackets while examples of static artefacts are rugby goal posts and fences in horse show-jumping.

How the artefact interacts with participants and other artefacts in the sport is of importance as it plays a role in how the artefact should be augmented. We define three main categories of interaction: impact by human body, impact by an artefact and continuous contact with a human.

Impact by a human body describes situations where an artefact is momentarily in contact with a human, and the human exerts some force on the artefact. Examples include punching a punch bag and throwing a rugby ball.

Impacted by an artefact categorises situations when an artefact comes into contact with another artefact. This type of interaction is evident in most sports with artefacts. Examples include two foils colliding in fencing and a golf club connecting with a golf ball. The significant difference between impacting with a human body and impacting with another artefact is centred around the complexity of augmenting the human body with respect to an artefact. It is possible to augment an artefact by constructing sensors and computational devices inside the artefact in addition to using methods in which the sensor is located off the artefact, but in the case when an artefact is being impacted by a human, it is harder to augment the human and so in general the artefact must be augmented or the sensing and computational mechanisms must be moved into the environment.

When the artefact is being held or otherwise attached to an athlete we describe it as being in continuous contact with a human. Examples include sports helmets, protective pads such as those worn in martial arts and boxing gloves.

From our example applications, Hawkeye has the most artefacts and we classify the relevant ones – from the point of view of the application – separately. The cricket ball is a communal object as it is shared among all the players. Its location is dynamic and it is impacted by humans (when thrown) and impacted by artefacts (when hit by the cricket bat). The bat is a personal object which is dynamic and in continuous contact with a human. It is also impacted by another artefact, namely the cricket ball. Lastly the cricket stumps are considered communal objects which are static and impacted by artefacts. It must be noted that there are many more artefacts that can be analysed in a similar way in a game of cricket, such as gloves, helmets and boundary ropes, so it is required to make some subjective decisions when considering what to include in the analysis.

Poseidon on the other hand is trivial with respect to artefacts as there are none apart from optional accessories, e.g., swimming hat, goggles and swimsuit.

Our classification of artefacts used in sport is intended to quickly describe the key characteristics of these artefacts and to highlight artefacts that are similar from an augmentation point of view. While it may appear trivial to point out that a squash racket is similar to a tennis racket, when using our taxonomy one can quickly recognise the similarities between a hockey stick, a golf club and a table tennis paddle. We can therefore identify other augmented sporting systems that may be similar when trying to augment an artefact that is close to another in the taxonomy.

## FORM

The notion of ubiquitous computing is adopted as a starting point also with regards to the form of the solutions and the associated technology implementations. A key philosophy in all ubiquitous computing work is the seamless integration of computing ability in the environment in which we live. Many of the new technologies used for computer-augmented sports systems, such as sensors, small computational devices and short-range wireless communication, are technologies that belong in the ubiquitous computing domain. Mark Weiser, who coined the term 'ubiquitous computing' to describe such environments, also proposed the notion of 'calm technology', meaning technology that operates in an unobtrusive manner, allowing the user to focus on their activity and not be distracted by the technology itself (Weiser, 1991). This unobtrusiveness is of particular relevance for applications intended to augment the sports domain because a minimum of disruption to the athlete is essential for the adoption of any sports technology.

In the context of computer-augmented sport, the design pressure to minimise disruption has clear effects on the form of the solutions. For example, a sensor and computation module used to computationally augment a golf club must have a small physical footprint in order not to interfere with the athlete's performance during swings. Hence, the size, weight and positioning of the system is critical. In this fashion, computer-augmented sport solutions are more demanding with regard to form than many other ubiquitous computing applications, such as office or entertainment applications.

The physical footprint is only one of several design parameters affected by the design pressure to minimise disruption, and performing augmentation of sports involves a range of other challenges. These challenges include supporting communication between various devices, ensuring computational robustness and fault tolerance of artefacts and ensuring good power utilisation.

In order to categorise the form of computer-augmented sports systems we sub-divide it into the categories of hardware, software and computer human interaction.

## Hardware

The category of hardware is used to describe the physical electronic objects that are used to realise the solution. These hardware components directly influence the system's capabilities.

As shown in Figure 3, we divide the category of hardware into organisation and characterisation. Organisation relates to the position of the physical components and specifically if they are located on an artefact (as defined in the sports domain section) or in the environment. In the characterisation category we analyse the hardware with respect to resource capabilities – which is used to describe how restricted the application is by the hardware capabilities of the system – and sensors and actuators used.

The category of organisation allows us to identify where the various parts of the solution are physically located. This is important to any implementation because it has implications for the degree of distribution of software used to co-ordinate the hardware and also has influence on what types of sports can be augmented with any given configuration of hardware. For example, if a solution requires hardware to be placed in the environment, it is better suited to augment indoor environments but considerably less suited for outdoor unbounded environments. As stated previously, the form factor of the artefact is also influenced by the organisation of the hardware, because addition of hardware to an artefact

*Figure 3. Hardware organisation and characterisation*

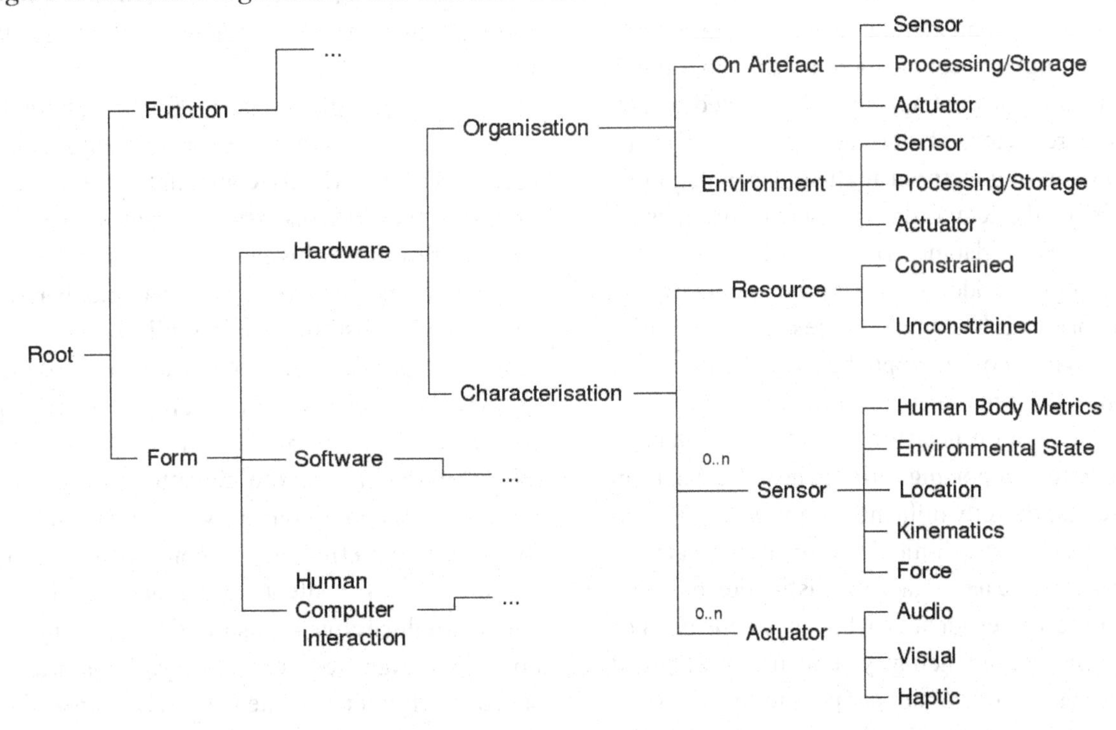

will increase its weight and possibly change its physical structure. In SensorHugo, the hardware is organised such that the force sensors used to detect blows are embedded in body protectors (on artefact), while Poseidon places all of its cameras around the pool (in the environment). Both systems also perform additional processing in the environment. SensorHogu correlates scores and handles real-time communication with the referee's handsets and the artefacts and Poseidon performs analysis of the images from the multiple cameras and controls the alerts and display functionality provided to the operators.

Where organisation deals with where the hardware is physically located, characterisation is concerned with the qualities of the hardware. We describe the availability of resources in the system under the general headings of resource constrained and unconstrained. This is a general indication of the amount of computational resources available to the application. To determine if the application is resource-constrained we ask the question: would more resources be beneficial in helping the system achieve its goals? Examples of relevant resources include battery power, processing speed, memory and network bandwidth. SensorHogu is categorised as unconstrained, because it does not appear to be limited by any hardware factors. Battery power which might be a constraining factor in applications with mobile augmented artefacts does not seem to be an issue with the SensorHogu, as it lasts for 4 days between charging. Poseidon is also in the unconstrained category as it has ample processing capabilities and a wired power supply because the hardware is embedded in the environment.

Sensors are crucial to any application intended to augment a sporting activity, because what can be sensed directly influences how the application works and what its limitations are. In our treatment of sensors, we have made the distinction between the type of sensor and what it is sensing. For example, some fencing systems use mechanical force sensors on the tips of épées to sense correct "touches". We consider this sensor to be in the same category as piezoelectric force sensors rather than categorising it with other mechanical sensors, which might be used to measure very different phenomena. By analysing the sensors with respect to what they are sensing as opposed to the type of sensor they are, we are able to categorise the area without needing to list every sensor available. SensorHogu is classified under this scheme as containing force sensors whilst Poseidon is classified as using kinematic sensors.

Actuators allow augmented systems sports systems to interact with the participants in the sport. We categorise actuators in a similar manner to our treatment of sensors in that we analyse the sensor modality used to transmit the feedback from the system to the interested parties. Therefore with reference to our example systems, Poseidon uses portable beepers carried by lifeguards while SensorHogu uses no actuators.

It is expected that as technological research progresses, advancements will mean that previously resource-constrained applications can be implemented in a non-resource-constrained fashion resulting in a better application. The category of resource is useful therefore in identifying applications that are promising with respect to future work.

An attempt might be made to further subdivide the resource constrained category by analysing the specific resource which is constraining the application. However, because of the inter-dependencies between resources, this easily becomes confusing and counter-productive. As an example, battery capacity can constrain certain applications, and in order to conserve battery power, a lower processing speed is often used as a means of conserving power. Wireless communication on the other hand can be extremely wasteful of battery power, and it is therefore common to use processor cycles to increase the efficiency of communication by compressing the content of communication. The desire to reduce power consumption results in two opposing design pressures with regard to processor speed. Such interdependencies make separating the concerns further very difficult.

## Software

We categorise the software elements of a system in a traditional manner by describing the functional and non-functional software features of the system. Functional features describe what the software system does while non-functional features describe qualities of the system. We choose these divisions to reduce the overhead in becoming familiar with the taxonomy and because they allow us to categorise the software along two orthogonal axes which are highly relevant to sports systems.

### Non-Functional Features

Non-Functional features of a software system are qualities of the system. They are prevalent throughout the system and are not explicitly concerned with the intended aims of the system, although the system as a whole may be unusable without the non-functional features. Many of these non-functional features exist by default in various software components, e.g., operating systems are often concerned with security, and therefore non-functional features may come "for free" in the implementation. In this section we attempt to capture if these non-functional features influenced the design and implementation of the system, e.g. was a real time operating system chosen over an alternative because it provided additional timing guarantees.

Security is the robustness of the system in the face of external influences, such as the ability of the system to function normally despite the existence of a malicious attacker. This takes the form of everything from robustness in the face of denial of service attacks through to malicious attackers attempting to exploit security flaws and viruses attacking the system. There are many

*Figure 4: Software features*

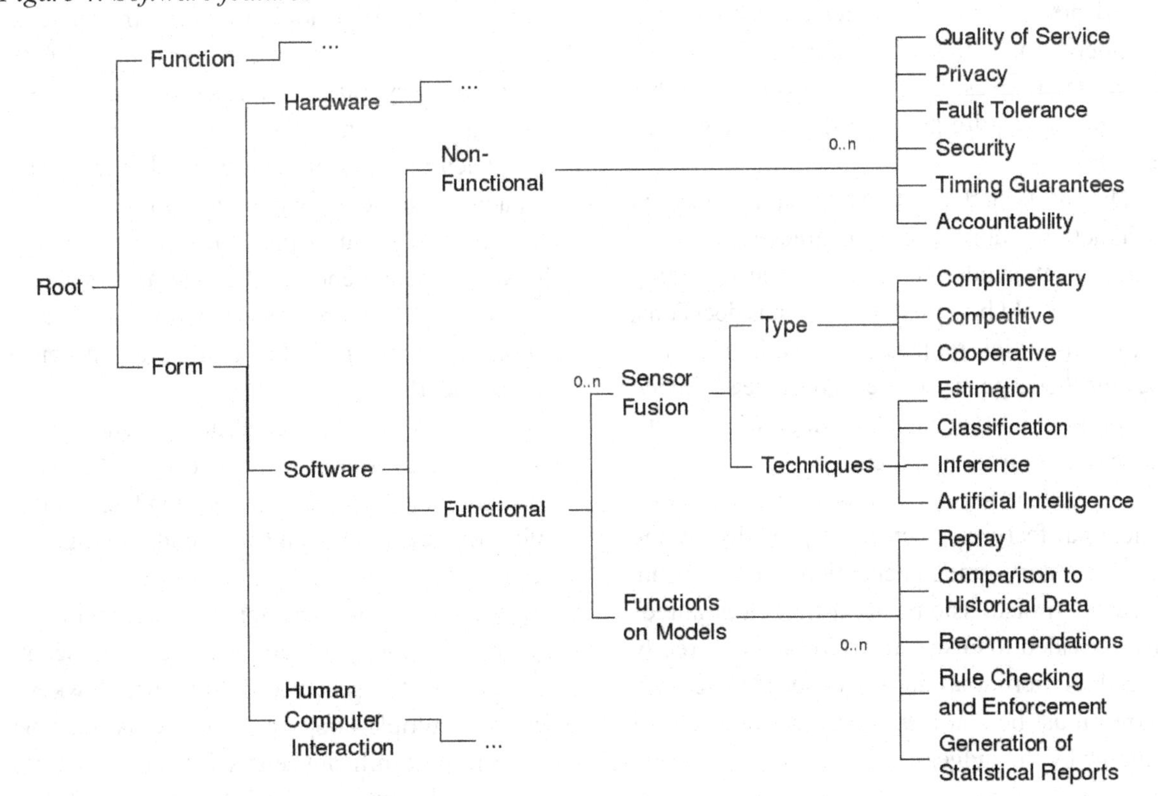

forms of security that can be employed to protect a system and a comprehensive analysis of these security features are outside the scope of this taxonomy.

Privacy is concerned with the confidentiality of the data and athletes involved in the system. Privacy is of crucial importance in most competitive sports and systems that do not provide sufficient privacy guarantees to their users may find it hard to gain acceptance. This is true across the spectrum of application scopes in this domain.

Fault tolerance describes the ability of the system to survive failures of components in the system be they software or hardware components. The robustness of the system is more important in some application scopes than others. One would expect for example that training applications have a lower need for robustness and that safety systems would require a very high degree of fault tolerance.

In order for applications to be used in some sporting settings timing guarantees will be required. Refereeing systems for example have a tendency to place high demands on timing considerations with some needing soft real-time timing guarantees. Safety systems on the other hand might require hard real-time timing guarantees.

The ability of the system to be able to account for its actions is an important characteristic in certain application domains. For example refereeing systems should be able to justify their decisions, e.g. by providing replays or by explaining how they reached a particular decision, at least during the initial adoption phase of deployment until people develop trust in them.

The SensorHogu system provides many non-functional features. Security is provided by the use of military grade encryption to encrypt the message transmissions. Real-time guarantees also provided to ensure that blows are correctly scored as there is a temporal aspect to scoring (blows must be confirmed within a one second window by the judges).

Poseidon too provides real-time image processing and redundant cameras for fault tolerance.

## Functional Features

Functional features of a software system are the aspects of the system which deal with implementing the application scope. In describing the functional features of an augmented sports system, we draw a distinction between features which refine sensor data to create some internal representation of the environment and to features that operate on these representations. The first category we class as sensor fusion and the later as functions on models.

Sensor fusion is a very important characteristic of any system relying on sensed data. It involves the merging of multiple inputs into a common representation and becomes important as soon as we have more than one sensor in the system. As the number of sensors and types of sensor used in the application increase the complexity of fusing this information also increases.

We divide the fusion into three different categories as defined by Durrant-Whyte et al (1988); complimentary fusion, competitive fusion and cooperative fusion. Complimentary fusion occurs when the environment that is being sensed by each sensor does not overlap. Competitive fusion occurs when more than one sensor is sensing the same phenomenon and cooperative sensing is when two or more sensors work together to sense a phenomenon that neither could sense independently.

We categorise the techniques used to fuse sensor data in four main categories. These are estimation techniques, classification techniques, inference techniques and artificial intelligence techniques.

Due to commercial sensitivities technical information for commercial systems is scant. Despite this it is safe to conclude that Hawkeye with its multiple cooperative cameras must be performing significant sensor fusion to generate a 3D model from multiple 2D images.

## Functions on Models

Many sports systems generate statistics to be displayed back to users. The Hawkeye system for example can generate bowling statistics on a bowler by bowler basis.

Many augmented sports systems provide a replay function to the users allowing a particular move or event to be replayed, in camera based systems this usually corresponds to replaying the time segment in question, but in other systems a representation of the internal model is generated and displayed to the user. From our example applications, Poseidon has the ability to replay dangerous situations after the event. Hawkeye too can display a model of ball flight which can then be rotated and viewed from any angle.

Some augmented sports systems, particularly those in the refereeing and safety scopes, monitor a sport and preform an action when certain rules are broken. This can be anything from creating alarms to awarding scores but usually is heavily domain dependant. For example the taekwondo refereeing system SensorHogu evaluates when certain scoring rules become true and award points based on these rule evaluations.

A particular form of analysis performed by many systems is the comparison to historical data of the refined sensor data. This can be used for example to compare golf swings to various professional golfers in order to provide some feedback in a golf training application. The Autonomous Sports Training from Visual Cues system compares users' swings with recorded swings from advanced players.

## Computer Human Interface

We include the computer human interface here as external to both hardware and software because although it is composed of a combination of both we consider it to be conceptually separate and a crucial element to the form of the system. As shown in Figure 5, we analyse computer human interaction under three distinct sub-categories to determine how the data is presented in terms of modality, when it is presented in terms of time and to whom the data is presented.

*Figure 5. Computer to human interface*

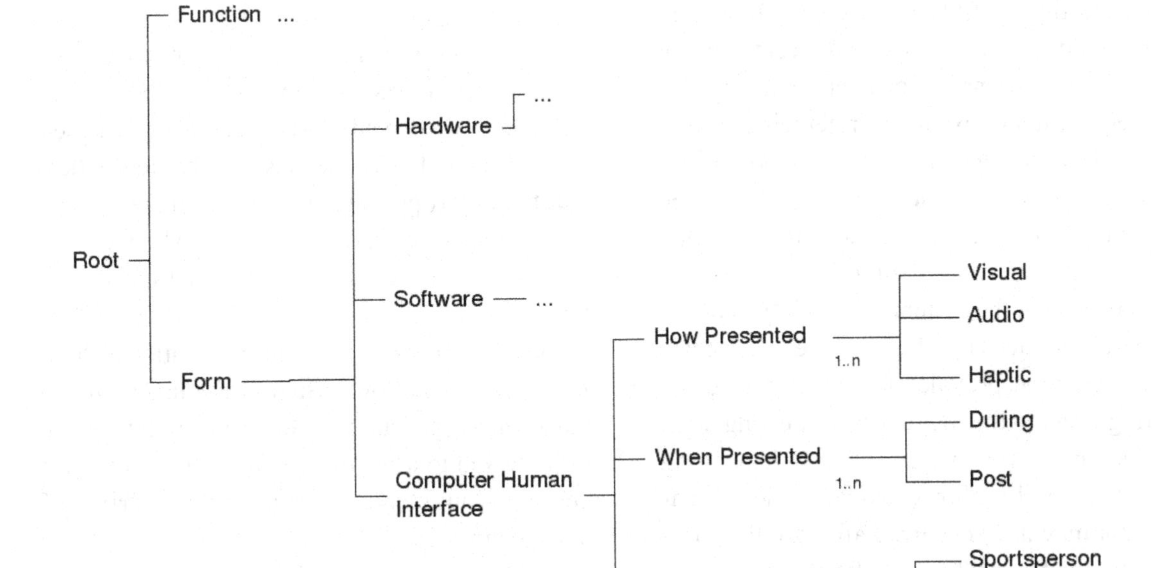

The first of these sub-categories concerns the sensory modality used to transmit the feedback from the system to the interested parties. For example a visual interface would generally not be suitable for a referee in a soccer match if the referee was expected to watch the match at the same time. A haptic interface or an audio interface might be deemed more suitable in such cases. In this fashion, the choice of sensory modality for the interface directly influences the capabilities of the system and how it can be used. Therefore, we partition the set of sensory interfaces into audio, haptic and visual interfaces. These particular modalities were selected because they are currently used in ubiquitous computing applications. If other modalities become available, the taxonomy can of course easily be extended. Using the Poseidon system as an example, it provides visual and audio interfaces to the system. It uses a monitor system with a flashing light and buzzer system to alert lifeguards. Taking the Autonomous Sports Training from Visual Cues system as our second example it provides a visual interface to users of the system through a computer display.

The second subcategory concerns the time at which the feedback is delivered to the interested parties. Data may be presented during and/or after a sporting event (e.g., a single golf swing). Presenting data during a sporting event is much harder than afterwards, however some applications require it. For example a refereeing system must operate with very strict time considerations (essentially soft real-time requirements), if the referee is to act immediately on situations that arise. These timing considerations can be relaxed if the purpose of the system is to supply analysis of a particular activity after the event has completed. Poseidon provides feedback during the sporting event to the lifeguards by alerting them and allowing them to view what is happening in the pool. It also allows people to view situations that may have occurred after the fact. The Autonomous Sports Training from Visual Cues system provides feedback after the event by allowing golfers to review their shot and compare it to other golfers.

The third subcategory concerns the recipient of the feedback. From the point of view of our analysis, the data can be presented to the sports person (i.e., the performing athlete) or to a spectator. In the category of spectator we include coaches, referees and other interested parties who may not normally be considered spectators in the traditional sense. Poseidon displays information to non-sportspeople (the lifeguards), while the Autonomous Sports Training from Visual Cues system displays information to the athlete and potentially additional non-sportspeople.

The method of interaction supported is one of the key attributes of the system and determines how it will be used. From this point of view it is imperative to classify this in some regard, and it is desirable that this description be independent of other concerns. For example the implementation details of a system might be identical with different interaction models, meaning the same infrastructure could be used to drive two very different applications.

## CONCLUSION

We have presented a taxonomy of computer-augmented sports systems which allows any computer-augmented sports system to be classified in terms of its key properties and characteristics. The taxonomy provides a clear map of the domain and defines a common vocabulary which can be used to discuss computer-augmented sports systems. We categorise these systems with relation to the problem they solve and the solutions which they use. This allows users of the taxonomy to place their particular problem in the taxonomy quickly and to draw out comparisons with other computer augmented systems over a variety of sub categories.

We have demonstrated the taxonomy by characterising various computer augmented sports systems with respect to the taxonomy. This both gives examples of how to use the taxonomy and illustrates the strength of the taxonomy.

As with all taxonomies and especially when trying to survey a new a expanding field it is inevitable that in the future some changes will need to be made to the structure of the taxonomy. We hope to have provided for this by making the taxonomy as modularised as possible and allowing various sections to be extended without requiring changes to be made to other areas of the taxonomy. We have deliberately taken a very high level approach to the categories in order to avoid discussing specific technologies which will inevitably become redundant in coming years.

## ACKNOWLEDGMENT

The authors are grateful to Enterprise Ireland for their generous support for the work presented in this chapter.

## REFERENCES

Durrant-Whyte, H.F. (1988). Sensor models and multisensor integration. *International Journal of Robotic Research, 7*, 97-113.

Ed, H., Chi, E.H., Song J., & Corbin, G. (2004). *Killer App of wearable computing: wireless force sensing body protectors for martial arts.* Paper presented at the 17th Annual ACM Symposium On User Interface Software And Technology, Santa Fe, NM.

Golftek (2008). Retrieved 13 May, 2008 from http://www.golftek.com/

Hawkeye (2008). Retrieved 13 May, 2008 from http://www.hawkeyeinnovations.co.uk/

Meier, R., & Cahill, V. (2005). Taxonomy of Distributed Event-Based Programming Systems. *Computing Journal, 48*, 602-626.

Smith, A.W.B., & Lovell, B.C. (2003). Autonomous Sports Training from Visual Cues. In B.C. Lovell, D.A. Campbell, C. E. Fookes, & A. Maeder (Eds.), *Proceedings of the Eighth Australian and New Zealand Intelligent Information Systems Conference* (pp. 171-175), Brisbane: The Australian Pattern Recognition Society.

Vision IQ (2008). Retrieved 13 May, 2008 from http://poseidon-tech.com/us/index.html

Weiser, M. (1991). The Computer for the 21st Century. *Scientific American,* 265, 94-104.

Yu, J., & Rajkumar, B. (2005). A taxonomy of scientific workflow systems for grid computing. *ACM SIGMOD Record, 34*, 44-49.

# Chapter III
# Dynamics and Simulation of General Human and Humanoid Motion in Sports

**Veljko Potkonjak**
*University of Belgrade, Serbia*

**Miomir Vukobratović**
*Institute "M. Pupin," Belgrade, Serbia*

**Kalman Babković**
*University of Novi Sad, Serbia*

**Branislav Borovac**
*University of Novi Sad, Serbia*

## ABSTRACT

*This chapter relates biomechanics to robotics. The mathematical models are derived to cover the kinematics and dynamics of virtually any motion of a human or a humanoid robot. Benefits for humanoid robots are seen in fully dynamic control and a general simulator for the purpose of system designing and motion planning. Biomechanics in sports and medicine can use these as a tool for mathematical analysis of motion and disorders. Better results in sports and improved diagnostics are foreseen. This work is a step towards the biologically-inspired robot control needed for a diversity of tasks expected in humanoids, and robotic assistive devices helping people to overcome disabilities or augment their physical potentials. This text deals mainly with examples coming from sports in order to justify this aspect of research.*

## INTRODUCTION

Currently, researchers in biomechanics and robotics are investigating many different problems in motion of humans and humanoid robots. Generalization is still missing. This general approach would be useful for several reasons. From a purely academic point of view, general methods are always seen as a final target. From a commercial point of view, a software package that can cover a diversity of motions is a more economic solution than several specialized packages. The particular argument comes from sports where a general model should cover a diversity of motion tasks imposed to a human athlete or a humanoid.

This work considers a new and generalized approach to the modeling of human and humanoid motion. In principle, modeling may follow an inductive approach or a deductive one. In the inductive approach, one analyzes different real situations like human or humanoid gait and running; playing tennis, soccer, or volleyball; gymnastics (exercises on the floor or by using some gymnastic apparatus); performing trampoline exercise; etc. Each problem needs a different model, appropriate to the situation–it should cover all the relevant effects. Once a number of situations are explored, one may try to make a generalization. However, there is no guarantee that it will be successful. In the deductive approach, one starts with considering a completely general problem. Once the general model is formulated, one may derive different real situations as being special cases. Such approach requires a serious effort to formulate a general model. This chapter is an attempt in this direction; it explains the principles, derives the general methodology and proves the feasibility and applicability on few examples. The initial results in this direction were published by Potkonjak, Vukobratovic, and their associates (Potkonjak & Vukobratovic, 2005; Potkonjak et. al., 2006; Vukobratovic et. al., 2007).

The new approach starts with an articulated system (e.g. a human body, a humanoid, or even an animal) that "flies" without constraints (meaning that it is not connected to the ground or to any object in its environment). We use the term *flier*. This situation is not uncommon in reality; it is present in running, jumping, trampoline exercise, etc. However, such motions are still less common than those where the system is in contact with the ground or some other supporting *object* in its environment.

A contact can be rigid or soft. With a rigid contact, one *LINK* (or more of them) is geometrically constrained in its motion. For instance, in the single-support phase of a bipedal gait, the foot (being a link of the system) is fixed to the support and does not move (or it moves in accordance with the motion of the support). With a soft contact, there is no geometric constraint imposed on the system motion, but the strong elastic forces between the contacted link and some external object make the link motion close to the object. Two examples of such contact are walking on a support covered with elastic layer, and a racket hitting a ball in tennis.

## MATHEMATICS

### Free-Flier Motion

We consider a flier as an articulated system consisting of the *basic body* (the torso) and several *branches* (head, arms and legs), as shown in Figure 1. Let there be $n$ independent joint motions described by joint-angles vector $\mathbf{q} = [q_1, \cdots, q_n]^T$ (the terms *joint coordinates* or *internal coordinates* are often used). The basic body needs six coordinates to describe its spatial position: $\mathbf{X} = [x, y, z, \theta, \varphi, \psi]^T$, where $x, y, z$ defines the position of the mass center and $\theta, \varphi, \psi$ are orientation angles (roll, pitch, and yaw). Now, the overall number of degrees of freedom (DOF) for the system is $N = 6 + n$, and the system position is defined by

*Figure 1. Unconstrained (free flier) and constrained system*

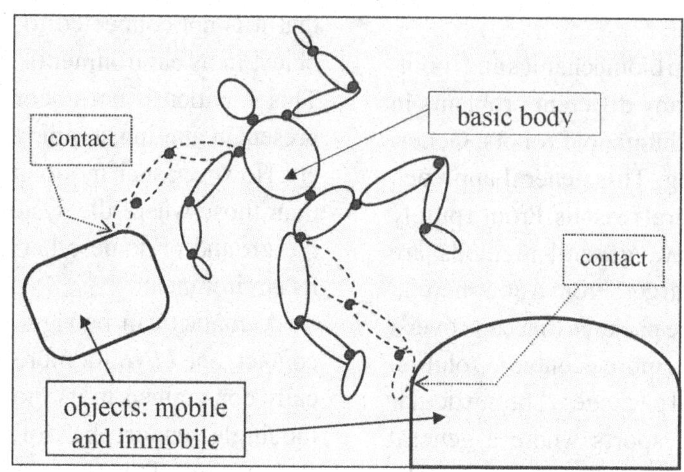


$$\mathbf{Q} = [\mathbf{X}^T, \mathbf{q}^T]^T = [x, y, z, \theta, \varphi, \psi, q_1, \cdots, q_n]^T. \quad (1)$$

We now consider the drives. It is assumed that each joint motion $q_j$ has its own drive – the torque $\tau_j$. Note that in this analysis there is no drive associated to the basic-body coordinates $\mathbf{X}$ (this is a real situation with humans and humanoids in "normal" activities, however, in space activities – actions like repairing a space station, etc. – reactive drives are added, attached to the torso; the proposed method for simulation can easily handle this situation). The vector of the joint drives is $\boldsymbol{\tau} = [\tau_1, \cdots, \tau_n]^T$, and the extended drive vector ($N$-dimensional) is $\mathbf{T} = [0_{1\times 6}, \boldsymbol{\tau}^T]^T = [0, \cdots, 0, \tau_1, \cdots, \tau_n]^T$; zeros stand for missing basic-body drives.

The dynamic model of the flier has the general form:

$$\mathbf{H}(\mathbf{Q})\ddot{\mathbf{Q}} + \mathbf{h}(\mathbf{Q}, \dot{\mathbf{Q}}) = \mathbf{T} \text{ or}$$

$$\begin{aligned} \mathbf{H}_{X,X}\ddot{\mathbf{X}} + \mathbf{H}_{X,q}\ddot{\mathbf{q}} + \mathbf{h}_X &= \mathbf{0} \\ \mathbf{H}_{q,X}\ddot{\mathbf{X}} + \mathbf{H}_{q,q}\ddot{\mathbf{q}} + \mathbf{h}_q &= \boldsymbol{\tau} \end{aligned} \quad (2)$$

Dimensions of the inertial matrix and its submatrices are: $\mathbf{H}(N \times N)$, $\mathbf{H}_{X,X}(6 \times 6)$, $\mathbf{H}_{X,q}(6 \times n)$, $\mathbf{H}_{q,X}(n \times 6)$, and $\mathbf{H}_{q,q}(n \times n)$. Dimensions of the column vectors containing centrifugal, Coriolis' and gravity effects are: $\mathbf{h}(N)$, $\mathbf{h}_X(6)$, and $\mathbf{h}_q(n)$.

## Contact Motion

Let us consider a *LINK* that has to establish a contact with some external *OBJECT*. In one example, it is the foot moving towards the ground and ready to land (in walking or running). In the next example, in a soccer game, one may consider the foot or the head moving towards a ball in order to hit. The external object may be immobile (like the ground), an individual moving body (like a ball), or a part of some other complex dynamic system (even another flier, like in a trapeze exercise in the circus). Note that the contact might be an inner one – involving two links of the considered system, an example being the tennis player holding the racket with both hands.

In order to express the coming contact mathematically, we introduce relative coordinates $\mathbf{s} = [s_1, \cdots, s_6]^T$ to define the position of the link with respect to the object to be contacted. Let us call them *functional coordinates* (often referred to as *s*-coordinates). Choice of coordinates in a particular example is made in accordance with the expected contact – to support its mathematical description. It is common to fix the *s*-frame to the object although it is sometimes more appropriate to fix it to the link and consider the position of the object relative to the link.

A consequence of the rigid (no deformation) link-object contact is that the link and the object perform some motions, along certain axes, together. These are *constrained* (*restricted*) *directions*. Let there be *m* such directions, *m* being a characteristic of a particular contact. Relative position along these axes does not change. Along other axes, relative displacement is possible. These are *unconstrained* (*free*) *directions*. In order to get a simple mathematical description of the contact, *s*-coordinates are introduced so as to describe relative position. Zero value of some coordinate indicates the contact along the corresponding axis. For a better understanding, we use some well-known examples (Figures. 2, 3, and 4). Example of Figure 2 comes from biped gait and shows the motion of the foot after the "heel strike". Case (a) is the motion of a woman's shoe with high heel. The foot-ground contact restricts three coordinates (coordinates $s_1 = x$, $s_2 = y$, $s_3 = z$), and leave the other three free (coordinates $s_4 = \theta$, $s_5 = \varphi$, $s_6 = \psi$); so, $m$=3. Case (b) is a rectangular robot foot where the contact restricts five coordinates ($s_1 = x$, $s_2 = y$, $s_3 = z$, $s_4 = \theta$, $s_6 = \psi$) and leave only one free ($s_5 = \varphi$); so, $m$=5. Later, the "flat foot" phase of the gait will constrain all six foot coordinates ($m$=6 in Figure 2c). However, if the foot starts to turn about its edge, like in the case of lost balance (Figure 2d), free coordinate will appear again (coordinate $\psi$ in the figure); and accordingly $m$ reduces to 5. The next example, the punch in the face, shown in Figure 3, restricts one coordinate ($s_1$) while other five are free ($s_2$, $s_3$, $s_4$, $s_5$, $s_6$). Finally, Figure 4 presents the coordinates used in the case of hand-pommel contact in a gymnastic exercise on a pommeled horse: $s_1$, $s_2$, $s_3$, $s_5$, $s_6$ are restricted while $s_4$ is free.

*Figure 2. Foot-ground contact in a bipedal gait: (a) heel strike phase – woman's shoe with high heel; (b) heel strike phase – rectangular robot foot; (c) "flat foot" phase of the gait; (d) foot overturning due to lost balance*

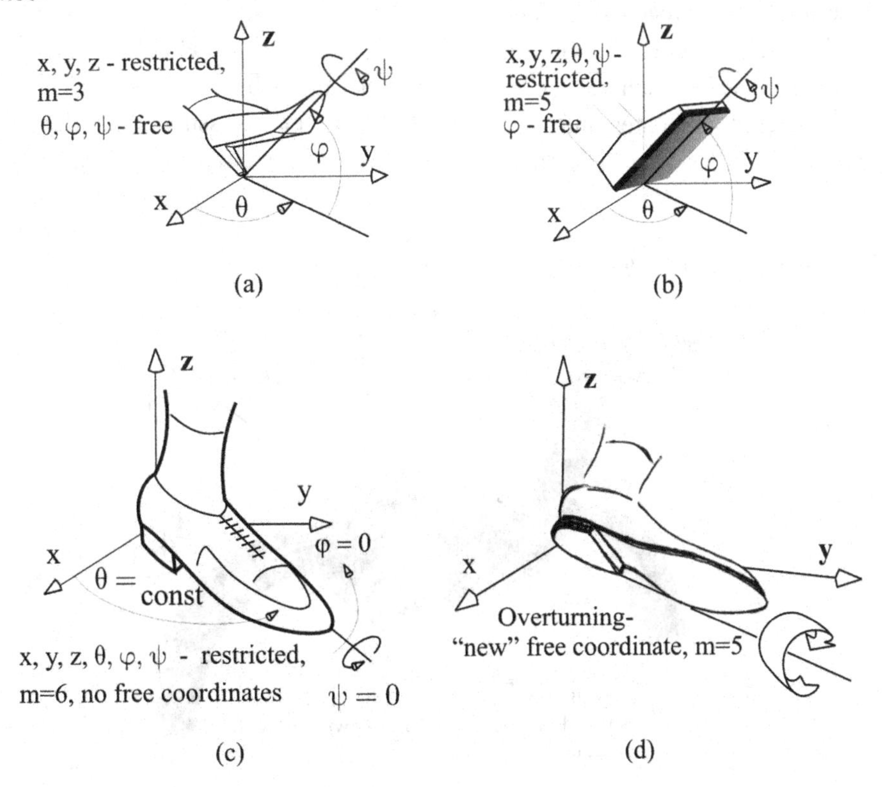

*Figure 3. Fist-head contact in a punch*

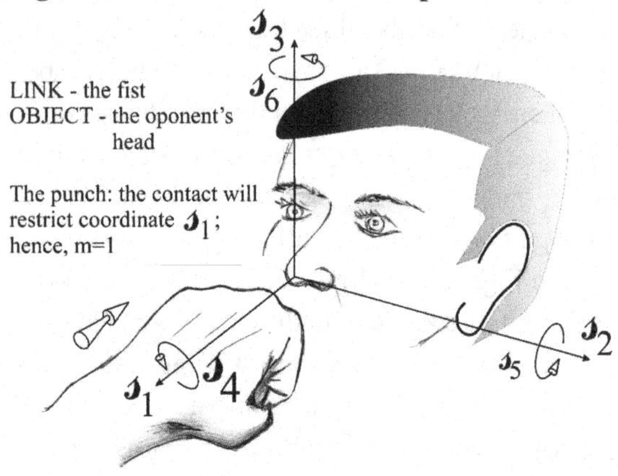


The motion of the external object has to be either known or calculable from the appropriate mathematical model, and then the *s*-frame fixed to the object is introduced to describe the relative position of the link in the most proper way. In the case of an inner contact (two links in contact), one link has to play the role of the object. Thus, in a general case, the *s*-frame is mobile. As the link is approaching the object, some of the *s*-coordinates reduce and finally reach zero. The zero value means that the contact is established. These functional coordinates (which reduce to zero) are called restricted coordinates and they form the subvector $s^c$ of dimension *m*. The other functional coordinates are free and they form the subvector $\mathbf{s}^f$ of dimension $6 - m$. Some examples are shown in Figure 2-4. Now one may write:

$$[\mathbf{s}^{c^T}, \mathbf{s}^{f^T}]^T = \mathbf{s} \text{ or } [\mathbf{s}^{c^T}, \mathbf{s}^{f^T}]^T = \mathbf{R}\,\mathbf{s} \qquad (3)$$

where **R** is a 6 × 6 matrix used, when necessary, to rearrange the functional coordinates (elements of the vector **s**) and bring the restricted ones to the first positions. Note that in different examples, the same link and object will make a different kind of contact. For instance, in handball, the hand grasps the ball, while in volleyball the hand hits the ball. Now, consider the ground (as an example of an immobile object); in walking, one type of contact of the foot and the ground exists, while in ice-skating the contact will be of a rather different kind.

Relative *s*-coordinates depend on link motion (thus on flier motion **Q**) and on object motion $\mathbf{Q}_b$ (index "*b*" will stand for "object"):

$$\mathbf{s} = \mathbf{s}(\mathbf{Q}, \mathbf{Q}_b). \qquad (4)$$

When speaking about a moving object one may distinguish two cases.

The first case assumes that the object motion is given and cannot be influenced by the flier

*Figure 4. Hand-pommel contact in the exercise on a gymnastic horse*

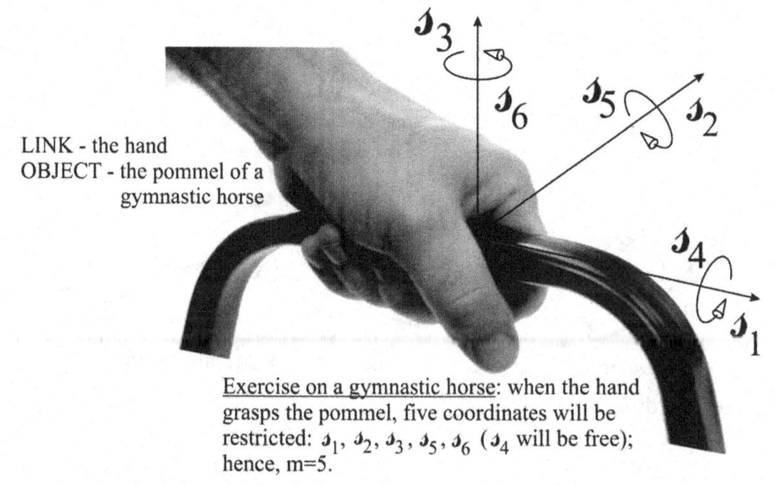

dynamics. The immobile object is included in this case. Such a situation appears if the object mass is considerably larger than the flier mass, or if the object is driven by such strong actuators that they can overcome any influence on the object motion. An example would be walking on the ground, i.e. contact between the foot and the ground. Ground is the large immobile object. So, one may say that $\mathbf{Q}_b(t)$ is known: either prescribed or calculated from the object model.

The other case refers to the object being a regular dynamic system, so that the flier dynamics can have an effect on it. The example is a handball player that catches the ball. The ball is an object strongly influenced by the player. Here, the time history of $\mathbf{Q}_b$ is a new unknown variable in the joint flier-object model.

Relation (4) can be written in the second-order Jacobian form (separation (3) included):

$$\ddot{\mathbf{s}}^c = \mathbf{J}_l^c(\mathbf{Q},\mathbf{Q}_b)\ddot{\mathbf{Q}} + \mathbf{J}_b^c(\mathbf{Q},\mathbf{Q}_b)\ddot{\mathbf{Q}}_b + \mathbf{A}^c \quad (5)$$

$$\ddot{\mathbf{s}}^f = \mathbf{J}_l^f(\mathbf{Q},\mathbf{Q}_b)\ddot{\mathbf{Q}} + \mathbf{J}_b^f(\mathbf{Q},\mathbf{Q}_b)\ddot{\mathbf{Q}}_b + \mathbf{A}^f \quad (6)$$

Dimensions of Jacobi matrices are: $\mathbf{J}_l^c(m\times N)$, $\mathbf{J}_b^c(m\times N_b)$, $\mathbf{J}_l^f((6-m)\times N)$, $\mathbf{J}_b^f((6-m)\times N_b)$ where $N_b$ is the number of DOF in the object, and the adjoint vectors dimensions are: $\mathbf{A}^c(m \times 1)$ and $\mathbf{A}^f((6-m) \times 1)$.

Classification of the contacts can be done based on the existence of the deformations in the contact zone. If there is no deformation, i.e. if the motions of the two bodies are equal in the restricted directions, then we talk about rigid contact. If deformation is possible, then the motions in the restricted directions will not be equal. Theoretically, they will be independent, but in reality they will be close to each other, due to the action of strong elastic forces. In this case, we talk about soft contact. It is clear that a rigid behavior is an approximation neglecting the always present deformation.

One class of contacts encompasses durable (lasting) contacts. This means that the two bodies (link and object), after they touched each other and when the impact is over, continue to move together for some finite time. The example is walking. When the foot lands on the ground, it keeps the contact for some time before it moves up again. The other class involves instantaneous and short-lived contacts. When the two bodies touch each other, a short impact occurs, and after that the bodies disconnect. A good example is the tennis (or volleyball) player who hits the ball. This latter class of contact is an approximation of the first one. It is clear that a general theory of impact, including the elastodynamic effects, can cover all the mentioned contacts (Vukobratovic, Potkonjak & Matijevic, 2003).

The contact-dynamics model is obtained by introducing contact reactions into the free-flier model (2). Reactions appear along the restricted directions $\mathbf{s}^c$. Reaction forces appear along the linear coordinates and reaction moments along the angular ones. In total, there are $m$ reactions forming the reaction vector $\mathbf{F}$. Introducing $\mathbf{F}$ into (2) one obtains:

$$\mathbf{H}(\mathbf{Q})\ddot{\mathbf{Q}} + \mathbf{h}(\mathbf{Q},\dot{\mathbf{Q}}) = \mathbf{T} + (\mathbf{J}_l^c(\mathbf{Q},\mathbf{Q}_b))^T\mathbf{F}. \quad (7)$$

Due to the $m$-component reaction $\mathbf{F}$, to solve the model one needs $m$ additional scalar conditions that describe the contact.

## Rigid-Contact Analysis

We first discuss rigid contact. It is described by the $m$-dimensional condition $\mathbf{s}^c = 0$. Starting from (5) this condition can be written in the Jacobi form:

$$\ddot{\mathbf{s}}^c = \mathbf{J}_l^c(\mathbf{Q},\mathbf{Q}_b)\ddot{\mathbf{Q}} + \mathbf{J}_b^c(\mathbf{Q},\mathbf{Q}_b)\ddot{\mathbf{Q}}_b + \mathbf{A}^c = 0. \quad (8)$$

Now, relations (7) and (8) involve $N+m$ scalar equations. Let us list the unknowns: $N$ scalar unknowns in $\ddot{\mathbf{Q}}$, $m$ in $\mathbf{F}$, and $N_b$ unknowns in $\ddot{\mathbf{Q}}_b$. So, the flier model can be solved if joined with the object model. To keep the discussion general, we adopt that the object dynamics is described by the $N_b$-dimensional model

$$\mathbf{M}(\mathbf{Q}_b)\ddot{\mathbf{Q}}_b + \mathbf{C}(\mathbf{Q}_b, \dot{\mathbf{Q}}_b) = \mathbf{T}_b + \mathbf{D}(\mathbf{Q}_b, \mathbf{Q})\mathbf{F}. \tag{9}$$

Now, (7), (8), and (9) describe the dynamics of a flier-plus-object contact motion, allowing one to calculate the accelerations $\ddot{\mathbf{Q}}$ and $\ddot{\mathbf{Q}}_b$, and reaction $\mathbf{F}$ (thus enabling integration of the differential equations and calculation of the system motion).

Under some condition, as discussed before, the object model (9) can be considered as independent of the flier dynamics. Then it can be solved separately and the solution used to resolve the flier model (7), (8).

### Soft-Contact Analysis

With the soft contact, it is $\mathbf{s}^c \neq 0$, and the condition (8) does not apply. Instead of it, the original form (5) is used. As a consequence, there are additional unknowns: $m$ components of $\ddot{\mathbf{s}}^c$. However, the contact forces (reactions $\mathbf{F}$) are not unknown any more. They are of elastic nature and can be written as:

$$\mathbf{F} = \mathbf{K}\,\mathbf{s}^c + \mathbf{B}\,\dot{\mathbf{s}}^c \tag{10}$$

where $\mathbf{K} = diag[k_j]$ is the stiffness matrix and $\mathbf{B} = diag[b_j]$ involves the damping coefficients. Note that, for the reason of simplicity, the elastodynamics of particles in the contact zone was not considered. It could be introduced in the way it was done in (Vukobratovic, Potkonjak & Matijevic, 2003; Potkonjak & Vukobratovic, 2005).

Now, the set of equations (7), (5), and (9), along with the expression (10), can be solved for $\ddot{\mathbf{Q}}, \ddot{\mathbf{Q}}_b$, and $\ddot{\mathbf{s}}^c$. The contact reactions follow from (10).

A generalization can now be made – a mixed system – allowing the selected directions to feature elasticity, and keeping the rigidity along the others. One can easily conclude that rigid directions introduce unknown scalar reactions $F_j, j = ...$, while the soft directions introduce unknown relative accelerations $\ddot{s}_i^c, i = ...$ The entire number of these unknowns is always equal to $m$. So, the total model is resolvable.

Solution to $\ddot{\mathbf{Q}}$ and $\ddot{\mathbf{Q}}_b$ allows simulation of the constrained system motion.

## IMPACT

One may recognize the three phases of the contact task (Vukobratovic, Potkonjak & Matijevic, 2003) as being: approaching, impact, and regular contact motion.

### Approaching

The link moves towards the object. From the standpoint of mathematics, approaching is an unconstrained (thus free) motion. Although all coordinates from the vector $\mathbf{s}$ are free, we use the separation (3) since the subvector $\mathbf{s}^c$ is intended to describe the coming contact and it gradually reduces to zero. Strictly speaking, the restricted coordinates (elements of $\mathbf{s}^c$) reach zero one by one. So, a complex contact is established as a series of simpler contact effects. Without loss of generality one may assume that all the coordinates $\mathbf{s}^c$ attain zero simultaneously and establish a complex contact instantaneously.

The flier dynamics during approaching is described by the model (2). The model represents the set of $N$ scalar equations that can be solved for $N$ scalar unknowns – acceleration vector $\ddot{\mathbf{Q}}$ (thus enabling the integration and calculation of the flier motion $\mathbf{Q}(t)$). The object motion is solved separately: first $\ddot{\mathbf{Q}}_b$ and then $\mathbf{Q}_b(t)$. Relations (5) and (6) allow to calculate the link functional trajectory: $\ddot{\mathbf{s}}$ and $\mathbf{s}(t)$.

Let us discuss the desired (reference) motion $\mathbf{s}^*(t)$ (sign * will be used to denote reference values of a variable). It might be planned so as to make a zero-velocity contact at the instant $t_c^*$: $\mathbf{s}^*(t_c^*) = 0$ and $\dot{\mathbf{s}}^c{}^*(t_c^*) = 0$. Example of such collision-free reference is in grasping a glass of wine. However, due to the control system, tracking error will appear and the actual motion $\mathbf{s}$ will differ from the reference, causing an undesired collision and impact. In other examples the reference motion of the link might be planned so as to hit the object (like in volleyball or in landing after a jump). In this case the collision is intentional. In any case, it is necessary to monitor the actual coordinates $\mathbf{s}^c$ and detect the contact as the instant $t'_c$ when $\mathbf{s}^c$ reduces to zero ($\mathbf{s}^c(t'_c) = 0$). It will be $t'_c \neq t_c^*$ Note that the actual contact velocity will not be zero: $\dot{\mathbf{s}}^c(t'_c) \neq 0$.

During the approaching phase, the integration of the system coordinates $\mathbf{Q}$ and $\mathbf{Q}_b$ is performed. So, at the instant of collision, there will be some system state: $\mathbf{Q}(t'_c), \dot{\mathbf{Q}}(t'_c)$, for the flier, and $\mathbf{Q}_b(t'_c), \dot{\mathbf{Q}}_b(t'_c)$, for the object.

## Soft Impact

The simulation of the soft collision and contact is rather simple. Once the contact is detected (instant $t'_c$ when $\mathbf{s}^c$ reduces to zero i.e. $\mathbf{s}^c(t'_c) = 0$), the integration procedure simply switches to model (7), (5), (9), (10) that describes the contact motion. The initial state for this new phase of motion is the final state reached in approaching: $\mathbf{Q}(t'_c), \dot{\mathbf{Q}}(t'_c)$, for the flier, and $\mathbf{Q}_b(t'_c), \dot{\mathbf{Q}}_b(t'_c)$, for the object. So, the impact will not be an instantaneous effect but a transient process until a steady state is reached. Extremely high contact forces are not likely (except for very high stiffness). The two bodies may disconnect after some time if the dynamics of the system dictates so.

## Rigid Impact

In this case, the situation is rather different. There exists a finite or infinitely short period while the system velocities change so as to meet the geometrical constraints imposed by the contact – it is the impact interval. Let it be $[t'_c, t''_c]$. At $t'_c$ approaching finishes and at $t''_c$ the regular contact motion described by (7), (8), and (9) starts. The shorter impact interval the higher impact forces. For this analysis we adopt the infinitely short impact: $\Delta t = t''_c - t'_c \rightarrow 0$. Hence, infinitely high impact forces are expected. Next, we assume a plastic impact (durable contact mentioned before) where the connected bodies keep the contact for some finite time, moving together. Finally, we assume that all the $m$ coordinates forming $\mathbf{s}^c$ reduce to zero simultaneously. The impact model means the equations that allow calculation of the change in the system velocities.

We now integrate the dynamic models (7) and (9) over the infinitely short impact interval $\Delta t$, thus obtaining:

$$\mathbf{H}\,\Delta\dot{\mathbf{Q}} = (\mathbf{J}_l^c)^T \mathbf{F}\Delta t \qquad (11)$$

and

$$\mathbf{M}\,\Delta\dot{\mathbf{Q}}_b = \mathbf{D}\,\mathbf{F}\Delta t, \qquad (12)$$

where

$$\Delta\dot{\mathbf{Q}} = \dot{\mathbf{Q}}(t''_c) - \dot{\mathbf{Q}}(t'_c) = \dot{\mathbf{Q}}'' - \dot{\mathbf{Q}}' \text{ and}$$
$$\Delta\dot{\mathbf{Q}}_b = \dot{\mathbf{Q}}''_b - \dot{\mathbf{Q}}'_b. \qquad (13)$$

The system state at the instant $t'_c$, i.e. $\mathbf{Q}', \dot{\mathbf{Q}}', \mathbf{Q}'_b, \dot{\mathbf{Q}}'_b$, is considered to be known. The position does not change during $\Delta t \rightarrow 0$, and hence: $\mathbf{Q}'' = \mathbf{Q}'$ and $\mathbf{Q}''_b = \mathbf{Q}'_b$. So, it is possible to calculate the model matrices $\mathbf{H}, \mathbf{J}_l^c, \mathbf{M}, \mathbf{D}$ in the equations (11) and (12).

Now, the model (11), (12) contains $N+N_b$ scalar equations with $N+N_b+m$ scalar unknowns: the change in velocities across the impact, $\Delta\dot{\mathbf{Q}}$ (of dimension $N$) and $\Delta\dot{\mathbf{Q}}_b$ (dimension $N_b$), and the impact momentum $\mathbf{F}\Delta t$ (dimension $m$). The additional $m$ equations needed to allow the solution are obtained by rewriting the constraint relation (8) into the first-order form and applying it for $t''_c$:

$$\mathbf{J}_l^c \dot{\mathbf{Q}}'' + \mathbf{J}_b^c \dot{\mathbf{Q}}_b'' = 0 \Rightarrow$$
$$\mathbf{J}_l^c \Delta\dot{\mathbf{Q}} + \mathbf{J}_b^c \Delta\dot{\mathbf{Q}}_b = -\mathbf{J}_l^c \dot{\mathbf{Q}}' - \mathbf{J}_b^c \dot{\mathbf{Q}}_b'. \tag{14}$$

The augmented set of equations, (11), (12) and (14), allow to solve the impact. The change $\Delta\dot{\mathbf{Q}}$, $\Delta\dot{\mathbf{Q}}_b$ is found and then (13) enables one to calculate the velocities after the impact (i.e. $\dot{\mathbf{Q}}''$ and $\dot{\mathbf{Q}}_b''$). So, the new state (in $t_c''$) is found starting from the known state in $t_c'$. The impact momentum $\mathbf{F}\Delta t$ is determined as well. The new state $\mathbf{Q}'', \dot{\mathbf{Q}}'', \mathbf{Q}_b'', \dot{\mathbf{Q}}_b''$ represents the initial condition for the third phase, the regular contact motion (explained previously).

## SIMULATION EXAMPLES

### Example 1: The Analysis of the Ground Contact and Dynamic Balance in the 3D Landing Motion

The first example demonstrates how the general model works, and particularly presents the analysis of the ground contact and the achieving and losing the dynamic balance in landing. This analysis of the contact will utilize the ZMP (zero moment point) theory (Vukobratovic & Juricic, 1969; Vukobratovic & Borovac, 2004).

#### Configuration and the Task

We consider a human/humanoid model having $n = 20$ DOFs at its joints, shown in Figure 5. This can be seen as a humanoid or an approximation of the human body. Definition of the joint (internal) coordinates, vector $q = [q_1, \cdots, q_n]^T$, is presented as well. Due to the limited space, we do not give the parameters of the human/humanoid body but we only mention that its total weight (mass) is 70 *kg*. The human or humanoid falls down from a height of 0.1034m (see Figure 6). This free (unconstrained) motion represents *Phase* 1 of the motion and it ends when the right foot touches the

*Figure 5. Adopted configuration of the human/humanoid*

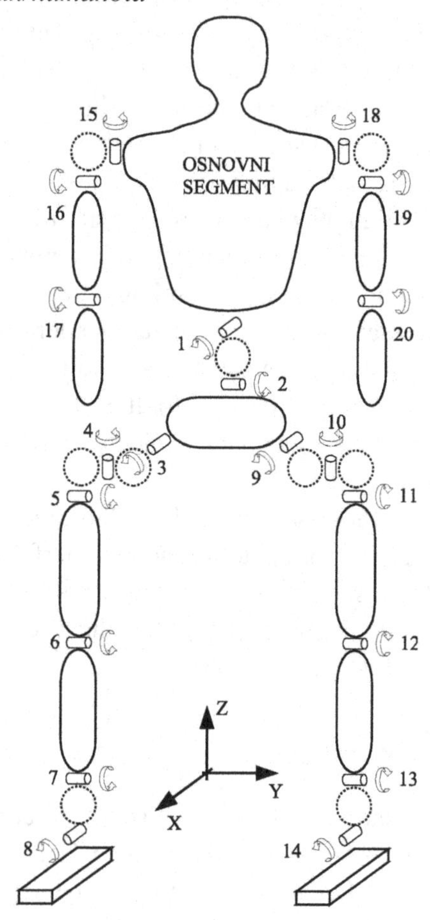


*Figure 6. Initial position of the robot with respect to the ground*

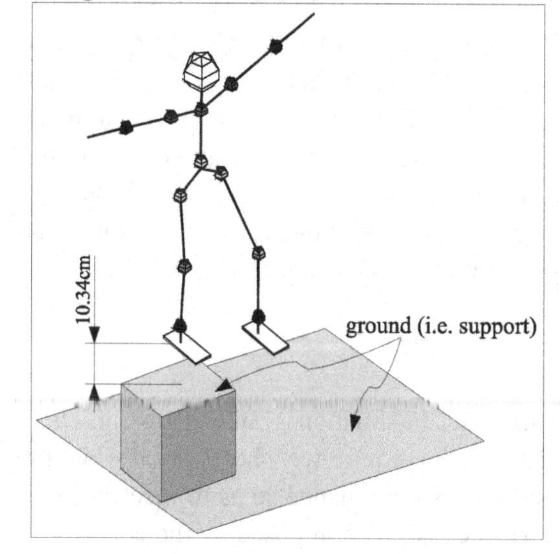

ground (support). Due to the configuration of the support, the left foot will never make contact.

The foot establishes the full contact with the ground (all six relative coordinates restricted). The infinitely short impact occurs at the instant of contact. The impact represents *Phase* 2.

After the impact, regular contact motion starts, constituting *Phase* 3. In this phase, the right foot has the full contact with the ground and the rectangular foot shape defines the support region. A single support is at stake since the left foot does not touch the ground. For some time ZMP will be inside the support region ensuring the dynamic balance. However, the large displacements of ZMP will be observed until ZMP finally reaches the edge of the support region. This is the end of *Phase* 3.

When ZMP reaches the edge, the dynamic balance will be lost and the foot will start rotating about the edge. This means that the humanoid overturns and falls down. This motion is considered as *Phase* 4.

During the motion, in all phases, the only control is the local linear feedback in each joint:

$$\tau_j = K_{Pj}(q_j{}^* - q_j) - K_{Dj}\dot{q}_j,\ j = 1, \cdots, n. \quad (15)$$

Sign * denotes the reference values, and $K_{Pj}$ and $K_{Dj}$ are feedback gains.

The general method is capable of modeling and simulating all four phases of the motion although they are rather different regarding the character of contact.

- In *Phase* 1, there is no contact – it is a free motion.
- *Phase* 2 and *Phase* 3 concern a full contact of the foot and the ground (durable contact is assumed). This means that all the six functional coordinates are restricted: $s^c = (s_1, s_2, s_3, s_4, s_5, s_6)$, $s^f$ does not exist, $m = 6$. This is shown in Figure 2c.
- *Phase* 4 considers a contact that allows one relative rotation, about the edge of the support region (i.e. the foot edge). This means that five coordinates are restricted. Thus, it holds that: $s^c = (s_1, s_2, s_3, s_4, s_5)$, $s^f = (s_6)$, $m = 5$. This is shown in Figure 2d.

The proposed method allows the calculation of all the relevant dynamic effects in all phases of the considered motion.

## Simulation Results

The software developed on the basis of the new general approach has been used to simulate the entire motion. When switching from one phase to the other, i.e., when the character of contact changes, the Jacobean matrix takes another form as a part of the general algorithm.

Let us discuss the results. Figure 7 shows the "movie" of the humanoid motion. Several time instants are selected and the position of the body presented. The characteristic instants and intervals are the following.

- ***Phase* 1** lasts for $t \in (t_0 = 0, t_1 = 0.145s)$. At $t_0$, the free falling down starts, and at $t_1$, the right foot touches the ground and the impact occurs. This is shown in Figure 7a-7b.
- ***Phase* 2** (the infinitely short impact) accomplishes at $t = t_1$ (Figure 7b).
- ***Phase* 3** lasts for $t \in (t_1 = 0.145s, t_2 = 0.714s)$. At $t_1$, the full-foot contact starts. At $t_2$, the Zero-Moment-Point (ZMP) reaches the edge of the rectangular shaped foot. This phase is shown in Figures 7b-7i. During this period, ZMP and Center of Pressure (CoP) are identical.
- ***Phase* 4** lasts for $t \in (t_2 = 0.714s, t_3 = 0.896s)$. At $t_2$, the dynamic balance is lost and the foot starts turning about its edge (we are talking about the left-side edge of the right foot), thus changing the character of the contact. The humanoid overturns and for such an unbalanced motion it is less com-

mon to talk about ZMP (ZMP is originally related to dynamically balanced gait) but rather about CoP only. CoP moves along the left-side edge and at $t_3 = 0.896s$ it reaches the front edge of the foot. This is shown in Figures 7i-7k.

- *Next phase* would take place for $t \geq t_3 = 0.896s$. At that instant, the foot would change the character of the contact again, starting to rotate about the left front corner of the rectangle. The initial instant of this phase, $t_3$, is shown in Figure 7k, but the simulation stops at that moment. Note that there would be no problem to continue simulation in this next phase.

*Figure 7. The "movie" of the human/humanoid motion. Characteristic instants defining the phases of motion are indicated*

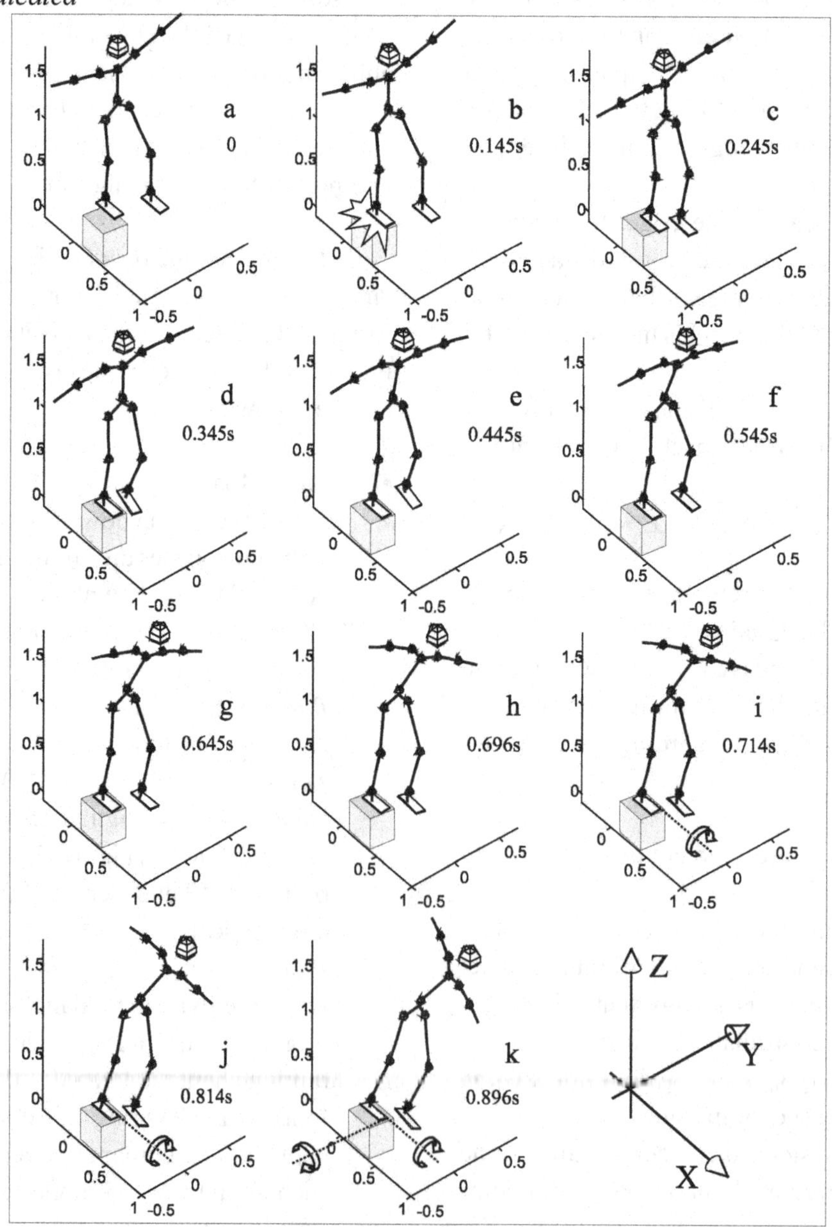

We now analyze the behavior of ZMP and CoP: the trajectory $(x^{ZMP=CoP}, y^{ZMP=CoP})$ inside the foot rectangle ($length \times width = 0.2m \times 0.1m$). Figure 8 shows the progress of this trajectory. Several characteristic time instants have been selected to show how ZMP and CoP moves. The "movie" starts at $t_1 = 0.145s$ (the instant of impact) and covers the contact motion of Phase 3 and Phase 4. Figure 8a presents the initial position of ZMP. Cases 8a-8g refer to the full-foot contact (Phase 3). During this period, at $t = 0.466s$, ZMP comes very close to the edge, almost touching it (compare 8d and 8e). However, the ZMP trajectory turns and ZMP moves back. Thus, the dynamic balance is preserved. Figure 8g (instant $t_2 = 0.714s$) shows the situation when ZMP finally reaches the edge with no intention to return. The balance is definitely lost and the overturning takes place.

*Figure 8. Trajectory of ZMP-CoP for the two phases of contact motion (full-foot-contact phase and overturning phase)*

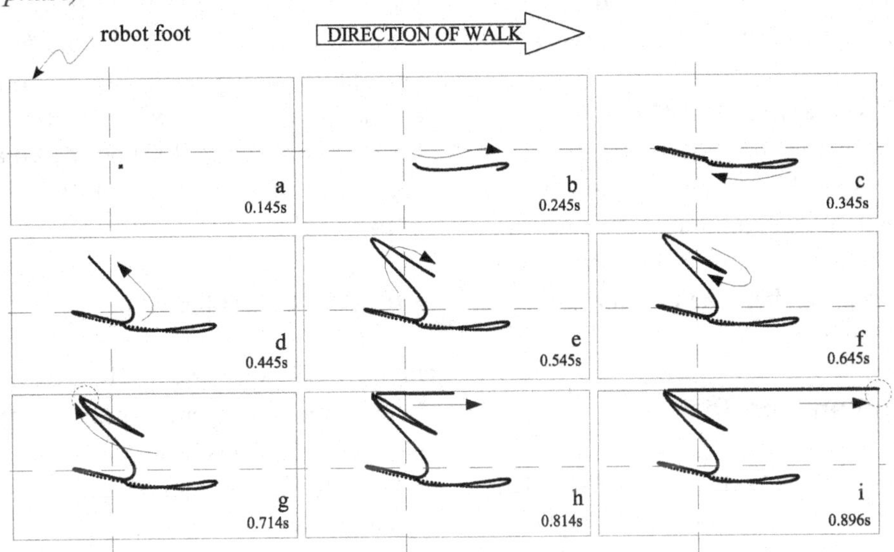


*Figure 9. Time histories of the reaction forces*

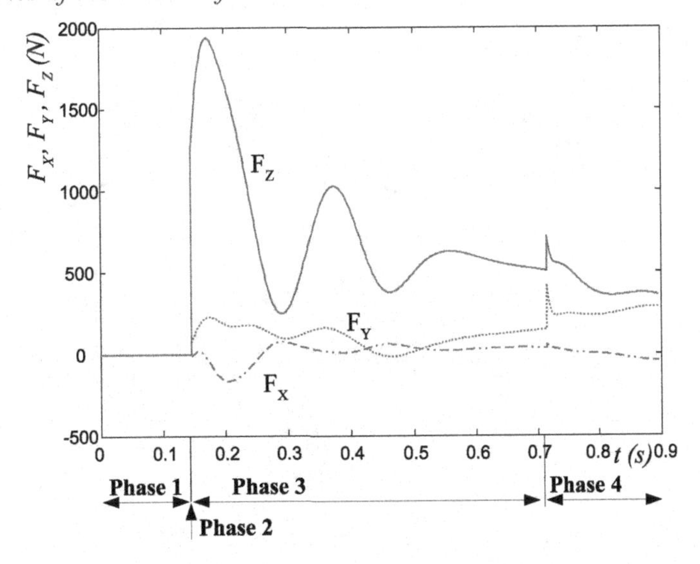

Cases 8g-8i refer to overturning (Phase 4). CoP point (as stated, we avoid using term ZMP for the unbalanced motion) moves along the left-side edge until it comes to the front edge ($t_3 = 0.896s$, Figure 8i). At that moment the overturning would become a more complex motion – rotation about the corner point. However, our simulation stops at $t_3$, although the method and the software support the simulation of such a next phase of motion.

Figure 9 shows the reaction forces: $F_x$, $F_y$ and $F_z$. During Phase 1, there was no reaction. The infinitely large values appearing in the instant of impact ($t_1 = 0.145s$) could not be presented. The diagrams effectively cover Phase 3 and Phase 4.

## Example 2: A Handball Action

This example considers a rather common action in handball playing–jumping with catching and throwing a ball. The example is elaborated in details in order to show the potentials of the proposed general method.

### Configuration and the Task

The configuration of the body is adopted equal to the one considered in the preceding example.

The task is selected as a specific action in handball. A player jumps, catches the ball with his right hand, swings, and throws the ball while still flying. The ball should hit the ground in front of the goalkeeper in order to trick him (Figure 10). In a realistic situation, this would be a learned pattern. So, we consider this player's motion as a reference, $\mathbf{q}^*(t)$, and try to realize it. For the purpose of simulation the reference motion was not measured but synthesized numerically. Here, we do not show the reference; the realized motion is considered sufficient information. The ball mass is $0.450kg$ and the diameter is $0.190m$.

The human/humanoid is equipped with actuators at its joints and the control system that will try to track the reference motion. Among different control strategies, we implemented local PD regulators described with the expression (15). Such a simple control is considered appropriate since dynamics, and not control, is the key point of this work.

### Analysis of the Results

Figure 11 shows the realized player's motion. His intention (and reference) was to jump, catch the ball, swing with his hand, and throw the ball. According to Figure 11, his attempt was successful. Let us analyze the player's motion in more detail. One can observe several phases of the motion:

*Figure 10. Required ball motion*

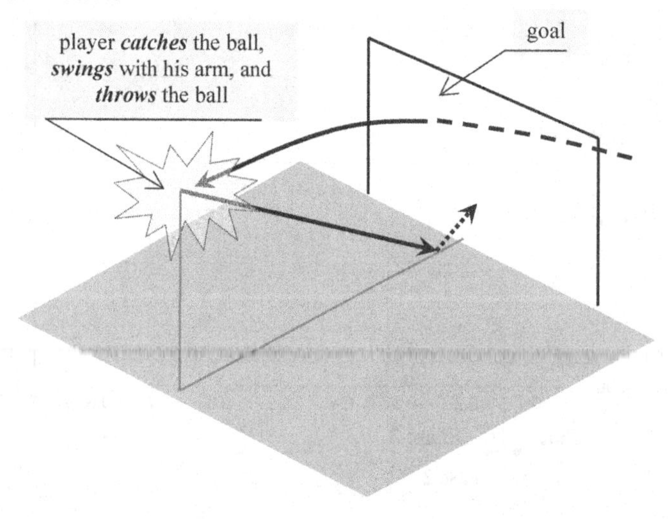

*Figure 11. The "movie" of the human/humanoid player motion. Characteristic instants defining the phases of motion are indicated*

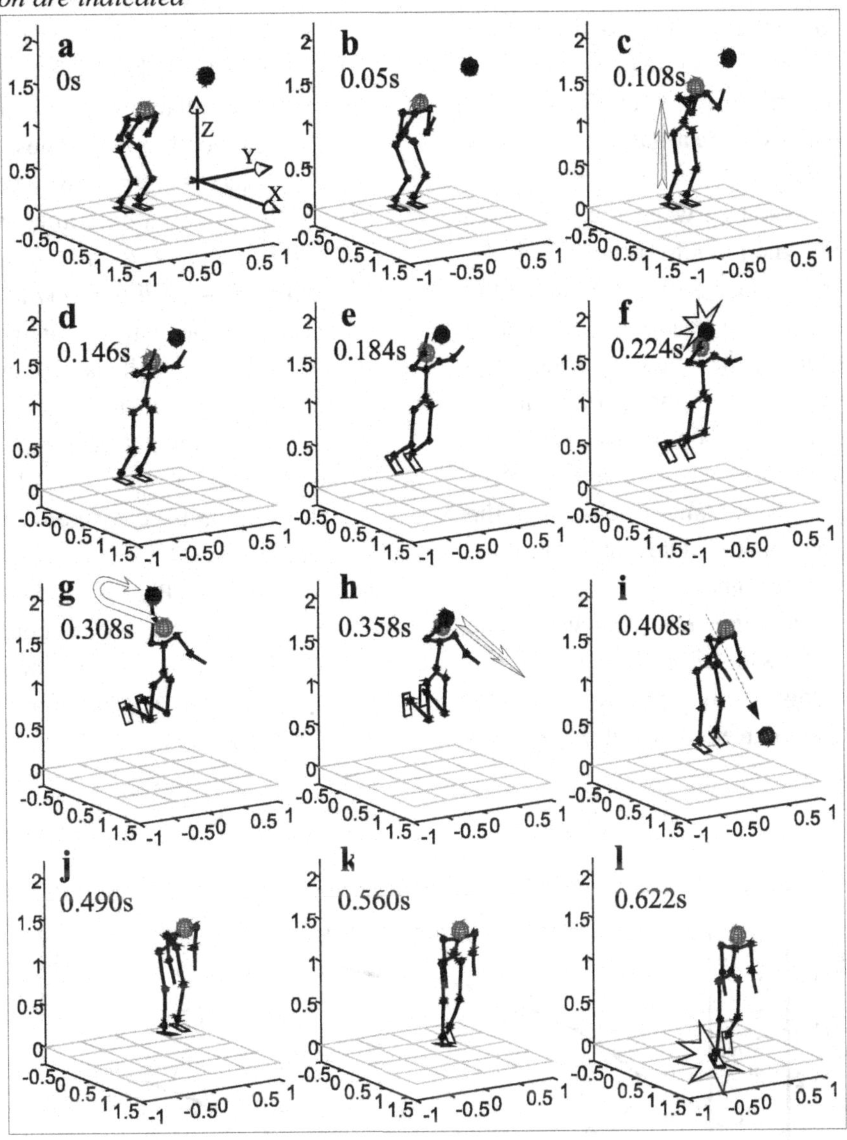


- ***Phase* 1 – *takeoff.*** Both legs are in contact with the ground accelerating the body upward (Figure 11a,b,c). This phase ends at 0.108*s*. At this instant the legs lose the contact with the ground (Figure 11c).
  - From the viewpoint of mechanics the flier (player in this example) has two contacts with an immobile object – both feet are on the ground. A closed chain is formed. Each contact restricts all six relative motions and accordingly creates six reaction forces/torques (twelve reactions in total).
- ***Phase* 2 – *free flight towards the ball.*** The player is moving (flying) up while expecting the contact with the ball (Figures 11d,e,f). The right hand makes the contact with the ball (catches the ball) at 0.224 *s* and Phase 2 ends (Figure 11f).
  - In this phase, there is no contact with any object; a true flier is considered.

The system has a tree structure (no closed chain).

- ***Phase 3 – impact.*** The contact of the player's hand and the ball always involves an impact effect. It occurs at 0.224 *s* (Figure 11f). For the current analysis we assume an infinitely short impact.
  - The collision between the hand motion and the motion of the object (the ball) will cause an impact. Since the contact will restrict all six relative coordinates, the impact is characterized by a six-component momentum. The infinitely short impact generates infinitely large reactions and, accordingly, the instantaneous change in the system velocities. The momentum still has a finite value.
- ***Phase 4 – swinging and throwing.*** The player swings with his right arm (Figure 11g) and throws the ball (Figure 11h). The ball leaves the player's hand at 0.358 *s* and Phase 4 ends.
  - From the point of view of mechanics, the contact is established between the flier's hand and the external mobile object (the ball). The contact restricts all six relative motions and produces six reaction forces/torques. The contacted object is a separate system (body) that has its own dynamics, now being influenced by the flier (and vice versa).
- ***Phase 5 – falling down.*** The ball and the player now move separately (Figures. 11i,j,k,l). The ball has its own trajectory, to hit the ground in front of the goal. The player falls down (free flying) until lands (he touches the ground at 0.622 s, Figure 11 l). The landing is performed correctly, so that the player will not overturn.
  - Since in Phase 5 there is no contact, a truly free flier is considered again. This lasts until a new impact makes the contact of the flier's right foot and the ground.

*Figure 12. Time histories of the main-body (torso) coordinates: x(t), y(t), z(t), θ(t), φ(t), ψ(t)*

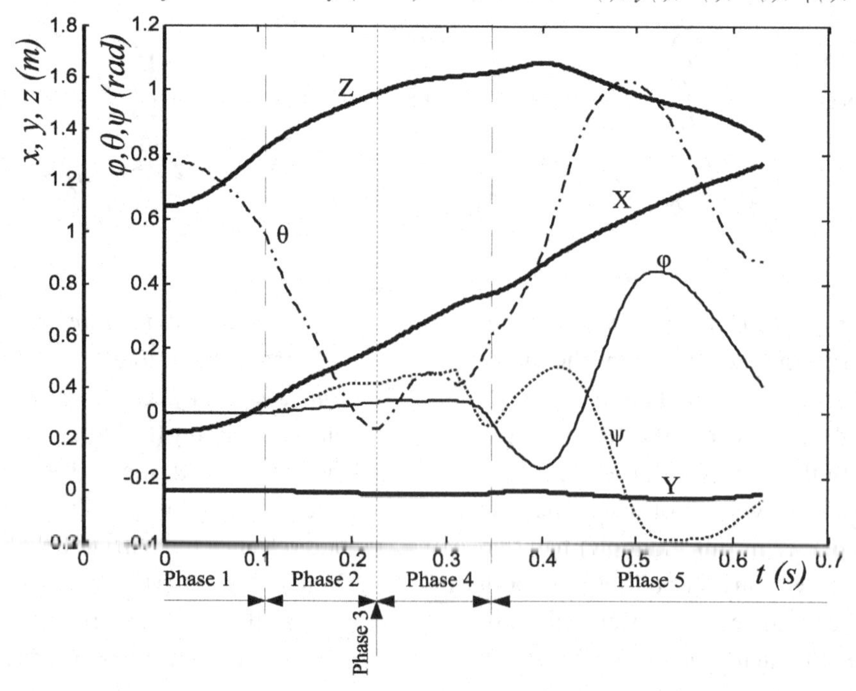

*Figure 13. Time histories of the joint coordinates: (a selected set)*

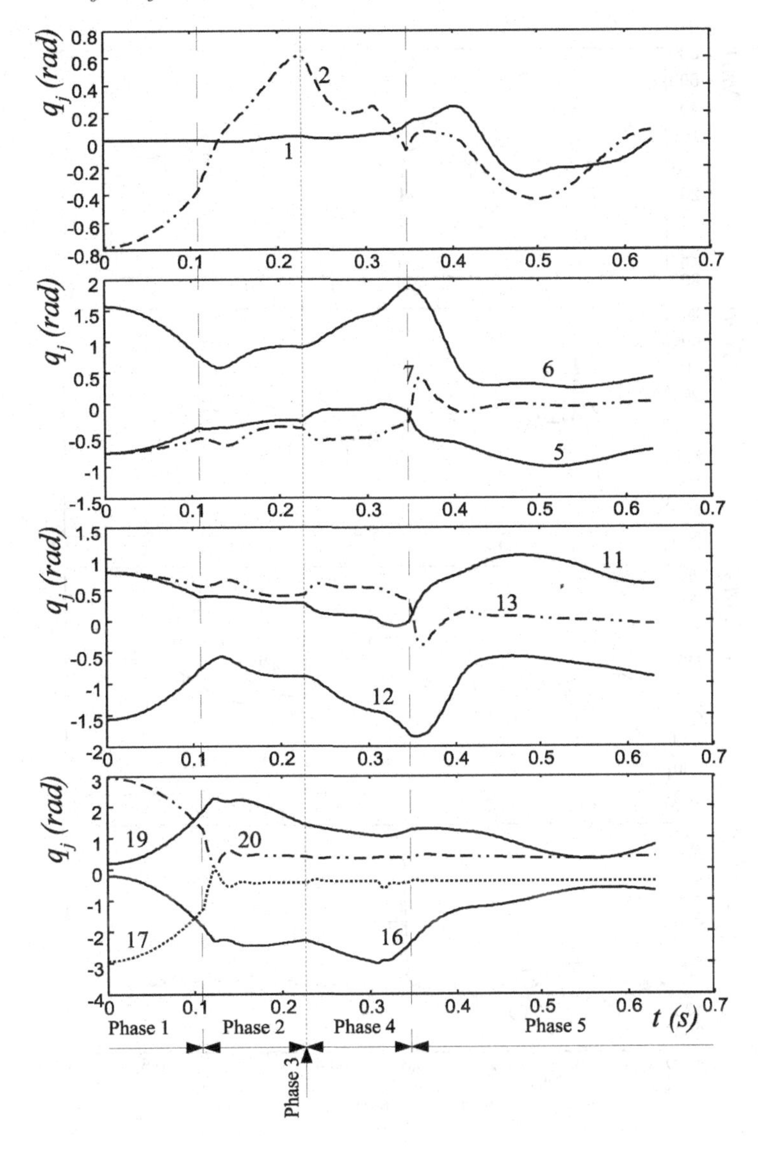


Our simulation stops when the player touches the ground.

Some characteristics of this motion deserve additional attention. Figures 12 and 13 show the time histories of realized player's motion. Figure 12 presents the motion of the main body – the torso. Time histories $X(t) = [x(t), y(t), z(t), \theta(t), \varphi(t), \psi(t)]$ are given. Figure 13 presents joint trajectories $q_j(t)$, $j$=1,.... A selected set of joints is given since not all trajectories are considered equally interesting. One can notice that from phase to phase different motions are dominant.

Figure 14 presents the torques generated at the player's joints: $\tau_j$, $j$ = 1,... . A selected set is given. One may see that high torques are present (at relevant joints) during Phase 1 (takeoff) and Phase 4 (swinging and throwing). In the swinging-throwing phase, it is obvious, as expected, that

*Figure 14. Time histories of the joint torques: (a selected set)*

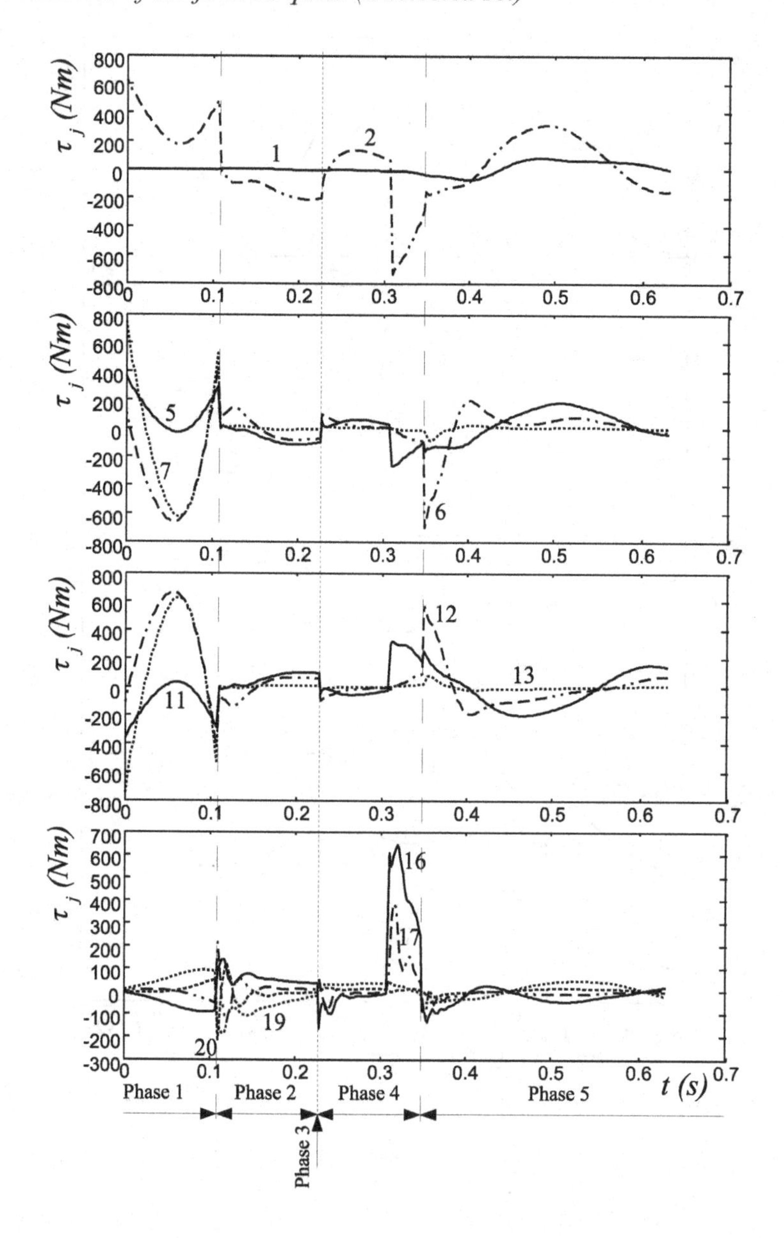


high torques are not generated all the time. While the arm (with the ball) is moving back, the torques are not high. The arm generates high torques when accelerating the ball in the forward direction in order to throw it strongly. It is also evident that almost the complete body (the majority of joints) is engaged in this demanding action (throwing). Driving torques (and total joint reactions that can also be calculated) offer the possibility of locating the dangerous situations that may cause injury – note that rather high picks are obvious in Figure 14.

We now present the force produced upon the ground during Phase 1 – takeoff. As mentioned,

*Figure 15. The launching force in Phase 1*

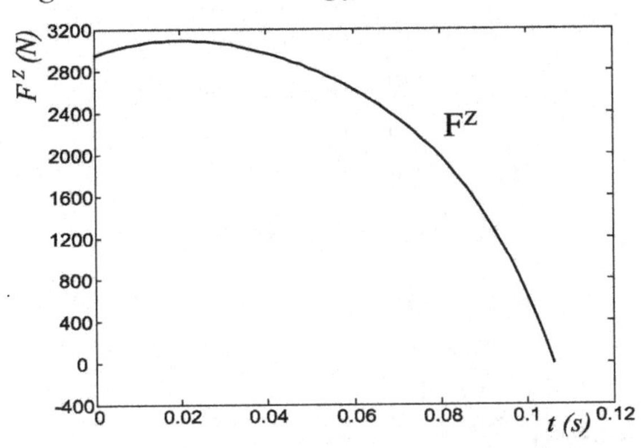


there is a six-component reaction between each foot and the ground. We present the total vertical component (sum for both feet) since it is the force that launches the flier up (Figure 15).

We now concentrate on the contact with the ball. The contact starts with an impact that has a six-component momentum. These components are given in Table 1. Since the impact is considered infinitely short ($\Delta t \to 0$), the infinitely large contact forces/torques will appear causing instantaneous changes in all system velocities. Table 2 shows these changes: first the changes in 20 joint velocities, $\Delta \dot{q}_j$, $j = 1,...,20$, and then in 6 velocities of the main body (torso), $\Delta \dot{x}, \Delta \dot{y}, \Delta \dot{z}, \Delta \dot{\theta}, \Delta \dot{\varphi}, \Delta \dot{\psi}$.

Figure 16 shows the three components of the contact force between the hand and the ball. In the figure, *x*, *y*, and *z* indicate the first three axes of the corresponding ball-to-hand *s*-frame ($s_1$, $s_2$, $s_3$). Contact moments have not appeared interesting. Strong forces appear in the second part of swinging-throwing phase (Phase 4), when the ball is accelerated in the forward direction. One can notice that the strongest force is generated in the *x*-direction and then in the *z*-direction. This comes out from the fact that the acceleration of the ball is mainly along the *x* axis, and then along negative *z*.

Figure 17 shows the motion of the ball. We have decided to present the three Cartesian velocities of the ball center. The immobile external Cartesian frame (*x*, *y*, *z*) is used. One can see that in Phase 1 the ball approaches moving according to a linear law along the horizontal axis *x* (in negative direction), meaning a constant negative velocity. Along the vertical axis *z*, there is a constant negative acceleration due to the gravity constant *g*. The same happens in Phase 5, but with the reverse

*Table 1. Impact between the hand and the ball – six components of the momentum*

| $k$ | 1 | 2 | 3 | 4 | 5 | 6 |
|---|---|---|---|---|---|---|
| $F_k \Delta t$ | 1.5291 | 0.2327 | 2.2001 | -0.0135 | 0.1723 | -0.0041 |

*Table 2. Impact between the hand and the ball – changes in system velocities*

| $\Delta\dot{q}_1$ | $\Delta\dot{q}_2$ | $\Delta\dot{q}_3$ | $\Delta\dot{q}_4$ | $\Delta\dot{q}_5$ | $\Delta\dot{q}_6$ | $\Delta\dot{q}_7$ | $\Delta\dot{q}_8$ |
|---|---|---|---|---|---|---|---|
| -0.3432 | 0.4858 - | 0.5460 | 0.5308 - | 0.7093 | 0.6356 - | 0.0425 | -0.6341 |
| $\Delta\dot{q}_9$ | $\Delta\dot{q}_{10}$ | $\Delta\dot{q}_{11}$ | $\Delta\dot{q}_{12}$ | $\Delta\dot{q}_{13}$ | $\Delta\dot{q}_{14}$ | $\Delta\dot{q}_{15}$ | $\Delta\dot{q}_{16}$ |
| -0.5380 | 0.5073 - | 0.3060 | 0.7014 - | 0.2141 | -0.6121 | 1.8907 - | 17.8599 |
| $\Delta\dot{q}_{17}$ | $\Delta\dot{q}_{18}$ | $\Delta\dot{q}_{19}$ | $\Delta\dot{q}_{20}$ | | | | |
| 3.8313 1 | .1954 - | 0.5187 | 1.0810 | | | | |
| $\Delta\dot{X}$ | $\Delta\dot{Y}$ | $\Delta\dot{Z}$ | $\Delta\dot{\theta}$ | $\Delta\dot{\varphi}$ | $\Delta\dot{\psi}$ | | |
| -0.0502 | -0.0166 | -0.0153 | 0.0746 - | 0.2328 | -1.1356 | | |

*Figure 16. Contact forces between the hand and the ball (x, y, z indicate the first three axes of the ball-to-hand s-frame)*

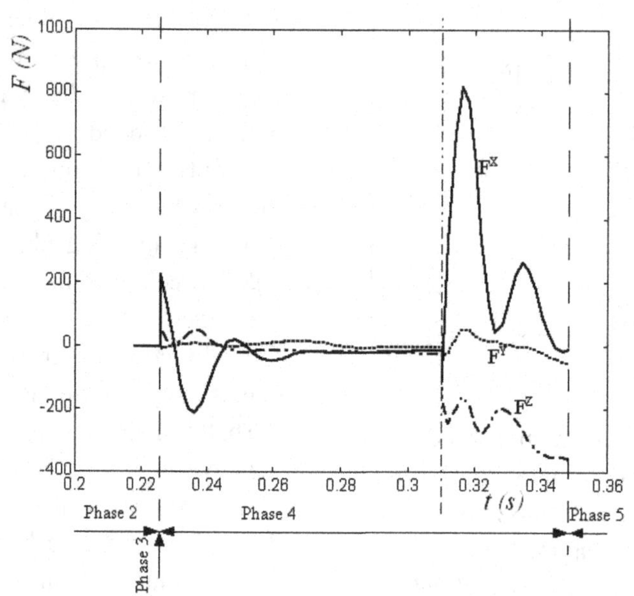


direction along $x$ (positive motion). Such motion laws are characteristic for the free motion (when there is no contact). In Phase 3 – impact – one can notice the sudden change in ball velocities. This is a consequence of the infinitely high contact forces that appear with the infinitely short impact (as previoulsy explained). Six components of the impact momentum were given in Table 1. In Phase 4 (swinging-throwing) high accelerations in forward ($x$) and downward ($z$) directions are

*Figure 17. Time histories of the ball velocity components (motion of its center)*

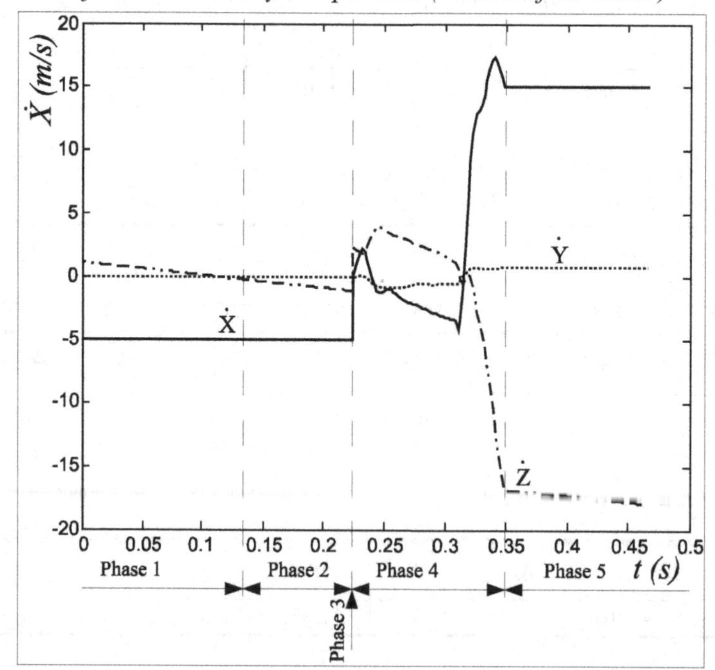

present in the second part of this phase before the ball is actually thrown. Remember that the ball should go forward and downward to hit the ground in front of the goal (Figure 10).

## Example 3: Soccer Goalkeeper

A more complex task, regarding the mechanical complexity of the entire system, is considered – the soccer goalkeeper jumping and catching a ball.

### Configuration and the Task

The player configuration used in Examples 1 and 2 has been modified by introducing more complex arms. The new configuration of arms is shown in Figure 18. So, the total number of DOFs at the player's joints is increased to $n = 28$. This complex arm configuration is needed to handle the problem of closed chain dynamics (when holding the ball with both hands). A simpler configuration would be inappropriate, as it would generate infinite contact reactions (forces/torques).

The task is selected as a specific action of a goalkeeper in soccer. The ball is moving towards the upper left corner of the goal. The goalkeeper jumps (diagonally across the goal), catches the ball with both hands, and finally lands while holding the ball. We consider this player's motion as a reference and try to realize it by using local PD regulators (as explained in previous examples and expression (15)). For this simulation the reference motion was not measured but synthesized numerically. Here, we will not show the reference but the realized motion only. The ball mass is $0.450 kg$ and the diameter $0.260 m$.

### Results

Figure 19 presents the results of simulation. Detailed analysis distinguishes the following phases of motion.

*Figure 18. The new configuration of the human/humanoid arms*

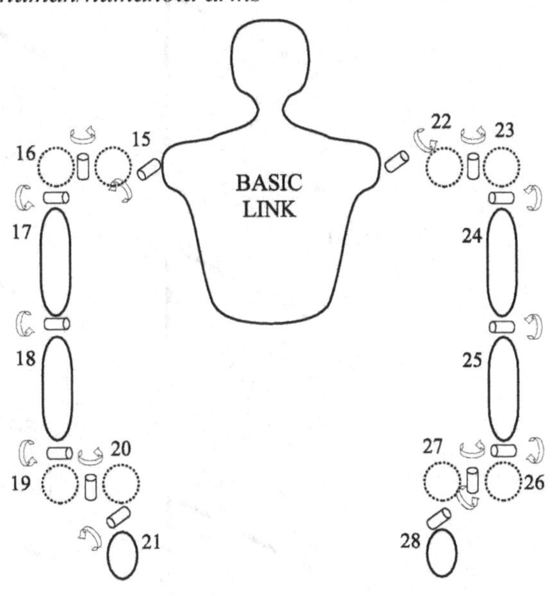


- In ***Phase* 1** the goalkeeper moves slowly to the left with both feet on the ground (Figure 19a-c).
- ***Phase* 2** starts when the right foot leaves the ground (at $t = 0.213s$, Figure 19d). So, the take-off is mainly up to the left leg.
- When the left foot leaves the ground (at $t = 0.2782s$, Figure 19f), ***Phase* 3**, a free flight towards the ball, starts.
- ***Phase* 4** is the impact (at $t = 0.722s$, Figure 19i). – the goalkeeper catches the ball.
- Falling down represents ***Phase* 5**.

All other dynamic characteristics could be calculated, like joints and main-body coordinates, joint driving torques, total reactions in joints, impact effects, power requirements, energy consumptions, etc.

## Example 4: Tennis Player

Tennis movements appear to be among the most complex. Due to the limited space we select one motion task and present only key facts and features.

*Figure 19. The "movie" of the human/humanoid goalkeeper motion*

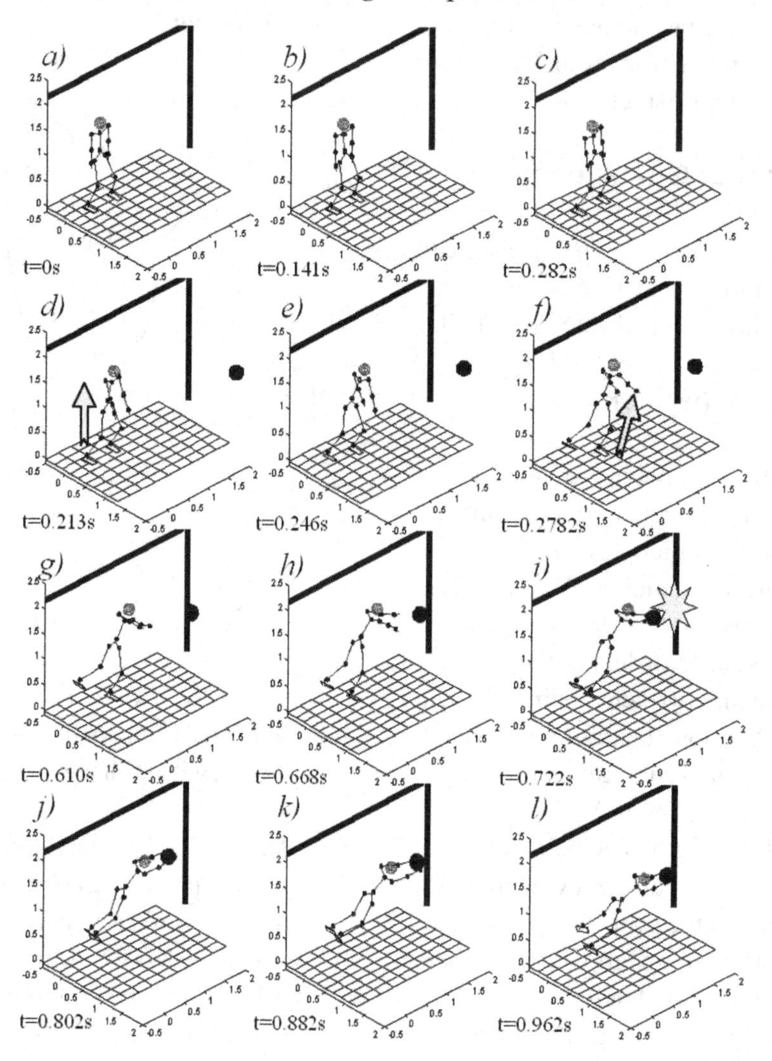


The simulation of a forehand hit is performed. The ball mass is $0.056kg$ and its diameter is $0.065m$. It approaches along the negative axis $x$. In modeling the player, the structure from the preceding example is used, with the racket added to the arm configuration. The hand-racket connection introduces two additional revolute DOFs – two angles about axes perpendicular to the racket length. These displacements are due to the hand tissue deformation and have limited range.

The key characteristic of this motion task is the presence of elasticity in the contact between the racket and the ball. It is important to note that elastic deformations appear on both sides, racket and ball. The string tension of the analyzed racket is about $22kg$. The elasticity and damping of the contact are verified by calculation and experiment: $k_1 = 20000N/m$ and $b_1 = 3.3$Ns/m.

In this example we will deal with a soft contact along one restricted coordinate and a rigid

*Figure 20. Ball-to-racket contact*

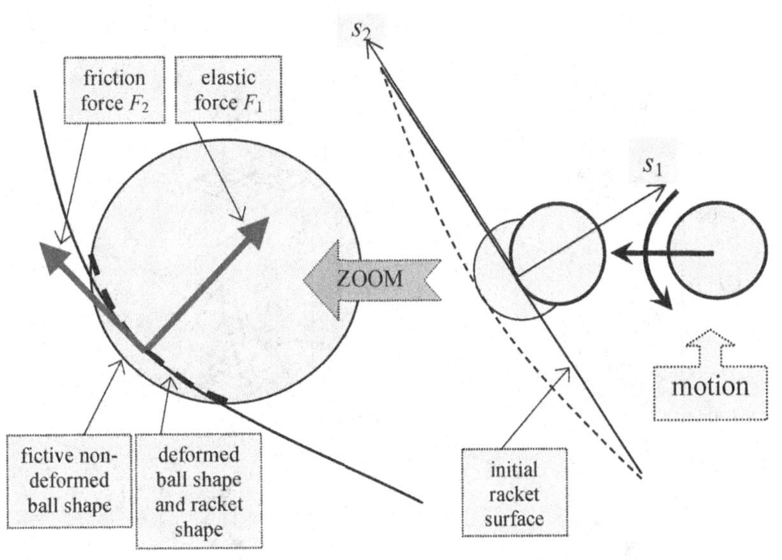


contact along the others. Modeling follows the general theory.

Figure 20 shows the planar view of the contact (the model works with three dimensions but this 2D presentation shows the key effects more clearly). The ball comes with some spin. When it touches with the racket, deformation starts. The coordinate $s_1$ being perpendicular to the racket measures the deformation. It is geometrically free but stays small due to elastic forces. The force along $s_1$, the elastic force (see eq. (10)), is $F_1 = -k_1 s_1 - b_1 \dot{s}_1$ (minus sign is due to the $s$-frame fixed to the link, racket this time). So, it is the soft contact along $s_1$. In direction $s_2$ (and in direction $s_3$ that is normal to $s_1$ and $s_2$ in 3D analysis), the contact point cannot slide due to strong friction forces ($F_2$ and $F_3$). So, we face the rigid contact in these directions. Rotations of the ball now reduce to the rolling motion. Under some conditions, the sliding force may overcome friction and sliding could start; however, this will not be considered.

Figure 21 presents the "movie" of the hit, obtained by simulation. Among different results available after the simulation study, we present the following. Figure 22 shows some effects of the ball spin. The ball approaches along the negative axis $x$; the input trajectory is shown by the shorter solid line. The upper diagram in the figure shows the motion in the horizontal $x$-$y$ plane and the influence of the input spin (rotation about vertical axis $z$) on the return angle. The case (a) means no input spin, (b) stands for the positive spin, 10 revolutions per second, and (c) for the negative value (-10*rps*). The lower diagram shows the vertical $x$-$z$ plane and the influence of the input spin about the horizontal axis $y$ on the return trajectory: (a) - no spin, (b) - positive, (c) - negative spin.

Figure 23 introduces the racket tilt (turning about the longitudinal axis) and discusses its influence on the return motion. The input trajectory is shown by the shorter solid line. The racket surface was first considered close to vertical (case (a)); then turned for 12 degrees up (case (b)); and finally turned for 12 degrees down (case (c)).

Figure 24 shows the deformation and the deformation force. As mentioned earlier and shown in Figure 20, the contact along the direction $s_1$ is soft and the coordinate $s_1$ describes the deformation (it

*Figure 21. The "movie" of the considered forehand hit*

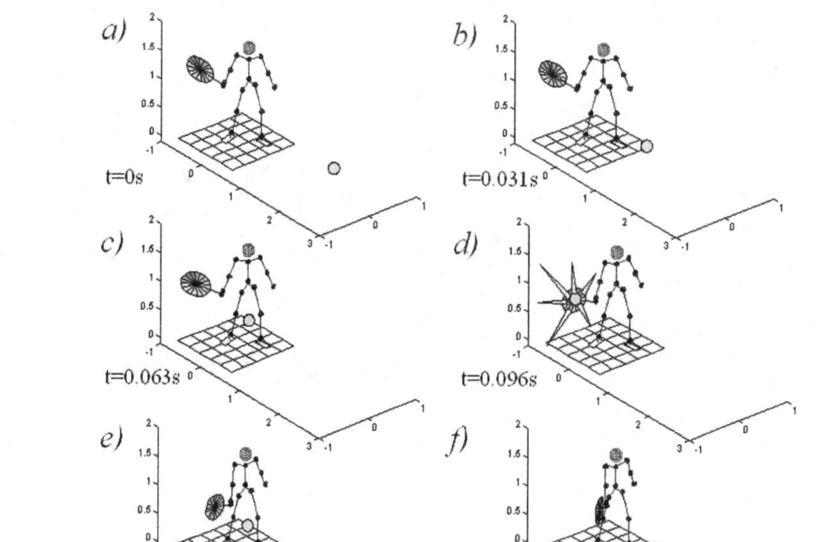


*Figure 22. Influence of the input spin on ball trajectory after impact. Upper diagram – return angles for the different spin about the vertical axis z: (a) - no input spin, (b) - positive spin, and (c) - negative spin. Lower diagram – the return motion for the different spin about the horizontal axis y: (a) - no input spin, (b) - positive spin, and (c) - negative spin*

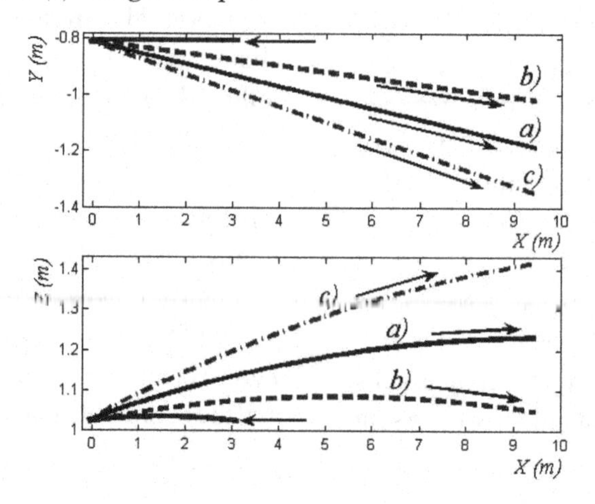

is negative all the time). The contact force along $s_1$ is of elastic nature, $F_1 = -k_1 s_1 - b_1 \dot{s}_1$. Figure 24 presents time histories of the deformation and the elastic force, for the interval of contact. The contact lasts for $5ms$. The maximum deformation is $70mm$, and the maximum force is $1417N$.

In order to show the numerical and computational complexity of the simulation model, we provide some data. Software was realized in MATLAB language (no specialized toolboxes). The example system to test the complexity and efficiency consisted of a tennis player having 37 DOFs, a ball having 6 DOFs, and two immobile objects supporting player's feet. The motion lasted for 0.3s. The simulation was based on Runge-Kutta integration (345 integration steps; model calculated 4 times in each step), by using PC Pentium 4 (CedarMill core, 3GHz, 1GB memory). The calculation took about 20 min (about 0.8s to calculate the model once). So long time was mainly due to the extremely slow calculation when programmed in MATLAB. The reprogramming to C++ is in progress and estimation (based on previous experience in dynamic modeling) is that the simulation will be accelerated about 1000 times (one calculation of the model would

*Figure 23. Influence of the racket tilt on the return motion: (a) - racket surface close to vertical; (b) - turned for 12 degrees up; and (c) - turned for 12 degrees down*

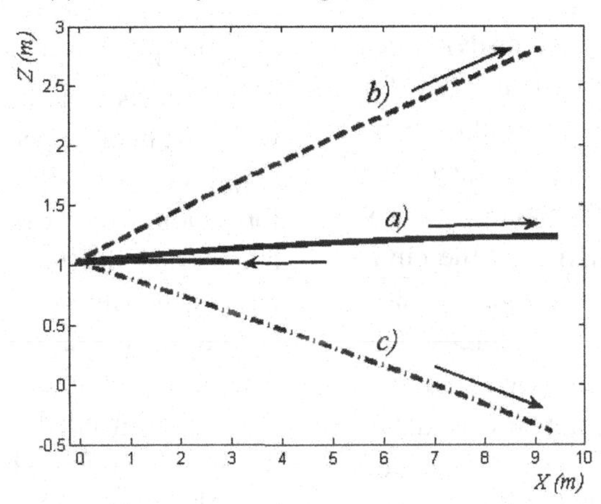


*Figure 24. Ball-to-racket contact: time histories of elastic deformation $s_1$, and the elastic force $F_1$*

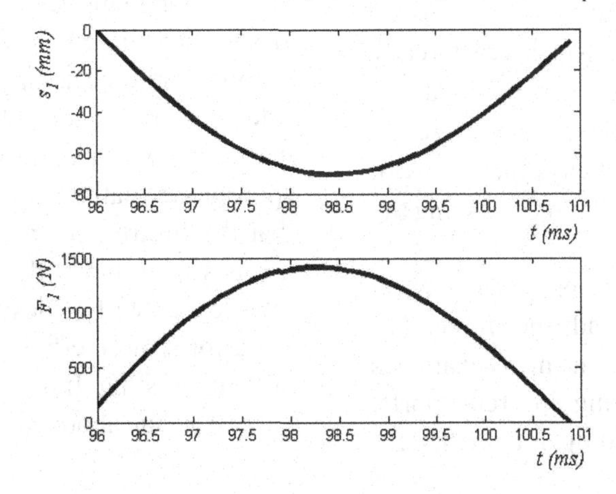

then take about $10^{-4}$s). This may be insufficient for real-time work on the applied computer but faster platforms will soon support the real time simulation.

## A LOOK TO THE FUTURE: SOME APPLICATIONS AND RESEARCH

Let us look to the future, foresee some interesting practical application of the proposed mathematical models, and think about the next steps and new research challenges. Of course, we focus our discussion on fields that are mainly (but not exclusively) related to sports. This means that the "classical" discussion on humanoid robotics will be avoided regardless of how important the dynamic models are in this field.

The mathematical models of human dynamics and the appropriate simulation systems allow for calculation of characteristics that could not be measured. Namely, the motion of an athlete can be captured by using cameras (or some on-body motion sensors). His interaction with the environment can be measured as well (e.g., a tension platform measures the ground contact forces). The other variables, like joint torques, power requirements, and total joint loads cannot be sensed. On the other hand, one notes that these "invisible" characteristics play the key role in achieving the better results (stronger hit, faster running, etc.) and the location of critical situations (regarding injuries and fatigue). Therefore, in the struggle for better results in sports, which are already close to the maximum of human potentials, a simulator could be a tool, means to analyze the current athlete's motion and then optimize it with the aim of improving the results and reducing the injury risks.

The next application of the proposed general mathematical model is seen in advanced computer graphics and animation. There are many examples of computer games, simulating different sports, where virtual athletes run, jump, hit a ball, fight and punch or perform other actions. Besides sports, one should mention games where soldiers (or knights, etc.) run, jump and fight. It is easy to notice that a motion of these virtual humans still appears rather "artificial" – it does not look like a motion of a real man. To make the motion realistic, it is necessary to model the dynamics of athletes (or soldier), and simulate the system on the basis of a full dynamic approach. Due to a diversity of motions, the applied model must be completely general. One should point out that this is not just the matter of computer games; any serious training systems (in sports and wider) based on virtual reality needs this approach.

Finally, we mention some possible future trends in sports (and wider), which might seem like close to science fiction. We have in mind the sophisticated robotic prosthesis and orthosis. The prosthesis would allow people with amputations to fully involve in sports (and other activities). The orthosis would allow the disabled to move while the healthy people would use it as a power amplifier which allows a faster and stronger performation. A medical application is the first target, but sport, industrial, and military potentials are obvious as well. If such robotic devices are designed and used as separate machatronic systems that closely interact with humans, mathematical models are needed to design and control them; due to interaction with humans, models have to cover both, machine and human dynamics. If the prosthesis or orthosis are fully integrated with the human body (one may even imagine a symbiosis that cannot be disassembled), the idea of *cyborg* arises. Born in SF novels, this concept might become reality. The importance of the sophisticated design and control based on full dynamics increases, and accordingly, the importance of appropriate mathematical models as well. Note that the future does not need to be such scary – it is simply likely that different mechatronic-robotic aids will be designed and used in sports to perform more efficiently and make the game more interesting. It is just the matter of rules i.e. the consensus about what to approve.

## CONCLUSION

This work tries to relate biomechanics to robotics, with the idea of benefit for both fields. The mathematical models have been derived to cover the kinematics and dynamics of virtually any motion of a human or a humanoid robot. All particular problems that arise could be derived from the proposed general model as its special cases.

Benefits for the robotics are seen in two directions: first, providing a tool for calculating dynamics of humanoids with the aim of formulating the fully dynamic feedforward control, and second, developing a general simulator for the purpose of system designing and motion planning.

Biomechanics in sports and medicine gets a tool for mathematical approaching the analysis of motion and disorders. Better results in sports and improved diagnostics are foreseen.

Furteher symbiosis of robotics and biomechanics in sports and medicine leads to the biologically-inspired robot control needed for a diversity of tasks expected in humanoids, and to robotic assistive devices helping people to overcome disabilities or augment their physical potentials.

This work dealt mainly with examples coming from sports in order to justify this aspect of research.

## REFERENCES

Bernstein, N. (1967). *The Co-ordination and Regulation of Movements*. Oxford, England, Pergamon Press.

Blajer, W., & Czaplicki, A. (2001). Modeling and Inverse Simulation of Somersaults on the Trampoline. *Journal of Biomechanics, 34*, 1619-1629.

Blajer, W., & Czaplicki, A. (2003). Contact Modeling and Identification of Planar Somersaults on the Trampoline. *Multibody System Dynamics, 10*, 289-312.

Bueng R.S., Bying-Ji, Y., & Sang-Rok, O., & Yound Soo, K. (2004). Landing Motion Analysis of Human-Body Model Considering Impact and ZMP condition. *In Proceedings of 2004 IEEE/RSJ Intl. Conf. Intelligent Robots and Systems*, Sendeai, Japan, (pp. 1972-1978).

Fukuda, T., Michelini, R., Potkonjak, V., Tzafestas, S., Valavanis, K., & Vukobratovic, M. (2001). How Far Away is “Artificial Man”. *IEEE Robotics and Automation Magazine, March issue*, (pp. 66-73).

Ken’ichiro, N., et al. (2004). Integrated Motion Control for Walking, Jumping and Running of a Small Bipedal Entertainment Robot. In *Proceedings of the IEEE International Conference on Robotics, & Automation*, (pp. 3189-3194).

Knuesel, H., Geyer, H., & Seyfarth, A. (2005). Influence of Swing Leg Movement on running stability. *Human Movement Science, 24*(4), 532-543

Morlier, J., & Mesnard, M. (2007). Influence of the Moment Exerted by the Athlete on the Pole in Pole-Vaulting Performance. *Journal of Biomechanics, 40*(10), 2261-2267.

Pei-Yan, Z., & Tian-Sheng, L. (2007). Real-Time Motion Planning for a Volleyball Robot Task Based on a Multi-Agent Technique. *J. of Intelligent Robot Systems, 49*, 355-366.

Potkonjak, V., & Vukobratovic, M (1979). Two New Methods for Computer Forming of Dynamic Equations of Active mechanisms. *J. Mechanism and Machine Theory, 14*(3).

Potkonjak, V., & Vukobratovic, M. (2005). A Generalized Approach to Modeling Dynamics of Human and Humanoid Motion. *Intl. Journal of Humanoid Robotics, 2*(1), 1-24.

Potkonjak, V., Vukobratovic, M., Babkovic, K., & Borovac, B. (2006). General Model of Dynamics of Human and Humanoid Motion: Feasibility, Potentials and Verification. *Intl. Journal of Humanoid Robotics, 3*(1), 21-48.

So., B.R., Yi, B.J., & Kim, W.K. (2002). Impulse Analysis and Its Applications to Dynamic Environment. *In Proc of ASME Biennial Mechanisms Conf.*, Montreal, Canada.

Stepanenko, Yu., & Vukobratovic, M. (1976). Dynamics of Articulated Open-Chain Active mechanisms. *Mathematical Biosciences, 28*(½).

Vanrenterghem, M., Lees, A., Lenor, M., Aerts, P., & De Clercq, D. (2004). Performing the Vertical Jump: Movement Adaptations for Sub Maximal Jumping. *Human Movement Science, 22*(6) (2004), 1713-1727.

Vukobratovic, M., & Juricic D. (1969). Contribution to the Synthesis of the Biped Gait. *IEEE Trans. on Bio-Medical Engineering, 16*(1), 1-6.

Vukobratovic, M., & Stemanenko, Yu. (1973). Mathematical Models of General Anthropomorphic Systems. *Mathematical Biosciences, 17*, 191-222.

Vukobratovic, M., Hristic, D., & Stojiljkovic, Z. (1974). Development of Active Anthropomorphic Exoskeletons. *Medical and Biological Engineering, 12*(1).

Vukobratovic, M., & Potkonjak, V. (1982). *Dynamics of Manipulation Robots*. Berlin, Heidelberg, Germany: Springer-Verlag.

Vukobratovic, M., Potkonjak, V., & Matijevic, V. (2003). *Dynamics of Robots with Contact Tasks*. Dordrecht, The Netherlands: Kluwer Academic Publishers.

Vukobratovic, M., & Borovac, B. (2004). Zero-Moment Point, Thirty-Five Years of Its Life. *Intl. J. Humanoid Robotics, 1*(1), 157-173.

Vukobratovic, M., Potkonjak, V., Babkovic, K., & Borovac, B. (2007). Simulation Model of General Human and Humanoid Motion. *Intl. J. Multibody System Dynamics, 17*(1), 71-96.

Vukobratovic, M., Herr, H., Borovac, B., Rakovic, M., Popovic, M., Hofmann, A., & Potkonjak, V. (in press). Biological Principles of Control Selection for a Humanoid Robot's Dynamic Balance Preservation. *Intl. J. Humanoid Robotics.*

Wakai, M., & Linthome, N. P. (2005). Optimum take-off angle in the standing long jump. *Human Movement Science, 24*(1), 81-96.

# Chapter IV
# Technologies for Monitoring Human Player Activity Within a Competition

**Brendan Burkett**
*University of the Sunshine Coast, Australia*

## ABSTRACT

*Monitoring of player activity within a competition is currently a reality within some high performance sporting teams, and the demand and level of sophistication for this information will continue to grow as players and coaches seek this knowledge. This new found scientific wisdom can guide the training and preparation of the athlete, with the aim of improving performance and reducing the likelihood of injuries. To date this information has been collected manually, which is time consuming and expensive. The challenges are to validate the accuracy of these systems and once this criterion is satisfied to expedite the analysis process to enable as close to real time feedback as possible. Outside of the coach and player arena there is growing demand from the other associated parties, such as the host broadcast media, referees and the spectator, all of whom are seeking new knowledge on "what is happening" during the play.*

## INTRODUCTION

Most of the world's professional and high performance sporting teams employ video and/or performance analysts to assist coaches with technical and tactical match analysis. The performance analysis packages currently in use range from simple statistical databases to high-end programs that incorporate video and extensive quantitative and qualitative analyses. Sport is now a professional pursuit and with this many changes have occurred. The amateur traditions where the local school teacher, also doubled as the star player have been replaced with competitive teams forced to continually explore all possible avenues to improve their game and obtain a competitive advantage over their rivals. No longer is it good enough to just employ the best players and coaching staff, teams are looking for that edge in other sectors. One interesting field of innovation is the use of technology from existing disciplines and modifying this to enable monitoring of player activity within a competition, this is known as player tracking. The information that can be obtained from player tracking systems are extensive, in particular key components of the sports performance can be objectively measured. These include the technical and tactical game data such as a spatiotemporal breakdown, distance covered by each player, velocity profile, work-to-rest ratios, and the proportion of time spent in other game related activities. With the continuous developments in technology and the evolution of sport specific user interfaces, sporting teams will be able to make immediate adjustments to their playing strategies to provide them with this competitive edge.

Presently the collection and subsequent analysis of sports performance data has required a high degree of human labour. For example, to code an 80 minute rugby match requires over 100 man hours and the input of between six and ten analysts. The time and human resources required is directly related to the amount of detail necessary for the coach and player - for example to code or track the number of times a player passed to their left versus their right, how many carries an individual had with the ball, number of times the ball was kicked from a certain place versus passed, number of missed tackles by each player, success rate of a particular move, etc… will all take extensive time to code and track. From this information more delayed descriptions of player activity can be quantified in areas such as the movement patterns, distances travelled, and velocity zones of each player in the game. Currently there is very little, if any, of this essential data available, however armed with this detailed information on actions performed by their team, as well as the opposition, the coaches would be able to react and make tactical changes. This could include adjustments to the team's strategy on the field, instruction to specifically focus and improve in certain aspects of the game, as well as instruction to target identified weaknesses in the opposing team. The ability to make these changes in real time or delayed time is dependent on the hardware, software, and the level of detail required for the coach. There are several other sectors who have an interest in knowing "what is happening" during the game, these end-viewers such as the media and the spectator who enjoys watching the game, are driving the commercial aspect of player tracking technology and the need to develop a system as close to real time as possible.

Given the intense labour input currently required to code games, the ability to semi-automatically code player movements in close to real time is seen as the "Holy Grail" of sports performance analysis. One of the leading systems, UK-based company ProZone® (ProZone Sports Ltd., Leeds, UK), offers this service at a considerable price. Complete installation of the system in a stadium is priced at approximately £300,000, and there is an analysis fee of approximately £2,500 per game (ProZone, 2008), thus highlighting the large financial resource requirements to monitor human player activity within competition. Once commissioned this system is fixed for that particular stadium and therefore cannot be transported from one venue to another, and there is an overnight turn-around time for feedback to players and coaches as the system requires manual tracking of players. The fact that sporting teams are willing to pay considerable sums for the knowledge that is uncovered in monitoring player activity within a competition demonstrates the importance of this new technology. Based on this new demand to "know what is happening" during the game from the player, coach, referee, media, and spectator sectors within the modern day sporting community, the objectives of this chapter are to:

review the current digital sport player tracking systems; review the fundamental mechanism of tracking objects; identify the typical sport science measures required by sporting teams; and discuss the issues relating to positional analysis of players; all with the aim of guiding the future developments in digital sport technology.

## BACKGROUND

Prior to the technological advances in the last decade, the tracking of a players movement throughout the game involved all team members watching the recorded video of the entire game with the head coach stoping and rewinding at time points within the game to identify movement patterns. As would be expected the analysis of an 80 minute game with all of the stoping, rewinding and associated discussion about what happened could take several hours. Naturally during this lengthy time period the players would lose interest as their concentration dwindled and the effectiveness of the process was lost. More recently human motion analysis has received increasing attention from computer vision researchers which has expedited this process and provided alternatives with the digitised video offering a faster mechanism for cueing video footage and subsequent playback. This improvement in video playback has subsequently identified the new problem of how to analyse this data and for many years the analysis of a sport event has been based on "observation sheets" partially filled-in during the match or from watching the recorded video (Moeslund, Hilton, & Krüger, 2006). Despite the new knowledge identified from this process, the motion acquisition and analysis were performed manually, a time consuming and tedious task. In the past, progress in introducing the computer vision technology to the team sports domain was slow, due to inadequate video and computational facilities, and the complexity of the tracking process itself. As players move rapidly, change direction unpredictably, and collide with one another, they violate the smooth motion assumption, on which many tracking algorithms are based. Players appear in the images as highly non-rigid forms. Many of the proposed approaches solved the motion acquisition and analysis problem only partially, and were therefore unable to provide an adequate solution to the sports experts, i.e. tracking every player on the entire field in every instance of time (Ekin, Teklap, & Mehrotra, 2003). To further understand the technologies required for player tracking background information is provided in the following three subsections: the types of device; the methods of tracking; and the typical measures.

### Types of Player Tracking Systems

In tracking player movements there are essentially two categories of systems: (i) a user worn system, where as the name suggests the user must wear a tracking or transmitting device; and (ii) a no device system, where the player's movement is tracked by using only external devices. As part of the no device system there is a sub category relating to portability, these systems are discussed in the following section.

#### User Worn Device

These tracking systems will track a player who is wearing the device "tag", within the designated field. Depending on the system these tags are generally smaller than a credit card and lighter than 20 grams, and the transponder reflects the signal sent from the transmitter and identifies each player in their exact location. The tracking of the player wearing a device has high accuracy and video update rates. This procedure is done simultaneously for all players, and is preferably repeated in every video frame. The tracking of each player is independent, and contingent upon the particular system in use; up to 50 players all on the same field, or just one player can be monitored. Generally these systems include four major components, (a) the main electronic unit to house most of the digital and radio frequency components, this also supplies the power, (b) the transmitter and receiving device and respective antennas, (c) the transponders which are worn by the players, and (d) the central base for data collection and analysis.

In the processing circuitry of each receiver there needs to be a mechanism for sorting incoming frequency signals according to their various shifts and also for determining the range of each shifted signal according to the time difference between a transmitted signal and each shifted received signal. Preferably each player is provided with a pair of transponders of different frequency shift and the receiver circuitry contains a mechanism for decoding the output of each pair of transponders to provide the direction of movement of a player. The transmitting antennas can be installed in various locations within the stadium, no line of sight is required to each player, however there should be no major physical obstacles between the antennas and the field (Dambreville, Rathi, & Tannenbaum, 2006). The player's position is calibrated relative to the field's centre. The signal is delivered from the transponders on the field and is reflected back to the receivers. Each receiver measures the distance to all the transponders. The distance from a transponder to both receivers, which are installed in a known position, is used to calculate the location of the player on the field. The identification of each player is accomplished through a predetermined frequency shift on the tag. Every tag receives the carrier frequency, transmitted from the antenna, shifts it by a unique frequency and sends the shifted signal back. In the receiver, the frequencies are sorted according to the different shifts, and every signal is used to determine the range. Thus, the player identification is extracted from the frequency sorting and the range of the tag, and from the time difference between the transmitted signal and the received one. Some systems require each player to wear two identification tags, as this enables the player's heading to be determined by calculating the difference in location between the two tags (Pitie, Berrani, Kokaram, & Dahyot, 2005). Using techniques such as high resolution, spread spectrum, and advanced signal processing algorithms on one hand, and state-of-the-art miniature microwave modules on the other, these systems can operate with an extremely low power and high precision.

Examples such as SporTrack® (Real Image Media Technologies Pvt. Ltd., Chennai, India) or Trak Performance® (Sportstec Australia, Warriewood, Australia), enable tracking of 100 players or more, simultaneously, in real-time for long periods of time. These systems are based on innovative state-of-the-art microwave modules and advanced tracking algorithms, to achieve an accurate, independent player location for each video frame. When the system is synchronised with a calibrated video camera, the exact location of each player can be displayed on the screen, thus allowing instant digital and virtual replay as well as any predefined statistical measures, a flow diagram is shown in Figure 1. Another example of a user worn player tracking system, designed only for training sessions, is inMotio® (Inmotio Object Tracking BV, Amsterdam Zuioost, The Netherlands), which has a high sampling frequency (1,000 Hz). This training session system will provide data in real time, however this reduces with the number of systems per players, so for example a team of 13 players, the real time frequency drops to 1000 divided by 13, or 77 Hz. These transmitters are about the size of a computer mouse and have other options like heart rate monitoring. Ideally this system work best for training environments rather than game situations.

## No Device Worn

There are a number of sports that do not allow the players to wear a transmitting device, these systems generally rely on a video analysis to

*Figure 1. Flow diagram for multiple transmitters*

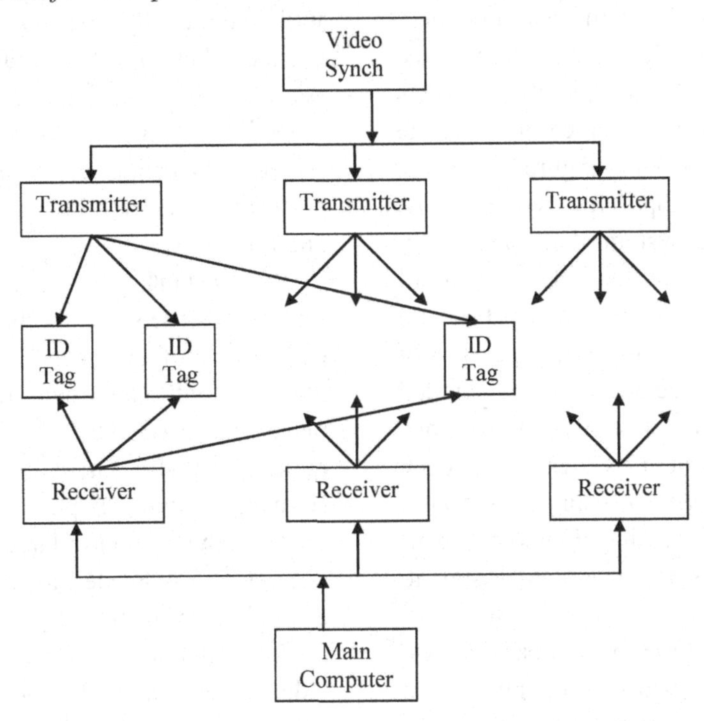


track the player's movements. Systems such as SportsCode® (Sportstec Australia, Warriewood, Australia), FairPlay® (Fair Play Pty Ltd., Jindalee, Australia), DartFish Team Pro® (Dartfish Australia, Sydney, Australia), or the SportVU® (SportVU Ltd., Kfar-Sava, Israel) analysis systems are essentially a video coaching tool that can collate the playing data by player, team, and/or by type of play. Outdoor sports such as football that take place on a pitch too large to be covered entirely by one camera have incorporated the use of fish-eye lens cameras to capture the required area (Pers & Kovacic, 2000).

In the systems such as SportVU "objects" are assigned; for example in a football game an object would include each player, the referee, as well as the ball. The system comprises of three off the shelf fixed cameras, positioned in one location on the pitch and the software tracks the relationship between the objects, where they are on the field, how far they have moved, at what speed, and the

*Figure 2. Typical image from FairPlay® software, the screen can display the video and video controls, and the tagged feature within the game, in this case the Own Goal circle option is highlighted. © 2009 Fair Play Pty Ltd. Used with permission.*

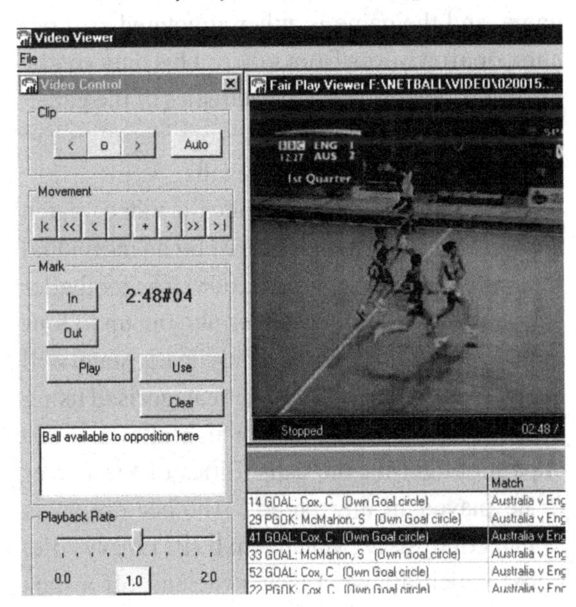

spaces and distances between each object throughout the game. Once these relationships are defined, then any combination of these factors (distance travelled, speed, etc...) that affect these objects can be analysed. A typical screen image of the FairPlay software is shown in Figure 2.

SoftSport Inc® (SoftSport Inc., Sylmar, USA) is a player tracking system that specifically concentrates on passes of the ball in a game of football. This company has two products on offer which analyses the full ninety minutes of the player's performance and provides detailed reports of completed and lost passes (SoftSport, 2008). The drawback of this system is the coach is required to enter all the data using the software, which can be time consuming. The software allows overhead images of player comparisons to be viewed, and presents the following statistics: Goals Scored, Shots On Goal, Impact Passes/Goals, Impact Passes/Shots, Total number of Passes, number of Completed Passes, number of Lost Passes, Completion Rate, and number of Intercepted Passes.

TrackSYS® offer a software package developed by Noldus Information Technology (Tracksys Ltd., Nottingham, UK) and is used for video annotation and presentation. This requires a set of events and players to be defined before the game, and the game is either annotated as it occurs, or afterwards from video. This data can be analysed to show the basic statistics of the game, as a time-event plot, or a series of video clips can be produced to display the results.

Sports Analytica® provides a video match analysis service which is geared for an individual footballer (SportsAnalytica, 2008). The footballer can send videos of themselves playing up to four matches, and the Sports Analytica® team will analyse the video. The game is categorised using a non-linear video system which allows video clips highlighting the capabilities of the player to be played in any order. Another option is Match Analysis® (Match Analysis, Emeryville, USA) which allows a coach to send off a narrated video of the game, and their specialists will methodically enter into the computer many events as they happen, this takes a couple of days. As previously mentioned ProZone® is another video based player tracking system in which the position of each player is manually marked; this generally takes over 34 hours to track 22 players and the ball.

SoccerMan® is as the name suggests a system that reconstructs the football game and based on the video from two views, creates a virtual 3D world with playground texture and textured player shapes (Bebie & Bieri, 1998; Xu, Orwell, & Jones, 2004). These images can be viewed from any virtual viewpoint. The main drawback of the system is that a large amount of manual intervention is needed to track the players, for example prior to tracking each player's head is identified manually in every tenth frame in both video streams, and a 3D spline is attached to this data as a basis for identifying the players' positions. Textures for each player are extracted, and textures that overlap need to be manually corrected. This ensures individual player texture representations are obtained and allows subsequent reconstruction of the game for visualisation from different viewpoints.

## Portable and Simple Player Tracking

Player tracking software can be installed on small portable devices such as PDA. These relatively cheap and small devices enable coaching staff to collect statistical and game data information for live coding. Some of the typical player statistics recorded is the number of shots at goal, or line breaks, or turn overs made. The user can create a template prior to the game to enhance the speed of data entry. There is no recorded video footage with this type of system, however it does allow the coach to quickly and cost effectively, capture player tracking data that can be referred to at any time during the game to assist in their decision making process. Depending on the software some

*Figure 3. Image of PDA screen for restricted live data coding © 2009 Sportstec Limited. Used with permission.*

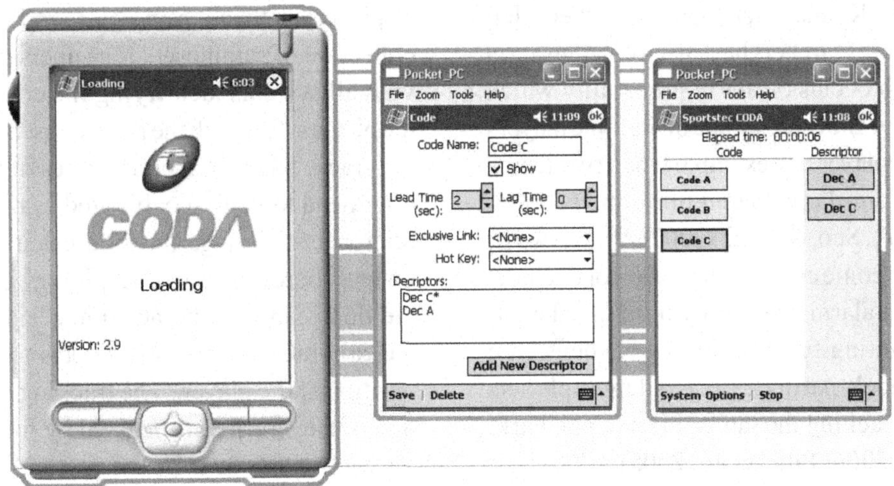


of this data can be exported back into larger programs such as SportsTec® either during or after the event. A typical image of the player tracking screen for football is shown in Figure 3.

## Mechanisms of Tracking

There are several mechanisms for tracking moving targets which can be applied in isolation, or in combination, and include active shape models, the Kalman filter, and condensation. Active shape models are flexible shape models that allow iterative refinements of estimates of the objects' pose, scale and shape (Cootes & Taylor, 1992; Dambreville, et al., 2006). One of the first people tracking systems developed was the Leeds People Tracker, designed for individual pedestrian surveillance, which employed contour tracking, active shape models, and Kalman filtering to track multiple people from a single camera (Siebel & Maybank, 2004). The computational efficiency of Kalman filters and its ability to estimate future states have resulted in several systems utilising this resource. The propagation of conditional densities over time, or condensation, has been used with contour tracking to track an object through cluttered scenes (Isard & Blake, 1996; Lu, Okuma, & Little, 2008). In this method developed for football, each player being tracked is independently fitted to a model, and the sampling probability for the group of samples is calculated as a function of the individual "goodness-of-fit" score. Estimates of position can be improved with the use of Kalman filters, which assists in tracking through occlusions. Problems arise with this system when tracking multiple targets, for example if several one-body trackers are employed that have the same tracking algorithm, then these single unit systems can merge together. The probabilistic exclusion principle was developed to address this issue (Czyz, Ristic, & Macq, 2007; MacCormick & Blake, 1999).

Some older systems such as Pfinder®, developed as a 'person-finder' which uses a multi-class statistical model of colour and shape to create a blob representation of a tracked person, can track the head and hands in real time (D'Orazio, et al., 2007; Wren, Azarbayejani, Darrell, & Pentland, 1997). This system will only work when there is a single person in the scene, but produces a more detailed model than is needed to obtain the position of sports players. By incorporating the colour variable to deal with shadows and occlusions the three layers of regions, people, and groups can be extracted (McKenna, Jabri, Duric, & Wechsler, 2000). Other researchers have used condensation tracking with occlusion alarm probability which aims to resolve the coalescence of multiple players into a single player, for example when two teams are wearing similar coloured uniforms (Choi & Seo, 2004; Ok, Seo, & Hong, 2002). When two or more players come close to each other, or occlude, the occlusion alarm probability used in calculating the weighting function for the condensation algorithm repels particles, to avoid multiple sets of particles tracking the same object (Choi, Park, Lee, & Seo, 2006; Pitie, et al., 2005).

Many attempts have been made at image segmentation from video by using background subtraction, adaptive background subtraction, and colour space models. Maintaining a temporal background model and performing background subtraction has been shown to be a fast and efficient method of extracting moving objects from a scene (Zivkovic, 2004). This performs best in relatively empty scenes through which objects are moving, but predominantly sporting activities do not fit into this category. In addition, if the player remains stationary for a short period of time, for example when they are away from the current action or in a tactical position, they can become incorporated into the background model. If a static background model is used to combat this, it may not allow for changes in the lighting conditions or small camera movements. Good, fast segmentation can be achieved by creating prior colour space models for foreground and background, in this case players and non-players. This method is robust to small camera jitters, and to stationary objects.

When tracking multiple objects a three step approach of condensation, propagation, and finally prediction is commonly adopted. An extra level of structure is required in the algorithms by tracking multiple objects rather than multiple single object tracking (Black & Ellis, 2006). Typically in this structure tracking of one player is described as a sample, tracking of a collection of players is then a sampleset, and tracking of a collection of samplesets is a supersampleset. As with all systems identifying the contact point of the players' feet with the floor is essential to ensure accuracy. The global coordinates of the field and floor need to be first calibrated as this forms the reference for image position calculations. The image coordinates of the players contact with the floor can then be accurately projected onto the calibrated global coordinates to identify the position of the player. The tracking of the collection of players are propagated by finding a score for each player within the particular collection of players that is similar, then the overall score for the collection of players is increased. If one or more of the collection of players is a poor fit, then the overall score is reduced. This aims to aid the propagation of the collection of players with the best overall fit of objects. The final stage is prediction, and this depends on the nature of the sport, for example, does this involve predictable straight line running compared to unpredictable agility manoeuvres?

## Typical Measures

There are various formats for the output of the player tracking data ranging from simple statistics that only list items such as the percentage of shots that were successful in a basketball game, through to the highly sophisticated database listed statistical data that is hyperlinked to video clips. With the continuous improvements in broadband transmissions a number of systems will enable players, coaches, and fans to view and analyse game data across the internet. Naturally the more sophisticated the data collection the greater time required for analysis, but the trade off is the ability to conduct comprehensive multi-levelled searches across a range of variables and scenarios. Most systems allow a template to be created beforehand to expedite the data collection and analysis, in some cases whilst the game is being played. Having a real time component of analysis thus allows the coach to compare both sides and make game-plan adjustments as the game is in progress. The typical data measured either during a training sessions and/or a competitive game includes the following:

a. Total distance travelled (m)
b. Mean velocity (m/s)
c. Work rates (J)
d. Heart rate (bpm)

From this fundamental data the following variables can be calculated such as the distances travelled at various velocities, breakdown velocity during specific time components (quarters, halves, or set intervals such as every 10 minutes…), and the relationship between heart rate and time or velocity etc… Coaches can use this to see how many short sprints the player runs in the game, how much time is spent stationary, or how quickly a change of pace occurs. An example of the distance travel and this relationship within the field of play for a wheelchair rugby player is shown in Figure 4. More advanced systems allow the operator to add graphics to recorded scenes to point out important information such as an instant replay. This is also attractive for host broadcasters as it can allow the commentators to do things such as put a circle around a particular player, highlighting and enhancing the scene.

## ISSUES, CONTROVERSIES, PROBLEMS

Positional analysis of sports players is of interest to coaches and trainers, as it allows a quantifiable method of analysing a players' performance, aids the design of better training regimes (also allowing quantifiable evaluation of such schemes on performance), and provides information for tactical analysis. It is important to obtain the positions of each player throughout the game to the highest possible level of accuracy and to a quantifiable level of accuracy. It is necessary to know the quality of the data obtained, and obtaining accurate data is dependent on the first four key issues listed as follows.

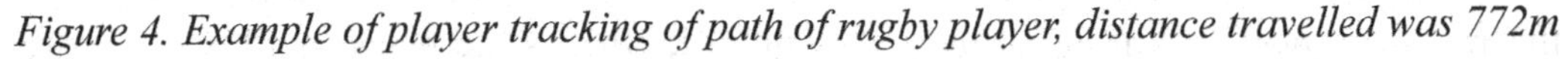
*Figure 4. Example of player tracking of path of rugby player, distance travelled was 772m*

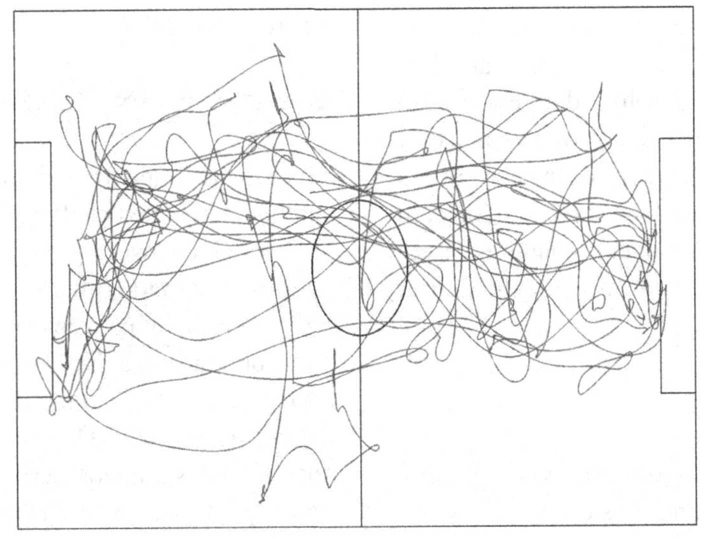

## Size of the Sport Field

To determine the trajectories of the players from the beginning to the end of the game, the objects in question have to be in the field of view for the duration of the whole match–and when considering the size of a sports field this is a very large capture window. If the system requires the video image to track the player, that is no device is worn, due to the perspective effects, the resolution of players on different parts of the image is significantly different and varies greatly between the nearest and the furthest parts of the oval or field. For example when using a 320 x 240 pixel image of a standard football field, if two vertically adjacent pixels on the image plane are projected onto the ground plane, then pixels in the nearest part of the image are approximately 3 cm apart, whereas those in the far goal mouth are approximately 45 cm apart. If the players are wearing the transmitting device however, the size of the sporting field is satisfactory for several transmitting and receiving devices.

## The Shape of the Players

The shape of sports players can vary significantly within one team and opposing teams. This variability creates a problem when using algorithms to automate the tracking of players. Furthermore the individuals player's shape can change quickly by raising their arms or stretching forward, laying on the ground, etc..., which does not happen in the original surveillance tracking systems for pedestrians or cars. Furthermore the players can move at variable speeds, have large variations in their silhouettes, often suddenly changing direction, all which makes their movements hard to predict.

## Occlusion

Sporting games are busy areas with players, crowds, mascots, cheerleaders etc all moving in and around the arena to create several occlusions. In contact sports it is common for players to obscure each other and/or for the players to merge together. This overlapping and subsequent reassigning of the images increases the complexity of the tracking. This emphasises the importance of considering the depth information in the image and the importance for the tracking to be performed using the ground plane coordinates, as this takes into account the distance that a player is physically able to cover over time. Knowledge of the players' position on the ground plane aids with resolving occlusions, particularly in scenes from such a perspective view, as often one player occludes part of another player when they are more than a metre away from each other.

## Evaluation

The fundamental requirement of any measurement system is the reliability and validity of the measurement. To quantify this requires the data measured via the new system to be compared to a known industry benchmark or gold standard. In the relatively green fields of player tracking in sport this extrapolation from the smaller confines of the laboratory is chartering into unknown territory. This highlights the critical issue of how player tracker systems should and/or can be evaluated; what is the bench mark to compare the reported data?

## Behaviour Modelling

Once the aforementioned issues on accuracy are addressed the next challenge is how to represent interactions between groups or individuals once the positional data is obtained? The large variability in movement patterns, movement status, and the sportspersons ability to quickly and erratically change their status make behaviour modelling almost impossible. The location and velocity of the ball plays a major role in the movements of sports players, along with the positioning of their team mates and the opposition players. The ball is relatively small and can move very fast, often not in contact with the ground as it travels in a complex 3D trajectory through the air. This is further complicated when the Magnus effect of the spinning ball produces a non-linear path. If the ball is the fundamental reference for players, then the accuracy of the position of the ball is decisive.

## Cost and the Time Frame for Collection and Analysis

As with most new technologies and developments the initial costs to have a system operational is relatively high, and as such raises the issue of equity and accessibility. The development and integration of a player tracking system must maintain the philosophy that at least the new raw knowledge on player movements is accessible to all participating teams to ensure fairness in the competition. What each team does with this raw data is another matter and this is where the minds of the coach and players come to fruition. In addition there is the current time frame issue for analysis of the collected data; at present the data analysis can take several hours, if not overnight, to be processed. Much attention is given to the integration of the tracking system graphical/statistical output with the video camera images. Successful integration means a synchronised system, enabling instant transformation from field coordinates to pixel location in every single frame. This synchronisation will require several establishment resources to setup the mechanism of collection and analysis. Currently TV broadcasts of sporting events use different graphic effects such as highlighting a player, marking a trail, or measuring position. In order to display the graphics on the screen in real-time, or instant replay, the object's position must be known. Current methods of tracking allow image tracking, but require working off-line to make this calculation, and most methods lack robustness. Thus, it takes a long time to track a player from a video clip, and the user must supervise the process, this prevents instant replays of interesting plays in a live broadcast with graphic effects, and time is money in the sports entertainment industry.

## Requirements and Restrictions

Finally the requirements and restrictions of the immediate users associated with player tracking of the sport all need to be addressed and abided by. For example the sporting governing bodies can have requirements that limit the player tracking options such as not allowing a device to be worn by the players as this may cause potential injury in a contact sport. Currently for contact sports such as rugby union, rugby league, and to a lesser extent Australian Football, transmitting devices are not appropriate. Sports such as soccer, netball, tennis and cricket however, would be some of the suitable options for a worn device to track player data. The governing body may also wish to restrict the real time ability of the player feedback as this new skill set may not be currently accessible to all sporting teams and would therefore be inequitable. On the other hand the coaches and players may have a different set of specifications for their sport, such as identifying the currently untapped knowledge of the players' status during various stages of the game – and the coach may seek this information in real time to enable tactical decisions to be made during the action. Other parties interested in accessing the new knowledge could include the media, who may use this information as part of the broadcast of the game and could also include some user-pay interactions, such as selecting the tracking of a particular individual or a different combination of game statistics. The referees or umpires would also have different requirements for this player tracking information, such as deciding if a player was in an off-side position during the play.

## Accuracy of Measure

Many different methods for tracking humans have been proposed in the past several years, yet surprisingly only a few authors examined the accuracy of the proposed systems. As the accuracy analysis is impossible without the well-defined ground truth, some kind of at least partially controlled environment is needed. Careful planning of image acquisition is crucial for the success of the whole project, inappropriately placed cameras can add a significant degree of difficulty and subsequent error to the tracking problem (Choi, et al., 2006). Analysis of an athlete's motion in a sporting game is well suited for that purpose, and it coincides with the need of the sport research community for accurate and reliable results of motion acquisition. Tracking of sports players has often involved "closed worlds", which involves giving context to the scene and relating that in a region of space and time. On the monitor screen specific pixels are associated with one object in the scene, for example each pixel must be associated with a player or field markings and from this video information the internal state of the closed world is used to identify the position of the players. This approach was established from the earlier work that identified actions from visual evidence in American football plays (Intille & Bobick, 1999), which was a different philosophy to the pedestrian tracking systems.

For the modelling of sports players' behaviour, zero error in players' positions would be ideal, although given the variability in human performance, an error of up to 0.5 m might be regarded as acceptable in hand-tracked data. It is expected that data will be usable for behaviour analysis if it is within 1 m of true position, so trajectories will be regarded as acceptable if they are within 1 m of the mean hand-tracked position. To demonstrate the changes in accuracy the following case study from Pers & Kovaic of a two-camera handball player tracking system, incorporating two complementary tracking algorithms is presented (Pers & Kovacic, 2000). The system was developed for simultaneously tracking several people in a large area of a handball court, using a sequence of 384-by-288 pixel images from fixed cameras. Three types of algorithms were evaluated and the count of human interventions required to obtain error-free tracking is shown in Table 1. This was followed by a test sequence, in which players were standing still. The measured distances therefore directly correspond to noise added by particular tracking method. To evaluate the effectiveness of the system the processing time per frame (based on standard PC equipped with 500 MHz Pentium III processor) was measured. Tests were performed on two video sequences of 30 and 50 seconds, respectively.

The advantages and disadvantages for each of the methods can be clearly seen, with the use of colour tracking in avoiding drift in player position. The motion detection system introduced little noise to the trajectories. On the other hand, more than one intervention per one second of playing time is required, which puts enormous pressure on human operator supervising the tracking process. Colour tracking, requiring a small number of interventions, was relatively fast, but

*Table 1. Comparison results for three different combinations of tracking methods (Pers & Kovacic, 2000)*

| Method | Interventions | Noise | Time |
|---|---|---|---|
| Motion detection algorithm | 45 | 80 meters | 0.424 sec/frame |
| Colour tracking algorithm | 12 | 249 meters | 0.175 sec/frame |
| Combination motion and colour algorithm | 14 | 55 meters | 0.229 sec/frame |

the trajectories contain extremely high amount of noise. The combination of both methods is nearly as fast as colour tracking alone, contains a low amount of noise and requires little intervention from the operator. Obviously the authors conclude that the combination system is the most suitable for use in an automated player tracker. Although the system was only tested in the handball match, we can assume that it would be equally useful in other team sports, provided that the cameras could be placed directly above the playing court.

## SOLUTIONS AND RECOMMENDATIONS

The first four issues raised on the size of the sporting field, the shape of the players, occlusions, and the subsequent evaluation of the player tracking system, all directly influence the accuracy of the system. Working through these challenges the size of the sport field is unlikely to change in the future as the dimensions of the game are specified. The perspective issues when capturing the large window of video data can be overcome with extra cameras placed around the field, but this increases the cost of hardware, the size of the data files as there are no extra video files, as well as increasing the processing time. Even if the players wear a device to transmit their position and subsequent movement patterns, the players and coaches also require a video image to actually "see" what happened, rather than looking at an animated model or stick figure representing their movement. Because of this challenge the preference is to develop a video based system that provides the real image of what happened during the play.

Similarly when using a video based system the recognition of the players is critical, and any change in shape of the players can further complicate the process. To allow for the possible change in shape of a player that can occur, for example when they raise arms to attract attention, or open stride when running, the height and width of the image being tracked will require modification - this is often done by adding Gaussian noise to the height and width pixels of the image. The limitation to apply a filter to any raw signal is that the original data then conforms to a more common profile, and this may result in occlusions of individual players. Continual improvements in the predictive modelling could be incorporated to reduce this possible confusion. Comparable issues arise with occlusion, and these issues are also unlikely to change as players will continue to merge together in contact sports. There are two mechanisms to address this issue, firstly to improve the unique image recognition of the individual player and to differentiate this from all other players. This could be done by recognising the player's number on their jersey along with the colour of hair, colour of their playing boots etc to provide a unique combination of attributes that recognise that player. This, along with an improvement in the behaviour modelling of the game will overcome the occlusion issues. The ultimate goal would be to robustly track the players of sports games and to learn their behaviour. Having created a behaviour model, this could be used to improve the performance of the multiple objects tracking system, by governing the expected dynamics of the players. A vicious circle exists, a robust tracker is needed to collect data in order to create a behaviour model *and* a behaviour model is needed in order to create a robust multi-object tracker.

The reference system for the model often revolves around the ball as the key component within the game, and the errors in accurately quantifying the position of this object from a 2D image (video), when the path of the ball is often in 3D highlights the need for multiple cameras to enable 3D tracking of the path of the ball. Others have tried various approaches for the task of tracking the ball, such as constraining motion in a single vertical by using a single moving camera, physics-based constraints (parabolic trajectories) and the shadow cast by the ball in soccer games (Kim, Seo, & Hong, 1998); or to address the occlusion of the ball by players in real footage by tracking the ball using colour and motion, assuming there are no players around the ball (Ohno, Miurs, & Sharai, 1999); or by using sets of pairs of cameras for tracking a tennis ball using monochrome cameras (Pingali, Opalach, & Jean, 2000). The last two of these are reported to have only been evaluated in a single experiment. The tennis ball tracking has been used in a number of broadcast tennis games to provide ball speeds and virtual replays. However, currently it is unable to handle player-ball interactions and occlusions, with which the other two systems also struggle. Due to the growing diversification within individual sports, team sports, and cyclic sports such as track cycling, there is a future need to develop specific player tracking systems to cater for these unique requirements.

As with technology based applications the cost and time for analysis can reduce with advances, or in some cases the simple adoption, of existing technology, to the required application. One of the key indicators on the time frame for collection and analysis is the output format. Currently these range from simple statistics that only list items such as the percentage of shots that were successful in a basketball game, through to the highly sophisticated database listed statistical data that is hyperlinked to video clips. With the continuous improvements in broadband transmissions, a number of systems will enable players, coaches, and fans to view and analyse game data across the internet.

To change the rules of the governing sporting bodies will require a combination of political requests from the sporting teams, coaches and media, to modify the current rules to allow player tracking. The valid issue of the safety of the players could be addressed with the reduction in the size of the user worn devices, as this will then reduce any potential injury when players make contact. Advances in microwave technology can allow a low power, lightweight tracking system due to precise digital components such as direct digital synthesiser, high frequency A/D converters and Digital Signal Processing. The main building block is an advanced, low power, spread spectrum radar that incorporates the advantages of Miniature Microwave Integrated Circuits components and power signal processors. These systems operate at low power (much lower than any ordinary mobile phone) and carefully chosen operating frequencies assures safe operation in any human or electromagnetic interference environment, and conforms to a number of current international safety standards. The end result is a small transponder worn by the player, similar to "smart cards" that are developed today around the world for many different applications. The transponders' small volume and weight make it almost unnoticeable to the player, and these devices could be sown into the player attire to further minimise the appearance.

The use of these microwave components allows the system not only to be small and lightweight, but also inexpensive. Due to the fact that the system is designed to operate at short distances, and with the use of spread spectrum techniques, the average and peak transmitting power is so low, that it is literally unaffected by electromagnetic interference and completely safe to humans. Carefully chosen transmitting frequencies and output power can enable these units to be well below any international radiation standards. The high update rate and the use of advanced waveforms enable large amounts of data to be processed in a very short time. The processor and these high rate algorithms are developed together to achieve the high efficiency needed for real-time operation, thus addressing the real time applications for coaches, media, and referees.

The combination of the issues listed before will culminate in improved accuracy of player tracking collection and analysis. The issue of video distortion and image rectification is commonly addressed in biomechanical data collection and analysis, but this discipline may or may not be involved in the validation and development of player tracking technology. As a relatively new discipline there are currently very few, if any, guidelines on the valid assessment of this data. The solution would be to introduce an accredited scheme where each system needs to demonstrate the calibration, operational, and mechanism of error correction, with their player tracking system. Similar national accreditation schemes have been created within Australia, and around the world, and the National Sport Science Quality Assurance Program developed by the Australian Institute of Sport is a good example of how the quality control for a number of laboratories around the country can be obtained (Australian Institute of Sport (Sport Science and Medicine), 2008). From the current developments in player tracking accuracy the combination of focusing on the position of the players feet, rather than their centre of mass, and the use of colour tracking appears to be the most suitable process for creating an accurate measure of position with the least amount of human interventions.

## FUTURE TRENDS

Monitoring of player activity within a competition is currently a reality within some high performance sporting teams, and the demand and level of sophistication for this information will continue to grow in the future as players and coaches seek this knowledge to enhance their sporting performance. This new found wisdom can add to the scientific understanding of human movement, thus increasing the level of performance, as well as making the activity safer. By quantifying the loads, stresses, and demands that actually occur during a real game situation, the coach and sport science staff can then accurately prescribe the intensity and duration that is required in the training environment to ensure the players are appropriately prepared for the real match scenarios. At present the demands of the game are estimated by observation or from past experience as a player. The future aspirations are to obtain this information in as close to real time as possible, as this data will enable the coach to make an informed tactical decision as the game is in play. This target is feasible within the currently available motion capture hardware in conjunction with the expansion of relevant sports specific data analysis software interfaces.

Outside of the coach and player arena, there is growing demand from the other associated parties, such as the host broadcast media and referees. Through the broadcast of the game the consumer who watches the sporting game is also seeking new knowledge on "what is happening" during the play. The ability to provide the media and consumer with new replay and statistical features is an attraction in the sports entertainment area. As with the coach and player, the media and viewer also seek the ability to view the game from their preferred angle, or to browse their preferred statistical measure as the game is in progress. A similar case is made by the referees and umpires of the sporting event, and with the increased knowledge of player, coach and media, the people who control the game will face greater scrutiny on the decisions that are made within the game. This cohort will also require a specific requirement on the player movement profile.

So where to next? The demands of the sport will require as close as possible real time feedback on the players' movement; this will need to be addressed in both worn devices, as well as in sports where wearing a device is not appropriate, yet. It is envisaged there will be a point in time when the demand for physiological measures such as heart rate, will outweigh the current rules and specifications of the game, and players will wear a transmitting device. This will be possible as the size of the identification tag decreases to the point where the possible injury from body or ground contact is well below the acceptable limits for any possible injury to the human. Other features that could then be added when the player wears a device include measuring the level of impact endured in a tackle. As before this new knowledge will objectively quantify the effectiveness of the attacking play or the appropriateness of the defensive technique, which can guide the off field training requirements, as well as providing an understanding of the actual loads absorbed in the body during contact which are necessary to establish proper safety guidelines. The continued development of a valid player tracking system could answer the following questions in the future:

- Can emergent behaviour be learned from observing real sports games?
- Can team membership be identified for each player?
- How do players interact with their team mates, and the opposition?
- Can the tactics be encapsulated in an understandable way?
- Can a generative model be created?
- Can the rules of the game be learned?

An exciting future trend for digital sport technology is the application of this game simulation that tracks the movement patterns of the social athlete or general public. The need to increase the levels of physical activity across all sectors of the community is well established, and this can be accomplished through physical activity, exercise, sport and fitness programs. Whilst scientific guidelines for the amount of physical activity exist, for example 30 minutes of moderate physical activity per day is recommended for an adult, as this equates to about 150 calories of energy expenditure, measuring the amount of physical activity in children has been identified as requiring urgent attention. New and novel methods of reversing the decline in physical activity are required across all age groups, with a priority focus required for children. One possible option is the development of interactive digital sport, in which the user participates or "plays" the simulated game. These systems could track the user's level of physical activity as they actually participate within a form of computer game, rather than being a passive observer of a computer animation. Technology is constantly expanding to "real life" scenarios, images, and experiences, with the most recent developments allowing increased human interaction when using digital sport technology. The ability to connect with the younger generation and increase their level of physical activity needs to be investigated further; digital sport could answer this problem.

The progression from essentially sedentary games such as PlayStation® to more interactive games such as the Nintendo Wii®, provide the platform for the operator to participate in the activity. The further development of player tracking in sport could provide another insight into the digital technology, with remote playing, interaction, and remote coaching applications. Already there is some application in the rehabilitation settings with physiotherapists creating "Wii-hab" settings within the hospital and nursing homes as part of the rehabilitation program and to keep residents mentally and physically stimulated.

## CONCLUSION

The growing demand from professional sport to know what is happening within the game are the key drivers for the continued research and development of player tracking systems. This relatively new technology is unlocking some of the myths about player activity, and this new found knowledge can provide fresh guidance to game tactics that can enhance the performance as well as allowing appropriate safety guidelines to be established. The quest from other key sectors such as the media, the spectator, and the referee, also place strong demand on the need for an accurate, valid, and close to real time feedback on player activity. By incorporating existing and new hardware technology in conjunction with evolving software interfaces this need can become reality. Often the development of technology begins with the high performance end of the market, such as the racing car industry, before this knowledge can flow down to the consumer end of the market. The same process applies with the tracking of player movement. In the future the development of this new technology could be extrapolated to the broader scope of the community for interactive games that increase the level of physical activity from actual player participation.

## REFERENCES

Australian Institute of Sport (Sport Science and Medicine) (2008). Retrieved 12 March, from http://www.ausport.gov.au/ais/sssm/quality_assurance

Bebie, T., & Bieri, H. (1998). *SoccerMan - reconstructing soccer games from video sequences.* Paper presented at the International Conference on Image Processing, Chicago, USA.

Black, J., & Ellis, T. (2006). Multi camera image tracking. *Image and Vision Computing, 24*(11), 1256-1267.

Choi, K., Park, B., Lee, S., & Seo, Y. (2006). Tracking the ball and players from multiple football videos. *International Journal of Information Acquisition, 3*(2), 121-129.

Choi, K., & Seo, Y. (2004, May). *Probabilistic tracking of the soccer ball.* Paper presented at the International Workshop on Statistical Methods in Video Processing, in conjunction with ECCV, Prague, Czech Republic.

Cootes, T., & Taylor, C. (1992). *Active shape models - 'smart snakes'.* Paper presented at the British Machine Vision Conference, Leeds, UK.

Czyz, J., Ristic, B., & Macq, B. (2007). A particle filter for joint detection and tracking of color objects. *Image and Vision Computing, 25*(8), 1271-1281.

D'Orazio, T., Leo, M., Spagnolo, P., Mazzeo, P.L., Mosca, N., & Nitti, M. (2007). *A Visual Tracking Algorithm for Real Time People Detection.* Paper presented at the Eighth International Workshop on Image Analysis for Multimedia Interactive Services. WIAMIS'07, Santorini, Greece.

Dambreville, S., Rathi, Y., & Tannenbaum, A. (2006). *Shape-based approach to robust image segmentation using kernel pca.* Paper presented at the IEEE Conference on Computer Vision and Pattern Recognition, New York, USA.

Ekin, A., Teklap, M., & Mehrotra, R. (2003). Automatic soccer video analysis and summarization. *IEEE Transactions on Image Processing, 12*(7), 796-807.

Intille, S., & Bobick, A. (1999). *A framework for recognizing multi-agent action from visual evidence.* Paper presented at the National Conference on Artificial Intelligence, Cambridge, UK.

Isard, M., & Blake, A. (1996). *Contour tracking by shochastic propagation of conditional density.* Paper presented at the 4th European Conference Computer Vision, Cambridge, UK.

Kim, T., Seo, Y., & Hong, K. (1998). *Physics-based 3D position analysis of a soccer ball from monocular image sequences.* Paper presented at the Sixth IEEE International Conference on Computer Vision, Bombay, India.

Lu, W. L., Okuma, K., & Little, J. J. (2008). Tracking and recognizing actions of multiple hockey players using the boosted particle filter. *Image and Vision Computing,* in press.

MacCormick, J., & Blake, A. (1999). *A probabilistic exclusion principle for tracking multiple objects.* Paper presented at the Int. Conf. on Computer Vision.

McKenna, S., Jabri, S., Duric, Z., & Wechsler, H. (2000). *Tracking interacting people.* Paper presented at the Fourth IEEE International Conference Automatic Face and Gesture Recognition, Washington, USA.

Moeslund, T.B., Hilton, A., & Krüger, V. (2006). A survey of advances in vision-based human motion capture and analysis. *Computer Vision and Image Understanding, 104*(2-3), 90-126.

Ohno, Y., Miurs, J., & Sharai, Y. (1999). *Tracking players and a ball in soccer games.* Paper presented at the IEEE International Conference on Multisensor Fusion and Integration for Intelligent Systems, Taipei, Taiwan.

Ok, H., Seo, Y., & Hong, K. (2002). *Multiple soccer players tracking by condensation with occlusion alarm probability.* Paper presented at the Statistical Methods in Video Processing Workshop, Copenhagen, Denmark.

Pers, J., & Kovacic, S. (2000, 06/14/2000 - 06/15/2000). *Computer vision system for tracking players in sports games.* Paper presented at the Image and Signal Processing and Analysis, IWISPA 2000. First International Workshop on Image and Signal Processing and Analysis. in conjunction with 22nd International Conference on Information Technology Interfaces., Pula, Croatia.

Pingali, G., Opalach, A., & Jean, Y. (2000). *Ball tracking and virtual replays for innovative tennis broadcasts.* Paper presented at the International Conference on Pattern Recognition, Barcelona, Spain.

Pitie, F., Berrani, S. A., Kokaram, A., & Dahyot, R. (2005). *Off-Line Multiple Object Tracking Using Candidate Selection and the Viterbi Algorithm.* Paper presented at the International Conference on Image Processing, ICIP 2005, Genova, Italy.

ProZone (2008). ProZone Sports Retrieved 12th May 2008, from http://www.pzfootball.co.uk/index.htm

Siebel, N., & Maybank, S. (2004, May). *The ADVISOR Visual Surveillance System.* Paper presented at the Applications of Computer Vision (ACV'04), Prague, Czech Republic.

SoftSport (2008). SoftSport Inc. Retrieved 12th May, 2008, from http://www.softsport.com/wc/

SportsAnalytica (2008). Retrieved 12th May, 2008, from http://www.briggspalmer.com/new/index.html

Wren, C., Azarbayejani, A., Darrell, T., & Pentland, A. (1997). Pfinder: real-time tracking of the human body. *IEEE Transactions on Pattern Analysis and Machine Intelligence, 19*, 780-785.

Xu, M., Orwell, J., & Jones, G. (2004). *Tracking football players with multiple cameras.* Paper presented at the International Conference on Image Processing (ICIP'04), Singapore, Singapore.

Zivkovic, Z. (2004). *Improved adaptive Gaussian mixture model for background subtraction.* Paper presented at the 17th International Conference on Pattern Recognition (ICPR'04), Cambridge, UK.

# Chapter V
# Video-Based Motion Capture for Measuring Human Movement

**Chee Kwang Quah**
*Republic Polytechnic, Singapore*

**Michael Koh**
*Republic Polytechnic, Singapore*

**Alex Ong**
*Republic Polytechnic, Singapore*

**Hock Soon Seah**
*Nanyang Technological University, Singapore*

**Andre Gagalowicz**
*INRIA, Le Chesnay, France*

## ABSTRACT

*Through the advancement of electronics technologies, human motion analysis applications span many domains. Existing commercially available magnetic, mechanical and optical systems for motion capture and analyses are far from being able to operate in natural scenarios and environments. The current shortcoming of requiring the subject to wear sensors and markers on the body has prompted development directed towards a marker-less setup using computer vision approaches. However, there are still many challenges and problems in computer vision methods such as inconsistency of illumination, occlusion and lack of understanding and representation of its operating scenario. The authors present a video-based marker-less motion capture method that has the potential to operate in natural scenarios such as occlusive and cluttered scenes. In specific applications in sports biomechanics and education, which are stimulated by the usage of interactive digital media and augmented reality, accurate and reliable capture of human motion are essential.*

## INTRODUCTION

Work on motion capture and analysis started as early as the 19th century, when Eadweard Muybridge began photographing horses to analyze their movement. During that period, a French physiologist Etienne-Jules Marey also embarked on his chronophotography work in studying the human performance filmed. Chronophotography is an application of the study of movement (science), and photography (art).The importance of motion capture (mocap in short), is motivated by its applications over a wide spectrum of areas. Tracking and following the movement of human joints over an image sequence, and recovering the 3D body posture and kinematics are especially useful for the study of the human locomotion and bio-mechanical applications (Corazza, 2006), such as gait analysis, injury prevention, rehabilitation and sports performance enhancement.

Since the mid-1970s, human motion analysis applications have made significant progress thanks largely to the tremendous advancement in digital electronics technologies. The measuring and analyzing of human movement have applications in many domains ranging from kinesiology, ergonomics, sports, 3D animation, 3D tele-presence, augmented reality, video surveillance, video data compression as well as medicine and clinical practice. Many, if not all, of these systems require the tracking of the motion trajectory of the subjects to yield kinematic information for further analysis.

There are many technologies developed to capture the human motion. They range from magnetic, mechanical and optical systems, for which the subject needs to wear sensors and markers on the body, to the non-intrusive one which is based purely on using video cameras. The kind of set-ups, methods and technologies that are used for motion capture are largely determined by their respective operational needs, that is, application specific. For example, it is a common practice in biomechanical studies for the subjects to wear reflective markers for movement analysis while operating in a very well-controlled environment.

In order to be suitable for human biomechanical applications, which is our main focus, the motion capture technique has to yield accurate and reliable quantitative information. To extend the applications, cluttered, outdoor and occlusive environments have to be considered. Also, the freedom of movement to the performer is an important consideration, which implies that wearing of sensors on the body has to be avoided. These issues are especially crucial if we are to consider a real scenario such as a sports tournament. Therefore, it has prompted many researchers to explore the use of video-based computer vision motion capture, which do not need any sensor or marker on the subject, for measuring the human movement.

Over the years, the use of video applications in the teaching of sports and physical activities have ranged from qualitative analyses to augment feedback in learning new sports skills (Koh & Anwari, 2004) to multimedia online environments with rich displays of animations and videos and embedded discussion forums to facilitate analysis and critique. These were designed to convey multiple representations of actual skilled sports performances and to facilitate the use of cognitive and social processes in learning and inquiry in a collaborative manner (Lim & Koh, 2006). However, efforts to use learning technologies are often hampered by prevailing attitudes and logistical issues. It is hoped that a marker-less motion capture system will circumvent the logistical issues and make for a seamless integration into novel pedagogical approaches in teaching sports and physical activities. Done well, this new form of learning technology has the potential to bring about a cultural change in the way we teach and learn.

Motivated by the enormous potential and widespread applications that lead to the pursuit of a video-based marker-less motion capture system, in this chapter we describe the application and technical advancements, future challenges and problems that lay ahead as we work towards a marker-less motion capture system. We will also present the work of our research team in the development of a marker-less motion capture system (see the following).

The outline of this chapter is as follows:

We will first review the existing technologies and methods. This is followed by a discussion on the challenges and outstanding problems. We will also present a video-based marker-less motion capture system proposed by our research team. The last few sections will illustrate the possible extension of our work with potential applications before the conclusion.

## EXISTING TECHNOLOGY

Existing commercially available motion capture (mocap) devices fall into 3 main categories: magnetic, mechanical and optical systems. All these methods require the subject to wear some form of devices or markers on the body. Attaching markers to the subjects poses a hindrance to the true movement of the performers and in some instances, may render the movement unnatural. There is therefore an emerging need to develop a system to operate in a setup where the subjects wear natural clothing and move in the natural context as opposed to laboratory conditions. A possible solution to answer this need is to explore the use of computer vision-based and graphics algorithms.

### Main Commercial Motion Capture Methods

Magnetic motion capture systems utilize sensors placed on the body to measure the low-frequency magnetic field generated by a transmitter source. The sensors and source are cabled to an electronic control unit that compiles their reported location within the measurement field. The electronic control unit is networked with a host computer that represents these positions and rotations in 3D space.

Mechanical motion capture systems require the performer to wear a mechanical armature fitted to their body. The fitting is usually done with elastic straps and belts, which hold plastic plates against the body of the performer. The sensors in a mechanical armature are usually variable resistance potentiometers or digital shaft encoders. These devices encode the rotation of a shaft as a varying voltage (potentiometer) or directly as digital value.

Existing commercial optical motion capture systems utilize reflective or pulse-LED (infra-red) markers attached to joints of the athletes' body. Multiple infra-red cameras are used to track the markers to obtain the movement of the subject. Currently, this kind of system is mainly used for biomechanical analysis. The main manufacturers of optical systems are Motion Analysis, Qualysis and Vicon.

In general, existing commercial mocap systems are only suitable for a very well controlled (mainly indoor) environment. Table 1 shows a comparison into the main features the different motion capture systems. Figure 1 shows an example of the motion capture setup using reflective markers carried out in our lab for analysis the walking pattern. This would be unrealistic and impossible to implement during a sport tournament for example. Furthermore, post-processing of motion capture data is tedious and time consuming e.g. re-establishing the correspondent markers across time due to the confusion of sensors in optical marker systems.

### Computer Vision-Based Motion Capture

The inputs to all vision-based system are purely images acquired from electro-optic sensors such as video cameras or infra-red cameras. Electro-

*Table 1. A comparison of different motion capture systems*

| Type of Systems | Advantages | Disadvantages |
|---|---|---|
| Magnetic | No sensor occlusion issue<br>Position and rotation are measured absolutely | Highly intrusive<br>Cumbersome for the performers and movement constriction<br>Sensors suffer from interference<br>Operates mainly in indoor and well-controlled scenarios |
| Mechanical | No sensor occlusion issue<br>Sensors do not suffer from interference | Highly intrusive<br>Cumbersome for the performers and movement constriction<br>Need frequent calibration of mechanism<br>Operates mainly in indoor and well-controlled scenarios environment |
| Optical (with Marker) | Low intrusive<br>Good realism of movements | Operates mainly in indoor and well-controlled scenarios<br>Markers suffers from occlusion, confusion and undesirable swapping<br>Highly priced (about US$ 200000 or more) |
| Computer vision-based | Non-intrusive<br>Natural freedom of movement for performer<br>Potential to perform in outdoor and cluttered environment | Technology is not fully matured, it needs more research and validation |

optic devices are popular largely owing to their non-intrusive and passive nature. Also, they are easy to set up and their prices are cheap. In order to get a good coverage of the performance, multi-cameras are usually used. Computer vision-based approaches for motion capturing, with their potential to operate in natural environments, have started to catch the attention of industrial motion capture manufacturers.

*Figure 1. Example of motion capture using reflective markers*

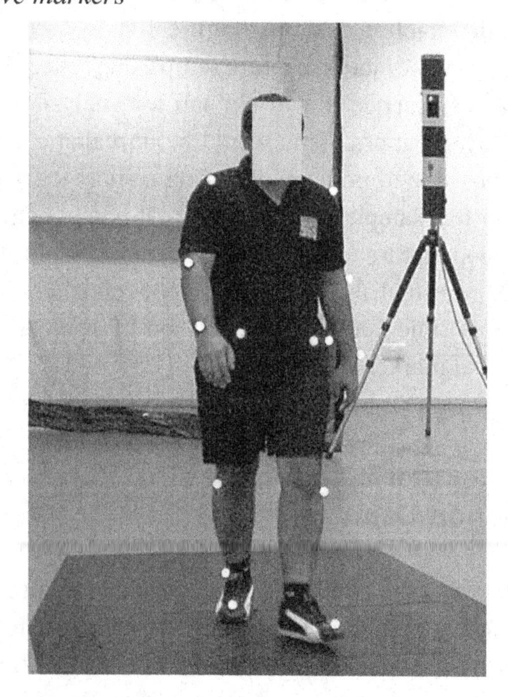

## Use of Computer Vision-Based Methods

Large numbers of work on the tracking and analysis of human motion using computer vision techniques had been proposed over the years. Reviews on computer vision-based motion capturing methods had been done in the years 1994 and 1999 (Aggarwal & Cai, 1994; Gavrila, 1999). In a more recent review, Moselund (2006) extended the earlier reviews of year 2001 (Moselund & Granum, 2001). The algorithms, technologies and equipment setups that have been used for motion capture are driven by the main application areas: surveillance, interaction/control, or analysis. In systems that require accuracy and reliability human models are usually used to facilitate the tracking (Carranza, 2003; Lerasle, 1999). Whereas in applications for visual surveillance and activities monitoring (McKenna, 2000), which require speed and robustness, such systems usually heavily involve low-level vision techniques such as image segmentation, thresholding, statistical and probabilistic formulations.

In the computer vision-based mocap survey by (Moselund & Granum, 2001), tracking is defined as establishing a coherent relation of subject and/or landmarks between frames. Pose estimation of subject is the process of identifying how the subject is configured in the scene e.g. posture of a human described by its kinematics at an instantaneous time. For 3D model-based motion capture approaches, the tracking and pose estimation are usually closely coupled with each other. The notion of motion tracking is used differently and loosely defined throughout the literature of visual analysis of human. In this article, since tracking and pose estimation are tightly coupled, and the aim is to quantify the human posture at each instantaneous time and follow it over a time sequence, the overall process refers to 3D human motion capture or 3D human motion acquisition.

Nearly every vision-based mocap follow the steps: (a) segmentation of subject from the rest of the image; (b) transformation of these segmented images into some kind of higher level representation to suit a particular tracking algorithm; and, (c) how the subject should be tracked from one frame to the next.

In these kinds of frameworks, many proposed algorithms have relied heavily on the image segmentation, which is a very crucial part of the system, and modeling of its result. In addition, some assumptions regarding the background scene had been used e.g. constant and low cluttered background. The information that is presented for segmentation can be either spatial or temporal image data.

## Common Motion Tracking Approaches

The use of silhouette is one popular approach where 3D volume data of human is built from multiple 2D silhouettes to yield the human posture that is tracked at each instantaneous time sample (Bottino & Laurentini, 2001; Cheung, 2003a; Mikic, 2003). This kind of approach is commonly known as the shape-from-silhouette (also called visual hull) method. It relies on calibrated still cameras, and segmenting the subject from the background by assuming constant and low cluttered background. The silhouette approach is a very mature topic of research in the domain of computer vision and image analysis.

Various motion tracking algorithms, frameworks and steps were proposed based on the visual hull approach. In Cheung (2003a), they extended their temporal shape-from-silhouette technique (Cheung, 2003b) to track the articulated rigid segments through the alignment of multiple-views silhouette poses across time. Then in Kehl & van Gool (2006), the authors proposed a *stochastic meta descent* minimization algorithm to fit a human model made up of super-ellipsoid parts to the volumetric data. In Nobuhara & Matsuyama (2006), the authors used a deformable mesh model with 'repulsive force' to guide the inter-frame 3D volumetric deformation and at the same time detecting collision. Then in Mikic (2003), they estimate the human movement parameters by using the extended Kalman filter after utilizing the Bayesian network to estimate the body part sizes from 3D voxel data. Also in Carranza (2003), they adapt an articulated 3D human model to the multi-view silhouette via Powell's minimization.

The extraction of motion and body structure through optical flow related method had also been proposed (Bregler, 2004). Other methods by Planker & Fua (2001) and Grest (2005) propose a recovery of motion through the estimation of disparity and depth map. Also there are suggestions (Kim & Pavlovic, 2007; Rosales & Sclaroff, 2006) to perform motion capture through the learning or training via a human motion database collected a prior. However, such a database may be incomplete and when the general human model dynamics are to be learned, the amount of data, model complexity and computational resources that are needed may be impractical.

### Estimation of Kinematics and Skeleton Structures

We need to acquire the true kinematics of the subject undergoing motion so that they can be useful for analysis. A human undergoing motion is mainly resulted from the contraction of muscles that drives the skeleton-bones, which can be represented via a kinematics chain. Thus, the visual appearance is an outcome of skin deformation caused by the skeleton movements and muscle actions.

Generally speaking, all the motion capture systems described before do not directly measure the movement of bones and skeletons. In the case of motion capture system that set up the subject to wear reflective markers, the reported markers location on the surface of the skin are used for estimating the kinematics and skeleton structures by using some form of mathematical least-square minimization techniques.

In medical applications, the human skeleton is obtained via X-ray. However its drawbacks are that X-rays are not easily available and it is impossible in the current technology that the human motion can be measured by X-ray. Lafortune (1992) attempted to explicitly measure human movement by inserting pins into the bones of the subject. Unfortunately, the procedure can result in the subject experiencing severe pain. Such a setup has very limited scope for application and is not practical in the general sport environment.

A common approach taken in video-based systems to estimate skeletons and joint locations is to use medial axis extraction algorithms based on mathematical morphology. Many techniques to compute 2D and 3D skeletons of object have been proposed over the years (Goh & Chan, 2007; Ma & Sonka, 1996). In Rosenhahn's (2005) work, skeletons were obtained from 2D images of the segmented subject. Wingbermuhle (2001) extracted the skeleton of the 3D model constructed by its visual hull. However, researchers have acknowledged that such approaches are very sensitive to noise and the uncertainties due to noise are not easily tractable. Other works attempt to improve the skeleton estimation by labeling body parts to kinematics chain of super-quadrics representing voxel data (Theobalt, 2004). It is also possible to derive a common kinematic structure through representing the shape of the subject in each time frame by using augmented Multiresolution Reeb Graph. However, we have to take note that the skeletons obtained mathematically are not the same as anatomic skeletons and do not take into consideration the skin deformations. Hence, numerical skeletons will not give the true anatomic joint locations e.g. the human's elbow joint is not at the centre of the surface skin.

## CHALLENGES AND OUTSTANDING PROBLEMS IN VISION-BASED METHOD

Despite the numerous research works on vision-based motion capture being done and published, there are still many challenges and problems to be resolved before they are ready to be accepted for wide usage.

### Computer Imaging Issues

In video images, their colour information is produced by many factors such as: (a) reflectance material of subject; (b) posture of subject; and, (c) light source. Hence, an image is a result of a highly dimensional combination, in which the lack of understanding and representation of its operating scenario is a main cause to the existing problems. Some of the researchers have termed this as the semantic gap (Dorai & Venkatesh, 2002). For example, the human posture undergoing motion is highly complex since they have many degree-of-freedoms (DOFs) and the skin is also deformable.

Most of the existing computer vision methods rely heavily on feature segmentation to be the entry point to their systems. Feature segmentation aims to extract the natural imaging features such as edges, highly textured points and corners from images. However, feature segmentation suffers severely from occlusion, lack of texture and foreshortening. There is no fool-proof method for image segmentation that can universally cater to all the environments. In addition, there is no consistent way to integrate different methods into a hybrid system.

Human undergoing motion will produce skin deformation, usually termed as skinning. The skinning issue has been studied in computer graphics; however it had not been properly reviewed in relation to any existing vision-based motion capture. Skinning is necessary for the geometric properties and color texture on the human skin to be warped correctly when undergoing motion. Some examples of human skin deformation include bending of elbow which causes the skin to stretch, the wrinkling of skin and the 180 degrees twisting of wrist which causes the skin at the wrist to twist about 180 degrees, while the skin near the elbow is twisted by only about 5 degrees. Other kind of deformations can arise from the bulging of muscle or the bending of the body. All these deformation factors are anatomically related. Therefore, the anatomical structure of 3D model must be able to mimic the subject closely.

One of the simplest and fastest ways to animate skinning is to consider a rigid skin model that is attached onto the articulation of the respective bones and to ignore the deformations near the joints. However, rigid skinning does not look realistic near the joints. The deformation of human skin is highly complex. By itself, it has attracted many research interests over the years, and mimicking the real human skin still remains a challenge.

When moving ahead towards a larger scale operating environment, there are issues such as the multiple interaction of people, the high occlusion and clutters as well as illumination uncertainties. These issues have yet to be resolved or properly tackled in the area computer imaging and artificial intelligence.

### Hardware Issues

The progress of computer technology has enabled very fast processing speeds of up to GigaHertz on a single terminal, and even possibilities of putting together multiple processors and computers to perform the computation. However, there are several hardware issues that we have to address before the motion capture system can be fully optimized to its potential. Considering a future system setup of many cameras with true RGB colours at high definition (HD) of 1920×1280 pixels, the physical properties of the hardware have to be ready for this challenge.

Bandwidth for data communication between the hardware devices and processors has always been one of the bottlenecks of the overall computational performance. When dealing with video data, the amount of data transferred between the devices poses a very serious issue. Although many video compression methods have been proposed, the compression ratio is always restricted to the entropy of the signal when near loss-less compression is needed for image analysis by the mocap system.

The size and the storage speed of the storage devices (in relation to the data bandwidth) of the computer is another key issue. These storage devices including the CPU and GPU memories must be able to cope with the amount of data whilst maintaining a real-time application. Last but not least, the CPU and GPU computing speed and parallelism of computation are also areas that cannot be ignored, considering the amount of information which the processors have to analyse.

## MOTION CAPTURE SYSTEM: A FRAMEWORK

This section describes a framework that we have proposed, and in which our research team is working on. The aim of our motion capture system is to create a framework and the realization towards a marker-less setup that can eventually be used in the cluttered and outdoor environments. The operating pipeline of our proposed framework is shown in Figure 2. The functionality of each block is modular and self-contained. We will also draw attention to some relevant methods that have been used in some other practices for each of the modules along the operation pipeline. And then, we will go on to show our results, which will illustrate the feasibility and potential of our approach.

Our methodology implements the model-based analysis-by-synthesis approach built on the concept of collaboration between computer vision and computer graphics to tackle our problems. The main concept has been proposed as early as 1990 (Gagalowicz, 1990), and has gained popularity (Terzopolus, 1999). The computer graphics will provide the knowledge-base for the 3D modeling, animation and synthesis, whereas, computer vision will supply the learning and analysis capabilities.

## Camera System

The advancement in imaging technologies over the past 10 years has seen video technology move from the interlacing sensor scanning of the NTSC or PAL to the progressive scanning in most of the current digital video devices, and towards a common high definition (HD) in the not far away future. To acquire fast movement, a high camera frame rate and fast shutter speed is needed. Our cameras run at a frame rate of 30Hz and shutter speed of at least 1/250 second.

The cameras are connected to a host computer. In our system, the data interface between the digital cameras and the computer is connected through the IEEE 1394 fireware for the ease of interfacing with computers. The IEEE 1394b interface is able to provide data bandwidth of up to 800Mbits per second at its current state, and with up to 6Gbits per second of data rate in the next few forthcoming years. Data repeaters could also be used to increase the distance of data transmission.

In a multi-camera system, all the cameras have to be synchronized so that all the images are acquired at the same time step. One common approach is through the use of an external triggering source, which is sent to all the cameras. Alternatively, the IEEE 1394 interface™ can be used to synchronize the cameras that are connected to the same PCI interface card on the computer.

Camera calibration is a software process that initializes the focal length, orientation and position of the cameras. This process also performs the correction in the images caused by lens distortion. The calibration of these parameters of a multiple camera system is to enable the recovery of 3D geometrical information through computing methods such as triangulation and back-projection of imaging rays.

Over the years, many camera calibration techniques have already been developed. Every camera calibration algorithm assumes that the

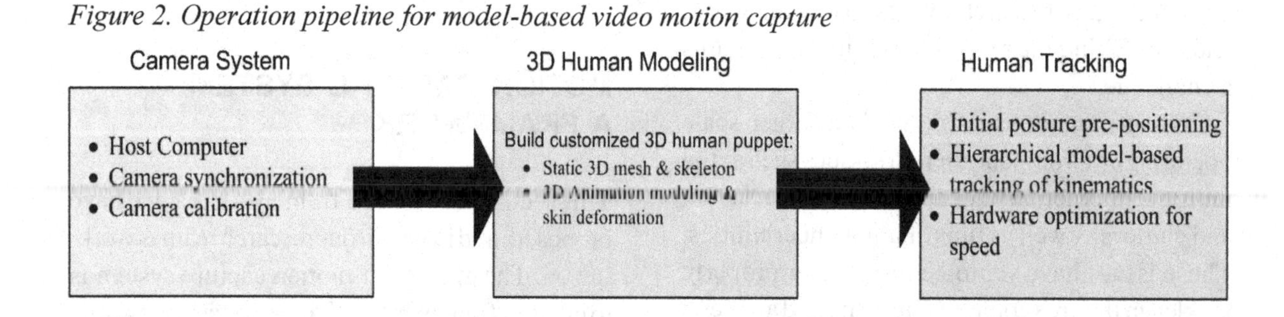


*Figure 2. Operation pipeline for model-based video motion capture*

3D object information and their corresponding features in the 2D images are already established. The main approaches are through: closed-form methods, numerical solutions, or relating linear camera model to the perspective model by iterative solutions. Readers who are interested in the technical algorithm in camera calibration may refer to the work of Bouguet, Dementhon & Davis (1995) or Zhang (2000).

## 3D Human Modeling

To have a 3D geometrical human model that closely resembles the size and shape of the subject, a very crucial component in the mocap system that will perform this function is the accuracy and reliability of tracking the subject in motion. Many studies (Gerard & Gagalowicz, 2003; Goncalves, 1995; Kakadiaris & Metaxas, 1998) have already indicated that the process of tracking is very sensitive to the shape parameters used. Thus, it is inappropriate to use, for example, a generic "averaging human" model for accurate and precise tracking of human with a different shape and size.

## Building the 3D Human Model

In order to make sure that the 3D human model can be used properly later on in the tracking process, a substantial amount of work has been spent on the research and development of the algorithm for 3D reconstruction of human model. Also, the software system for building this 3D human should allow only minimal to moderate amount of human interaction such as clicking on a few feature points and drawing a few curves. This is to ensure that the whole 3D reconstruction process does not take more than several minutes.

Representation of the customized 3D human model comprises the following steps:

i. The surface skin made up of regular triangle mesh.
ii. Bones made up of triangle mesh as well. Alternative, for simplicity, they can be 3D curves/lines formed by the medial axes of these bones. The bones of different body parts are also linked through a forward kinematics chain, the root of which starts from the pelvis. During 3D animation, the bones will drive the surface skin by manipulating its kinematics, which deform the surface.
iii. Given also the calibrated cameras parameters and real images of the subject, the texture coordinate of each vertex on the 3D model can be computed by back-projecting its 3D position through the camera parameters onto the 2D image coordinate i.e. (u, v) coordinate. This texture coordinate is used for shading and rendering of the 3D model during the motion animation.

This scheme of representing our 3D model using triangle mesh and texture were chosen since they are easy representations for image synthesis and are well supported in many existing graphics hardware.

There are many systems and methods that have been developed to construct the 3D human model. They usually construct the surface skin of the model, and their methods fall into two categories: (1) 3D scanner-based system, and (2) passive multi-camera system.

The scanner-based and active systems are available from many commercial companies such as Cyberware®, Hamamatsu and (TC)$^{2}$®. Typically, they use the triangulation principle, and with laser light or pattern projection method and a CCD camera. These systems usually capture the shape of the entire human body in about 15 to 20 seconds. The output of such a system is a set of highly dense point cloud representing the surface of the model. Before this model can be used, the point cloud needs to be furnished into a regular surface 3D triangular mesh by using 3D geometrical software with some manual cleaning up of possible spurious data points. Drawbacks of a full-body scanner-based system include its high pricing and the requirement for the subject to stay still and rigid for the whole duration of scanning, about 15 seconds for full body coverage, which is quite constrictive and not very practical in certain applications.

The passive multi-camera approach is much cheaper; also video cameras are more easily available and their set-ups are much simpler. Once the synchronized cameras are calibrated, we are able to reconstruct the 3D model of the subject when the features points and curves of the subject seen in the images are properly defined.

## A Proposed Method for 3D Human Reconstruction

In our project, we used the passive multi-camera approach since it is cheaper and easier to set up. Here, we will describe our general setup and approach for modeling the 3D human. Readers who are interested in our 3D human reconstruction method can refer to Gagalowicz & Mathieu (2006) or Quah (2005). Other works related to 3D human reconstruction can be found in Matsuyama (2004) or Starck & Hilton (2003).

The method that we proposed uses multi-cameras to build the external surface skin of the subject, and at the same time estimate its underlying skeleton. We make use of a 3D generic human model with skeleton (top of Figure 3) and adapt it to the real subject seen in the multiple images, thus yielding the customized 3D human model of the subject with its skeleton automatically positioned inside the external skin.

Initialization of the 3D human reconstruction begins by establishing the 3D anatomically related characteristic points on the surface of the generic model and their 2D correspondents in the respective images (bottom of Figure 3). This is achieved by using an interactive point-matching tool that we have developed. Next, our algorithm builds the 3D human model through the use of camera calibration and 3D ray triangulation. The feature

*Figure 3. (top) Generic surface model with skeleton, and (bottom) example of characteristic points selection on an image of a real golf player that is corresponded to points on surface of the generic model*

points reconstruction yields the global geometry of the subject (Figure 4). However, these are insufficient to give the local body geometry. The local body contours of the subject are refined through the matching of silhouette curves (Figure 5). The skeleton of the subject is estimated based on the information of changes from generic model to the customized model. Figure 6 shows the result of 3D human of our subject with its estimated skeleton.

## Human Tracking

We propose a model-based analysis-by-synthesis framework for human motion capturing (see block diagram on Figure 7). The concept of this work is built on the work by Gagalowicz & Gerard (2000) for tracking rigid objects. Prior to this framework, a textured 3D puppet model that closely resembles the subject was built (from the previous section)

*Figure 4. 3D reconstruction via feature points that yield the global geometry of the subject, but inadequate for the local body contours*

*Figure 5. Matching of silhouette curves for refinement via local contour*

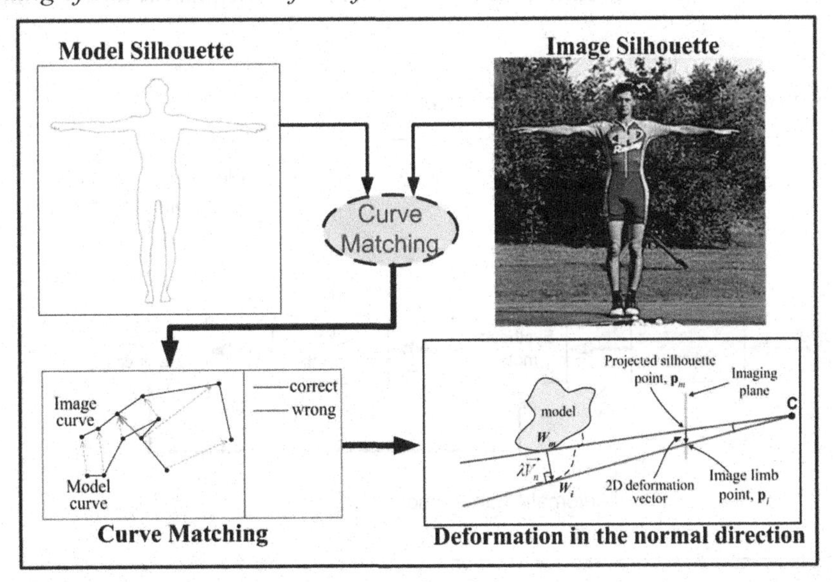

*Figure 6. Final result of a close-up of the surface skin with its skeleton for the subject superimposed onto the image*

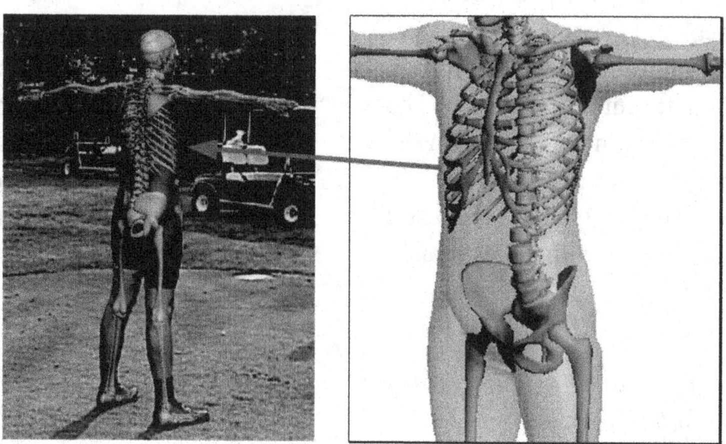

to facilitate the motion tracking. Our method does not use explicit or standard image background segmentation or *skeletonization*, which is very sensitive to noise. A more detailed explanation of our algorithm could be found in Quah (2007).

Our algorithm starts with automatic realization of the colour/texture onto the puppet model from its initial pre-positioned posture. The pre-positioning was done using a 3D modeling software just for the initializing to the first image. Next, our computation synthesizes the 3D puppet movements such that they minimize the differences between the synthesized movements and real athlete's motion in the subsequent images. This is achieved through the use of the simulated-annealing algorithm to search through the various probable degree-of-freedoms of the joint kinematics. The joint kinematics then drives the skin of the model puppet. The solution to the joint kinematics is taken when the skin puppet posture yields the smallest error when superimposed onto the video containing the subject. The colour texturing onto the puppet is analyzed and updated once every few frames so that the synthesis is not influenced by the changing articulated posture and illumination variations. In this project, we implemented the

*Figure 7. Model-based human motion tracking framework*

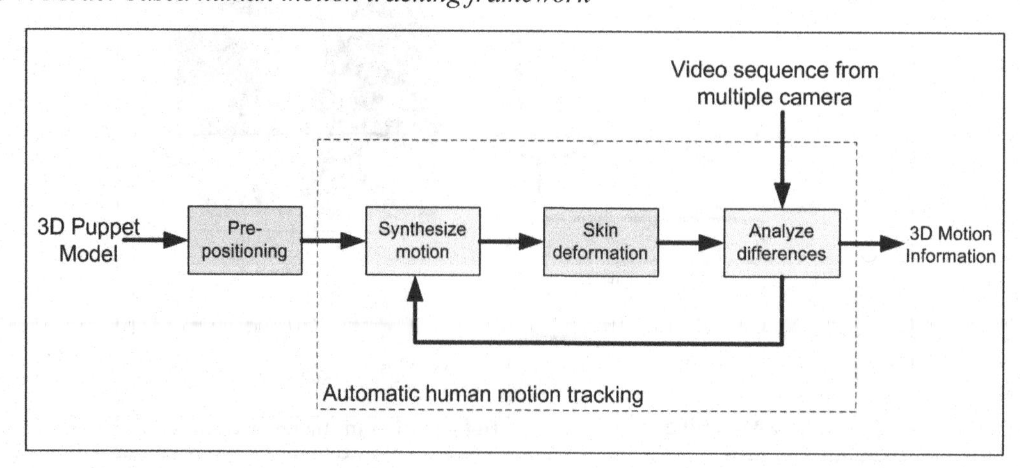

rigid skin deformation. The rendering for motion synthesis is the most computational intensive module and it is sped up by using the graphics processor unit (GPU).

### Results & Discussions

In this section, we present some of our typical tracking results. The meshes of the upper-arm and forearm are superimposed onto sequences of images to visualize the tracking results. The top images show the results for the camera view that we used for tracking. The bottom images show the results on another camera from the same videoing scene that we used for verification and visualization. From the results, we can see that our algorithm is able to track the human arms in environments with clutter and occlusion (Figure 8). The 3D meshes of the arms are superimposed onto the sequence of images to visualize the tracking quality. They are overlaid onto 2 different camera views to validate their 3D depth validities. The left images are from camera 1 and the images on the right are from camera 2.

## FUTURE WORKS

This section describes the future direction of our research team.

### Human Skin Deformation and Shading

In practice, the human undergoing motion will cause the skin to deform in non-rigid manners. The accuracy and reliability of human tracking can be improved if these non-rigid deformations can be

*Figure 8. Results of tracking in clutter and self-occlusion setting*

closely modelled. In computer graphics animation terms, the surface geometries of the animated characters could have the attributes of: twisting, bulging, bending, stretching and squashing. The scientific principles of these attributes could be used to derive and synthesize the proper surface geometries for use in motion tracking.

The shading for the synthesized image must also not be overlooked. Although very realistic images had been synthesized before (Boivin & Gagalowicz, 2001), to put it to use in practice together with proper skin deformation, different lighting condition and computationally efficient is still a big challenge.

### Automatic Posture Pre-Positioning

In this article, pre-positioning had been done interactively by using 3D graphical tools. To initiate the tracking automatically, the initial posture of the subject needs to be recognized by the machine vision system. The 3D geometries of the postured model and subject images have to be properly registered. We also have to take into account the skin deformation while manipulating the kinematics of 3D model at the stanza posture to the initialized posture. One suggestion to do this is through machine learning of example postures from multiple subjects, and then synthesizes them during an automatic pre-positioning operation.

### Full-Body and Multiple People Tracking

A natural extension to the work from this chapter is to perform full-body and eventually multiple people tracking in cluttered environments. For full-body tracking of a single subject, the complete 3D body model of the subject will be used, thus, more degrees-of-freedom and possibly more accurate skinning function have to be used. For extension to multiple people tracking, we have to consider occlusion of subjects and detecting collision and interactions between subjects. A higher level data topological relationship may also be needed if there are many subjects in the scene.

### Computational Efficiency

Computational speed is a factor of consideration if we are to add in more sophisticated operations involving the likes of skin deformation, automatic posture pre-position and multiple people tracking. Specific dedication is needed to make sure that the algorithms that manipulate the model geometries, data synthesis and matching take place efficiently so that the whole tracking process can yield the desired results in a realistic time-frame. Efficient data streaming is also required, since the large amount of data transfer within the module of the tracking system can be a bottleneck.

## IMPACT OF VISION-BASED MOTION CAPTURE AND PRACTICAL APPLICATION FOR THE FUTURE

This section describes the impact of a purely vision-based motion capture and its implication toward the domain of sports kinesiology. We also give a few practical applications that uses this system for the future.

The innovation of these technologies will potentially benefit the biomechanics community through the utilization of marker-less motion capture methods by minimizing skin-markers artifacts and equipment cost (as such configurations require only video-based cameras). Additionally such technologies may enable user-friendly setups accessible by the masses, particularly for the sports science, sports coaching, sports retailing, animation arts industry and main-stream education communities. With a wide spectrum of users, the push for the advancement of such technologies will be imminent.

There are many potential applications of Interactive Digital Media (IDM) that require capturing and analysis of human movements in the fields of sports coaching, sports science, education, medicine and animation domains. We have chosen two examples to illustrate the permeability of its application.

## Educationally Contextualized Interactive Digital Media

In typical main stream schools, the learning of academic subjects like physics, mathematics, biology, physical education and the coaching of sports skills are usually taught predominantly within each discipline's domain. There are common inter-related concepts like mechanics in physics, mensuration in mathematics, kinesiology in biology and simple technique analysis in physical education or sports coaching which can be learnt within a common educational framework. The relevant connectedness between these academic and sports coaching domains are usually not perceived by the learner. To develop the holistic thinking student, one of the strategies is to have the connectedness between these domains, which forms a significant part of the formal main stream school curriculum, located within the learner. The learning of interdisciplinary academic concepts incorporating experiential physical activities in realistic conditions provides a potential platform for establishing the connectedness. Traditional classroom or laboratory strategies have their limitations in that the concepts taught are usually not fully contextualized to real situations involving actual activities. Therefore, this may not facilitate deep learning. Interactive digital Media (IDM) technology offers a very attractive tool to address this problem by providing contextualized augmented feedback, infused with interdisciplinary information, in activities performed in real settings. Such situations and activities are best facilitated using imaging technologies especially in three dimensions (3D). Objective and augmented feedback through interactive 3D video-based approaches using Augmented Reality (AR) technologies can, thus, be used to enhance the teaching, learning and coaching processes. We will illustrate three pedagogical areas that our project may have potential contributions.

## Inter-Disciplinary & Holistic Learning

Using IDM in learning within a common theme may involve various inter-related academic disciplines. For example, in the example of a golf swing, the angles, speed, time concepts in mathematics can be explored together with physics principles like kinetic energy, potential energy and momentum under the domain of skills acquisition or physiology focusing on the musculoskeletal forces and moment arms. Also, these concepts can be taught experientially whilst the students actually perform the swing in actual realistic environment. Benefits of experiential and interdisciplinary` learning environment in education are well documented (Badenoch, 2005; Redman, 2006).

Additionally, as computing hardware—both wired and wireless—approaches ubiquity, new opportunities emerge to use technology to enrich individuals' experiences of objects and places. Because all areas of academic inquiry potentially benefit from background and context, AR has the possibility of enhancing education across the curriculum. By holistically exposing students to an experiential, explorative, and authentic model of learning early in their higher education experience, AR has the potential to help shift modes of learning from students' simply being recipients of content to their taking an active role in gathering and processing information, thereby creating effective knowledge.

## 3D Spatial Understanding

In order to solve three dimensional mathematical and especially geometrical problems, spatial abilities are an important prerequisite (Brown & Wheatley, 1989; Fennema & Sherman, 1977;

Guay & McDaniel, 1977). Many students have difficulties solving tasks that require spatial visualization skills and spatial thinking. They may use strategies such as learning construction steps by heart without fully understanding spatial problems and their solutions in 3D space. Spatial abilities can be improved by well-designed training (e.g. Souvignier, 2001) and a number of training studies have shown the usefulness of AR in training spatial ability (Durlach *et al.*, 2000; Larson & Van Rooyen, 1998).

## Future Research: Extension of Developed Technology into Related Applications

Exploiting the developed new IDM technologies that aim to be faster and more accurate, there is great potential to use them, in future research, to develop 'Immersive Virtual Environments' and 'Education Games' applications where the participants' movement can be detected more accurately and faster and hence, more realistically contributing to a more enhanced experience. In sports coaching, automatic augmented objective feedback may be provided to both coaches and athletes where immediate feedback on speed, angle and distance information amongst others can be presented contributing to an effective and engaging coaching session. This will significantly facilitate the application of sports biomechanics for sporting excellence in line with the ethos of the Committee on Sporting Singapore (CoSS) (2000) encompassing various activities like SEA games, Asian games, Olympics and normal physical activates.

Also, there is potential for future research in the area of 'Digital Literacies and Modalities of Representation' where this project will lend itself to be a potential case study for understanding the issues related to pedagogy, learning, design and persuasion using multi-modal forms of electronic representations and media (e.g. interactive art, digital textbooks) in IDM environments. Also, in the 'Research on Impact of IDM use among Students' domain, this project will provide a framework for understanding, in future studies, the impact of widespread use of IDM on students' social, emotional and cultural development in a longitudinal context. The research activities and products (including achieved aims) in this project, potentially, provide the catalyst for more research using IDM technologies within learning environments through journal publications, conference/ poster presentations and various professional sharing sessions. These will potentially generate more questions requiring further research.

## Sports Analysis

In sports research, sports biomechanics typically provide the scientific basis for movement intervention by means of analyses and review of movement performance data. The use of such technology with a suitable user interface will empower the coach to gain insights into existing performance levels more readily and conveniently. This in turn can enable coaches to be better able to identify key movement attributes that are impeding performance. Feedback to learners can be near real time and contextualized so that new learning interventions introduced can be assessed for its efficacy and appropriateness in terms of the kinematics of the movement or that of the performance outcomes. Certainly, such a platform to augment feedback can provide a further impetus to motivate learners to improve themselves and to optimize their training time. Current approaches in sports coaching that involve the use of video technology range from qualitative analysis of sports techniques to more complicated and time consuming controlled environments for empirical movement analysis. Both methods are limited in their application either through a lack of sophistication or detail in the case of qualitative analyses or cost and availability in the latter environment. The advent of a marker-less system would help overcome this limitation and promote new pedagogical approaches to coaching sports.

## CONCLUSION

AR technologies are technological tools that must be developed and contextualized before their benefits can be fully manifested. Currently, the Republic Polytechnic (Singapore) and Nanyang Technological University (Singapore) are examples of two institutions that have been awarded government research grants for research and development in this area. In this chapter, we reviewed the existing systems for human motion capture, which have many limitations. Looking ahead to the future and challenges, we present a model-based approach for marker-less motion capture that will widely extend its scope of applications in many domains of practices. From the results, we demonstrate its feasibilities and potentials to operate in challenging scenarios. We also illustrated that our system is particularly important in the sports and educational domain, with examples of ongoing development through the use of video and interactive digital media.

In closing, we are only limited by our imagination while we attempt to advance technology. Technology will not replace us but we may be replaced by those who can exploit technologies in the right context. We will continually advance knowledge; the more we acquire, the more is our desire; the more we learn, the more we are capable of learning.

## ACKNOWLEDGMENT

This project was funded by the National Research Foundation of Singapore grant NRF2007IDM-IDM003-052. We would like to thank the INRIA golf stream project team for the filming of golfers. We would also like to thank Mr. Ta Huynh Duy Nguyen for assisting in the image acquisition. Also we would like to thank Saharuddin Omar and Ian Harris Sujae for the optical mocap setups.

## REFERENCES

Aggarwal, J. K., & Cai, Q. (1994). Human Motion Analysis: A Review. *Proceedings of the IEEE Workshop on Motion of Non-Rigid and Articulated Objects,* (pp. 2-14), Austin, Texas: IEEE.

Badenoch, D. (2005). *Interdisciplinary Curriculum Design and Teaching for Integrative Learning and Development of Graduate Qualities: A case study of reconceptualising Physical Education: University of South Australia.* Paper presented to the Division of Education, Arts & Social Sciences' Teaching & Learning Colloquium, University of South Australia.

Boivin, S., & Gagalowicz, A. (2001). Image-based Rendering of Diffuse, Specular and Glossy Surface from a Single Image. *SIGGRAPH,* (pp. 107-116).

Bottino, A., & Laurentini, A. (2001). A Silhouette Based Technique for the Reconstruction of Human Movement. *Computer Vision and Image Understanding, 83,* 75-95.

Bouguet, J-Y. (nd). *Camera calibration toolkit for Matlab and in OpenCV.* Retrieved May 12, 2008, from http://www.vision.caltech.edu/bouguetj/calib_doc/

Bregler, C., Malik, J., & Pullen, K. (2004). Twist Based Acquisition and Tracking of Animal and Human kinematics. *International Journal of Computer Vision, 56,* 179-194.

Brown, D.L., & Wheatley, G.H. (1989).Relationship between spatial Knowledge. In C. Maher, G. Goldin, & R. Davis (Eds.), *Proceedings of the 11th Annual Meeting, North America Chapter of the International Group for the Psychology of Mathematic Education* (pp. 143-148). New Brunswick, NJ: International Group for the Psychology of Mathematics Education.

Carranza, J., Theobalt, C., Magnor, M., & Seidel, H-P. (2003). Free-Viewpoint Video of Human Actors. *SIGGRAPH*, (pp. 569-577).

Cheung, G., Baker, S., & Kanade, T. (2003a). Shape-from-silhouette for articulated objects and its use for human body kinematics estimation and motion capture. *Proceedings of the IEEE Conference on Computer Vision and Pattern Recognition*, (pp. 77-84), Madison, Wisconsin: IEEE.

Cheung, G., Baker, S., & Kanade, T. (2003b). Visual Hull Alignment and Refinement Across Time: A 3D Reconstruction Algorithm Combining Silhouette with Stereo. *Proceedings of the IEEE Conference on Computer Vision and Pattern Recognition*, (pp. 375-382), Madison, Wisconsin: IEEE.

Committee on Sporting Singapore (CoSS) (2000). Retrieved May 12, 2008, from http://app.mcys.gov.sg/web/sprt_towards_committeesporting.asp

Corazza, S., Mundermann, L., & Andriacchi, T. (2006). The Evolution of Methods for the Capture of Human Movement Leading to Markerless Motion Capture for Biomechanical Applications. *Journal of Neuroengineering and Rehabilitation*, *3*, 6.

Cyberware (n.d.). Retrieved May 12, 2008, from http://www.cyberware.com

Dementhon, D.F., & Davis, L. (1995). Model-based Object Pose in 25 Lines of Code. *Intl. Journal of Computer Vision*, *15*, 123-141.

Dorai, C., & Venkatesh, S. (2002). Media Computing: Computational Media Aesthetics (The International Series in Video Computing). *Springer*, Edition 1.

Durlach, N., Allen, G., Darken, R., Garnett, R.L., Loomis, J., Templeman, J., & von Wiegand, T. E. (2000). Virtual environments and the enhancement of spatial behavior: Towards a comprehensive research agenda. *PRESENCE-Teleoperators and Virtual Environments*, *9*, 593-615.

Fennema, E., & Sherman, J. A. (1977). Sex-related differences in mathematics achievement, spatial visualization, and affective factors. *American Educational Research Journal*, *23*(1), 51-71.

Gagalowicz, A. (1990). Collaboration between Computer Graphics and Computer Vision. *Proceedings of the International Conference on Computer Vision*, (pp. 733-737), Osaka, Japan: IEEE.

Gagalowicz, A., & Gerard, P. (2000). Three Dimensional Object Tracking using Analysis/Synthesis Techniques. In A. Leonardis, F. Solina, R. Bajcsy, & F. Solina (Eds.), *Confluence of Computer Vision and Computer Graphics* (pp. 307-330), Dordrecht, NL: Kluwer.

Gagalowicz, A., & Mathieu, M. (2006). Modeling 3D Humans from Uncalibrated Wide Baseline Views. Paper presented at *SAE Digital Human Modeling Conference*, Lyon, France.

Gavrila, D.M. (1999). The Visual Analysis of Human Movement: A Survey. *Computer Vision and Image Understanding*, *73*, 82-98.

Gerard, P., & Gagalowicz, A. (2003). Human Body Tracking using a 3D Generic Model Applied to Golf Swing Analysis. Paper presented at *MIRAGE 2003Conference*, Rocquanecort, France.

Goh, W.B., & Chan, K.Y. (2007). The Multiresolution Gradient Vector Field Skeleton. *Pattern Recognition*, *40*, 1255-1269.

Goncalves, L., Di Bernardom, E., Ursella, E., & Perona, P. (1995). Monocular Tracking of the Human Arm in 3D. *Proceedings of the International Conference on Computer Vision*, (pp. 764-770), Cambridge, Massachusetts: IEEE.

Grest, D., Woetzel, J., & Koch, R. (2005). Nonlinear Body Pose Estimation from Depth Images. In W. Kropatsch (Ed.), *Proceedings of the DAGM - German Association for Pattern*

*Recognition Conference*, (pp. 285-292). Vienna, Austria: DAGM.

Guay, R.B., & McDaniel, E. (1977). The relationship between mathematics achievement and spatial abilities among elementary school children. *Journal for Research in Mathematics Education*, *8*(2), 211-215.

Hamamatsu (n.d.). Retrieved May 12, 2008, from http://usa.hamamatsu.com

Kakadiaris, I.A., & Metaxas, D. (1998). 3D Human Body Acquisition from Multiple Views. *International Journal of Computer Vision, 30*, 191-218.

Kehl, R., & van Gool, L. (2006). Markerless Tracking of Complex Human Motions from Multiple Views. *Computer Vision and Image Understanding, 104*, 190-209.

Koh, M., & Anwari, K. (2004). Integrating video and computer technology in teaching – an example in gymnastics initial PE teacher-training in Singapore. *British Journal of Teaching Physical Education*, 35(3), 43 -46.

Lafortune, M.A., Cavanagh, P.R., Sommer, H.J., & Kalenak, A. (1992). Three-Dimensional Kinematics of the Human Knee during Walking. *Journal of Biomechanics, 25*, 347-357.

Larson, P., & Van Rooyen, A. (1998). The Virtual Reality Mental Rotation Spatial Skills Project. *CyberPsychology and Behavior, 1*, 113-120.

Lerasle, F., Rives, G., & Dhome, M. (1999). Tracking of Human Limbs by Multicular Vision. *Computer Vision and Image Understanding, 75*, 229-246.

Lim, W.Y., & Koh, M. (2006). Effectiveness of Learning Technologies in the Teaching and Learning of Gymnastics. *Pacific Asian Education, 18*(2), 69–77.

Ma, C. M., & Sonka, M. (1996). A Fully Parallel 3D Thinning Algorithm and Its Applications. *Computer Vision and Image Understanding*, 64, 420-433.

Matsuyama, T., Wu, X., Takai, T., & Nobuhara, S. (2004). Real-time 3D Shape Reconstruction, Dynamic 3D Mesh Deformation, and High Fidelity Visualization for 3D Video. *Computer Vision and Image Understanding, 96*, 393-434.

McKenna, S. J., Jabri, S., Duric, Z., Rosenfeld, A., & Wechsler, H. (2000). Tracking Groups of People. *Computer Vision and Image Understanding, 80*, 42-56.

Mikic, I., Trivedi, M., Hunter, E., & Cosman, P. (2003). Human Body Model Acquisition and Tracking using Voxel Data. *International Journal of Computer Vision, 53*, 199-223.

Moeslund, T.B., & Granum, E. (2001). A Survey of Computer Vision-based Human Motion Capture. *Computer Vision and Image Understanding, 81*, 231-268.

Moselund, T.B., Hilton, A., & Kruger, V. (2006). A Survey of Advances in Vision-based Human Motion Capture and Analysis. *Computer Vision and Image Understanding, 104*, 90-126.

Motion Analysis (nd). Retrieved May 12, 2008, from http://www.motionanalysis.com

Nobuhara, S., & Matsuyama, T. (2006). Deformable Mesh Model for Complex Multi-Object 3D Motion Estimation from Multi-Viewpoint Video. In M. Pollefeys (Ed.), *Proceedings of the 3rd Intl. Symposium on 3D Processing, Visualization and Transmission (3DPVT 2006)*, (pp. 264-271). Chapel Hill: University of North Carolina.

Peak Performance. (nd). Retrieved May 12, 2008, from http://www.peakperform.com

Planker, R., & Fua, P. (2001). Tracking and Modeling People in Video Sequences. *Computer Vision and Image Understanding, 81*, 285-302.

Quah, C. K., Gagalowicz, A., Roussel, R., & Seah, H S. (2005) 3D modeling of humans with skeletons from uncalibrated wide baseline views. In A.Gagalowicz & W. Philips (Eds.), *Proceedings of the International Conference on Computer Analysis of Images and Patterns,* (pp. 379-389), Versailles, France: Springer.

Quah, C.K., Gagalowicz, A., & Seah, H.S. Markerless 3D Video Motion Capture in Cluttered Environments. In F.K. Fuss (Ed.), *Proceedings of the Asia-Pacific Congress on Sports Technology,* (pp. 121-126). Singapore: Nanyang Technological University.

Qualysis. (nd). Retrieved May 12, 2008, from http://www.qualysis.com

Redman, R.W. (2006). The challenge of interdisciplinary teams. *Research and Theory for Nursing Practice, 20*(2), 105-107.

Rosales, R., & Sclaroff, S. (2006). Combining Generative and Discriminative Models in a Framework for Articulated Pose Estimation. *International Journal of Computer Vision, 63*, 251-276.

Rosenhahn, B., He, L., & Klette, R. (2005). Automatic Human Model Generation. In A.Gagalowicz & W. Philips (Eds.), *Proceedings of the International Conference on Computer Analysis of Images and Patterns,* (pp. 41-48), Versailles, France: Springer.

Souvignier, E. (2001). Training räumlicher Fähigkeiten. [Training spatial abilities.]. In K.J. Klauer (Ed.), *Handbuch Kognitives Training* (pp. 293-319). Göttingen: Hogrefe.

Starck, J., & Hilton, A. (2003). Model-based Multiple View Reconstruction of People. *Proceedings of the International Conference on Computer Vision,* (pp. 915-922), Nice, France: IEEE.

$(TC)^2$. (nd). Retrieved May 12, 2008, from http://www.tc2.com

Terzopolus, D. (1999). Visual Modeling for Computer Animation: Graphics with a Vision. *Computer Graphics, 33*, 42-45.

Theobalt, C., Aguiar, E., Magnor, M., Theisel, H., & Seidel H-P. (2004). Marker-free Kinematic Skeleton Esimation from Sequence of Voume Data. In R. Lad & G. Baciu (Eds.), *Proceedings of the ACM Virtual Reality Software and Technology Conference,* (pp. 57-64), Hong Kong: ACM.

Vicon. (nd). Retrieved May 12, 2008, from http://www.vicon.com

Wingbermuhle, J., Liedtke, C-E., & Solodenko, J. (2001). Automated Acquistion of Lifelike 3D Human Models from Multiple Posture Data. In W. Skarbek (Ed.), *Proceedings of the International Conference on Computer Analysis of Images and Patterns,* (pp. 400-409), Warsaw, Poland: Springer.

Zhang, Z. (2000). A Flexible New Technique for Camera Calibration. *IEEE Trans. Pattern Analysis and Machine Intelligence, 22*, 1330-1334.

# Chapter VI
# Technology to Monitor and Enhance the Performance of a Tennis Player

**Amin Ahmadi**
*Griffith University, Australia & Queensland Academy of Sport (Centre of Excellence), Australia*

**David D. Rowlands**
*Griffith University, Australia*

**Daniel A. James**
*Griffith University, Australia & Queensland Academy of Sport (Centre of Excellence), Australia*

## ABSTRACT

*Tennis is a popular game played and viewed by millions of people around the world. There is a large impetus for players to improve their game and technology is becoming an important tool in doing this. This chapter discusses the current technology used in tennis and also discusses the biomechanics of the various strokes, so that the application of the technology can be better understood. Since the serve is a crucial part of a player's game, this chapter focuses on the serve, but still discusses the other tennis strokes. The chapter is divided into 2 parts: the biomechanics of the strokes and the technology used to monitor tennis. The technology section details some of the major tools to monitor and analyze the tennis swing, including high speed digital cameras, marker-based optical systems, and inertial sensors. Examples are provided of how these technologies can be applied. Finally, a small discussion is presented, which gives an idea of future directions in tennis monitoring.*

## INTRODUCTION

Tennis is ranked as one of the most popular sports in the world. According to their website, the International Tennis Federation has 144 member nations, from Algeria to Zambia. Tennis is played at all levels by a diverse cross-section of society. It is played at both a social and professional (elite) level and encompasses tournaments based upon ability, age, gender, and disability. Tennis is a fast, competitive game which requires a great deal of skill from the athlete, whether it is in serving or in the ground strokes. All aspects of the game require power and finesse. Since the game relies heavily on the skill of the athlete to defeat the opponent, then all players are looking to improve their skill level.

At the professional level, the major tournaments are televised and viewed around the world by millions of people. This means that the game of tennis has become big business, with prize money and advertising revenue worth millions of dollars every year. The Grand Slam tournaments are considered to be one of the most highly prestigious tournaments in the world and include the Australian Open, the French Open, Wimbledon, and the US Open. Usually these consist of 32 seeded players and others ranked in the top 100 qualifiers in the world. The winners of the grand slam tournaments can collect large prize funds and a significant number of ranking points. For instance, in 2008, both men's and women's singles champions received 750,000 pounds as part of a total prize fund of 11,812,000 pounds. Any improvement in the skill level of the professional athlete means that the athlete can improve in the world ranking and greatly improve their income, as well as their personal and national prestige.

In order for the player to improve their skill level, they need to monitor and analyze all aspects of their game. Traditionally, they had to perform some self assessment, or get instructions from a coach to indicate the areas of possible improvement. Now, technology offers reliable tools for quantitative assessment. Technology has made this more possible, since all aspects of the game can now be analyzed. From the media point of view, technology would enable online real-time feedback, which can be broadcasted all over the world during the televised tennis competitions. Hawk-eye technology is an example of one of those technologies that media have used to capture some statistics, such as speed of the ball at any point of a rally; service comparisons (i.e. direction and depth of serves, placement of 1st and 2nd serves etc.); bounce points of the ball and so on.

A major challenge faced by coaches and sports scientists is to be able to accurately measure what athletes are doing in the training and competition environments. Such information is important, because it has potential to provide insight into physical activity levels associated with performance, as well as the techniques involved in the activity. Video cameras and digital optical marker-based systems are currently used to perform movement analysis. However, they are primarily laboratory-based, expensive, not readily accessible, and often require considerable processing time, due to the large amounts of data that they generate. Such tools are often impractical for athletes during training and competition. For example, the cameras require lighting conditions and field of view, which are more readily available in a laboratory. The development of methods that provide meaningful information about athletic performance, without laboratory assessment and long processing delays, has the potential to significantly assist in improving skills. In this chapter, we introduce the use of laboratory, video based and remote sensing technology to evaluate swing performance.

Due to the popularity and importance of tennis, there is an increasing need to monitor a tennis player on the field. This chapter examines tennis action and the use of the technology. It first presents an introduction to tennis and explains the actions and importance of the various strokes in tennis. The biomechanics of the strokes is then discussed. After an understanding of the strokes is gained, the technology to monitor the strokes is introduced. Due to the importance of the serve, this chapter focuses mainly on the serve action, but does discuss the other actions as well.

## BACKGROUND

Tennis is a game that is naturally composed of two parts, consisting of the service in which the player starts by putting the ball into play, and the ground strokes in which the player returns the ball. Therefore, improving a tennis game involves enhancing the service, ground strokes, as well as the correct court and body positioning for each stroke. Ground strokes can be divided into three main strokes: forehand, backhand and volley.

Forehand is usually known as the most natural stroke and is the easiest one to master. The forehand involves hitting the ball with the palm facing in the direction of the stroke. Key principles in generating a powerful and accurate forehand are the grip on the racquet, the footwork to get into position, the position on the court, the stroke where the ball is hit, and finally the follow through. This type of stroke is heavily affected by the style of the grip used by a player, which can have an influence on the energy coordination transfer to hit the ball. Foot positioning, foot movement, and timing can also contribute differently to the energy transfer of the stroke. However, the speed and angle of the racquet at impact, in addition to the path of the racquet prior to impact, play a significant role to determine the speed and spin of the ball.

Backhand is another main ground stroke and can either be a one-handed or two-handed stroke. The backhand involves hitting the ball with the back of the hand facing outwards and the swing brought across the body. The backhand is generally considered to be more difficult to master and lacks the consistency and power of the forehand. Many beginners and amateur players have trouble hitting the ball using the backhand due to lack of strength. There are even a number of advanced players who have a significantly better forehand than backhand. Both one-handed and two-handed backhands follow the same key principles needed to produce a proper forehand stroke, which are grip, footwork, position, and follow through.

The volley can be considered as an offensive stroke in tennis. The volley involves striking the ball without allowing it to bounce. Usually the player uses the volley to finish the point by approaching the net. This reduces the reaction time for the opponent and gives a wide range of angles for the player to land the ball in the opponent's court. Also, it can help the player to reduce the risk of unwanted bounces on improper courts. Volley techniques are more focused on accuracy rather than speed, and are therefore based more on the range of motion and optimal projection principles of tennis biomechanics. Similar to the forehand and the backhand, the key principles of the forehand and backhand volley are exactly the same as those mentioned for the aforementioned two ground strokes.

According to Bahamonde (2000), the serve is the most important and critical stroke in tennis. The serve is used to enter the ball into play and is the principal offensive weapon. It is known that fast serves can dominant the game at the elite levels (Girard, Micallef & Millet, 2005; Knudson, 2006). Therefore, for a player to be successful in a tennis match, the player must fully master the serve. The serve is the only stroke that a player has full control over. However, it is also the most difficult stroke amongst all the other strokes to master, because it involves many complicated motions, which need to happen at exactly the right time (Bahamonde, 2000). Many high level tennis players use a fast, powerful serve as their first serve, which imparts to the ball maximum velocity with minimum top or side spin. The higher the velocity of the serve the more probable it is to win the point (Chow et al., 2003). However, there is a greater risk for error as well. If the first serve fails, then most high level players will use the slice serve as their second service, which imparts lower velocity to the ball but gives it more spin.

It is apparent that a good understanding of the biomechanics of the player's game is essential in improving the technical aspects of a player's game. In contrast, a lack of knowledge of the coordination of the complex motions involved in the game can result in a poor performance.

## SPORTS BIOMECHANICS

Any meaningful discussion of tennis techniques requires knowledge of tennis biomechanics. Biomechanics is the field of study that focuses on the motion and the causes of motion in the human body. More precisely, it is a mechanical study of the human body based upon the fundamental disciplines found in physics and science. The goal of the biomechanical study is to improve the human's motion. Hence different groups of scientists, such as anatomists, biomedical engineers, physical therapists and coaches, are interested and involved with biomechanics principals. Sports biomechanics focuses on how the human creates a wide range of motions during a sporting activity. The biomechanics of the tennis game has been studied in the literature. In this section we will introduce the biomechanical principles in tennis and then focus on the biomechanics of the first and the second serve.

It is not the authors' intention to give a detailed explanation of the biomechanical principles of tennis. The aim of this section is to allow the reader to understand a little of the body motions of interest so that the applications of the technology can be better understood and be more relevant to the case study presented later.

## BIOMECHANICS PRINCIPLES IN TENNIS

Using the principles of biomechanics makes it possible to understand the movement of a body during any of the strokes used in tennis. In fact, it is vital to know what body motions were employed and how they were created in order to analyze and modify the motion to improve the performance of the player.

In 2003, Knudson showed that there are about nine principles of biomechanics required to describe human body movements. He also stated that only six of them are relevant to the movement of the body and the ball trajectory during the tennis game. Although each of these principles can be investigated in more detail for each stroke, we will just briefly introduce them so that the rest of this chapter becomes more understandable.

According to Knudson (2003), the six principles of biomechanics in tennis are:

1. **Force and time:** This principle states that modifying the motion of any segment of the body is the function of force and time. In other words, a greater motion is the result of greater force over a shorter period of time, or a smaller force over a larger period of time. In tennis, most of the movements are based on a greater force over a shorter time.
2. **Coordination and transfer of energy:** This principle explains the source and the origin of energy for body and segment motions in tennis. The source of the energy can come from the outside world or can be between the segments of the body, which are linked together. This principle explains that a great force can be generated in the tennis racquet, due to the fact that energy can be transferred from the ground up through the lower trunk, hip, upper trunk and arm. This means that the force at one segment can be transferred to another segment, since the segments are linked in the body.
3. **Balance and inertia:** Balance is the ability to control the motion of the body during any activity. Balance is becoming more important in sports that require both high accuracy and fast movements, such as tennis. The balance and inertia principle states that balance is a compromise between the motion and stability during the tennis game.
4. **Range of motion:** Range of motion is the principle which can be generally observed by coaches. This is an overall distance either linear or angular that a body or body segment must travel to create the motion. This principle also states that the less body segments involved in creating a movement, the more accurate the movement will be. On the other

hand, the more body segments that are used can lead to a higher speed movement at the end of the linked segment system. There is no specific optimal range of motion for the strokes of the tennis player.

5. **Optimal projection:** This principle states that for each stroke there is a window for optimal ball flight in tennis. Considering environmental effects such as air resistance, gravity and size of the tennis court, among others, this principle makes it possible to estimate the optimal projection for the various tennis strokes.
6. **Spin:** In tennis, spin is used to control the flight path as well the bounce of the ball on the court. This is called the spin principle. Topspin strokes can reduce the risk of hitting the ball too far and increase the margin height from the net. Drop shots are good examples of backspin to modify the ball bounce.

## Biomechanics of the Tennis Serve

The serve has been identified as a major contributor to game performance (Bahamonde, 2000). Previously we have introduced the biomechanics principles of general tennis, but will now concentrate on the tennis serve. The serve is a highly constrained motion from rest that limits external interaction and therefore is well suited to investigation. We are interested in the biomechanical principles in tennis serves to be able to understand the segment motions during the service.

Kinematics refers to the study of the geometry of the motion. It describes the motion in terms of time, displacement, velocity and acceleration. If a motion is occurring in a straight line, then it is called linear kinematic and if the motion is occurring about a fixed point, then it is called angular kinematic. As we will see later, studying the tennis serve includes examining both the linear and angular kinematics. It is important to remember that kinematics is only concerned with the mathematical analysis of the motion and does not investigate the forces required to generate that motion.

It has been shown that the head speed of the racquet contributes to the velocity of the serve and therefore it is important to examine the motions that contribute to the racquet head velocity. The racquet head velocity is the endpoint result of the kinematic chain. The aim is to determine the linear racquet head velocity as a result of the kinematic chain.

A simplified kinematic model used to describe the racquet head velocity of the tennis racquet during the serve can be broken down into the following three basic motions (Gordon & Dapena, 2006):

- Motion of the lower trunk relative to the ground
- Motion of the upper trunk relative to the lower trunk
- Motion of the arm segment (shoulder, elbow and wrist) relative to the upper trunk

Upper trunk, lower trunk and arm segments are shown in Figure 1.

*Figure 1. Lower trunk, upper trunk, upper arm, mid hip of a subject, as well as the racquet face center are shown*

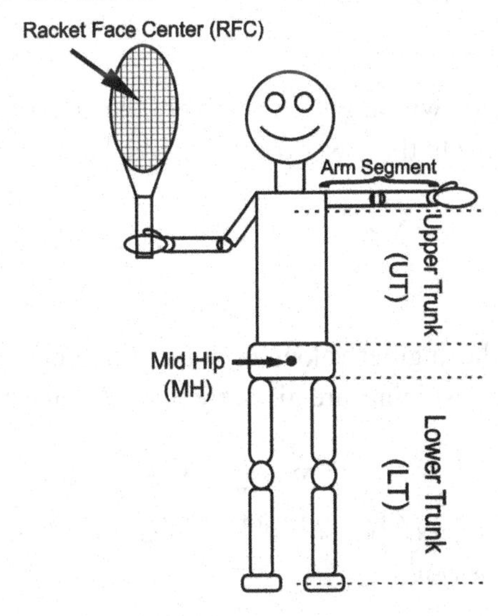

Arm segment contains the shoulder, the elbow and the wrist components. These motions each contribute a velocity component to the racquet head linear velocity. The racquet face center linear velocity ($V_{RFC}$) can be defined as the sum of six factors (Gordon & Dapena, 2006).

$$V_{RFC} = \mathbf{V}_{MH} + \mathbf{V}\,\boldsymbol{\omega}_{UT/LT} + \mathbf{V}_{SH/UT} + \mathbf{V}\,\boldsymbol{\omega}_{SH} + \mathbf{V}\,\boldsymbol{\omega}_{EL} + \mathbf{V}\,\boldsymbol{\omega}_{WR} \tag{1}$$

Where

$\mathbf{V}_{MH}$ is the linear velocity of the mid-hip (MH)
$\mathbf{V}\,\boldsymbol{\omega}_{UT/LT}$ is the twist angular velocity of the upper trunk (UT) relative to the lower trunk (LT)
$\mathbf{V}_{SH/UT}$ is the linear velocity of the shoulder joint (SH) relative to the upper trunk (UT)
$\boldsymbol{\omega}_{SH}$, $\boldsymbol{\omega}_{EL}$ and $\boldsymbol{\omega}_{WR}$ are the angular velocities at the shoulder (SH), elbow (EL) and wrist (WR) joints respectively.

The angular velocity of the lower trunk relative to the ground is given by

$$V\,\boldsymbol{\omega}_{LT} = V\,\boldsymbol{\omega}_{LT\text{-}X} + V\,\boldsymbol{\omega}_{LT\text{-}Y} + V\,\boldsymbol{\omega}_{LT\text{-}Z} = (\boldsymbol{\omega}_{LT\text{-}X} \times \mathbf{r}_{RFC/MH}) + (\boldsymbol{\omega}_{LT\text{-}Y} \times \mathbf{r}_{RFC/MH}) + (\boldsymbol{\omega}_{LT\text{-}Z} \times \mathbf{r}_{RFC/MH}) \tag{2}$$

The twist angular velocity of the upper trunk relative to the lower trunk is given by

$$\mathbf{V}\,\boldsymbol{\omega}_{UT/LT} = \mathbf{V}\,\boldsymbol{\omega}_{UT/LT\text{-}Z} = (\mathbf{V}\,\boldsymbol{\omega}_{UT/LTz} \times \mathbf{r}_{RFC/MH}) \tag{3}$$

The angular velocities at the shoulder, elbow and wrist joints are given by the following

$$V\,\omega_{SH} = V\,\omega_{SH\text{-}X} + V\,\omega_{SH\text{-}Y} + V\,\omega_{SH\text{-}Z} = (\boldsymbol{\omega}_{SH\text{-}X} \times \mathbf{r}_{RFC/SH}) + (\boldsymbol{\omega}_{SH\text{-}Y} \times \mathbf{r}_{RFC/SH}) + (\boldsymbol{\omega}_{SH\text{-}Z} \times \mathbf{r}_{RFC/SH}) \tag{4}$$

$$\mathbf{V}\,\boldsymbol{\omega}_{EL} = \mathbf{V}\,\boldsymbol{\omega}_{EL\text{-}X} + \mathbf{V}\,\boldsymbol{\omega}_{EL\text{-}Y} + \mathbf{V}\,\boldsymbol{\omega}_{EL\text{-}Z} = (\boldsymbol{\omega}_{EL\text{-}X} \times \mathbf{r}_{RFC/EL}) + (\boldsymbol{\omega}_{EL\text{-}Y} \times \mathbf{r}_{RFC/EL}) + (\boldsymbol{\omega}_{EL\text{-}Z} \times \mathbf{r}_{RFC/EL}) \tag{5}$$

$$V\,\omega_{WR} = V\,\omega_{WR\text{-}X} + V\,\omega_{WR\text{-}Y} + V\,\omega_{WR\text{-}Z} = (\boldsymbol{\omega}_{WR\text{-}X} \times \mathbf{r}_{RFC/WR}) + (\boldsymbol{\omega}_{WR\text{-}Y} \times \mathbf{r}_{RFC/WR}) + (\boldsymbol{\omega}_{WR\text{-}Z} \times \mathbf{r}_{RFC/WR}) \tag{6}$$

Where

$\mathbf{r}_{RFC/MH}$, $\mathbf{r}_{RFC/SH}$, $\mathbf{r}_{RFC/EL}$ and $\mathbf{r}_{RFC/WR}$ are vectors pointing from the mid-hip, shoulder, elbow and wrist, to the racquet face center respectively. In this model, the motion of the hitting upperarm relative to the ground was separated into motion of the lower trunk relative to the ground, motion of the upper trunk relative to the lower trunk, and motion of the upper arm relative to the upper trunk.

Some studies have been conducted (Bahamonde, 1994; Elliott, Marsh & Blanksby, 1986; Elliott, Marshall & Noffal, 1995; Elliott & Wood, 1983; Legnani & Marshall, 1993; Payne, 1978; Smith, 1979; Sprigings, Marshall, Elliott & Jennings, 1994; Van Gheluwe, De Ruysscher & Craenhals, 1987; Van Gheluwe & Hebbelinck, 1985), which indicate that the pattern of the overarm swing motion during the serve is very similar to other sports, such as badminton or baseball. However, the motion is constrained based on the rules of tennis. It is possible to create different kind of serves by the proper adjustment of the arm and the racquet motion during the impact of the ball. In later sections, we will discuss the coordination and energy transfer as methods to maximize the racquet speed during the serve stroke.

## Biomechanics of the First Serve

According to Knudson (2006), the first service is the serve intended to produce the greatest ball velocity with the least ball spin. Hence it is also called the power serve. The aim of this serve is to maximize the tennis racquet at impact time. Note that there is always some topspin and side spin components in the serve, due to the upward and sideward motions of the body during the serve. The first service is normally a flat powerful stroke where the ball is hit very close to the top of the net. However, the margin for error is very small, so as a result, the fault rate of the first serve is higher than that of the second serve. On the other hand, there is a chance to win the point by performing an ace on the first serve. It is worth remembering that the service is the only stroke in which the player has full control of the ball and the game (Bahamonde, 2000).

The key concept in the tennis serve is to increase the racquet head speed through correct coordination and the correct transfer of energy. A high-speed, accurate serve requires a sequential coordination pattern involving many body segments. This kinematic chain starts from the ground, to the legs, to the trunk and the higher trunk, and finally to the racquet arm (Elliott, 2006). In fact, the energy transfers from the segments with larger muscles, such as the leg, to the segments with smaller muscles, such as the arm (Knudson, 2006). The approximate sequential order of main contributors to racquet speed, between the maximum knee flexion and final impact with the ball, are listed (Gordon & Dapena, 2006):

1. Shoulder external rotation
2. Wrist extension (bend up)
3. Twist rotation of the lower trunk
4. Twist rotation of the upper trunk relative to the lower trunk
5. Shoulder abduction
6. Elbow extension
7. Forearm rotation (pronation)
8. A second twist rotation of the upper trunk relative to the lower trunk
9. Wrist flexion (bend down)

It is now important to determine which factors contribute the most to increasing the racquet head speed during the serve. According to some studies, both the elbow extension and wrist flexion contributions are large (Gordon & Dapena, 2006). However, other studies indicate that the upper arm internal rotation and wrist flexion are important, and have 54.2% and 30.6% contribution respectively to the linear racquet head velocity (Marshall & Elliott, 2000). It is very important to be aware of the fact that these approximate figures are related only at impact time, and do not indicate the importance of trunk rotation and leg drive in the early stages (Elliott, 2006). The forearm pronation has also been shown to make a negative contribution on the serve (Elliott, 2006). Other segments can also contribute in positioning the body. There are also some other sequencing of the hips, trunk, and upper extremity joint actions in the serve. Regarding the classical sequential coordination, the correct sequence consists of the pelvic rotation followed by the trunk rotation after the weight shift. This is called the "differential rotation". However, not all the tennis players employ differential rotation. Also, since there are approximately seven degrees of freedom in the serve stroke, it is quite complex to identify the sequence for every serve (Knudson, 2006). Therefore, players with different strengths and flexibilities in the trunk or shoulder might have slightly different sequences.

In the serve, like most of the other strokes in tennis, leg drive against the ground produces the force to create a powerful stroke (Elliott, 2006). There are two main footwork techniques that tennis players employ to serve a ball. They are known as foot-up (pinpoint) and foot-back (platform). Both techniques have a different effect on the direction and size of the created force to hit the serve. In the foot-up technique, the rear foot moves to the front foot during the toss phase of the serve. In contrast, the foot-back technique keeps the rear foot back from the front foot during the same phase. The foot-up technique is suitable to create a fast serve with a considerable spin component, whereas the foot-back technique is more suitable for the serve-volley play, due to the wider stance and hence a greater horizontal ground reaction.

The main focus has been the biomechanics of the first serve. However, the second service is also important and the biomechanics are discussed in the rest of this section.

### Biomechanics of the Second Serve

The second serve in tennis is usually either a slice or a twist serve employing spin at the sacrifice of speed (Chow et al., 2003). The slice serve is the combination of an angled racquet face and sideward racquet path during the impact. This gives the ball side spin, which causes it to curve leftwards and then skid leftwards when it bounces. What makes the twist serve different from the slice is its greater topspin, which causes it to curve leftwards and then skid rightwards when it bounces. Both of these serves can be used to draw a player from his/her stationary position, which in turn makes it harder for the player to return the ball or get back into position for the next shot. In elite tennis players, slice/twist serves may have ball speeds between 70 to 80 percent of the flat serve. The reason is that the racquet travels through a circular back swing to brush the ball on the side and the top during the impact. Although the ball speed decreases by around 20 to 30 percent, the racquet speed is almost 90 to 95 percent of that during the flat serve (Knudson, 2006). This implies that a combination of joint motions is used both before and during impact. So a "wrist flick" in a particular direction at the moment of impact to produce a slice/twist serve does not work.

In contrast to the flat serve, there are not many studies focusing on the timing and the coordination of upper joints used to create the second serve. The most consistent difference between the flat and the slice/twist serves is the racquet path through the impact. Comparing the flat serve and the slice serve, the racquet path to the ball is 10 to 20 degrees to the right of the path for the flat serve (Chow et al., 2003). There is also an angled racquet face of between one to four degrees to cause the required spin component. Comparing the twist serve and the flat serve, there is a further 20 degrees of knee bend (Lo, Wang, Wu & Su, 2004). This extra bending is due to the different ball toss for the twist serve. Similar to the slice serve, the timing and the motion of joints of the arm and the upper body, and how they contribute to create the twist serve, has not yet been investigated thoroughly. However, there is some support for the idea that greater upward and sideward motion of the racquet might be created by increased elbow extension, wrist flexion and ulnar deviation. Also, there is greater elbow extensor muscle activation in the wrist over that of the flat serve (Coutinho, Pezarat & Veloso, 2004).

## AVAILABLE TECHNOLOGIES TO MONITOR A TENNIS PLAYER

In the rest of this chapter, we will introduce the current available technologies to monitor the biomechanics of a tennis player during a serve or during the ground strokes. Firstly, we will examine digital video cameras and marker-based optimal monitoring systems. Secondly, we will discuss the networked inertial sensors as a tool to capture the motion of athletes, especially tennis players. Examples and methodologies are presented throughout this section to make the technologies more understandable.

### Video Cameras and Optical Measurements

Generally there are two different approaches to monitoring high speed motion. Both approaches rely on markers placed on the moving object of interest, with the position and motion of the markers recorded. One approach is to only capture the markers and nothing else. This is known as a Digital Optical System. In a Digital Optical System the markers are illuminated by, and reflect, infra-red light. A detector records the position of the marker from which the marker positions and trajectories can be determined. These systems usually provide two or three-dimensional data in near real-time. It has been widely used in different areas such as life sciences, animation and engineering, among others. The other approach is to use specialized software to extract the marker positions from a captured video sequence. This is known as a Digital Video System. The markers are normally just a colored disc. In a Digital Video System the video sequences are captured with a normal or high speed video and digitally stored on a computer. Once the data is stored, various motion analysis software can be applied to track the marker positions in each frame and calculate a corresponding location in each frame.

Digital optical systems are restricted to indoor activities, which in most cases is the laboratory. Underwater or outdoor activities during training or competition cannot be monitored due to issues with lighting and field of view. One way to conduct in-situ monitoring is to use a Digital Video System, which relies upon the use of digital video cameras. Digital video cameras can be categorized to digital video cameras and high speed digital video cameras.

It is the nature of each activity to be monitored, which determines whether a digital camera or a high speed digital camera is required. Most digital cameras have a capture speed between 30 fps (frame per second) to 200 fps at a full resolution of 720 (Horizontal) x 576 (Vertical) for standard definition cameras. For high definition digital video cameras, full resolution can be as high as 1920 (Horizontal) x 1080 (Vertical). Therefore, if the frequency of the desired movement is not more than 200 Hz, then digital cameras can still be a good option. High speed sequences must be recorded by using high speed digital cameras. Most of the cameras in this category are capable of recording 1000 fps in full resolution, which is usually about 1280 (Horizontal) x 1024 (Vertical) pixels. By reducing the resolution, some high speed cameras can achieve a capture speed of up to 675000 fps. Normally these cameras are equipped with different range of memory, varying between 2GB (Gigabyte) to 32GB. A 32GB memory can provide about 25 seconds of 1024 x1024 resolutions at 1000 fps. It is clear that reducing the resolutions can lead to a faster recording time. It should also be noted that the amount of available memory, as well as the capture speed and resolution, all put a limit on how long the subject can be videoed. Obviously the more memory available means a longer possible time for motion capture.

It is obvious that both systems will operate differently and therefore each system will require different methodologies to use them. In order to understand the process of using and analyzing the data from these systems, we present a summarized methodology of each. To better understand the Digital Optical System, references to the Vicon capture monitoring system will be used. The Vicon capture monitoring system referred to consists of near infra-red cameras (MX cameras), data collection units (MX Net), and analysis software (Nexus). It is worth noting at this point, that different commercial systems might use slightly different methodologies to process motion than the methodologies given proceed, but they will follow the same general methods.

**Digital Video System Methodology**

1. **Setting up and calibration:** It is possible to compute 2-D coordinates from one camera only and to compute 3-D coordinates from two or more cameras. Similar to the digital optical analysis, the cameras need to be positioned around the performance area on tripods. Once the cameras are set in position, the system can be calibrated.
2. **Preparing the subject:** Contrast markers must be attached to the area of interest on

the subject and tracked from the video images by using video tracking software. If it is not possible to attach the markers, then data can be collected first and manually digitized. Note that care should be taken when attaching the markers and the investigator must ensure they are firmly attached. If they are attached too loosely, such as on baggy or floppy clothing, then the markers will move around too much, which will make it difficult to see the actual motion intended to be monitored. Also, the investigator should be careful that there is nothing which may be falsely interpreted as a marker.

3. **Recording the video:** As soon as the subject starts to perform a task, the subject can be recorded. The video must be transferred to the computer for further analysis.
4. **Capturing the coordinates:** The video for the section of interest must be presented to the software reconstruction algorithm, which can then determine the 3-D coordinates and trajectories of the markers. Care has to be taken in the synchronization of the video inputs to the software.
5. **Analyzing and presenting:** Analysis such as calculating angles, velocities and so on can be performed once the 3-D data coordinates are produced. It is common to present the resultant parameters in graphs and create an animation from the collected data for presentation.

**Digital Optical System Methodology**

1. **Choosing / designing an appropriate model:** This is the first step of the analysis. Each manufacturer of the equipment typically supplies a biomechanical modeling script, which allows the user to define some properties such as joint centers, orientations and so on. This script is known as the *model* and there are normally different models supplied for different types of monitoring. The model relies on markers in set positions

*Figure 2. Reflective markers placement on the subject's body and racquet*

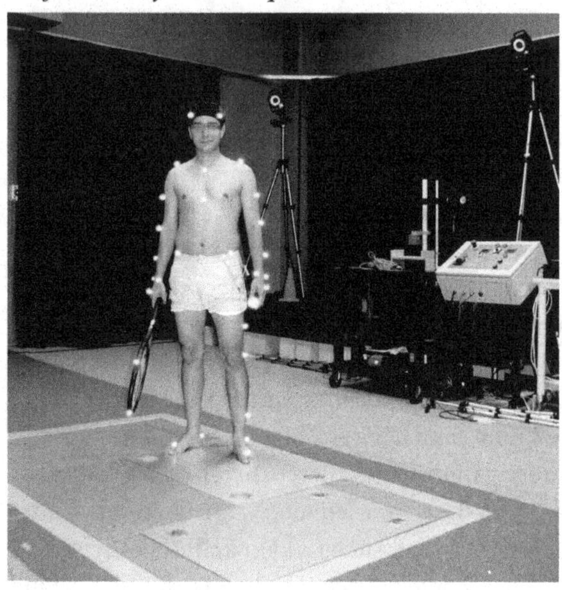

on the body to work properly. Note that the user may need to adapt an existing model, or even create a new model depending upon the motion of interest. Figure 2 shows the marker placements as defined in a typical model.

2. **Setting up the cameras and calibrating the system:** In this step, we need to set up the cameras around the region of interest where the motion will occur. Part of a typical camera set up can be seen in Figure 2. The cameras can be seen mounted securely on tripods around the volume of interest. Once the cameras are switched on, they need to be synchronized with the system. During the cameras' set up, the Frame rate (sampling frequency) of the cameras can be set. Each camera needs to be moved physically to get the best view of the volume. Also, overlapping the camera views is essential for 3-D analysis. For the highest quality of marker detection, it is useful to design a camera placement and volume specific for each individual study. It should be noted

that moving a camera closer or further from the subject will alter the focus point for the camera, and may result in some blurring of the marker edges. Clearly, the more cameras the better, since the visibility of markers will be increased under the range of motion. Any unwanted reflections, such as the strobe of other cameras in the view, must be masked.

Once the cameras are set up, the system needs to be calibrated. The aim of calibrating the system is to let the system calculate the position of markers inside the volume as accurately as possible. The better the system is calibrated, the more accurate the system can calculate the position of the markers during the data collection trials. Calibrating the cameras can be slightly different from company to company. One solution to perform this task is to swing a T-shape wand in front of the cameras inside the volume. The number of samples per camera that will be collected for calibration is called *Refinement frames*. In order to calibrate over larger volumes, the refinement frames samples should be increased. Also during camera calibration, it is useful to decrease the sampling frequency to collect more calibrating data and to change it back to the desired sampling frequency for data capture. Moving the wand in a figure of eight pattern is a useful method of dynamic calibration, as this motion includes x, y and z components of motion. It is also beneficial to move around inside the volume whilst turning and spinning, since this guarantees that all cameras will get to see the wand. The investigator should ensure that the calibration was successful. In the Vicon system™, an indication of the errors associated with each camera is given. Next, the origin needs to be set by putting a special set of markers on the origin to establish the global coordinate system. In the Vicon system™ this set of markers is contained on an L shaped frame. The 3-D positions of the markers are reported to the system using the infra-red light of the camera strobes.

3. **Preparing the subject:** There are usually two different approaches to prepare the subject. One approach is to attach the markers straight on to a person's skin or clothes and record the 3-D positions of the markers. The second approach is to wear a suit with all the markers already attached to it. Figure 2 shows the reflective markers attached to the body and the racquet using the first approach. They are attached using double sided tape. Again, the investigator must ensure the markers are firmly attached and that the actual motion intended to be monitored can be seen. Also, in the case of reflective markers, the investigator should check that there is nothing reflective which may falsely be interpreted as a marker.

   Depending upon the model chosen in the first step, the analysis software will need measurements of some of the physical characteristics of the body. The particular characteristics required depend upon the model chosen. For example in the Vicon Plug-in Gait model, the subject's mass, height (without shoes), left and right leg length (ASIS to medial malleolus), knee width (from KNEE marker to medial axis of rotation), and ankle width (from ANKLE marker to medial malleolus) need to be measured and entered into the analysis software.

4. **Performing the measurement:** In order to perform the measurement, the marker tracking software must be run, which will enable the cameras to send the marker data in real-time to the main engine. The engine is used to calculate the 3-D positions of the markers, as well as any other calculations required by the defined model in the first step. A typical measurement consists of a static trial and many dynamic trials. The

static trial allows the software to collect the required information to process the dynamic trials.

The static trials require the subject to step into the middle of the capture volume and assume a stationary T-pose. Two to three seconds of static data capture is sufficient. Once the static data is captured, then each marker must be labeled as required by the model chosen in the first step. The analysis software generally contains a visualization component, which shows the markers in 3-D space and allows them to be labeled.

Once the markers are labeled, then the dynamic trials can be performed and captured. The labeled data can then be visualized in the analysis software. An example of a dynamic trial of a person performing a tennis serve is shown in Figure 3.

5. **Processing and exporting data:** One major step in processing the data is to check if there is any gap between the marker's trajectories throughout the entire trial. For small gaps, an automatic gap-filling algorithm based upon spline fitting can be used. However, for large gaps, a manual gap filling strategy is more practical.

   Any calculations on the captured data, such as velocity, angles and so on, can be performed when needed. Once the calculations are performed, it is also possible to transfer the calculated biomechanical information to other software for more analysis or presentation if required. For instance, in a coaching role, the coach may only want certain information displayed in a user friendly manner.

## Examples Using Digital Video and Digital Optical Systems

The integration of a high-speed digital video camera with motion analysis software makes possible a wide range of measurement applications. In this section, we will present some examples from the literature to explain how the described methods can be applied to measure the performance of tennis players.

To start with, we can briefly explore how some researchers (Elliott, Fleisig, Nicholls & Escamilia, 2003) collected data to measure the kinematics used by tennis players to produce high velocity serves and the effects on upper limb loading in the tennis serve. The data was collected from professional tennis players by two electronically synchronized high-speed video cameras capturing at 200 Hz. The cameras were setup in a way to be able to capture the side and front view of a tennis player during the entire service motion. A 2 x 1.9 x 1.6m calibration frame was employed (from Peak Performance Technologies Inc., Englewood, CO, USA) and each player was videotaped

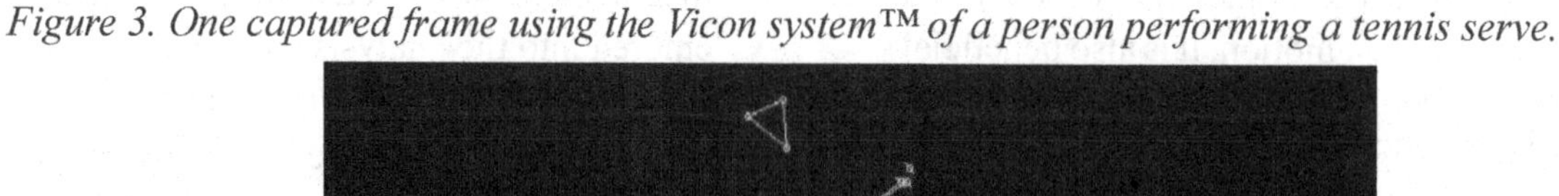

*Figure 3. One captured frame using the Vicon system™ of a person performing a tennis serve.*

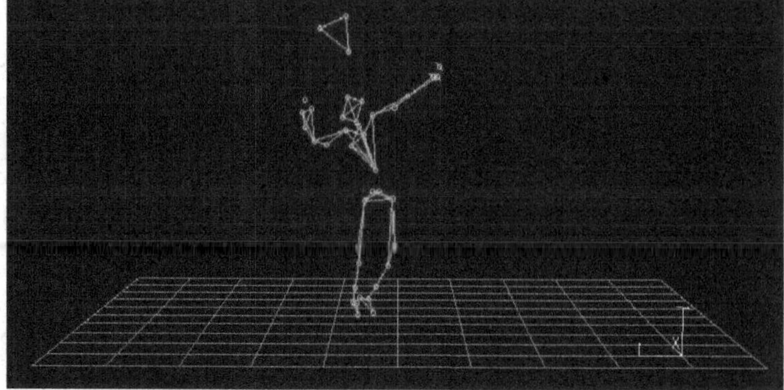

at the end of the court, both before and after each match. The position of the frame encompassed the space used by players during the entire serve. The volume was large enough to contain most of the tennis serve that was digitized. Once the data were collected, only the three fastest successful serves from each tennis player were picked for analysis. The chosen serves were manually digitized with the Peak Motus system (Peak Performance Technologies Inc., Englewood, CO, USA). For this particular experiment, the model contained 20 points, including the center of the mid-toes, ankle, knee, hips, shoulders, elbows, wrists, head, racquet grip, and racquet head. Each of these 20 points was digitized in each frame. The starting frame was when the ball was tossed and the final frame was until a few frames after the ball was hit (impact time). Using the Peak Motus system, the 2-D digitized images and the direct linear transformation, a 3-D coordinate data was calculated. The kinematics parameters were extracted and calculated using a computer program specifically written for this purpose.

Another example is from Gordon and Dapena (2006) who were trying to identify the contributions of joint rotations to racquet speed in the tennis serve. In this study, 13 markers were attached on each subject, as well as the racquet. Ten markers were placed on the subject's racquet arm, with the remaining on the racquet. Of the ten markers, two markers were attached anterior and posterior to the shoulder, and the other eight were positioned to the elbow and wrist area. The markers were placed in a way that one marker was aligned with either a wrist marker set or an elbow market set at intervals of approximately 90 degrees. Each subject was asked to warm up thoroughly first, before placing the markers on the subject as well as the racquet. The subject was then asked to perform several serves to get used to the markers prior to hitting three fast services. All serves were recorded by using three Locam motion picture cameras (Redlake Corporation, Morgan Hill, CA, USA) set to a 100 Hz frame rate. One of the cameras was positioned to the right of the tennis player, with the other one behind and to the left of the server, and the last one near the left net post. All the cameras were placed at the height of approximately 1.5 meters. Only the serves nearest to the center line of the service box were selected for analysis. A Vanguard projection head (Vanguard Instrument Corporation, Melville, NY, USA) was employed to project the film images on to the COMPLOT digitizer (Houston Instrument Division, Bausch & Lomb, Austin, TX, USA). The locations of all the body landmarks were digitized manually in each frame. All frames between the maximum knee flexion and impact time were digitized. As the cameras were not completely synchronized, it was necessary to establish the correspondence between the frames of all three cameras by using the methods of Pourcelot, Audigie, Degueurce, Geiger and Denoix (2000). Also, the quintic spline functions developed by Woltring (1986) were used to fit the data from each camera. For instance, intermediate values between frames were interpolated using the quintic spline functions of the cameras. Finally, the direct linear transformation (DLT) method, which was developed by Abdel-Aziz and Karara (1971) and Walton (1981), was utilized to compute the 3-D coordinates of the landmarks from the digitized data.

As an example of using digital optical monitoring systems, in 2007, the biomechanical laboratory located at Griffith University, Gold Coast campus was used to collect some tennis serve data. The lab is equipped with eight MX13 infra-red cameras. The first two cameras were set approximately at the height of 1.75 meters from the ground, and the other six were mounted on two tracks at the height of 3 meters, facing downward toward the subject. Each camera supports 1.3 million pixels and can record 1280 x 1024 greyscale pixels at 484 fps (up to 10 000 fps at lower resolution). The experiment was set to collect data from a sub elite tennis player inside the lab. For the highest quality of the marker detection, a proper arrangement for camera placement and a volume specific to the study were designed. Two cameras were faced to the front and back of the subject, and the other six were placed at the sides (three cameras at each side). After calibration, 37 markers were placed all over the subject according to the full-body plug-in-gait model. As part of the plug-in-gait model requirements, measurements such as the subject's mass, height, leg length, knee and ankle width were provided to the system's software. Nexus 1.1 was used as the data acquisition and processing platform during the whole experiment. Two to three seconds of data were captured as static trial, prior to capturing the dynamic trials. A static trial was performed to generate required data to run the model in Nexus. During each dynamic trial the subject was asked to perform a first serve. Finally, Fill Gaps (Woltring, 1986) was applied on the collected data through the Nexus pipeline. The trajectory of markers were extracted from the Vicon software for further upper arm rotation, forearm pronation, and kinematic chain analysis in MATLAB®.

In the following section, we will see how inertial sensors can be employed to measure some biomechanical parameters in the tennis serve.

## Inertial Sensors

Accelerometers which measure acceleration, and gyroscopes which measure rotation motion, are the main inertial sensor families that can be used for human motion analysis. Accelerometers are sensors and instruments for measuring, displaying and analyzing acceleration and vibration. They can be used on a stand-alone basis, or in conjunction with a data acquisition system. Accelerometers can have one axis to three axes of measurement. Three main features that must be considered when selecting accelerometers are: amplitude range, frequency range, and ambient conditions. Acceleration amplitude range is measured in Gs, which is based upon the acceleration due to gravity. For the ambient conditions, such things as temperature should be considered, as well as the maximum shock and vibration that the accelerometers will be able to handle. The sensor measures both the static (gravitational) and dynamic (inertial) accelerations. However, in dynamic situations where the axes of sensitivity of the accelerometer tilts relative to gravitational force vector, it is not possible to separate the static and dynamic acceleration, unless using the appropriate high pass filter (Ohgi, Yasumura, Ichikawa & Miyaji, 2000). In relation to a swinging object, it will also be necessary to detect the rotational movements, as the swinging instrument can move in a 3-D environment. Rotary information can be achieved by the use of gyroscopes. To obtain absolute velocity and acceleration information, it is necessary to know the boundary conditions, which are easy to predict in a large number of swing movements. In this type of analysis it is also important to limit the time intervals over which integration is performed, to minimize the accumulation of integration error.

Accelerometer technology has some advantages over the previously described video methods. Using accelerometers eliminates the need for experiments to be carried out solely inside the laboratory, which means that experiments can be run in the real environment. Accelerometers are small and light enough to be placed on any part or segment of the body without hindering performance. They are also very cheap compared to the other technologies. Within many sports applications, accelerometers are now used to measure and classify activity and effort levels. For instance, the key characteristics of the basic swing of Japanese swordsmanship can be extracted using accelerometers. James, Gibson and Uroda (2005) mounted four accelerometers on a sword to record data from a basic swing. This demonstrated a high correlation between athlete skill and the recorded measures. They proposed that this can be used as a biofeedback tool to aid athletes in correcting swing characteristics in a variety of sports. Accelerometers have also been proven to be capable of detecting the kinematic chain in the golf lateral swing (Ohgi & Baba, 2005). In that study, golfers with different levels of proficiency were categorized once the swing was performed. Using the same technology, it is possible to study the tennis serve motion.

The purpose of this section is to show that body mounted sensors such as accelerometers can be used to monitor the performance of tennis players. Since the first serve has been reported to be the most important tennis stroke, the following example will demonstrate the use of inertial sensors to perform skill assessment.

## Example Using Inertial Sensors

Research investigating the translational and rotational motion of the swing has been performed using accelerometers and applied to tennis (Ahmadi, Rowlands & James, 2006). As part of this research, an accelerometer measurement system comprising of three accelerometer nodes was used to identify the correlation between the skill level and the characteristics of the first serve swing in tennis. This system included a microcontroller board; Bluetooth® connection (wireless connection); two different types of tri-axial accelerometers to capture the slower and faster type of motion (ADXL202, ADXL210, Analog Devices Incorporation, Norwood, MA, USA); and a Visual Basic (VB) client running on a laptop (James, Gibson & Uroda, 2005). The VB client on the laptop receives the wireless signals and saves them for further analysis.

*Table 1. Hardware specification of the sensor monitoring system*

| Feature | Value |
|---|---|
| Resolution (Accelerometers) | 10 bits |
| Sampling Rate @ 3 channels | Up to 500 Hz per channel |
| Accelerometer Range | ADXL202 : ± 2g<br>ADXL210 : ± 10g |
| Data Transmission Type | Wireless (Bluetooth® 1.1) |
| Transmission Range | Up to 250 meter |

*Figure 4. Location of sensors and processing box on (A) the athlete's wrist, (B) the athlete's waist and (C) the athlete's knee. The acceleration coordinates are also shown when the athlete is facing the tennis court. A sensor node containing two accelerometers is shown in the top right corner.*

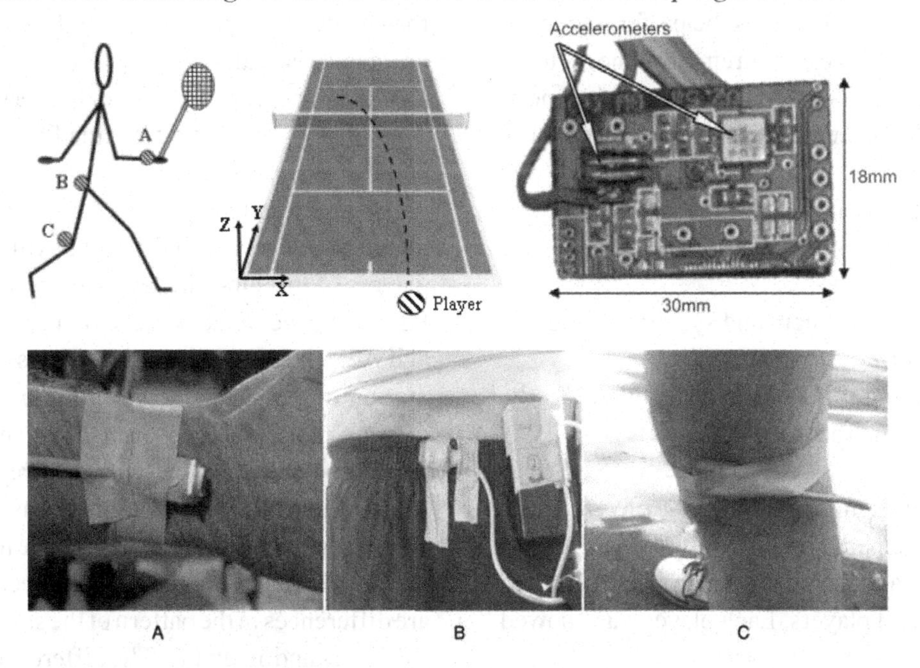

The specification of the system is summarized in Table 1.

In Table 1, the "Accelerometer Range" parameter refers to the acceleration range that can be measured by the accelerometers. For instance, the ADXL210 accelerometer can sense the range of acceleration from -10g to +10g, which is equivalent to -98 m/s$^2$ to +98 m/s$^2$ where 1g is 9.8 m/s$^2$.

Three distinct accelerometer nodes were employed to capture the kinematic chain during the serve. The position and coordination of sensors when a player is facing the court is shown in Figure 4. The accelerometer nodes are attached on the knee, waist and wrist of the tennis player. The processing box is attached on the player's shorts using a clip and the accelerometers are attached to the area of interest using strapping tape. The attachments must be firm to eliminate excess movement of the sensors.

Low-g accelerometers (± 2g) were applied on the knee and the waist, and a higher-g accelerometer (± 10g) was attached on the wrist due to the higher acceleration of the racquet-arm movement compared to the other segments.

In any investigation it is very important to develop a proper protocol before the data collection session. The protocol typically consists of a list of instructions that must be performed prior, during, and after the measurements. The protocol is important in that it standardizes each experiment and guarantees that the investigator does not forget any details. In this study, a protocol was developed that consisted of five major parts:

1. Camera set up
2. Sensors placement and synchronizatio
3. Monitoring
4. Detaching the sensors from athlete's body
5. Saving the recorded data

Four male tennis players with different levels of skill participated in this study. Two were sub-elite players and the other two were recreational/developmental players. Each player was allowed to take as much time as needed to warm up. Once the player was ready, the accelerometer system was mounted on three different positions (knee, waist and wrist) of the player's body. The player was then asked to perform approximately 10 first serves to ensure that the player felt comfortable with the system and that the system was not limiting the player's movements. Each player was then asked to serve from the deuce side and return to this position after each serve. The player was asked to delay for approximately 5 seconds before starting the next serve. This allowed the investigators to recognize the serve's signals from signals due to other movements and helped the player to focus on the next serve. Two digital video cameras (30fps) were also recording the serve to video tape in order to capture the movement. The accelerometer nodes were synchronized with the recorded images to determine the kinematic action, so that specific events could be identified and then a model developed from the acceleration data.

Average acceleration values for each of the sensors for each athlete were used to determine the trends, so that the effects of variations between individual serves were minimized. The average value was determined from the 10 serves performed by the athlete as part of the experimental protocol. The standard deviation was also used as an indicator of repeatability of the service action. The impact time was determined from the video and the sensor data.

Figures 5 and 6 show selected sensor measurements for players with different skill levels. The players of the sub-elite level are denoted P1 and P2, and at the amateur-developmental level, A1 and A2. The selected measurements are the side motion of the hand (shown in Figure 5) and the forward motion of the waist (shown in Figure 6). As can be seen, players with different levels of skill exhibit different swing patterns.

The most indicative motion of the hand can be seen in the side motions. It can be seen that there are differences in the pattern of the side motion for sub-elites and amateurs. The differences are most apparent around the impact. Since the amateur player does not employ his waist to generate power to the serve, his side motion pattern is completely different from that of the sub-elite player. There is also a significant difference in acceleration magnitude between the amateurs and sub-elite players around impact time.

The forward waist motion also exhibits a difference between the sub-elite and the amateur. This is shown in Figure 6. The forward motion of the waist is linked to the rotation of the waist in the service motion. The rotation pattern varies from the sub-elite players to the developmental players. It shows that for the developmental players, there is either no rotation or some rotation at the wrong time. In contrast, the full rotation occurs around the impact time for the sub-elite players.

Repeatability is another key feature that may be used to classify the athletes. The grey lines in Figures 5 and 6 show the standard deviation for each player. As can be seen, the sub-elite players have less variability through this whole swing than the developmental players. This indicates that the sub-elite players are more consistent in their service and their action is more controlled. It should also be noted that the standard deviation alone cannot be used to detect the correct motion, as a player might consistently repeat a bad habit. Therefore, both the standard deviation and the average of the swing pattern can be used for an indication of the proficiency level of an athlete.

The kinematic model for the first serve was observed. Furthermore, this study revealed that side-forward motion of the hand along with the forward motion of the waist of an athlete can be used as indicator to assess the athlete's skill level. Overall, this has shown that accelerometers can be used to monitor and give indicators towards assessing an athlete's skill level.

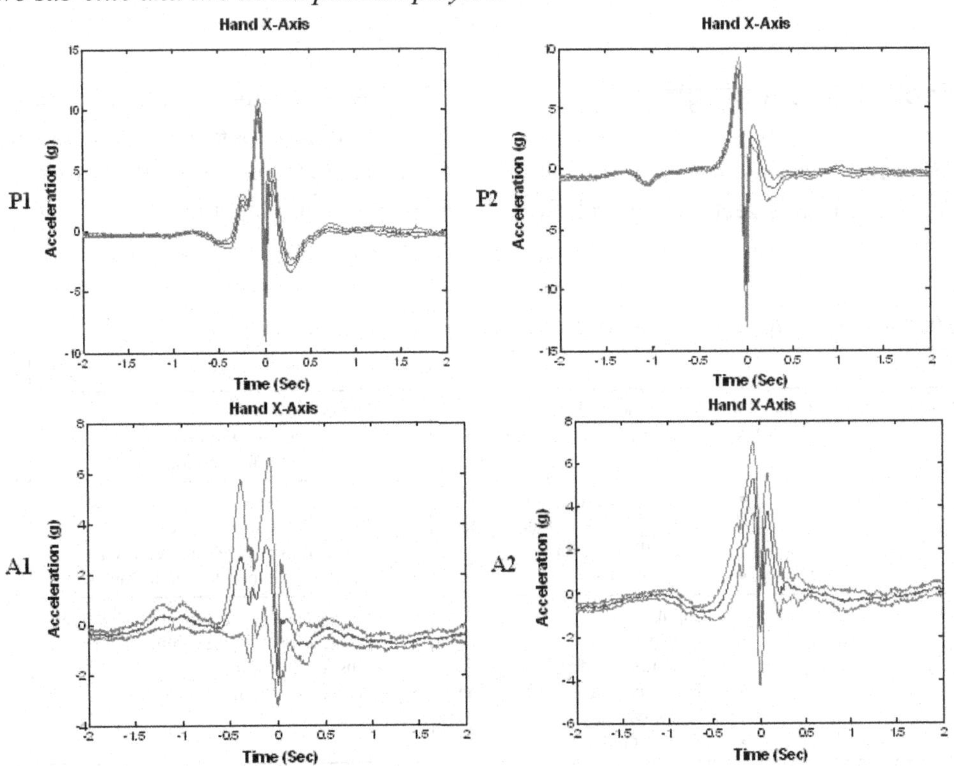


*Figure 5. Average (middle line) and average ± standard deviation (outer line) of the hand forward motion for two sub-elite and two developmental players.*

*Figure 6. Average (middle line) and average ± standard deviation (outer line) of the waist forward motion for two sub-elite and two developmental players.*

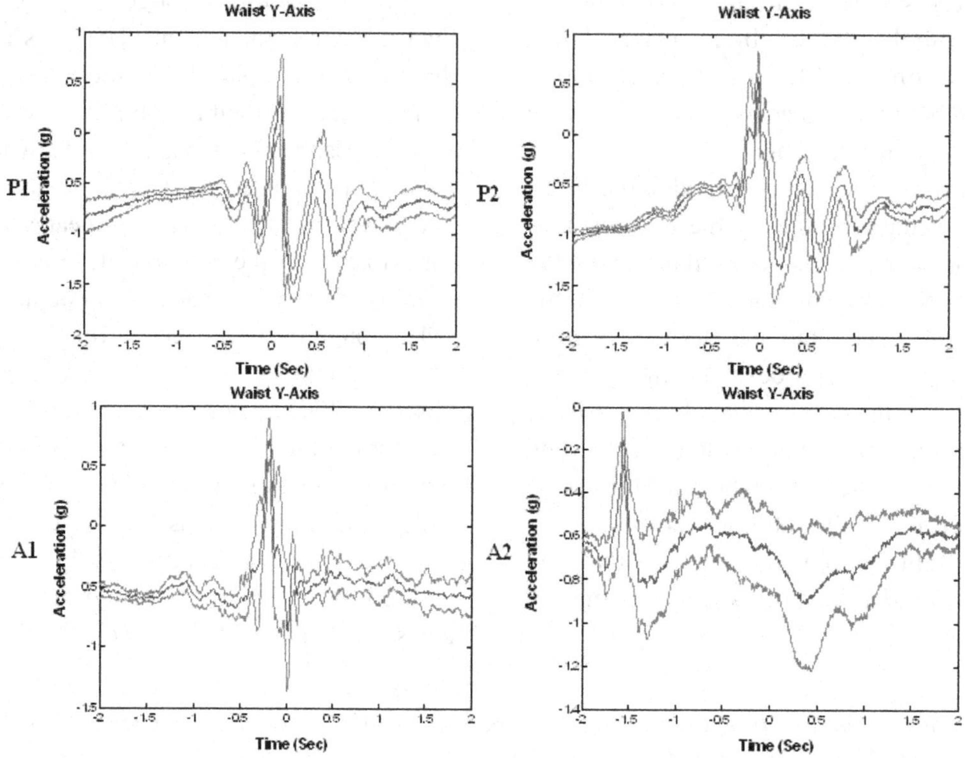

## Comparison of Technologies

Table 2 summarizes and compares the three technologies discussed in this section. The relative advantages, disadvantages, and cost factors are given for digital video cameras, optical markers, and inertial sensors. The table was developed with respect to tennis monitoring.

*Table 2. Advantages and disadvantages of using the technologies in tennis*

| Technology | Pro | Con | Cost |
|---|---|---|---|
| Digital Video Camera Systems | • Real-time feedback<br>• Real environment monitoring | • Not cheap<br>• Post processing required<br>• Limited time for motion capture | Medium |
| Marker-based Digital Optical Systems | • Provide trajectory data<br>• Powerful tools to manipulate the collected data<br>• Non-invasive | • Expensive<br>• Mainly laboratory-based<br>• Post and pre processing required<br>• Need to be expert to run the lab | High |
| Inertial Sensors | • Light<br>• Cheap<br>• Real-time feedback<br>• Real environment monitoring<br>• Non-invasive<br>• Self-test possible | • Sensor and recorder must be attached<br>• Possible issues with cabling from sensor to recorder<br>• Power must be from batteries | Low |

## SUMMARY

Tennis is a very popular sport that is played by millions of people around the world at all levels from amateur to professional. There is a demand for players to improve their game by enhancing their stroke play. Technology allows a coach to monitor the player's action and take appropriate action. This chapter focused on some of the main technologies to monitor the action of a tennis player.

In order to understand the data from the technology then knowledge of the biomechanics is essential. This chapter explained some of the various strokes in tennis and discussed the importance of each during the tennis match. Due to the importance of the serve in a tennis match then the biomechanics of the serve was examined in detail.

This chapter also detailed some of the major technologies to monitor and analyse the tennis swing. Those systems were the high speed digital cameras, marker-based optical systems, and inertial sensors. The advantages and disadvantages of each technology to monitor the tennis player was summarised in Table 2.

## FUTURE TRENDS

This chapter focused on some of the main technologies to monitor the action of a tennis player. The technology to monitor tennis will continue to improve in the future, but there are a few issues that still need to be resolved before the current technology becomes more useful.

In the marker based systems, there is a lack of suitable models specifically aimed at tennis and the different components of tennis. For instance, at the moment there is no specific model that can be used to extract biomechanical measures of the tennis serve. Newer models are currently under development by the research community.

There will be an increasing trend towards using a combined inertial sensors platform to capture the translational and rotational components. It is expected that these platforms will be integrated and miniaturized so that they do not interfere with the tennis player. It is also expected that these sensors will become more dependent upon wireless technologies and that they will form part of a wireless network cluster.

The swing is a component of a number of different sports. Once linear and rotational motions have been analyzed, the same technology and analysis routines as used in tennis can be applied to many more sports that are based on swing action. For example, the same analysis routine to monitor the upper arm rotation or forearm pronation during a tennis serve can be used to calculate the same kinematic parameters during bowling in cricket, or putting and driving in golf.

The use of technology to monitor athletes leads to large volumes of data being collected. It is not expected that the final users of the technology such as coaches and athletes will understand the data. Therefore, there is a need to present the data in a meaningful way to the end user. This means that there will be a trend towards greater emphasis on data visualization, visualization technology and storage in the future.

Overall, the trend will be towards using more technology in tennis so that the players and coaches can get a more objective analysis of the player's techniques. It is also expected that technology will form a greater part of the coaching regime.

## REFERENCES

Abdel-Aziz, Y.I., & Karara, H.M. (1971). Direct linear transformation from computer coordinates into object coordinates in close-range photogrammetry. In *Proceedings of the ASP Symposium on Close-Range Photogrammetry* (pp.1-18). Falls Church, VA: American Society of Photogrammetry.

Ahmadi, A., Rowlands, D.D., & James, D. (2006). Investigating the translational and rotational motion of the swing using accelerometers for athlete skill assessment. In *Proceedings of the 5th International IEEE Senors Conference* (pp. 980-983). Daegu Exhibition & Convention Centre (EXPO), Daegu, Korea.

Bahamonde, R.E. (1994). *Biomechanical analysis of serving during the performance of flat and slice tennis serves.* Unpublished doctoral dissertation, Indiana University, USA.

Bahamonde, R. (2000). Changes in angular momentum during the tennis serve. *Journal of Sports Sciences, 18*, 579-592.

Chow, J., Carlton, L., Lim, Y., Chae, W., Shim, J., Kuenster, A.F., & Kokuban, K. (2003). Comparing the pre- and post-impact ball and racquet kinematics of elite tennis players' first and second serves: A preliminary study. *Journal of Sports Sciences, 21*, 529-537.

Coutinho, C., Pezarat, P., & Veloso, A. (2004). EMG patterns of the upper limb muscles in the first (flat) and second (topspin) serve. *Medicine and Science in Tennis, 9*(3), 14-15.

Elliott, B. (2006). Biomechanics and tennis. *British Journal of Sports Medicine, 40*, 392-396.

Elliott, B., Fleisig, G., Nicholls, R., & Escamilia, R. (2003). Technique effects on upper limb loading in the tennis serve. *Journal of Science and Medicine in Sport, 6*(1), 76-87.

Elliott, B., Marsh, A., & Blanksby, B. (1986). A three-dimensional cinematographic analysis of the tennis serve. *Journal of Applied Biomechanics, 2*(4), 260-271.

Elliott, B., Marshall, R., & Noffal, G. (1995). Contributions of upper limb segment rotations during the power serve in tennis. *Journal of Applied Biomechanics, 11*(4), 433-442.

Elliott, B., & Wood, G. (1983). The biomechanics of the foot-up and foot-back tennis serves techniques. *Australian Journal of Sports Sciences, 3*(2), 3-6.

Girard, O., Micallef, J., & Millet, G.P. (2005). Lower-limb activity during the power serve in tennis: Effects of performance level. *Medicine and Science in Sports and Exercise, 37*, 1021-1029.

Gordon, B.J., & Dapena, J. (2006). Contributions of joint rotations to racquet speed in the tennis serve. *Journal of Sports Sciences, 24*(1), 31-49.

James, D., Gibson, T., & Uroda, W. (2005). Dynamics of a swing: A study of classical Japanese swordsmanship using accelerometers. In A. Subic & S. Ujihashi (Ed.), *The impact of technology on sport* (pp. 355-360). Melbourne, Australia: Australasian Sports Technology Alliance.

Knudson, D. (2003). *Fundamentals of biobmechanics.* New York: Kluwer Academic Plenum Publishers.

Knudson, D. (2006). *Biomechanical principles of tennis technique using science to improve your strokes.* California: Racquet Tech Publishing.

Legnani, G., & Marshall, R.N. (1993). Evaluation of the joint torques during the tennis serve: Analysis of the experimental data and simulations. In *Proceedings of the VIth International Symposium on Computer simulation in Biomechanics* (pp. 8-11). Paris: International Society of the Biomechanics, Technical Group on Computer Simulation.

Lo, K., Wang, L., Wu, C., & Su, F. (2004). Biomechanical analysis of trunk and lower extremity in the tennis serve. In *Proceedings of the XXII International Symposium of Biomechanics in Sport* (pp. 261-264). Ottawa: University of Ottawa.

Marshall, R.N., & Elliott, B.C. (2000). Long-axis rotation: The missing link in proximal-to- distal segmental sequencing. *Journal of Sports Sciences, 18*(4), 247-254.

Ohgi, Y., & Baba, T. (2005). Uncock timing in driver swing motion. In A. Subic & S. Ujihashi

(Ed.), *The impact of technology on sport* (pp. 349-354). Melbourne, Australia: Australasian Sports Technology Alliance.

Ohgi, Y., Yasumura, M., Ichikawa, H., & Miyaji, C. (2000). Analysis of stroke technique using acceleration sensor IC in freestyle swimming. *The Engineering of Sport*, (pp. 503-511).

Payne, A.H. (1978). Comparison of the ground reaction forces in golf drive and tennis service. *Aggressologie, 19*, 53-54.

Pourcelot, P., Audigie, F., Degueurce, C., Geiger, D., & Denoix, J.M. (2000). A method to synchronise cameras using the direct linear transformation technique. *Journal of Biomechanics, 33*(12), 1751-1754.

Smith, S.L. (1979). *Comparison of selected kinematic and kinetic parameters associated with the flat and slice serves of male intercollegiate tennis players.* Unpublished doctoral dissertation, Indiana University, USA.

Sprigings, E., Marshall, R., Elliott, B., & Jennings, L. (1994). A three-dimensional kinematic method for determining the effectiveness of arm segment rotations in producing racquet-head speed. *Journal of Biomechanics, 27*(3), 245-254.

Van Gheluwe, B., De Ruysscher, I., & Craenhals, J. (1987). Pronation and endorotation of the racket arm in a tennis serve. In B. Johnsson (Ed.), *Biomechanics X-B* (pp. 667-672). Champaign, IL: Human Kinetics.

Van Gheluwe, B., & Hebbelinck, M. (1985). The kinematics of the service movement in tennis: A three-dimensional cinematographical approach. In B. Johnsson (Ed.), *Biomechanics IX-B* (pp. 521-525). Champaign, IL: Human Kinetics.

Walton, J.S. (1981). *Close-range cine-photogrammetry: A generalized technique for quantifying gross human motion.* Unpublished doctoral dissertation, Pennsylvania State University, University Park, PA.

Woltring, H.J. (1986). A Fortran package for generalized, cross-validation spline smoothing and differentiation. *Advanced Engineering Software, 8*, 104-113.

# Chapter VII
# Quantitative Assessment of Physical Activity Using Inertial Sensors

**Daniel A. James**
*Griffith University, Australia & Queensland Academy of Sport (Centre of Excellence), Australia*

**Andrew Busch**
*Griffith University, Australia*

**Yuji Ohgi**
*Keio University, Japan*

## ABSTRACT

*The testing and monitoring of elite athletes in their natural training and performance environment is a relatively new area of development that has been facilitated by advancements in microelectronics and other micro technologies. The advantages of monitoring in the natural environment are that they more closely mimic the performance environment and thus can take into account environmental variables; and where used in competition, can monitor the athlete's performance under competitive stresses. Whilst it is a logical progression to take laboratory equipment and miniaturize it for the training and competition environment, it introduces a number of considerations that need to be addressed. This chapter introduces the use and application of inertial sensors as ideal candidates for the portable environment, describing the technical challenges of making the transition from the lab to the field, using a number of case studies as examples.*

## INTRODUCTION

Athletic and clinical testing for performance analysis and enhancement has tradition-ally been performed in the laboratory, where the required instrumentation is available and environmental conditions can be easily controlled. In this environment, dynamic characteristics of athletes are assessed using simulators such as treadmills, rowing and cycling machines, and flumes for swimmers. In general, these machines allow for the monitoring of athletes using instrumentation that cannot be used easily in the training environment, but instead requires the athlete to remain quasi static, thus enabling a constant field of view for optical devices and relatively constant proximity for tethered electronic sensors, breath gas analysis, and so on. Today however, by taking advantage of the advancements in microelectronics and other micro technologies, it is possible to build instrumentation that is small enough to be unobtrusive for a number of sporting and clinical applications (James, Davey & Rice, 2004). One such technology that has seen rapid development in recent years is in the area of inertial sensors. These sensors respond to minute changes in inertia in the linear and radial directions. These are known as accelerometers and rate gyroscopes respectively. This work will focus on the use of accelerometers, though in recent years rate gyroscopes are becoming more popular as they achieve mass-market penetration, benefiting from increased availability and decreases in cost and device size.

Accelerometers have in recent years shrunk dramatically in size, as well as in cost (~$US20). This has been due chiefly to the adoption by industries such as the automobile industry where they are deployed in airbag systems to detect crashes. Micro electromechanical systems (MEMS) based accelerometers like the ADXLxxx series from Analog Devices (Weinberg, 1999) are today widely available at low cost.

The use of accelerometers to measure activity levels for sporting (Montoye et al., 1983), health and gait analysis (Moe-Nilssen & Helbostad, 2004) is emerging as a popular method of biomechanical quantification of health and sporting activity. This is set to increase with the availability of portable computing, storage and battery power, available due to the development of consumer products like cell phones and portable music players.

Researchers have also used accelerometers for determining physical activity and effort undertaken by subjects. These kinematic systems have been able to offer com-parable results to expensive optical based systems (Mayagoitia, Nene & Veltink, 2002). Rate gyroscopes, a close relative of the accelerometer, measure angular acceleration about a single axis and are also used to determine orientation in an angular coordinate system, although these suffer from not being able to determine angular position, in the same way accelerometers have trouble with absolute position. Additionally, many physical movements, such as lower limb movement in sprinting, exceed the maximum specifications in commercially available units that are sufficiently small and inexpensive for such applications.

This chapter introduces the technological basis of inertial sensor technology, including its construction and theory of application, together with how the data might be acquired, stored and collected by a physical device. A number of applications that have penetrated the mass market are introduced, with the focus being on emerging applications. This range of sporting applications cover physiological event detectors, integration with other sensors, and data mining for complex phenomena such as aerodynamic forces on athletes.

## INERTIAL SENSOR TECHNOLOGY

### Description and Construction

Accelerometers measure linear acceleration at the location of sensor itself, typically in one or more axis. Today's sensors are millimeters (e.g. 5 $mm^2$) in size with multi-axial versions available in an integrated package. In general, a suspended mass is created in the design and has at least one degree of freedom. The suspended inertial mass is thus susceptible to displacement in at least one plane of movement (Figure 1). These displacements arise from changes in inertia and thus any acceleration in this direction. Construction of these devices vary, but typically use a suspended silicon mass on the end of a silicon arm that has been acid etched away from the main body of silicon. The force on the silicon arm can be measured with piezoresistive elements embedded in the arm. In recent years, multiple accelerometers have been packaged together orthogonally to offer multi-axis accelerometry (Weinberg, 1999). The attached piezoresistor's dimensions are elongated or contracted depending on the movement of the inertial mass. Thus as this mass responds to changes in acceleration, so too does the electrical properties of the piezoresistor, which can be monitored.

Alternatively, the arm's position can be monitored as a variable capacitor that changes with arm position or with SAW techniques (Surface Acoustic Waves). In addition, the position of the arm can be maintained in place by an applied electric field; this compensatory applied field is then a measure of the acceleration of the sensor (Kloeck & de Rooij, 1994; Motamedi & White, 1994).

*Figure 1. Piezoresistive method is shown for measuring accelerations felt by a suspended inertial mass*

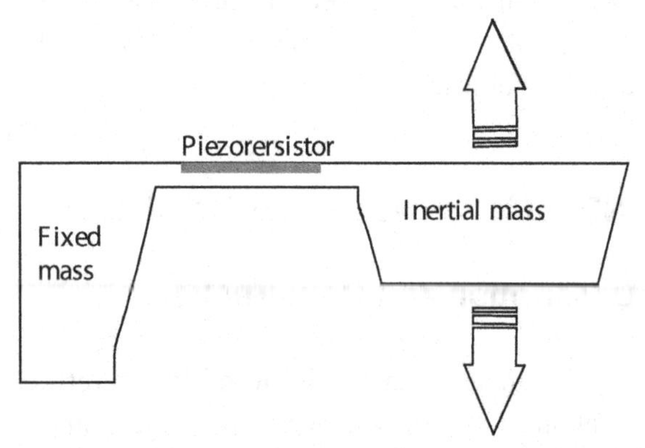


## Function and Theory of Application

Accelerometers measure linear acceleration. Gyroscopes measure angular acceleration. Thus acceleration of body $\mathbf{A_b}$ can be described as the sum of a number of components:

$$A_b = \ddot{\mathbf{p}} + \mathbf{g} + \dot{\boldsymbol{\omega}} \times \mathbf{r} + \boldsymbol{\omega} \times (\boldsymbol{\omega} \times \mathbf{r}) \quad (1)$$

Where **p** is the displacement vector of the origin of the local coordinate system with respect to the global coordinate system, **g** is the gravity, **ω** is the angular velocity of the specified point **p**, and **r** is the radius of the rotation (Ohta, Ohgi, Kimura & Hirotsu, 2005). In practice, each of these four components may or may not be relevant to a particular activity, or can be accounted for as a constant for a sufficiently short period of interest.

Acceleration is the time derivative of velocity and velocity is the time derivative of position. Thus accelerometers can measure these quantities through numerical integration to fully quantify the dynamics of motion and potentially position as well. It is well understood though that the determination of position from acceleration alone is a difficult and complex task (Davey, James & Anderson, 2004), where cumulative errors from the double integration approximation and orientation changes can obscure the desired signal.

Because of this, accelerometers are often used for short-term navigation and the detection of fine movement signatures and features (such as limb movement). Accelerometers can be used to determine orientation with respect to the earth's gravity, as components of gravity are aligned orthogonal to the accelerometer axis. In the dynamic sports environment, complex physical parameters are measured and observed in relation to running and stride characteristics (Herren, Sparti, Aminian & Shultz, 1999), and in the determination of gait (Williamson & Andrews, 2001).

The mechanical coupling of inertial sensors to the region of interest also requires a number of considerations. Firstly, within the musculoskeletal system it is usually the dynamics of joint and bone segments that are of primary interest. However, it is rarely possible to mount the sensors directly to these segments. Instead, sensors are mounted directly onto the skin surface and the tissue beneath it necessarily separates itself from the bone of interest in usually non-linear coupling that is susceptible to stretching and distortion generated during movement. Subject comfort and the neutrality of the sensor is also important; if the subject is uncomfortable in any way the performance characteristics of their movement may be subtly different. Finally, attachment of the acceleration sensors also has the potential to affect the very dynamics of what is to be measured, by the addition of additional mass and surface area on the site of interest. For example, introducing added mass creates drag in monitoring swimmers, and skin based artifacts when monitoring leg or shank movement. Fortunately accelerometers today weigh only a few grams and can be packaged simply using epoxy or similar compounds. The advent of body hugging clothing and suits in many sports allows for the incorporation of sufficiently small sensors and accompanying instrumentation relatively easily, using sewn pouches for example.

## Sampling Theory

Accelerometer output is typically recorded digitally. Depending on the application, sample rate and resolution are important factors. Thus careful adherence to sampling theory is required to ensure that aliasing due to under sampling does not occur (Cutmore & James, 1999). This is particularly relevant to studies based on human motion. Whilst the peak of human activity occurs below 20 Hz, sampling at much higher rates is required to capture the full detail of the motion or if short term navigation is sought, as there are significant higher order terms in the data that contain important information. Sample rates such as 1000 Hz with 12-bit resolution generally preserves data quality for most applications. Of course sampling at these rates for multi-axial devices generates large data sets very quickly. Recent research (Lai, James, Hayes & Harvey, 2003) suggests that in many cases the resolution can be reduced to 8 or 6 bits without loss of information, though preserving the high data rate is important. It is argued that error minimization through shaping the measurement error in time and space, mindful of the noise characteristics of the device (such as the noise floor of the sampling circuit and the bandwidth response of the devices themselves), are more important than the number of bits used in sampling.

## Systems Integration

Utilization of a sensor platform greatly simplifies the process of measuring a physical event via a sensor and turning it into something useful, by providing support for the sensor, basic signal conditioning, storage, and application specific processing of the data. In the case of sporting applications, a modular approach allows rapid customization and modification, as the technical expertise, understanding and expectations of sport scientists can develop rapidly along with prototype systems developed for testing. Accelerometers, when combined in such a system, enable the recording of athlete activity and the storage and/or transmission of data.

Thus a sensor platform should support the acquisition and processing of accelerometers and gyroscopes an order of magnitude above the frequency of interest (typically 100 Hz or greater). On board data acquisition, signal conditioning, basic processing storage, and data transmission characteristics will depend on the intended application.

A sport specific or generalized sensor platform is then packaged according to the environmental demands of the sport, and where appropriate, near real-time transmission or display of the data is available for coach and athlete feedback.

A number of sporting activities have been investigated using accelerometers by the authors. Central to these investigations has been the development of a modular sensor system that can be customized for the intended sport. This system (James et al., 2004) was initially applied to both rowing and swimming, before being trialed in a number of other sports, and commercial prototypes produced. In each case, the system was packaged separately and allowed additional modules to be added as required to facilitate data storage; RF (radio frequency) communications in near real-time and post event IR (infra-red) data download; and sealing for aquatic applications. Figure 2 shows a diagrammatic layout for a modular sensor system encompassing batteries, power controller, processing, sensor and communications via IR or RF.

## Applications

### Off the Shelf Applications

The humble pedometer, which could be considered to be a single axis acceleration threshold counter, today is seeing increased popularity in programs such as the 10, 000 step program. Enter recreational and sporting goods manufacturers like Polar™ and Nike® who today offer a variety of accelerometer based devices that can be mounted on the body and often the foot. With the concurrent development of training software, the running enthusiast can have their daily training regime logged for historical review, to set personal training goals, and interact with a portable music player (e.g. the Apple iPod®). HangTimer™ has recently emerged as a wearable device for winter sport enthusiasts, measuring time spent in the air for land-based sports that have airborne components. Video game manufactures looking for alternate UI (user

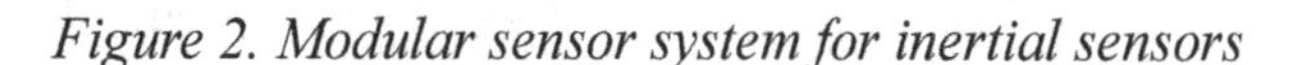
*Figure 2. Modular sensor system for inertial sensors*

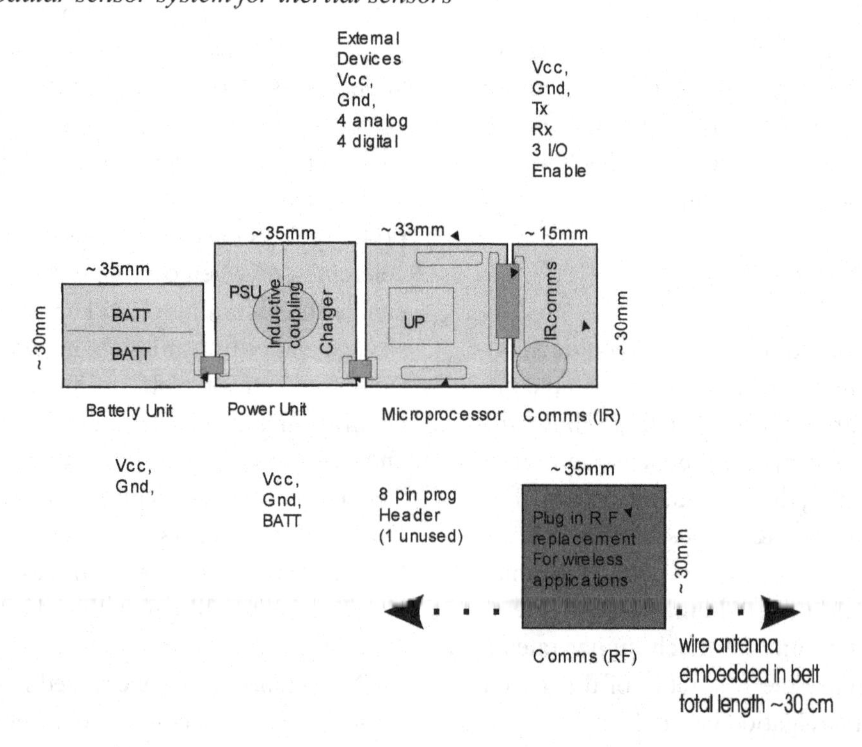

interfaces) have also turned to inertial sensors as a method of game control. Many mobile phones and computers now contain accelerometers (often for other purposes such as hard drive emergency shut down) that are the main UI for the control of simple games. Nintendo's recent Wii™ platform features inertial sensors as part of the hand held controller, being used predominately in sporting simulation games.

Recently inertial sensors have been combined with other sensors, such as magnetometers (for absolute orientation to the earths magnetic field) to offer an alternative to visually based marker systems. These systems are gaining acceptance and are finding increasing applications in the sporting and video game development applications.

## Emerging Applications

Feature detection on the torso, wrist and club head in the sport of golf allows the movements of amateur and professional golfers to be compared in a quantitative way. In rowing, instrumentation of the scull (and optionally oars) allows the phases of the rowing motion to be tracked over the length of the course; statistics such as these allow for development of race strategy and efficacy of training drills to be assessed. Skill acquisition of athletes is investigated in the sport of cricket, where both bowling and batting activities can benefit from on-field monitoring using inertial sensors. In the fourth application, ski jump, the use of inertial sensors has been applied to calculate the timing of events, as an aid to synchronize other sensors and video data, but also as an aid in examining the aerodynamics of flight.

While these applications are not exhaustive, they demonstrate the emergence of these sensors as a viable tool across a broad range of sporting activities at the elite level.

### Golf

Golf swing for the improvement and accuracy of golf ball placement is of great interest to golf players, professional and amateur alike. Trunk and the lower extremities contribution towards the kinematical power and the rotational motion of the upper extremities describe the energy transfer mechanism. Jorgensen (1999) qualitatively explained the golfer's lateral sway motion, which had been believed to utilize the trunk energy by using a simple 2-linked rod model. This is well known among golfers as "left wall of the driver swing". The driver should accelerate during the first stage of the down swing, between the top of swing to the time when the shaft would be horizontal. Then it should decelerate to the opposite direction to the target line. A common instruction to the golfer is that the horizontal acceleration by sway motion should be switched during down swing.

There have been a number experimental and theoretical approaches for the swing and subsequent "uncock" technique, and the wrist–club shaft interface (Inoue, 1997; Umegaki et al., 1998). According to Nagao and Sawada's (1977) research, the optimum uncock timing should be performed at -100ms before the impact. Inoue (1997) examined the wrist turn qualitatively using a 2-linked rigid body model and a non-linear spring between two links, and then proposed an optimum timing of the wrist turn by their theoretical analysis. Thus, the golfer's performance on their driver carry is strongly affected by both their energy production and the transfer mechanism. Accelerometers are useful tools for measuring activity with such fine temporal resolutions, and thus can reveal a golfer's skill by its miniaturization and shockproof structure. Figure 3(a) shows a swing monitoring application tool using a number of accelerometers. Three different ranges of piezoresistive MEMS accelerometers were attached on the trunk ($\pm$29.4m/s$^2$), left forearm ($\pm$490m/s$^2$), and driver head ($\pm$980m/s$^2$), as shown in Figure 3(b).

Experiments on various levels of golfers show clear differences between professionals and amateurs. When the down swing began on all professional golfers, their X-axis acceleration at

*Figure 3. (a) Golf swing diagnosing application (b) Acceleration sensor location*

(a)

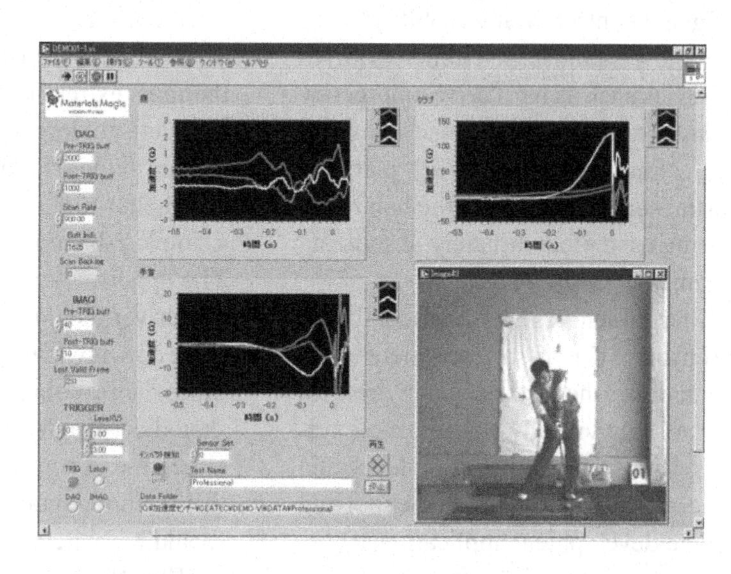

(b)

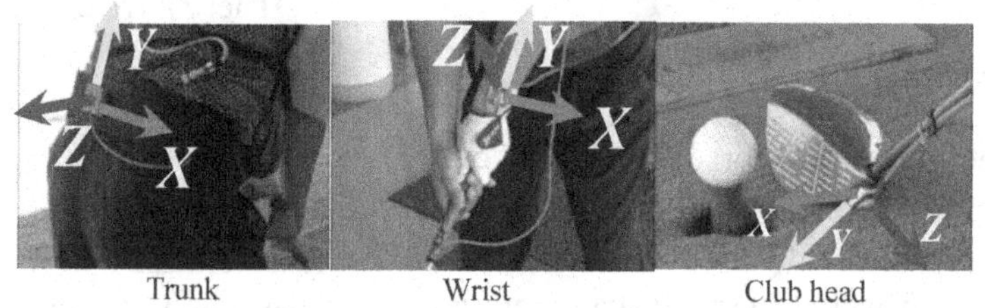


Trunk Wrist Club head

the hip decreased and kept negative. Then, -70 to -30 ms before the impact, it changes to positive. The reason why X-axis acceleration of the hip changed to positive value is that the professional golfers might decelerate in a backward direction in order to prevent their body moving forward. On the other hand, with the amateur golfers, the transition timing from negative to positive fluctuated depending on their each shot between -40ms to -15ms. Late stop motion makes transition time delay. According to Jorgensen (1999), such delay of the golfer's deceleration of sway motion is likely to cause insufficient energy transfer from the trunk to both the upper extremity and the club.

Inoue (1997) and Umeguki et al. (1998) argued that golfer's uncock leads the kinetic energy to transfer into the golf club during the down swing. The rotational velocity of the upper extremities decreases steeply as a consequence. In the wrist accelerometer (attached onto the distal end of the forearm), the Y-axis acceleration of the wrist, which was along with the longitudinal axis, had a centrifugal acceleration as a major component. Under this configuration and in a local coordinate system, the centrifugal acceleration indicates negative value. Therefore, a decrease of the rotational angular velocity of the forearm causes the increase of the Y-axis wrist acceleration. Our experiment validated these postulates. In all professional golfers, their wrist Y-axis acceleration during the down swing phase had a local minimum at -100ms before impact. On the other hand, the amateur golfer's Y-axis acceleration at the wrist kept decreasing until the impact. Thus uncock operation was unexecuted so that upper extremity and the club were tightly connected and rotated together. It is suggested that the wrist joint torque was likely to be kept even after the appropriate uncocking timing in the down swing phase. The most important skill of this energy transfer is that the uncock must be carried out with appropriate timing. Therefore, the local minimum of the Y-axis acceleration at the wrist could be used for the indication of the release of wrist cocking.

## Rowing

The use of under boat impellers to record rowing scull velocity is common practice, though the additional drag is of concern to rowers, especially in competition. More recently, 3-D accelerometers have been used to aid more precise stroke phase determination, and GPS has also been employed for more precise position determination. Output from the platform to a palm top computer is facilitated using a serial port, enabling immediate visual feedback to both coach and athlete.

Key metrics like stroke rate can be derived from real-time data and displayed on the palm top computer for immediate feedback to the athlete. It can also be viewed using a telemetry module for immediate (and post event) use by coaches and sports scientists. Post activity analysis and stroke diagnosis has been useful for teams to compare inter team differences, thus helping to refine and develop race strategies. Figure 4 shows basic accelerometer traces with the release and catch phases clearly identified. Statistical information derived from these phases of motion are useful benchmarks for coaches; they have traditionally been difficult to obtain without purpose instrumented boats and/or video analysis.

*Figure 4. Acceleration activity data recorded during rowing activity*

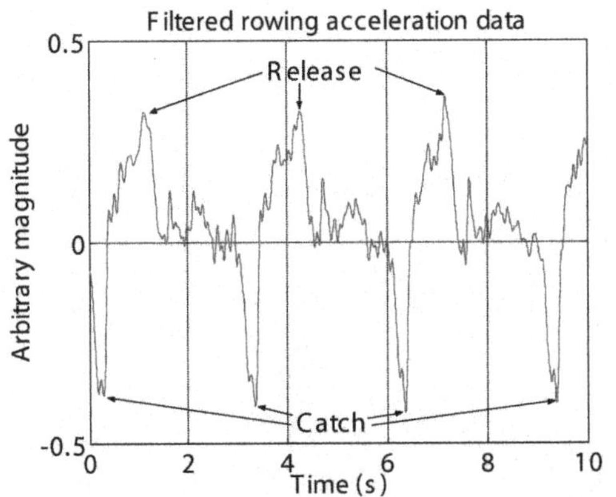


## Cricket (Bowling and Batting)

Cricket, a sport with a long history and great tradition, has only recently embraced the use of technology for the purpose of enhancing the performance of both players and officials. In the last 20-30 years, technological innovations have entered the sport range, from the relatively simplistic use of video replays to adjudicate in run-out decisions and the adjustment of target scores in rain affected matches (Duckworth & Lewis, 1998), to detailed biomechanical analyses of the legality of bowling actions (Lloyd, Alderson & Elliott, 2000).

In the area of performance assessment and enhancement, a considerable amount of research has been undertaken using 3-D tracking technology, such as the Vicon system™, which enables the accurate localization of athletes and their equipment by means of wearable visual markers, in conjunction with a system of high-speed cameras. Whilst extremely accurate, the use of such a system is limited in that it can only be operated in a controlled laboratory setting, and not for performance monitoring of athletes in competitive situations. The use of inertial sensors, which are small, lightweight and relatively unobtrusive, can bridge this gap, allowing cricket players to be monitored in both training and match environments.

**Batting Analysis**

Striking a cricket ball effectively is a complex movement involving many parameters, and has traditionally been evaluated in a purely subjective manner based on the visual assessment by expert observers. Such metrics are prone to errors due to differences in observers, the inability of the human visual system to accurately perceive high-speed motion, and other human factors. It is also not unusual for different batsman to have markedly different batting techniques, which further reduces the accuracy of human assessment. Inertial sensors allow this complicated motion to be analyzed objectively, and a number of useful parameters to be assessed, such as the time taken in various parts of the swing; the power of the shot; the direction of the shot; and the quality of the swing (Busch & James, 2007).

Tri-axial accelerometers, when placed at multiple positions on the bat itself, allow for calculation of a number of useful parameters, such as rotational velocity at any point in the swing; time of impact with the ball; and the angle of the bat blade, which has a strong correlation with the final direction of the shot. Figure 5 shows the typical output of such a sensor, and the events that can be readily extracted from the data. When collected over the duration of a training session, such information allows players and coaches to analyze a batsman's technique, identify problems which are unnoticeable to the human eye, and provide a quantitative way of measuring improvement over time.

By adding additional sensors to the bat, it is also possible to accurately determine the position on the bat struck by the ball: a very useful metric for assessing the ability of players in various conditions (Busch & James, 2007). This is accomplished by timing the arrival of the shock wave created by the ball's impact at each of the sensor locations, and using the differences between them to assess the approximate position. Accuracies of up to 2cm have been reported using this technique, although in practice such precision is largely unnecessary, as impact zones are generally placed into broad categories, such as "high on the bat", "low on the bat", and "middle of the bat".

**Bowling Analysis**

The analysis of the bowling actions of cricketers has come under heavy scrutiny in recent times, with many international level players being accused of using actions that contravene the laws of

*Figure 5. Data extracted from a typical shot. Note the small dip in acceleration in the x and z axes (solid and dotted lines respectively) at the tie of impact, and the large spike during the swing. Twisting of the bat at the point of impact and follow-through is noted in the y axis (dashed line).*

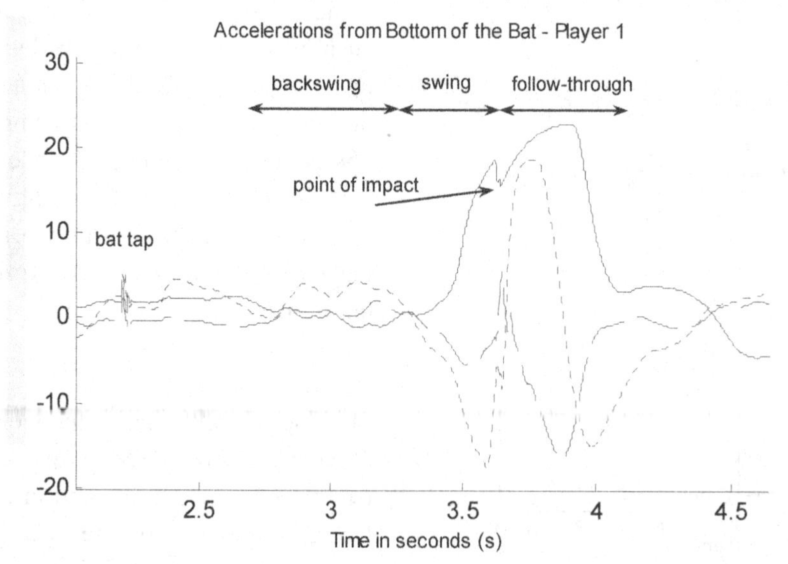

the game. The response of cricket's administrators to this problem has been varied over the years. Traditionally, the on-field umpires were the sole judges of a bowler's action, and would indicate such by calling "no-ball" when, in their opinion, the bowler's arm straightened perceptibly during ball delivery. The highly subjective nature of this approach lead to great inconsistency, and due to this the International Cricket Council (ICC) introduced a new testing procedure, whereby bowlers whose actions were identified as suspect by umpires or match referees were instructed to undertake additional, off-field biomechanical testing. Such testing is typically carried out using motion capture systems in an artificial environment, and allow the bowler a limited degree of elbow extension (at the time of publication this limit is fixed at 15 degrees). While such testing is accurate, it is time-consuming, expensive, and does not allow for bowlers to be tested during match conditions. Inertial sensors are currently being investigated as a possible solution to this problem, as they are lightweight enough to be worn by a bowler without hindrance, and accurate enough to provide a good estimate of elbow extension during delivery. With a suitable wireless solution, a complete online system is also possible, enabling instantaneous feedback to officials on the legality of each ball bowled.

## Ski Jump

Ski jumping has recently attracted fluid dynamics scientists, because of its highly complex flight phase and potential for aiding performance. A number of kinematic and kinetic methods have been developed as aids for field studies (Babiel, Hartmann, Spitzenpfeil & Mester, 1997; Kaps, Schwameder & Engstler, 1997; Komi & Virmavirta, 1997; Yamanobe & Watanabe, 1999). A popular method for kinematics is image analysis; further force platform data analysis during the last few seconds on the in-run slope has been examined. For the flight phase, experimental wind tunnel tests have also provided useful information (Seo, Murakami & Yoshida, 2004; Seo, Watanabe & Murakami, 2004). Most of these studies were conducted on the slope or in the air. Total kinematical and kinetic analysis from the start, in-run, take-off, steady flight, to landing is something that has been sought, but difficult to achieve in the literature. Inertial sensors have recently been applied to this problem to provide further information (Ohgi, Seo, Hirai & Murakami, 2006, 2007).

Figure 6 shows an example of the inertia sensor data in ski jump. Each event such as start (ST), take off (TO) and landing (LA) are indicated. After 0.5 second, the accelerations $A_x$ and $A_z$ were very stable, and the landing event is very clear. $A_x$, which is along with jumper's spine, was almost zero. On the other hand, $A_z$, which is along with the jumper's front to back, was kept 7 to 8 $m/s^2$ during his flight phase. The angular velocity $\omega_y$ changed negative, then it changed positive at the take off moment. Then, it kept small positive value during the flight phase around 20 to 30 deg/s. Before the landing, it changed rapidly positive to negative for the preparation to the telemark position. From the video analysis, the initial trunk angle and jumper's flight trajectory were obtained. This initial trunk lean angle $\theta_0$ was used for the time integration of the angular velocity.

In order to understand fluid dynamics of ski jumping, both the global and local coordinate systems must be defined. In addition, a simple mathematical model should be adopted in these coordinate systems. In Figure 7, U indicates a velocity vector of the center of gravity of the jumper-ski model. For our convenience, we assumed the hip joint to be a center of gravity. And our sensor unit was also assumed to be attached on the same location. $\theta$ is the trunk angle to the horizontal plane, $\alpha$ is the angle between trunk and the velocity vector U. Then, flight angle $\beta$ is defined by $(\alpha—\theta)$.

Aerodynamic forces acting on the ski jumper are drag and lift. These forces can be described

*Figure 6. Inertia sensor data, accelerations $A_x$ and $A_z$ and the angular velocity $\omega_y$*

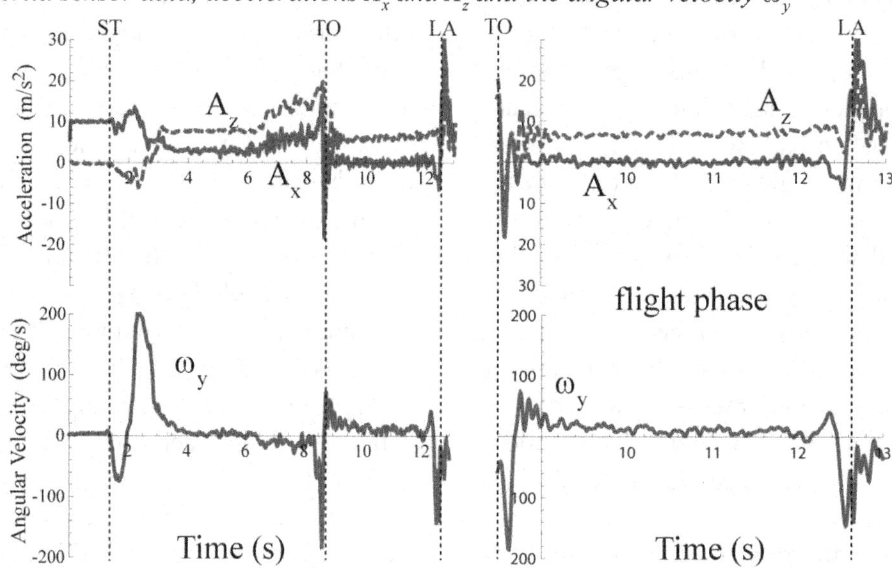


on both the global or local coordinate systems. We can express them on the global coordinate system as follows:

$$D = -m(-\ddot{x}\cos\beta + \ddot{z}\sin\beta + g\sin\beta) \quad (2)$$

$$L = m(\ddot{x}\sin\beta + \ddot{z}\cos\beta + g\cos\beta) \quad (3)$$

These equations can then be transformed onto a local coordinate system. A number of assumptions for the obtained acceleration data make this transformation easier, particularly $A_x$ and $A_z$. We suppose that the center of gravity is equivalent to the hip joint and our co-located sensor device. Thus the radius of rotation on the local coordinate system is assumed to be zero (**r** = **0**). Hence we hypothesize that obtained acceleration has only translational and gravitational accelerations as its components. The forward tilting angle, θ is obtained by the time integration of the gyroscope, the period of integration is sufficiently short to avoid integration error. Therefore, obtained acceleration is formulated as follows:

$$A_x = \ddot{x}\cos\theta + \ddot{z}\sin\theta - g\sin\theta \quad (4)$$

$$A_z = -\ddot{x}\sin\theta + \ddot{z}\cos\theta - g\cos\theta \quad (5)$$

Aerodynamic forces, drag and lift are then obtained by using eq.(1) and eq.(2). Figure 8 shows an example of estimated drag and lift forces by the inertia sensors, when compared to video image analysis methods. The sample rates from the inertial sensors are much higher and so the data shows considerably more detail.

*Figure 7. Ski jumper-ski model and its coordinate system*

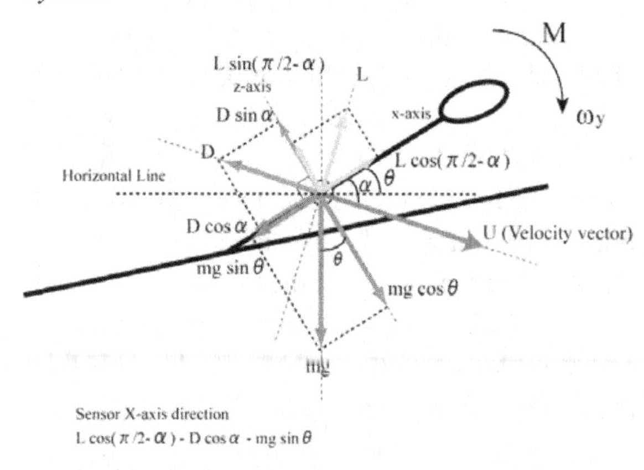

*Figure 8. Estimated aerodynamic forces by the inertia sensor method and the video analysis method*

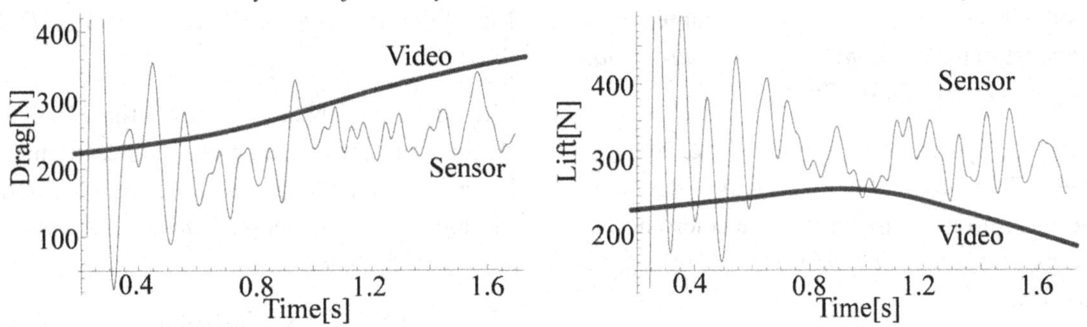


## CONCLUSION

This chapter has introduced the technology of inertial sensors, its function and implementation, together with its use in the popular market place and a number of emerging sporting applications. As a valid technique in their own right, inertial sensors are still very much in the research and early adopter phase, though commercial and pre commercial devices are emerging. These sensors represent another tool in the bag of the sports scientist, not to replace other tools, but as a compliment to them. The opportunities for athlete assessment through the application of these wearable sensors in the training and competition environments are still being discovered.

Whilst a brute force approach using inertial sensors for sporting applications has yielded mixed success, the application of signal processing of the data and the use of sport specific knowledge has allowed the extraction of key features in the data, which can then be interpreted in a useful manner. Critical to the success of the sample applications has been to ensure that the use and development of the sensors is in direct partnership with key stakeholders, including engineers, athletes, coaches and sport scientists. This has kept the technology development and interpretation firmly grounded on providing useful outputs that directly benefit athletes, which is critical to enhancing sporting activity.

## ACKNOWLEDGMENT

"Investigating the Australia-Japan Connection" is supported by the Commonwealth through the Australia-Japan Foundation of the Department of Foreign Affairs and Trade.

## REFERENCES

Babiel, S., Hartmann, U., Spitzenpfeil, P., & Mester, J. (1997). Ground reaction forces in alpine skiing, cross-country skiing and ski jumping. In E. Muller, H. Schwameder, E. Kornexl & C. Raschner (Eds.), *Science and skiing* (pp. 200-207). USA: Taylor & Francis.

Busch, A., & James, D.A. (2007). Analysis of cricket shots using inertial sensors. In F. Fuss, A. Subic & S. Ujihashi (Eds.), *The impact of technology on sport II* (pp. 317-322). Netherlands: Taylor & Francis.

Cutmore, T.R.H., & James, D.A. (1999). Identifying and reducing noise in psychophysiological recordings. *International Journal of Psychophysiology, 32*(2), 129-150.

Davey, N.P., James, D.A., & Anderson, M.E. (2004). Signal analysis of accelerometry data using gravity based modeling. In *Proceedings of SPIE: Vol. 5274* (pp. 362-370). USA: SPIE.

Duckworth, F.C., & Lewis, A.J. (1998). A fair method of resetting the target in interrupted one-day cricket matches. *Journal of the Operational Research Society, 49*(3), 220-227.

Herren, R., Sparti, A., Aminian, K., & Schutz, Y. (1999). The prediction of speed and incline in outdoor running in humans using accelerometry. *Medicine & Science in Sports & Exercise, 31*(7), 1053-1059.

Inoue, Y. (1997). A study on dynamics of golf swing. In *Proceedings of Dynamics and Design Conference* (pp. 99-103). Japan: Japanese Society of Mechanical Engineering.

James, D.A., Davey, N., & Rice, T. (2004). An accelerometer based sensor platform for insitu elite athlete performance analysis. In *IEEE Sensor Proceedings*. Vienna: IEEE.

Jorgensen, T.P. (1999). *The physics of golf.* USA: Springer.

Kaps, P., Schwameder, H., & Engstler, G. (1997). Inverse dynamic analysis of take-off in ski jumping. In E. Muller, H. Schwameder, E. Kornexl, & C. Raschner (Eds.), *Science and skiing* (pp. 72-83). USA: Taylor & Francis.

Kloeck, B., & de Rooij, N. (1994). Mechanical sensors. In S.M. Sze (Ed.), *Semiconductor sensors* (pp. 153-204). New York: Wiley.

Komi, P.V., & Virmavirta, M. (1997). Ski-jumping take-off performance: Determining factors and methodological advances. In E. Muller, H. Schwameder, E. Kornexl & C. Raschner (Eds.), *Science and skiing* (pp. 3-26). USA: Taylor & Francis.

Lai, A., James, D.A., Hayes, J.P., & Harvey, E.C. (2003). Application of triaxial accelerometers in rowing kinematics measurement. In *Proceedings of SPIE: Vol. 5274* (pp. 531-542). Bellingham, WA: SPIE.

Lloyd, D.G., Alderson, J., & Elliott, B.C. (2000). An upper limb kinematic model for the examination of cricket bowling: A case study of Muttiah Muralitharan. *Journal of Sports Sciences, 18*(12), 975-982.

Mayagoitia, R., Nene, A., & Veltink, P. (2002). Accelerometer and rate gyroscope measurement of kinematics: An inexpensive alternative to optical motion analysis systems. *Journal of Biomechanics, 35*(4), 537-542.

Moe-Nilssen, R., & Helbostad, J. (2004). Estimation of gait cycle characteristics by trunk accelerometry. *Journal of Biomechanics, 37*(1), 121-126.

Montoye, H.J., Washburn, R., Servais, S., Ertl, A., Webster, J.G., & Nagle, F.J. (1983). Estimation of energy expenditure by a portable accelerometer. *Medicine & Science in Sports & Exercise, 15*(5), 403-407.

Motamedi, M.E., & White, R.M. (1994). Acoustic sensors. In S.M. Sze (Ed.), *Semiconductor sensors* (pp. 97-151). New York: Wiley.

Nagao, N., & Sawada, Y. (1977). A kinematic analysis of the golf swing by means of fast motion picture in connection with wrist action. Journal of Sports Medicine & Physical Fitness, *17*(4), 413–419.

Ohgi, Y., Seo, K., Hirai, N., & Murakami, M. (2006). Measurement of jumper's body motion in ski jumping. In E. Morriz & S. Haake (Eds.), *The Engineering of Sports, 6*, 275-280. USA: Springer.

Ohgi, Y., Seo, K., Hirai, N., & Murakami, M. (2007). Aerodynamic forces acting in ski jumping. *Journal of Biomechanics, 40*, S402.

Ohta, K., Ohgi, Y., Kimura, H., & Hirotsu, N. (2005). *Sports data.* Japan: Kyoritsu Publishing.

Seo, K., Murakami, M., & Yoshida, K. (2004). Optimal flight technique for V-style ski jumping. *Sports Engineering, 7*(2), 97-103.

Seo, K., Watanabe, I., & Murakami, M. (2004). Aerodynamic force data for a V-style ski jumping flight. *Sports Engineering, 7*(1), 31-39.

Umegaki, K. et al., (1998). Influences of magnitude of vertical rotational torques and timing of grip torque generation on golf swing motion. In *Proceedings of the Symposium on Sports Engineering* (pp. 111-115). Japan: Japanese Society of Mechanical Engineering.

Weinberg, H. (1999). Dual axis, low g, fully integrated accelerometers. *Analog Dialogue, 33*(1), 1-2.

Williamson, R., & Andrews, B.J. (2001). Detecting absolute human knee angle and angular velocity using accelerometers and rate gyroscopes. *Medical and Biological Engineering and Computing, 39*(3), 294-302.

Yamanobe, K., & Watanabe, K. (1999). Measurement of take-off forces in ski jumping competition. *Japanese Journal of Biomechanics in Sports and Exercise, 3*(4), 277-286.

# Chapter VIII
# Computer Supported Collaborative Sports: An Emerging Paradigm

**Volker Wulf**
*University of Siegen, Germany*

**Florian 'Floyd' Mueller**
*The University of Melbourne, Australia*

**Eckehard F. Moritz**
*SPORTKREATIVWERKSTATT, Germany*

**Gunnar Stevens**
*Fraunhofer FIT, Germany*

**Martin R. Gibbs**
*The University of Melbourne, Australia*

## ABSTRACT

*Augmenting existing sports experiences with computing technology is increasingly gaining attention due to its potential for performance enhancement. However, most of these approaches focus on existing single-user activities. The authors are presenting the newly emerging field of Computer Supported Collaborative Sports (CSCS) to draw attention to the social aspect of sport and its potential to support novel experiences for players that are not available in traditional sports environments. They discuss important dimensions in the design space of CSCS by detailing two example applications and lay out further research directions for the design of collaborative technologies in computer augmented sports.*

## INTRODUCTION

Computer games have turned into a popular form of entertainment. An increasing number of people are playing computer games, making it one of the most rapidly growing leisure activities. When asked for the most fun entertainment activities, 35% of Americans mentioned computer and video games outranking alternatives such as watching television, surfing the World Wide Web, reading books, or going to the cinema (IDSA). Since their introduction, computer games have fascinated its users and drew people's attention. However, the success of computer games has been watched critically. Controversial game content, social isolation of players and the promotion of sedentary lifestyles are major concerns with regards to computer games.

Quite a number of computer games deal with shooting or killing activities. An often expressed criticism in regard to this type of games is based on the assumption that killing activities within games will lead to an increased aggressive behavior in daily life (Rauterberg, 2003). While empirical investigations with regard to this hypothesis show heterogeneous results (Fritz & Fehr, 1997), the design of ethically less questionable, but equally fascinating game content can be a challenge.

Critics have pointed out that intense use of computer games may lead to social isolation of the players (Provenzo, 1991). However, social arrangements such as playing single user games in a group or LAN (Local Area Network) parties where multi-user games are played in physical proximity can compensate for this problem. Some computer games address this issue by allowing playing together across geographical distances.

Another problematic issue with regard to computer games is the lack of physical activity when playing – in stark contrast to the 'physical' content of many games: most game content involves muscled heroes who perform intense exerting physical activity, quite different to the player in front of the screen. The typical input devices of computer games are game pads, keyboards and mice, unsuitable for promoting physical activity. Output is typically provided to the players by auditory and graphical means (e.g. loudspeakers and screens). The research area of Ubiquitous Computing has begun to introduce new input and output technologies which are also applicable for games (Björk, Holopainen, Ljungstrand, & Mandryk, 2002). Some approaches have taken sportive activities like skateboarding and karate as a platform and augmented them with information technology. By doing so, existing sports activities can experience an additional 'game content' (Ishii, Wisneski, Orbanes, Chun, & Paradiso, 1999; Mokka, Väätänen, & Välkkynen, 2003; F. Mueller, Agamanolis, & Picard, 2003).

With our contribution, we want to get one step beyond by further integrating computer games and computer augmented sports. We postulate the approach of Computer Supported Cooperative Sports (CSCS). By leveraging innovative input and output technologies we believe we can offer users new experiences in shared computationally augmented game environments.

## OVERVIEW

This article is structured as follows: First, we will present related work in computer games that use augmented sportive interfaces. Then we will outline the concept of Computer Supported Collaborative Sports. Two prototypes of this design paradigm will be presented: the FlyGuy offers flight experiences in shared 3D spaces and Table Tennis for Three offers tangible game play in a mixed-reality environment for three distributed players. We will conclude by discussing our findings in regards to future applications of the design space and the role of CSCS for emerging distributed sports activities.

### Ubiquitous Games and Computer Augmented Sports

Ubiquitous computing offers a relatively new approach of interacting with computers through real world objects and spaces, which can provide novel opportunities for innovative games and physical experiences. For example, the 'STARS' environment offers a platform to implement different board games on a computer augmented table. Real world objects, such as chess figures, can be moved on the board and their positions can be tracked. Based on this input, a game engine can compute appropriate output behaviour (Magerkurth & Stenzel, 2003). Based on similar input technologies, Harvard and Lovind (Harvard & Løvind, 2002) have developed toys based on a rather different conceptual idea. They try to encourage storytelling by moving away from the computer screen and take physical objects (typically simple plastic toys) as an interface that permits the exploration of the quirks of a story. Stories can be recorded and attached to different toys and their actual position.

A different approach is taken by Sanneblad and Holmquist (Sanneblad & Holmquist, 2003). They distribute a game area onto several handheld computers in a way that the whole area can only be seen by means of all the different displays. The players have to move towards each other to perform gaming activities, e.g. controlling Pac-Man in the classic arcade game in those parts of the game area which are not represented on their personal handhelds. In this case physical activities of the players result from the need to see the entire game area.

Other approaches record human movements in order to navigate in virtual environments. Humphrey II, developed by the Futurelab in Linz, is a flight simulator where the user emerges into a 3D virtual space by means of a head mounted display. The behaviour of an avatar representing the users can be controlled by means of arm movements. In the Virtual Fitness Center (Virku) an exercise bicycle is positioned in front of a video screen. The physical movements conducted on the exercise bicycle are used as input to modify the representation of 3D virtual environments from map information. Reversely, the map information affects the pedaling efforts. In an early implementation the players move this way along a hilly landscape in Finish Lapland (Mokka et al., 2003).

Other approaches address collaborative sporting activities explicitly. They can be understood as early instances of CSCS research. AR2 is an augmented reality airhockey table with a virtual puck. The two players wear head-mounted displays to see a virtual puck on the table in front of them (Ohshima, Satoh, Yamamoto, & Tamura, 1998).

Airhockey over a Distance (F. Mueller, Cole, O'Brien, & Walmink, 2006) uses a physical instead of a virtual puck for distributed gameplay: the puck is shot out at the remote end by puck cannons whenever the player hits the local puck across the middle line.

The Wii® game console comes with a controller that contains accelerometers to support physical activities in its games, and force-feedback is provided through subtle vibration in the controller. Although such exertion games are achieving commercial success, they have been criticized for not being comparable to the sports activities they are simulating (Graves, Stratton, Ridgers, & Cable, 2007). For example, Wii Tennis does not facilitate the same energy expenditure and therefore similar physical health benefits than a traditional game of tennis. However, computationally augmenting such activities can offer novel experiences, such as supporting distributed participants.

Dance Dance Revolution (DDR) is a physical game that requires players to follow dance instructions on a screen. The players' movements are detected by sensors embedded in the 'dance platform' that forms the stage the players are performing on. This game can be very exhausting, and early investigations indicate that it can contribute to an understanding of music-based characteristics in CSCS applications (Behrenshausen, 2007).

PingPongPlus is a system which augments traditional table tennis by means of a tracking device for the ball and a video projector. Different applications have been designed which project images on the table according to the location where the ball hits the table. When a ball hits the table in the "*water ripple mode*", an image of a water ripple appears from the spot the ball landed (Ishii et al., 1999). Mueller et al. (2003) have developed a system called "*Breakout for Two*" which allows players to interact remotely through a life-size videoconference screen using a regular soccer ball as an input device. Both players kick the ball against their local wall on which an audio and video connection with the other player is displayed. By tracking the position where the ball hits the wall various games can be added on each player's side via an overlay technique.

While there are quite some interesting developments in the ubiquitous and entertainment computing fields, the sports engineering community has not captured the full potential of computer-augmented sport devices, we believe. Most research is still restrained to analyse and model traditional sport devices or aspects of the human body (for a good summary see Subic & Haake, 2000; Ujihashi & Haake, 2002). Respective contributions are often only to be found in training science, with a specific purpose to use the computing technology to achieve particular training objectives, or in rehabilitation in which the technology is used to support regaining specific physical capabilities (Powell, 2008).

## COMPUTER SUPPORTED COOPERATIVE SPORTS

Computer Supported Cooperative Sports investigates the design of computer applications which require sportive input activities to gain collective game experiences (F. Mueller & Gibbs, 2007; F. Mueller, Stevens, Thorogood, O'Brian, & Wulf, 2007). It is an interdisciplinary research field where sports engineers, computer scientists, designers, sport scientists, and social scientists need to cooperate, guided by a systematic design approach (Moritz, 2004). In the following section, we elaborate on the concept and discuss important aspects of the design space for such CSCS applications.

### Integrating Sports and Games

In the following section, the hermeneutic and practical core of sports and games and their implications will be identified and related to one another. Sports in a traditional understanding has been defined as "*organized play that is accompanied by physical exertion, guided by a formal structure, organized within the context of formal and explicit rules of behaviour and procedures, and observed by spectators*" (Anshel, 1991, p.143). Still widely spread, this formalizing definition coerces sports into a specific scheme and strangely strangles the scope for innovation with respect to social and individual use value. However, there are also more context-sensitive approaches, defining sports as a "*specific expression of human movement behaviour*" (Haag, 1996, p.8) that becomes "sports" only by "*a situation-specific reception and an attribution of meaning*" (Heinemann, 1998, p.34). Eventually it is the purpose an individual assigns to a movement which she/he considers being sportive (which in many cases encompasses 'physical exertion'), that defines sports. Reasons to do sports include fun, health, socializing with others, maintaining fitness, and compensating for sedentary occupation (Meyer, 1992; Moritz & Steffen, 2003).

'Doing' sports and playing games have many similarities, especially the voluntary character of the activities motivated by a perception of fun. In the domain of computer games, sport genres have already been utilized: players can simulate sport competitions, such as soccer championships, on their computer. The aim of CSCS, however, is not to simulate sports activities, but to offer the opportunity of 'doing' sports.

### Input and Output Devices for Sports Activities

An important dimension in the design space of CSCS is the type of sports activity which shapes the input and output interface to the computer augmented environment. If sport is defined by the meaning individuals assign to the involved body movements, it is possible to imagine a wide scope of different activities. If we either presuppose the objective of increasing fitness levels or at least aim to minimize long-term physical harm, one of the essential requirements would be the balancing of external load distribution, for example not to require an over-utilization of the biceps while offering no stimulation to the triceps. Practical technical and bio-mechanical considerations and the wish to monitor progress suggest a reduction of movement complexity to a simple combination of translational and rotational movements – in which, however, one might have to compromise a 'natural' feeling while moving around in a virtual world.

With regard to the design of the input interface the question arises how to register sport activities appropriately. If this cannot be done by monitoring movements and forces within the device directly, e.g. the actual engine torque, then sensors would need to become an essential part of the design. These sensors can either measure the movements of the human body (e.g. stirring and pedaling an exercise bike) or of different types of sport tools (e.g. the ball in table tennis or the stick in hockey).

With regard to the design of the output interface in a distributed game environment one has to think of how to represent the activities of other actors and the physical texture of virtual space. This can either happen merely visually or also physically by means of forced feedback. For instance, in the Virku environment the physical texture of the virtual landscape translates into different levels of required pedaling efforts.

## Collaboration

The concept of collaboration in CSCS environments requires some discussion. Sports, like many game genres, seem to imply competition either among individuals or among teams. However, in dancing or acrobatics it is the feeling of being together in combination with (joint) movements that people are aiming at. So, in principle, CSCS can be centred on cooperation or competition. Hence, the meaning of collaboration in CSCS can span the whole spectrum from multi-user competitive settings (e.g. computer-augmented table tennis or a bicycle race in a virtual 3D environment), to settings of mere co-presence (e.g. playing soccer individually in a shared audio and video space or riding bicycles together in a virtual space) and settings where cooperation is needed to achieve the common goals (e.g. moving in a game area distributed via different handhelds or producing output loads that are converted into a stimulating input for the partner at a remote location).

From a computer science perspective, collaborative settings can be classified along the time-space dichotomy (Johansen, 1988). With regard to the design space, players in CSCS applications can either interact in the same place or at remote locations. With regard to time, most of the applications in the field of entertainment computing are synchronous in the sense that the players interact with each other at the same time. However, asynchronous applications such as community systems may help to shape social relationships among players. Seay, Jerome, Sang Lee, & Kraut (2003) and Friedl (2003) describe how synchronous and asynchronous computer mediated communication such as chat and email can be integrated into Massive Multiplayer Online Games. Friedl (2003) stresses the importance of asynchronous features. Web pages allow, for example, the displaying of information about player's performances in past games that is available at any time.

Another important dimension with regard to collaboration is the question of whether the players know each other beforehand or whether they form a social bond within the game environment. In the latter case specific technical features may be needed to introduce or match human actors (Al-Zubaidi & Stevens, 2004). Friedl (2003) points out that personal information and information about players' performance can stimulate social interactions.

## Objectives and Vision

CSCS emerges in an interesting intersection of sports, game and innovative technologies. It may help to tackle problems which are of imminent importance to individuals and the society as a whole:

- **Animated fitness equipment:** Has the potential to enhance motivational factors to improve health and fitness, and to maintain such commitment by combining exertion with diversion (and diversity).
- **Animated fitness worlds:** Could combine play, sports, and fitness: A leisure attraction may create an opportunity to get kids away from stationary computer gaming, and thus to fight obesity and social isolation.
- **Computer controlled sports equipment:** Could allow monitoring movements and performance, adapting training and rehabilitation, and enable remote supervision.
- **Computer enhanced sports equipment:** May offer further understanding of the realms of emotions and feelings in sports, especially through combining movements and visual displays, in contrast to purely mechanical sports equipment.
- **Computer supported collaborative sports equipment:** Could link people together to engage in collaborative physical activities. This could enhance motivation and open up new social channels for friends, strangers or even distributed teams.

To arrive at these objectives, however, high demand is put on how to conduct respective research and development projects. A project team heterogeneously assembled with engineers, computer scientists and sports experts will have to combine their competencies, guided by a systematic approach to innovation in sports, and backed up by a distributed project management. Initial pilot projects in this area have been conducted and will be reported upon in the next sections.

## THE FLYGUY APPROACH

We have developed a concept called FlyGuy for an innovative CSCS device which combines fitness training with playful challenges, social interaction, and versatile entertainment. The work was conducted in a multidisciplinary design team which consisted of researchers and

*Figure 1. First sketches of the FlyGuy and design alternatives of the frame which holds the human actor when flying. © Springer-Verlag Berlin Heidelberg 2004. Used with Permission.*

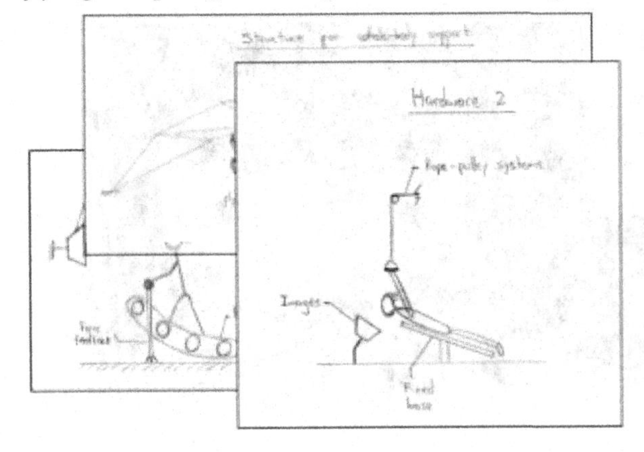


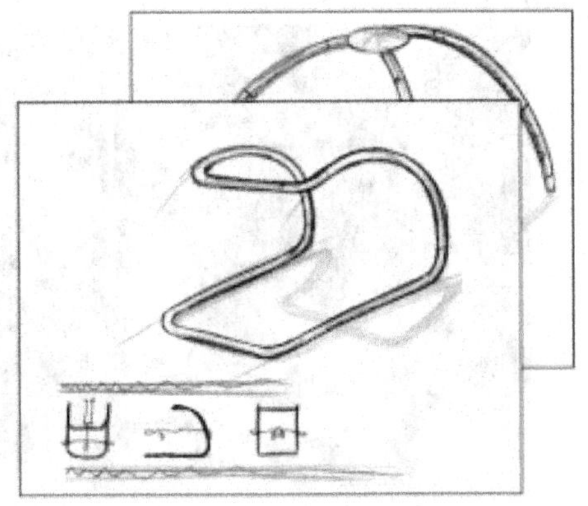

students from Germany, Japan, Mexico, and the United States.

In contrast to PingPongPlus (Ishii et al., 1999) or Breakout for Two (F. Mueller et al., 2003), we wanted to design a collaborative environment for new physical experiences and sports activities. In a first face-to-face meeting of the project team, flying was identified as an interesting sports activity because humans can only experience this in a computer augmented environment or by means of specific avionic devices such as hang gliders.

We have therefore created the following concept: the player immerses via a head mounted display into a 3D virtual environment. She/he controls a flight simulation through her/his body motions. In a first explorative implementation two handles need to be grabbed by the hands; the flight direction can be changed via rotation of the torso, the height by pulling or pushing a lever horizontally (Figure 1). One of the reasons we chose to realize the flight movement in this 'starfighter' fashion was that it appeared to be the most natural way to the test persons we asked to 'fly' on a small table structure. A training effect is intensified by providing resistance for both concentric and exocentric movements; thus it is possible to realize extreme intensity and quick exhaustion. In further stages, we plan to include leg movements for acceleration and deceleration. The motions are captured by sensors located in the joints of the lever structure and transformed into electrical signals which are then being transmitted to a microcontroller and a PC. The data is used as input to control the flight simulation which is perceived by the player via a head mounted display. The player is hanging in a frame made of aluminum similar to the frame of a hang glider.

In the virtual space, the player has the possibility to solve different flying tasks and meet other persons and fly and exercise with them, even if they are in a geographically distant location in the real world. Whenever the players reach certain proximity in the virtual space, an audio channel is opened to allow for communication.

For creating the virtual environment we explored different popular 3D game engines and opted to tailor an existing game like Half-Life II for our purpose. This also supports addressing the need to arrive at a sufficient user base for efficient usage of the system, as it makes it easier to integrate other players which do not have the FlyGuy device, but can play with conventional hardware.

After detailing the concept, the team separated again and worked on its realization (mechanics, mechatronics, network structure, virtual environment, output devices, biomechanics, game plan, sports scientific aspects, etc.) in a distributed fashion.

In a second face-to-face meeting a functional prototype was assembled and tested. This prototype was built to explore technical design issues

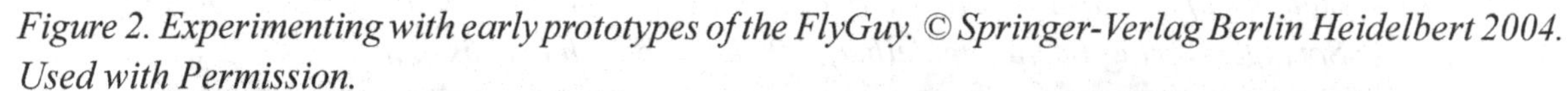
*Figure 2. Experimenting with early prototypes of the FlyGuy. © Springer-Verlag Berlin Heidelbert 2004. Used with Permission.*

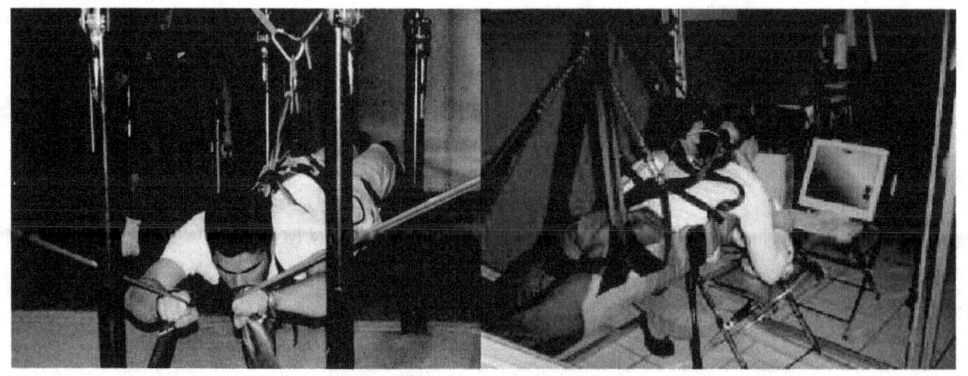

such as the location of the hanging anchor point and the fixture of the lever structure. Further functional design issues were explored such as the steering mechanisms (Figure 2). We also investigated which degrees of freedom and what kind of support is necessary to provide the basis for efficient and safe training.

We evaluated our concepts internally during the design process. It turned out that some aspects of the prototype's design worked out better than others. For example the horizontal flight posture and the steering of the flight simulation were rated positively and intuitive, while the usage of a fixed and stiff lever structure was regarded suboptimal because it does not match the idea of 'free' flight motions. Other aspects which need to be improved are the overly complicated access into the device and the lack of adaptability with regard to different user anthropometries.

## TABLE TENNIS FOR THREE

Another prototype of the CSCS paradigm is 'Table Tennis for Three'. Table Tennis for Three is a tangible game that uses a real ball, bat and table but supports players in geographically distant locations. It is aimed at providing a health benefit by encouraging physical activity and training reflexes as well as hand-eye coordination. Just like table tennis, it is easy to learn and supports a sense of achievement quickly. Through the inclusion of a videoconference, the aim is to support similar benefits known from traditional casual table tennis play such as exercise, enjoyment and bringing people together to socialize.

Table Tennis for Three does not only overcome the need for collocation between participants, it also demonstrates the scaling opportunity of the CSCS concept by supporting three players simultaneously in three different locations, offering another example of a novel sports experience.

### Gameplay

Each player has a ball, a paddle and a table tennis table (Figure 3). Game play involves hitting the table tennis ball with the paddle against a backboard. This backboard is one half of a table tennis table, which is usually pushed together with the other half to create the playing surface. By tipping one of these halves from the horizontal to the vertical position it is possible for players to play the ball against the backboard created. This setup is familiar to table tennis players who have

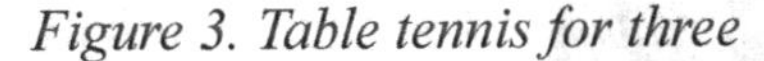

*Figure 3. Table tennis for three*

practiced by themselves by repeatedly bouncing the ball off the backboard with their paddle.

This backboard has projected images of eight large 'bricks' on it. These bricks are identical for all players, i.e. they are synchronized across all three stations. These bricks are semi-transparent and are projected onto the backboard with a projector mounted to the ceiling. In addition to the bricks, it also projects two video streams of the other players in the game (Figure 4). One player is positioned on the left of the backboard, and the other on the right. Each table has a set of loud speakers and each player wears a microphone so the three participants can converse with each other.

The backboard is equipped with sensors mounted on the back that detect when and which brick the players are hitting. These bricks 'break' when hit by the ball because the sensors register the location of the impact. All three players see the same brick layout and the same brick status. If a brick is hit once, it cracks a little. If it is hit again (regardless by which player), it cracks more. The crack appears on all three stations (Figure 5). If hit three times, it 'breaks' and is removed from play, revealing more of the underlying videoconferencing: the player 'broke' through to the remote player. However, only the player that hits the brick the third and final time receives the point. This helps to make the game more interesting

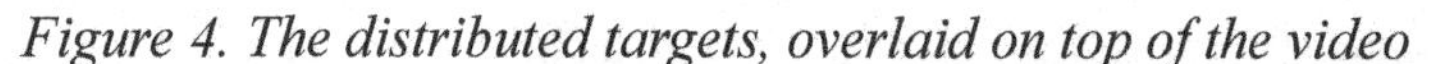
*Figure 4. The distributed targets, overlaid on top of the video*

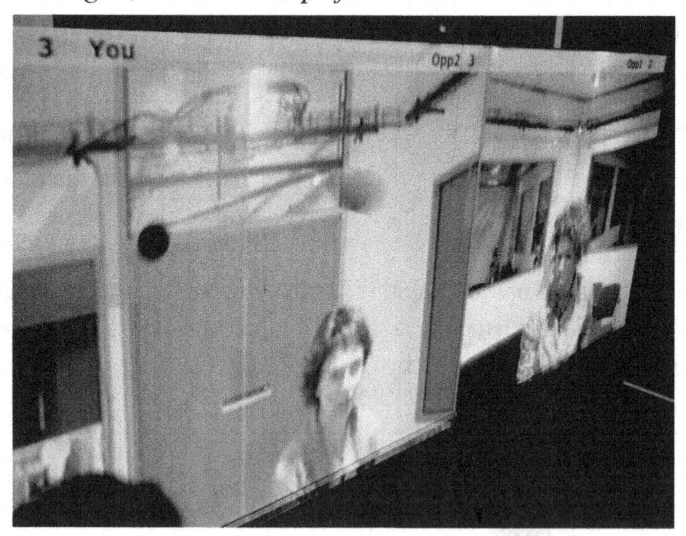


*Figure 5. The blocks are shared across the stations, a hit is visible to all players*

because it offers players a number of strategies for winning the game. They can either try to crack as many bricks as possible by placing the ball quickly or they can poach points from other players by waiting for the opportunity to snatch away points through hitting bricks that have been already hit twice by the others.

Each brick that is completely broken scores one point, and the running score is displayed along the top end of the projection.

Play continues until all bricks have been cracked three times and been removed from play. At this point the player who has scored the most points is announced as the winner and after a delay of 15 seconds, the game resets all the bricks and play can recommence.

## Technical Implementation

We operated Table Tennis for Three in three separate rooms connected via a LAN network connection. The backboards were instrumented so that the time and approximate location of a ball striking the table could be detected. Eight piezoelectric sensors were attached to the rear of the backboard in locations corresponding to the gameplay blocks projected on the front of the backboard (Figure 6). The sensors detect the sound vibrations in the wooden board created by a ball striking it. This approach is similar to the system described by Ishii et al. (1999), however, we were not able to achieve a highly accurate system with four sensors (which should cover the entire surface through interpolation), and therefore opted for the use of eight sensors.

The one sensor that receives the vibration signal first, exceeding a certain threshold, determines the location of the impact. After an analogue to digital (A/D) conversion and data acquisition, software concludes which of the bricks should be cracked. This information is sent to software that updates all other stations using client-server architecture. Each station then updates the graphical content accordingly, and synchronizes game data such as the score. A camera was placed in the centre of the upper edge of each backboard. This camera was used to capture and send video conferencing streams from each play station to the other two tables.

The videoconferencing implementation is deliberately kept independent from the technical gameplay component in order to provide an optimal videoconference experience. Developing a videoconferencing system is not a trivial task, and

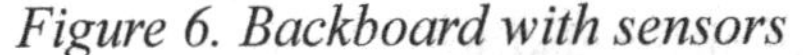
*Figure 6. Backboard with sensors*

many open-source and commercial systems claim to offer the best compromise between bandwidth limitations and image and audio quality. These software (some of them are hardware) implementations balance the most effective compression codecs with en- and decoding CPU requirements, deal with varying network throughputs, provide circumventions for firewall issues, and minimize noise- and echo effects. In order to be able to always utilize the latest advances in videoconferencing technology, we implemented the Table Tennis for Three gameplay independently and placed it on top as a separate half-transparent layer. This ensures that any researcher who wants to recreate the system can take advantage of their existing videoconferencing infrastructure and is not locked into a proprietary system that might be outdated quickly.

## User Experiences

An evaluation with 41 participants using observational data, questionnaires and interviews indicated that the participants enjoyed playing Table Tennis for Three and that they could see such a CSCS game being helpful in facilitating rapport between people who are physically apart but want to stay in touch (F. Mueller & Gibbs, 2007 ). In particular, they expressed a strong sense of "*playing together*" and commented on the fact that it "*gave them something to talk about*". The physicality of the game allowed participants to quickly engage and interact, and most players reported that they had fun, considered it a workout, forgot the world around them when playing, and wanted to play again. This user study strengthened our approach that the CSCS concept can contribute to a sense of social bond between geographically distant players while offering increased fitness incentives. Designers of CSCS games might be interested in knowing that our participants liked to practice their skills beforehand, showing 'practice' behavior comparable to traditional sports. However, at least two participants of Table Tennis for Three also reported on a negative experience. Both players mentioned they had trouble understanding the other players over the audio channel which was probably one factor that affected their experience. Such results shed light on future work needed on the CSCS concept, such as investigating the role of verbal support amongst participants. Such type of research will lead to further design recommendations for applications which support distributed sports experiences across multiple locations.

## CONCLUSION

We have described the concept of Computer Supported Collaborative Sports and presented two prototypes that highlight the sportive and collaborative aspects of such computer augmented activities. The attempt to transfer the excitement of computer games to motivational aspects for fitness training and thereby allowing joint physical activities by partners far apart might mark an important trend in the future of entertainment computing and fitness-oriented sports.

- The introduction of computers into cooperative sports equipment does not only offer new areas of application for computers in entertainment but also opens up new dimensions in sports and fitness:
- There is a whole array of novel means to increase motivation to participate in sportive or health-sustaining activities.
- Linked via the internet, people in different locations can do sports, share physical fun or follow rehabilitation exercises together.
- The development of virtual worlds and connected input-output devices could offer sensory and emotional sensations that cannot be experienced otherwise; 'flying' being just one example.

To explore the design space out-lined in this paper, we need to implement a variety of different CSCS devices. The more we depart from sport activities given already in the physical world, the more effort need to be spent on the design process. While the input activities in the case of Table Tennis for Three were still rather close to their origin, in case of the FlyGuy a new repertoire of sportive movement had to be invented.

Moreover, we need more profound empirical evaluations of CSCS applications. We have collected so far mainly data on the devices' short term appropriation. However, long term data is needed to better understand how motivation develops and whether the intended health effects can be detected. We also need to better understand how different types of players appropriate these applications.

## ACKNOWLEDGMENT

The work on the FlyGuy was supported by the Japanese Ministry of Education in the project Tele Athletics New Experience (TANE). The work was conducted at the Technical University of Munich (Germany), Stanford University (USA), University of Siegen (Germany), Tokyo Institute of Technology (Japan), Tsukuba University (Japan), Universidad de las Americas (Puebla, Mexico), and University of Magdeburg (Germany). The Table Tennis for Three system is based on work initially supported by a University of Melbourne and CSIRO Collaborative Research Support Scheme. We thank Shannon O'Brien, Matt Adcock, Bo Kampmann Walther, Keith Gibbs, Irma Lindt, the IDEAS Lab, the Interaction Design Group and Ivo Widjaja. The still pictures of Table Tennis for Three were taken by Kerin Bryant.

## REFERENCES

Al-Zubaidi, K., & Stevens, G. (2004). *CSCP at Work.* Paper presented at the Proceedings of the Conference Mensch und Computer 2004 (MC 2004).

Anshel, M. A. (Ed.). (1991). *Dictionary of the Sports and Exercise Sciences.* Illinois.

Behrenshausen, B. G. (2007). Toward a (Kin) Aesthetic of Video Gaming: The Case of Dance Dance Revolution. *Games and Culture, 2*(4).

Björk, S., Holopainen, J., Ljungstrand, P., & Mandryk, R. (2002). Special Issue on Ubiquitous Games. *Personal Ubiquitous Comput., 6*(5-6), 358-361.

Friedl, M. (2003). *Online Game Interactivity Theory.* Hingham: Charles River Media.

Fritz, J., & Fehr, W. (Eds.). (1997). *Handbuch Medien und Computer-Spiele.* Bonn.

Graves, L., Stratton, G., Ridgers, N. D., & Cable, N.T. (2007). Comparison of energy expenditure in adolescents when playing new generation and sedentary computer games: cross sectional study. *BMJ, 335*(7633), 1282-1284.

Haag, H. (Ed.). (1996). *Sportphilosophie.* Schorndorf: Verlag Karl Hofmann.

Harvard, Å., & Løvind, S. (2002). "Psst"-ipatory Design. Involving artists, technologists, students and children in the design of narrative toys. In T. Binder, J. Gregory & I. Wagner (Eds.), *Proceedings of the PDC 2002 Participatory Design Conference.* Malmö.

Heinemann, K. (1998). *Einführung in die Soziologie des Sports.* Schorndorf: Verlag Karl Hofmann.

IDSA. Interactive Digital Software Association. Retrieved at http://www.idsa.com

Ishii, H., Wisneski, C., Orbanes, J., Chun, B., & Paradiso, J. (1999). PingPongPlus: Design of an Athletic-Tangible Interface for Computer-Supported Cooperative Play. In *Proceedings of CHI'99* (pp. 394-401). New York: ACM-Press.

Johansen, R. (1988). Current User Approaches to Groupware. In R. Johansen (Ed.), *Groupware* (pp. 12-44). New York: Freepress.

Magerkurth, C., & Stenzel, R. (2003). Computerunterstutztes Kooperartives Spielen -- Die Zukunft des Spieltisches. In J. Ziegler & G. Szwillus (Eds.), *Proceedings of Mensch & Computer 2003 (MC ,03), Stuttgart*: Teubner.

Meyer, M. (1992). *Zur Entwicklung der Sportbedürfnisse.* Unpublished Dissertation, German Sports University, Cologne.

Mokka, S., Väätänen, A., & Välkkynen, P. (2003). Fitness Computer Games with a Bodily User Interface, *Proceedings of the Second International Conference on Entertainment Computing* (pp. 1-3). Pittsburgh, Pennsylvania: ACM International Conference Proceeding Series.

Moritz, E. F. (2004). *Systematic Innovation in Popular Sports.* Paper presented at the 5th Conference of the International Sports Engineering Association.

Moritz, E. F., & Steffen, J. (2003). Test For Fun – ein Konzept für einen nutzerorientierten Sportgerätetest. In K.v. Roemer, J. Edelmann-Nusser, K. Witte & E.F. Moritz (Eds.), *Sporttechnologie zwischen Theorie und Praxis* (pp. 43-63). Aachen.

Mueller, F., Agamanolis, S., & Picard, R. (2003). *Exertion interfaces: Sports over a Distance for Social Bonding and Fun.* Paper presented at the CHI 2003.

Mueller, F., Cole, L., O'Brien, S., & Walmink, W. (2006). *Airhockey Over a Distance – A Networked Physical Game to Support Social Interactions.* Paper presented at the Advances in Computer Entertainment Technology ACE 2006.

Mueller, F., & Gibbs, M. (2007). *A Physical Three-Way Interactive Game Based on Table Tennis.* Paper presented at the IE '07: Proceedings of the 4th Australasian conference on Interactive entertainment (2007), RMIT University, Melbourne, Australia.

Mueller, F., & Gibbs, M.R. (2007). *Evaluating a distributed physical leisure game for three players.* Paper presented at the Proc. of Australia on Computer-Human interaction OZCHI '07.

Mueller, F., Stevens, G., Thorogood, A., O'Brian, S., & Wulf, V. (2007). Sports over a Distance. *Personal and Ubiquitous Computing (PUC), 11*(8), 633–645.

Nintendo. WiiSports, http://wii.nintendo.com/software_wii_sports.html

Ohshima, T., Satoh, K., Yamamoto, H., & Tamura, H. (1998). AR2 Hockey, *Conference Abstracts and Applications of SIGGRAPH'98* (pp. 268). New York: ACM-Press.

Powell, W. (2008). *Virtually Walking? Developing Exertion Interfaces for Locomotor Rehabilitation.* Paper presented at the Paper presented at the CHI 2008. Workshop submission to "Exertion Interfaces". Retrieved from http://workshopchi.pbwiki.com/f/PowellW_Exertion08V1.2.pdf

Provenzo, E.F. (1991). *Video Kids: Making Sense of Nintendo.* MA: Harvard University Press Cambridge.

Rauterberg, M. (2003). *Emotional Aspects of Shooting Activities: 'Real' versus 'Virtual' Actions and Targets.* Paper presented at the 2nd International Conference on Entertainment Computing (ICEC 2003), Pittsburgh, PA.

Sanneblad, J., & Holmquist, L.E. (2003). *Designing collaborative games on handheld computers.* Paper presented at the SIGGRAPH 2003 Conference on Sketches & Applications.

Seay, A., Jerome, W., Sang Lee, K., & Kraut, R. (2003). *Project Massive 1.0: Organizational Commitment, Sociability and Extraversion in Massively Multiplayer Online Games.* Paper presented at the LEVEL UP Digital Games Research Conference.

Subic, A.J., & Haake, S.J. (Eds.). (2000). *The Engineering of Sport 3*. Cambridge.

Ujihashi, S., & Haake, S.J. (Eds.). (2002). *The Engineering of Sport 4*. Cambridge.

# Chapter IX
# Digital Sport:
## Merging Gaming with Sports to Enhance Physical Activities Such as Jogging

**Florian 'Floyd' Mueller**
*The University of Melbourne, Australia*

## ABSTRACT

*Recent advances in computing technology have contributed to a new trend that merges digital gaming with physical sports activities and combines the advantages of both; such as contributing a health benefit and supporting distributed participants. This chapter describes prominent examples, and their underlying theoretical concepts and perspectives. In particular, it presents a design prototype, "Jogging over a Distance," which offers social joggers the opportunity to run together, although being in two different locations. This approach demonstrates the potential for the merging of computer gaming technology with sports activities, to offer combined effects that have traditionally been limited to each respective domain. Future work on enhancing existing sports and gaming activities will support novel experiences previously not possible. This exciting new field has the potential to enhance users' lives by making positive health contributions.*

## INTRODUCTION

Digital computer games, from their early beginnings in arcades to the multi-million dollar titles of today, follow rule-sets; often allow for interactions; and are predominantly high-score focused. Furthermore, digital games encourage their players to improve upon their skills in a competitive fashion. Very similar characteristics are attributed to sports. Also, gamers can become members of internationally organized clubs and participate in prestigious gaming tournaments, similar to the ones that exist in professional sports (Pedersen, 2002). Such activities have been labeled "online sports" or "e-sports". Further, there is a voluntary characteristic associated with both sports and computer games, and both are believed to be motivated by a perception of fun (Wulf, Moritz,

Henneke, Al-Zubaidi & Stevens, 2004). It appears that physical sports and computer gaming have many aspects in common.

However, what differentiates computer gaming from traditional sports is its lack of support for physical exertion. Unlike sports, computer games do not encourage the use of gross motor skills, as their focus is mostly on fine motor skills. They do not intend to exert players physically, because they do not demand physical effort to reach the game's goal. This is very different to sport's focus on training and mastering the human body, which dates back to the ancient Greek's celebration of the body and its movements. Computer games involve predominantly the development of cognitive skills and neglect involving the body in its supported activities. Players are only required to press buttons on a keyboard, game pad, or mouse in order to control an avatar's movements, even though these avatars perform actions that, in reality, would require great strength and endurance. The mapping of users' actions is not proportional to the represented activities. This lack of inclusion of physical activities is characteristic for computer games, but that means gamers also miss out on the benefits of physical activity.

Sports have many advantages, including physical, social and mental health benefits. From a physical perspective, sports can contribute to a healthier body by reducing the risk of obesity, cardiovascular disease and diabetes (Pate et al., 1995). From a social and mental health viewpoint, sport is believed to teach social skills (Morris, Sallybanks & Willis, 2003), encourage team-building, and support individual growth, as well as community development (Gratton & Henry, 2001). Some argue sport can foster social integration and personal enjoyment (Long & Sanderson, 2001; Wankel & Bonnie, 1990); provide opportunities to meet and communicate with other people; and contribute positively to self-esteem and well-being (Bailey, 2005). These are benefits that also contribute to the growth of social capital (Huysman & Wulf, 2004; Putnam, 2000). Sports activities can facilitate bonds between people, resulting in loyalty and team spirit. Team sports, in particular, are considered to be character building. Sports clubs not only function as a place to exercise, but also as a social space (Putnam, 2000). International sporting events also demonstrate that sports have the ability to overcome the language barrier and bring people together from various cultural backgrounds.

In summary, sports offer many benefits. The health and social benefits are of particular focus here. Computer games share some of these concepts, but fall short in offering a fitness aspect, due to interaction mechanisms that focus on fine-motor skills. Their role in facilitating social support is ambivalent. However, computer game technologies have now emerged that allow for novel social experiences, such as participating in shared activities between geographically distant players.

The objective of this chapter is to introduce the reader to an emerging approach in the amalgamation of digital games technology and sports activities. By combining the unique advantages from computer gaming technologies (such as supporting geographically distant participants) and the health benefits of sports, new experiences can be facilitated that offer mutual benefits to users, previously only available in each area.

Computing technology has been used in sports applications before. However, such work has mainly focused on supporting performance enhancement of professional athletes. Less work has been undertaken on using computing technology to enhance the experience of very differently fit sports people, and to support the social benefits associated with sports activities. The work presented here aims to demonstrate that there is potential for computer gaming technology to support a wide range of sports participants and even offer new experiences.

To highlight the feasibility of this approach, a system called "*Jogging over a Distance*" is described. It demonstrates how gaming technologies can be combined with sports activities, and in particular, how the benefits of networking advances can be utilized in sports settings to enable athletes to participate in shared sports activities, although being geographically apart.

## BACKGROUND

From a computer gaming technology perspective, players recently have seen an increase in what has been labeled "exergaming": a portmanteau of "exercise" and "gaming". Computer games that make the player sweat have emerged, labeled exertion games. Game companies have discovered the potential of embracing physical activities in their games and using it to entice players to buy new hardware. Systems such as the EyeToy™ (EyeToy Kinetic, 2008) use advances in camera and vision analysis technology to detect body movements of the participants, who perform bodily actions in front of a screen to control an avatar. The Wii™ game console from Nintendo® comes with a controller that contains accelerometers to support physical activities in its games, and force-feedback is provided through subtle vibration in the controller (Nintendo, 2008a).

Exertion games are believed to be able to work against the prevailing computer gaming image of facilitating a modern world's sedentary lifestyle. These games are attracting new audiences that have previously not been catered for, including novices outside the traditional "young and male" gamer demographic (Snider, 2008). It has also been reported that gamers have appropriated existing technology, such as Nintendo's Wii™Sports, to supplement their fitness goals. However, without health progress reports, gamers have been using blogs and household scales to supplement their gaming equipment, in order to track their health improvements and combat obesity issues. New fitness products for the living room have emerged that promise more effective and motivating workouts through interactive status reports, with personalized feedback and progress report functionality (Nintendo, 2008b).

The convergence of sport and gaming has challenged our understanding of what is a sport. The Dance Dance Revolution® game, which requires jumping on dance pads according to screen-based instructions, has been approved by the International Dance Organization as an official dance sport discipline named "machine dance" (IDO – MachineDance European Championships, 2006). Dance Dance Revolution® is probably the most successful exertion game worldwide; it has even been integrated into physical education curricula in the USA and fitness gyms (Behrenshausen, 2007).

Other commercial exertion game systems can be plugged into game consoles to replace the joystick or game pad interaction. Typical examples are stationary exercise bikes or treadmills with a connection to a computer. Sensors measure the speed of the wheel (or belt) on the fitness equipment, and the computer converts this data into commands typically performed with a traditional game console controller (Gamebike, 2008). So the faster the user pedals on the bike, the faster the car in the computer game drives. Other similar devices utilize existing exercise equipment, such as the Peak Training System Sensor Kit (2008) that can be retrofitted to any rowing machine or treadmill. Some of these interfaces take a simplistic approach to encouraging exercise: only if the user engages in physical activity (regardless of speed), the computer game becomes active and is controlled with a traditional game pad, but if the user stops exercising, the game pauses (Gamersize, 2008). The Bodypad™ (2008) supports exerting body activity as input control through pressure sensors that are strapped onto the hands and legs, replacing the button presses. The Powergrid (Interaction Laboratories, 2008) uses isometric force, in contrast to pushing and pulling weights most other exercise equipment affords. The user exhibits force against a sturdy metal pole, which does not move, however sensors inside measure the applied intensity and convert it into game control commands. The exercise bike in "Virku" takes the exercise bike-systems a step further by adding a feedback channel: the digital world reacts to the physical effort the user puts in, but the virtual world also affects the exertion activity in return. By pedaling, the user travels through a virtual version of a rural park area. The pedaling speed affects the player's traveling speed in the environment, but the game environment also affects the pedaling effort e.g. riding uphill increases the required pedaling effort and downhill decreases it (Mokka, Väätänen & Välkkynen, 2003).

Academic prototypes and commercial products have been developed that consider social aspects in motivating users to exercise and aim to leverage people's relations with one another for an enhanced experience. For example, websites such as Fitsync.com support athletes in tracking their exercise data, but also offer to share the results with other sportspeople for comparison purposes (Fitsync, 2008). Social networking websites such as Bodyspace™ aim to leverage the motivational power of social connections (Bodyspace, 2008). Sports that emphasize the visual appearance of the body, such as bodybuilding, can benefit from such sites, as they offer self-presentation capabilities: they allow members to upload pictures of themselves performing in their sport. However, these websites are limited by the existence of a temporal difference between the sports activity and its technological support. In other words, only after the workout is finished and the pictures taken can the body-builder upload the images and exchange messages with peers, usually hours later, in a different location and context. Fitsync.com allows for the use of PDAs and mobile phones during a gym workout to reduce the time between exercise and digital support mechanisms. However, there is still a conceptual separation between the exercise and the persuasive element, as the physical activity does not experience any technological augmentation, and the social augmented aspect is, although influenced, not dependant upon the physical activity.

The Nike+ system (Apple, 2008) on the other hand offers a continuous augmentation of the physical activity: it adds entertainment technology to a jogging experience and combines it with a performance tracking feedback channel. A sensor in the jogger's shoe tracks running pace and sends it to the user's iPod®, where it is stored until it is synced with a computer. Through connection to a website, the results can be shared with thousands of other runners worldwide. Virtual runs can be arranged and other runners challenged for specific time or distance goals, offering a form of asynchronous running competition. However, the pace data is also available during the run, and the software in the iPod is programmed so that it uses this performance measure to influence the runner's workout on-the-fly: the entertainment usually provided via music through the headphones fades out if the runner reaches a personal running record, and the recorded voice from Lance Armstrong compliments the jogger on reaching their goal. Pressing a button on the iPod® not only displays the runner's speed; a voice, augmenting the entertainment channel, also informs the runner about the remaining time, distance and calories burnt, giving instant feedback anytime during the activity.

Other pace measurement devices exist that claim to be motivational through the use of peer support, or other computing technology. The *MPTrain* is a mobile device that monitors heart rate and speed of a jogger to select music, with a particular tempo delivered through headphones to encourage the user to slow down, speed up, or keep pace (Oliver & Flores-Mangas, 2006). Other prototypes that use social interaction to encourage walking or jogging include *Houston* (Consolvo, Everitt, Smith & Landay, 2006) and *Chick Clique* (Toscos, Faber, Shunying & Mona Praful, 2006). *Houston* is a mobile phone application that monitors step count and displays it alongside the step count of friends. *Chick Clique* is a similar mobile phone application for sharing step count. This social peer pressure approach focuses on teenage girls and uses instant messaging to keep the social group connected and aware of each other's progress. *Shakra* (Barkhuus et al., 2006) supports physical activity awareness in a mobile setting, and the authors report on the beneficial aspect of competitive progress exchange as encouragement to exercise more. *Melodious Walkabout* is a headphone based system that assists joggers in finding their way by using directional audio. It plays music files to guide the user in the right direction using GPS data (Etter & Specht, 2005). Another device that incorporates the user's activity to affect their audio is the "Are We There Yet?" system (Adcock, 2008), which modifies the playback speed of audio books according to how much travel time remains for the user. If the user increases his or her speed, resulting in the estimated time to destination to be sooner than anticipated, the playback of the audio book increases so that the read-out story ends just when the user arrives.

## THEORETICAL FOUNDATIONS

The theoretical foundations for this merger of physical actions from sports with computer gaming technologies has been influenced by psychological investigations in emotional experience and bodily signs of emotion, which has been used to explain exergaming phenomena (Lehrer, 2006). One of the early questions under investigation in this field was whether emotions (such as fear and excitement) follow bodily signs (such as facial expressions and tears), or whether it is the other way round (Lehrer, 2006). Understanding this can inform the design of interactive physical experiences. A phenomenological approach to understanding the body's role has led to recent investigations in proprioception, suggesting a continuous bi-directional feedback loop between the mind and the body. Damasio (2000) calls it an "embodied mind", in contrast to the exclusively embrained mind: a concept that emerged out of an earlier, rationalistic tradition of understanding interactive technology (Winograd & Flores, 1987). Merleau-Ponty (2002), in his investigations on phenomenological thinking, described the consciousness, the world and the human body as a perceiving thing that is intertwined and mutually "engaged". This view considers a correlate of the body and its sensory functions. Having a theoretical lens like this implies that a manager for exertion games should not ask one designer to be responsible for the gameplay, while another designer is assigned the interface, but rather these two tasks go hand-in-hand and should be considered as one entity.

Such an approach has further implications. "An emotion begins as the perception of the bodily change" (Lehrer, 2006). Therefore, moving around in an exercise activity can generate the perception of increased heart rate and sweating. This means in order to prepare a participant for an upcoming exertion interaction, the brain triggers a wave of changes in the physical viscera, such as a quickening of the pulse and the releasing of adrenaline. Once such a physical experience has started, these effects are exaggerated, because the muscles need oxygenated blood (Lehrer, 2006). Supporting this constant interaction or loop between brain and body is believed to create a more emotionally engaging experience for exertion games, compared to traditional screen-based games (Lehrer, 2006). Designers who are aiming for an emotional experience can leverage these findings in their designs. A more general recommendation is the inference that the mind-body loop calls for a more holistic approach in our understanding of combined sports and computer game experiences.

## OPPORTUNITIES

An emerging research field has begun to leverage the advantages that can result from a holistic approach to considering the body and the mind in computer based activities. Sports actions are augmented with pervasive computing technologies and borrow from computer game principles to help users improve their skills. Computer game activities are supplemented by involving bodily actions to control and immerse players in their environments, while simultaneously creating more emotionally rich experiences that have a significant impact on players' lives. This approach goes beyond adding an exertion component to existing computer games, but rather aims to create physical experiences that are not possible without the computer augmentation, similar to the "beyond-being-there" approach in videoconferencing. This work by Hollan and Stornetta (1992) realized that videoconferencing equipment, until then, was mostly designed with the aim of recreating a collocated experience as closely as possible. They proposed that an augmented tele-experience would never be as good as face-to-face, and therefore designers should not try to emulate this, but rather focus on creating compelling experiences that offer benefits "beyond" what is possible face-to-face. The same concept underlies the approach in augmenting exertion interactions: it is not designed to replace existing sports activities, but rather offer opportunities that are not possible without the technological addition.

This chapter argues that by creating awareness and sensitivity to the unique opportunities sports activities can offer, better interactive experiences can be designed. Novel experiences in augmented online environments can be offered, instead of trying to simulate existing ones. For example, researchers and practitioners have the opportunity to retain the benefits of physical exertion in interactive environments, but also capitalize on the ability of computer games to support geographically distant participants, through telecommunication advances. The chapter now turns to one example, an original prototype, which follows this framework of utilizing gaming technologies and principles to support sports activities. It is a jogging support system for casual joggers who value the motivational and social benefits of jogging together, but are separated by distance.

## JOGGING

### Social Jogging

Jogging is one of the easiest ways of participating in sports activity in terms of requirements, because the jogger does not need any special equipment and running can be done almost everywhere. Jogging can also be a beneficial health exercise for participants of all levels. In particular, joining others to jog is often recommended to increase and sustain participation rates. Jogging in teams allows for socializing, and runners use each other's commitment for motivation. For example, joggers motivate one another to run faster and further by challenging each other's strength. This contributes to more effective exercise.

Ubiquitous computing technology to motivate joggers already exists. For example, heart rate monitors, pedometers and MP3 players are commonly worn by joggers to augment their experience. However, these devices focus on the individual and provide the individual a way of tracking progress. Here I present a prototype called "*Jogging over a Distance*" that offers experiences traditionally known from gaming such as distributed player support and spatialized audio, delivered via entertainment means, but applied to a sports setting. The conceptual themes and applied frameworks that led to the described prototype are detailed, along with the technological challenges faced in using gaming technologies for sports use. These included consideration of computing devices suitable for rugged outdoor use, programming efficiency and audio spatialization techniques for voice communication in un-tethered environments, as well as reliable acquisition and adequate tracking of performance data in order to offer a satisfying augmented sports experience.

The system, *Jogging over a Distance*, is aimed at runners who want to run together but are geographically apart. It is facilitated by a ubiquitous distributed audio environment that retains the outdoor exercise experience, while aiming to support the social and motivational benefits. It is acknowledged that some joggers prefer running alone; however, in this work, the focus is on joggers who enjoy running with others.

Internet forums indicate that joggers often run with others (O'Brien, Mueller & Thorogood, 2007). A survey conducted with 77 joggers confirms this, with 57% of respondents stating they run with at least one other person. The top reasons for running with others include socializing (83%), motivation to run faster (78%), motivation to participate (53%), and to have fun (53%). Many social joggers value the ability to have conversations with their partners and use their exercise sessions as a way to stay in touch with their friends. One respondent noted,

*About twice a month I run with some of the girls I went to college with. It's a great time to chat and catch up! Even though we see each other and chat regularly, we always seem to talk more openly while we run." Another participant gave an example of the benefits he received from running with a partner: "I ran on Sunday with another runner, and she wanted to add a little more distance to the route. We talked about it as we ran and agreed where to run. I ran more than I would have if I ran by myself. After the run, I was glad that I did the extra mileage. Also, my running companion ran faster than I would have in the early part of the run (I actually had to ask her to slow down a little for the first mile), and I think I pushed her at the end of the run. It was mutually beneficial.*

A frustration participants have with social jogging is finding the "right" jogging partner: one who can meet them at the same location and who jogs at roughly the same pace. This challenge of finding a partner often results when people move away, or when one partner becomes faster than the other due to training. One jogger explained that he only has one friend with whom he can run, but his friend moved across the country and "*now I know of no one my age who runs the way I do... many run longer and a lot run shorter... I still wish I knew people to run with to shake things up a bit.*" Another recently re-located runner stated, "*I run alone, [but] I wish I could find a couple of people to run with but haven't had much luck in finding a running partner since I moved two years ago.*"

For casual joggers, being able to hold a conversation can be an indicator that they are running at a suitable pace: not too fast and not too slow for an optimal health benefit. This is often referred to as the "Talk Test" (Porcari et al., 2002). While social jogging can motivate people to run faster and further than solo jogging, partners should have roughly the same physical capabilities in regard to both speed and distance. Some runners may run alone, simply because they have yet to find a jogging partner.

An important yet challenging aspect of social jogging is therefore finding jogging partners who run at the same pace and who live nearby. An opportunity exists to overcome this challenge and enhance the sports participation of joggers.

## Mobile Support

To support social communication between joggers, which is motivational for their exercise, a solution featuring an audio connection between the joggers was selected. An audio interface suits a mobile, outdoor environment: it is simple, lightweight, and allows users to visually focus on their environment. Furthermore, an audio interface supports outdoor running; unlike treadmill-based systems, the proposed system is wearable and un-tethered.

## Preliminary Experiment

To investigate the feasibility of an audio channel to support joggers and their social interactions, an experiment was conducted, in which a jogger was equipped with a small laptop in a backpack; a headset; and a wireless internet connection over a mobile 3G data network. The runner used Voice over IP (VoIP) to transmit his speech. The other person did not run, but sat in her office about 800 kilometers away, connected via the VoIP link. Of initial interest was testing the technical feasibility of the proposed setup; it was found that it is feasible to use a commercial wireless system to support a fast-moving jogger across a large area. But what was intriguing was the remote participant's report that she had a sense of sharing the other person's running experience, and through the conversation and noise from the outside environment, was able to visualize what the other runner was possibly seeing. This indicated that running with a remote person has the potential to be engaging.

## Mobile Phone Experiment

In order to understand the experience joggers would have if they could communicate with a remote running partner through an audio channel (Figure 1), 18 volunteers were asked to go running at the same time, but in opposite directions, equipped with a mobile phone and a Bluetooth® headset (O'Brien et al., 2007). The vivid sense of presence the audio conveyed to the participants, confirmed the first experiment: participants not only mentioned hearing the other person's voice, but also the wind, the noise of the footsteps depending on the different ground surfaces, and the breathing of the remote jogger. As a result, the combined acoustical information created a social and enjoyable experience.

Considering the system was simply applying existing technology to a new context, it was surprising how much participants enjoyed their run. On a scale of 0-100 (with 100 being best) participants ranked, on average, their enjoyment level of running alone as 55, their enjoyment of jogging side-by-side as 79, and their enjoyment with the mobile phones as 75. Although this is not a reliable measure, the high rating is promising.

*Figure 1. Jogging with a mobile phone connection*

One participant explained, "*There were times when I was just jogging along like I always jogged and chatting away like I always chat away and it was more or less exactly the same as running with someone.*" Another participant stated, "*It had the advantage of running with the other person and I could run where and how I wanted to run [...] you had almost the same experience because you were constantly communicating with them.*"

Knowing how fast they and their partner were going was important for half of the participants. For one participant, this kept her running. She explained, "*There's some pride that you don't want to stop. I thought about stopping a bit today, and that would have been easier, because [my partner] wasn't there, but I didn't know if she could tell over the phone, so I didn't try.*" One participant suggested the partners could carry some tracking devices and then verbally tell each other their speeds, which he felt would greatly improve his experience. Participants seemed interested in knowing their partner's pace in order to help them endure the distance of the jog, by either cheering them on to go faster, or motivating them to keep going.

These results prompted the development of a prototype to push further the idea of jogging "together" with geographically distant partners. This system is called "*Jogging over a Distance*".

## JOGGING OVER A DISTANCE

One possible solution to finding social jogging partners is to enable people to jog with remote friends and other remote joggers (Figure 2). With *Jogging over a Distance*, jogging partners could live in opposite parts of the world, yet share the experience of jogging together. By meeting at the same time in separate locations, long distance friends could become, or stay, social jogging partners.

The prototype created not only supports conversation, but also uses audio to communicate pace. The *Jogging over a Distance* system therefore supports the motivational aspect of being able to compare each other's pace. Similar to jogging side-by-side and adjusting pace with one's partner, the *Jogging over a Distance* prototype transforms the conversation into spatialized audio to simulate hearing one's partner in front, to the side, or behind. The idea is that the information can contribute to an increased awareness of the other person's presence, hence creating a shared sportive experience.

*Figure 2. Jogging together although geographically apart*

### Design Goals

The prototype aims to allow two joggers to experience a run "together", similar to the experience of jogging side-by-side. The design is focused on the following points:

- To retain the experience of jogging, including its health and social aspects, the system is mobile and can be used outside. The workout as well as the social experience is greatly enhanced by being outside. Indeed, participants in the experiment indicated they jog outside, many to experience nature.
- One aim was to keep the technology as non-intrusive to the jogging experience as possible, and therefore to build a system that, if commercially deployed, could take the form-factor of a mobile phone or MP3 player. The current design is heavier than an iPod® and uses larger headphones, however, future technology advances will allow for the functionality to fit into a mobile phone. This way the user would have to carry only one device plus the headphones.
- The decision was made to restrict the modality to an audio-only experience, based on the fact that many joggers are familiar and comfortable with taking an MP3 player on their run. The participants will therefore experience a social run delivered through an audio-only channel, not distracting their visual focus by requiring them to look at a display.
- Joggers already exert themselves physically and their mental focus is occupied with a strenuous run, therefore it is important the experience does not require any button presses. In fact, the interface should be controlled explicitly, not implicitly. Therefore, the decision was made to let the jog become the Exertion Interface (Mueller, Agamanolis & Picard, 2003) to enhance the experience.

## Adding Pace Awareness

By increasing the sense of presence of the other person, it is possible to enhance the experience of jogging together for geographically distant jogging partners, beyond the previous audio-only approach (discussed earlier). Adding pace awareness could be beneficial for an increased sense of presence between the jogging partners, which in turn could contribute to the motivational and social effects of jogging together. A mobile audio system was therefore designed that connects two joggers over a distance, supporting their conversations while using spatialized sound to communicate pace awareness.

## Experience

In the current design, the two joggers, who run in different locations, wear microphones to speak to each other. They also have headphones that deliver the conversations exchanged with the remote partner, but the audio is positioned on a virtual 2D horizontal plane around the user's head. While they jog, speed data is collected and used to position the audio of their conversations and environment in a 2D sound environment. As one jogger speaks, their partner hears the localized audio and is able to detect whether the other person is going faster, same pace, or slower, and thus is in front, to the side, or behind. Similar to a collocated setting, the audio cues runners when to speed up or slow down in order to "stay" with their partner. This approach supports joggers' desire for socializing and motivation to keep pace, as indicated in the survey and forums. The joggers can discuss running routes, encourage each other, or simply listen to the environment noises of the other location.

The system is per default optimized for a condition in which the two jogging partners run roughly the same pace: the joggers should experience the other person's voice sometimes coming from the front, sometimes coming from the back, but mainly coming closely from the center position. If one jogger is constantly faster than their

*Figure 3. Technical implementation*

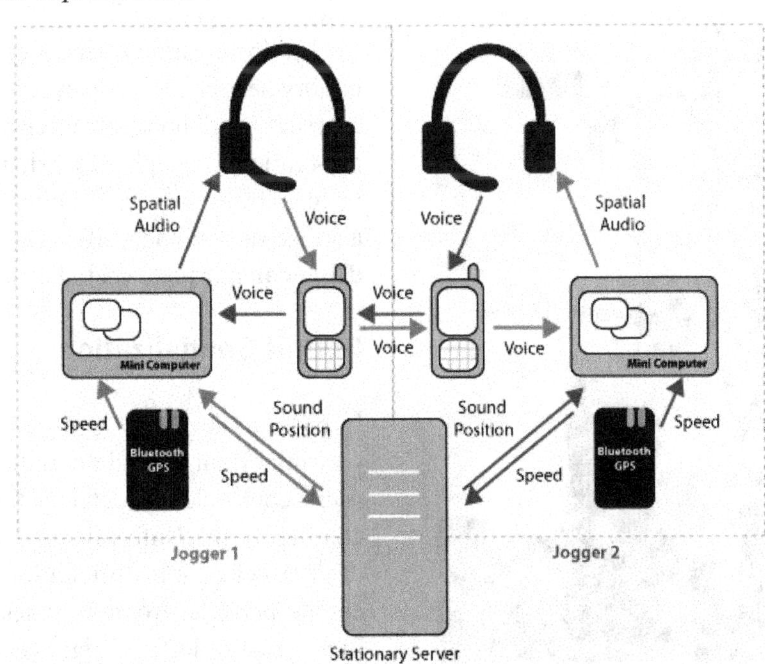

partner, the sound would always appear coming from the back, which might therefore seem not to be very reactive. For this kind of scenario, an offset in the code can compensate and adjust the two joggers' speeds virtually. This enables two joggers to jog together although they would not enjoy jogging together side-by-side: one would be too exhausted trying to keep up, the other frustrated by the slow running speed. Finding a partner that has the same jogging speed is a frequent problem reported by participants, but *Jogging over a Distance* can help accommodate these joggers.

*Jogging over a Distance* consists of two identical systems, each with a miniature computer; a Bluetooth® GPS device; a wireless modem; a mobile phone; and a headset (Figure 3) (Mueller, O'Brien & Thorogood, 2007). Each system is carried in a small, close fitting bag while the user jogs (Figure 4).

The mini PC runs Java™ and Processing software, which determines the participants' speed from the GPS data. Each computer is connected to a commercial wireless broadband service, which covers the major urban parks where joggers run. Speed and time data from the GPS device is received by the mini computer via Bluetooth®. The computer then transmits this data wirelessly over a 3G network to a server, which calculates the speed difference and adjusts for GPS inconsistencies. The server determines how fast each jogger is running in relation to his or her partner. As a result of this, an algorithm calculates a sound position value for each jogger. As each jogger talks, their voice is picked up by a microphone and the audio is transmitted via a conventional mobile phone. VoIP technology was initially used, as in the preliminary experiment, but the lag and reliability was insufficient for the current purpose. Before routing the incoming audio from the remote jogger's mobile phone to the headphones, the mini computer applies a spatialization algorithm to the sound source. The mini computer uses the sound position value received from the server to transform the audio data into spatial 2D audio by placing the sound source onto an imaginary plane around the jogger's head, delivered via the headphones. The result is that the jogger hears their partner's voice coming from a certain direction.

*Figure 4. The prototype*

The localization effect is enhanced by high-quality headphones, however it is acknowledged that such headphones are often bigger in size than conventional headphones and, therefore, may not be liked by all joggers. Smaller headphones have also been tested, and while not as superior, they still deliver an adequate spatial sound experience.

## Sound Spatialization

In order to develop *Jogging over a Distance*, it was important to find an audio setup in which users could clearly detect where the sound is coming from. Unfortunately, without the use of visual cues, it is difficult for people to differentiate between front and back sound sources, in contrast to left and right (Burgess, 1992). In addition, mobility has been found to decrease audio target accuracy by 20 percent (Marentakis & Brewster, 2006). Fortunately, target accuracy for the current application does not need to be very precise. However, the user needs to be able to clearly differentiate if the other person's voice is coming from the front or the back. To find a solution for communicating sound location while in a mobile environment, different headphones and audio spatialization implementations were evaluated.

In an informal experiment, five participants were recruited to jog on a treadmill at a public gym, while listening to spatially positioned audio cues with various headset models. These headsets were off-the-shelf surround sound headphones, internally designed surround sound headphones, and regular headphones. Some of the spatialization implementations used HRTFs (head related transfer functions), while others relied solely on filtering frequency. Jogging makes sound localization difficult due to the participant's exhaustion level and the moving around of the body and the head. An intensification approach for the prototype was therefore adopted. Instead of positioning the remote sound on an imaginary axis from 12 o'clock to 6 o'clock (from a birds-eye perspective, with the person being in the center of the clock, looking at 12 o'clock), it is proposed that the sound is positioned on an axis from 1:30 to 7:30 (Figure 5). This exploits a person's ability to easily distinguish between left and right audio sources, while simultaneously conveying an experience of hearing sound appearing from the front or back. Initial experiments confirmed that this design greatly improved the sound localization ability of participants, while still creating the impression that the other person is talking either "from behind or in front".

*Figure 5. Bird's eye view of spatialized sound*

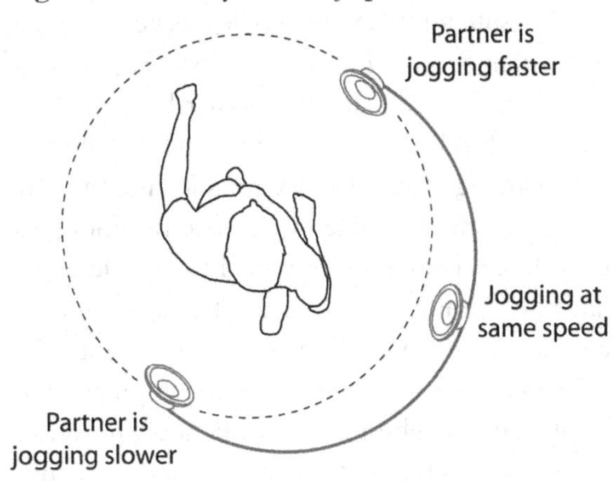


### Limitations

In its current implementation, the system does not take into account any elevation data and assumes that both joggers run on a flat surface in order to effectively compare pace data. Also, the joggers should ideally run for the same amount of time and start simultaneously.

## DISCUSSION

*Jogging over a Distance* targets social, casual joggers who enjoy jogging with others and like to converse during their runs. Not all joggers talk during their exercise: six percent of surveyed joggers who run with others replied that they do not talk while running, and some enjoy the tranquility of running alone. However, the design of *Jogging over a Distance* is based on feedback from participants who claim they jog with others for social and motivational reasons. *Jogging over a Distance* is also not aiming to replace the traditional social "jogging together" experience. Rather, it provides a novel experience and enables an activity that is not possible without the technical augmentation, for joggers who want to run with friends or family but who are located elsewhere. Furthermore, the ability of processing pace data via computational means before delivery to a sports partner allows for modification to improve accessibility. By adjusting exercise intensity data to a suitable level for the involved exercise partner, it is possible to create sports experiences that compensate for different skill levels between the participants. In *Jogging over a Distance*, runners of different speeds can run together, because their exercise level can be computationally adjusted before being utilized on the remote end, thereby providing an enjoyable experience for both joggers. Such a situation in which the participants have different physical capabilities can often be unsatisfactory in a collocated environment. The joggers can wear weights or other equipment that limits their capabilities, or deliberately decrease pace to that of the slower partner, but these experiences are often associated with feelings of not being challenged enough by the faster runner, and notions of inadequacy and over-exhaustion by the slower one. *Jogging over a Distance* could also modify the baseline variables on-the-fly based on the players physiological conditions. This adjustment could occur concealed from the participants to facilitate implicit motivational support. For example coaches could use this to push their athletes as part of a training plan to reveal the athlete's limits.

Future work will include a comparative study to measure whether spatial audio delivery has an effect on the perceived presence of the other person. Furthermore, it could be interesting to investigate a greater sense of presence can serve as a motivational tool to encourage people to run faster, further and more often. The results of this research will inform future designs that aim to support social interactions between geographically distant participants.

## FUTURE WORK

Although prototypes and even commercial products have emerged that combine computer gaming technologies and sports, this field is only in its infancy and far from having achieved its goals. Although exertion games have been shown to offer physical health benefits (Lanningham-Foster et al., 2006), their cardiovascular energy expenditure is often limited, and has therefore been compared to only moderate traditional activities such as brisk walks (Maddison et al., 2007). Incorporating more strenuous activity to support richer health benefits will therefore be one of the next challenges to consider in future games that are aimed at addressing the obesity issue. Furthermore, although it is acknowledged that exertion games can help attract new audiences (Bianchi-Berthouze, Kim & Patel, 2007) to participate in fitness exercises, they are generally considered to be only facilitators for future participation in sports clubs and gyms, as opposed to being distinct sports activities on their own. Turning such exertion games into distinct sports experiences could be a rewarding opportunity for future research.

Future designs and evaluations of new prototypes will help uncover the potential computers can contribute to the sports domain, and in which ways athletes, from the beginner to the professional, can benefit from computer game technology advantages. Creating a conceptual framework around the topic is a remaining issue. Such a theoretical model could provide guidance for the development of future systems. In particular, it should be able to consolidate conceptual themes from the two core areas – digital play and sport – and hence has potential to contribute towards a beneficial balance that considers the benefits of each domain in this merging field.

Many technological advances have been supporting professional athletes in recent years, however the market for hobby sportspeople has a growing commercial potential. With an increasing awareness of the health issues our sedentary lifestyles support, people could become more self-conscious and more involved in sportive activity. Supporting these amateur, but passionate sportspeople has created a growing financial market, and might be an increasingly important area for future developments. In particular, the design of cheap, robust technology that can support health goals will be a central challenge.

Technology can supplement participants' motivational efforts by providing them with social support mechanisms through their peers. Empirical investigations should be undertaken to investigate how these mechanisms need to be designed for maximum effectiveness. In regards to the previously outlined design issues, a few elementary questions emerge. First, can technology be an influencing factor when it comes to the size of these social support networks? Do these networks have the same opportunity to scale effortlessly, as is seen in digital games? Does the communication support need to be synchronous, or what effect does an asynchronous channel have? Finally, does a high degree of anonymity facilitate or hinder social support mechanisms? Knowledge from online communities in social network games could contribute insights here. Answering these questions will benefit designs in the merging area of digital games and sport.

## CONCLUSION

With this contribution, selected aspects of a growing trend in merging digital game technology with physical sports activities have been highlighted. In particular, the goal of this chapter was to sensitize the reader to the many opportunities offered by the application of digital gaming technologies to bodily activities. This approach is exemplified by describing a system that offers advantages known from computer gaming (i.e. supporting players in geographically distant locations by utilizing networking technologies) to engage users in a physical activity together, despite being apart. It has been shown how this system, called *Jogging over a Distance*, can address the issue of finding the right jogging partner: one that is roughly running at the same pace and available at a mutually convenient time. This is often difficult to achieve in a traditional, collocated setting. Other research prototypes exist that augment sports with digital technologies to support performance enhancement, and prominent examples were described. Commercial systems have already entered households across the world, and users have appropriated them for their health goals. However, many of these systems that have their origin in a gaming tradition offer only limited support for energy expenditure. With a simple example of using mobile phones for jogging activity, this chapter shows how technological advances can support traditional sports activity, while being accessible and affordable to the recreational athlete. This study showed how powerful existing technology can be when applied to a new context such as sport, as indicated by the enthusiastic comments from participants. It is hoped this inspires more work in the area.

This work on networked sports activities should not suggest that the benefits of introducing digital gaming technologies to fitness applications are limited to supporting distributed participants. Offering enhanced training effects; supporting teams with players of different capabilities; and allowing athletes to work out with others at different times are all scenarios that could be envisioned by the utilization of technological advances. Computer gaming technologies can also serve as persuasive tools to motivate people to start a healthier, physically active lifestyle and thus address a growing obesity issue. Online systems, often in combination with smart sports equipment, have demonstrated this. Users benefit from such approaches by being offered more engaging experiences, and they can use these systems to support a healthier lifestyle, as well as increase their mental, social and physical well-being. However, participants also need to be aware that some of these advances cannot and should not replace traditional sports activities. For example, participating in sports clubs has many advantages beyond the obvious health benefit, such as contributing to social capital, and users should not forget that technological solutions might not be able to support all of these aspects equally well. A sportsperson should always weigh up the advantages and disadvantages a new technological advance offers. It might be helpful to consider these opportunities as complementary to existing sports opportunities, instead of as a replacement. Also, some design solutions offer experiences that are not just simulations of existing activities, but rather aim to go "beyond" by enabling functionality traditionally difficult or impossible to obtain. *Jogging over a Distance* exemplifies such an approach by supporting sportspeople of different physical capabilities. The consideration of such opportunities in design will enable athletes to participate in sports activities that offer exciting possibilities for their well-being in the future.

Digital gaming in combination with sports activities now has the potential to work against its past criticism of supporting a sedentary lifestyle and diminishing physical fitness. Instead, it is making a significant leap towards contributing positively to health effects and enabling novel well-being experiences previously difficult to achieve. Seizing the opportunities is a task that lies ahead for all of us in this area, whether we are designers, researchers, developers, practitioners, decision makers or users, and collectively, a rewarding contribution can be achieved that contributes to the well-being of many people worldwide.

## ACKNOWLEDGMENT

Thank you goes to the CSIRO, where early work inspired parts of the research, and in particular Shannon O'Brien and Alex Thorogood for their ongoing support of the development and Wouter Walmink for help with some of the design work.

## REFERENCES

Adcock, M. (2008). *Are we there yet?* Retrieved April 30, 2008, from http://web.media.mit.edu/~matta/projects.html.

Apple. (2008). *Apple – Nike + iPod.* Retrieved April 30, 2008, from http://www.apple.com/ipod/nike/.

Bailey, R. (2005). Evaluating the relationship between physical education, sport and social inclusion. *Educational Review, 57*(1), 71-90.

Barkhuus, L., Maitland, J., Anderson, I., Sherwood, S., Hall, M., & Chalmers, M. (2006). Shakra: Sharing and motivating awareness of everyday activity. In *Ubicomp 2006*. California, USA: ACM Press.

Behrenshausen, B.G. (2007). Toward a (kin) aesthetic of video gaming. *Games and Culture, 2*(4), 335-354.

Bianchi-Berthouze, N., Kim, W., & Patel, D. (2007). Does body movement engage you more in digital game play? And why? In *Affective computing and intelligent interaction* (pp. 102-113). (vol. 4738). Berlin: Springer.

Bodypad. (2008). *Home*. Retrieved April 30, 2008, from http://www.bodypad.com.

Bodyspace. (2008). *Home*. Retrieved April 30, 2008, from http://bodyspace.com.

Burgess, D. (1992). Techniques for low-cost spatial audio. In *Proceedings of UIST'92*. California, USA: ACM.

Consolvo, S., Everitt, K., Smith, I., & Landay, J.A. (2006). Design requirements for technologies that encourage physical activity. In *Proceedings of CHI 2006* (pp. 457-466). Quebec, Canada: ACM Press.

Damasio, A. (2000). *The feeling of what happens: Body, emotion and the making of consciousness*. Washington: Harvest Books.

Etter, R., & Specht, M. (2005). Melodious Walkabout – Implicit navigation with contextualized personal audio contents. In *Adjunct Proceedings of the Third International Conference on Pervasive Computing* (pp. 43-49). (vol. 3468). Germany: Springer.

EyeToy Kinetic. (2008). *Home*. Retrieved April 30, 2008, from http://www.eyetoykinetic.com

Fitsync. (2008). *Products*. Retrieved April 30, 2008, from http://fitsync.com/website/pages/productsPersonal.html

Gamebike. (2008). *Home*. Retrieved April 30, 2008, from http://www.cateyefitness.com/GameBike/index.html

Gamersize. (2008). *Home*. Retrieved April 30, 2008, from http://www.gamercize.net/

Gratton, C., & Henry, I. (Eds.). (2001). *Sport in the city: The role of sport in economic and social regeneration*. UK: Routledge.

Hollan, J., & Stornetta, S. (1992). Beyond being there. In *CHI 1992: Proceedings of the SIGCHI conference on Human factors in computing systems* (pp. 119-125). New York: ACM Press.

Huysman, M., & Wulf, V. (Eds.). (2004). *Social capital and information technology*. London: MIT Press.

IDO – MachineDance European Championships. (2006). *Machine Dance*. Retrieved April 30, 2008, from http://www.machinedance.nl/arrowdance/uploadedmedia/ec_machinedance_information_260506.pdf

Interaction Laboratories. (2008). *Powergrid*. Retrieved April 30, 2008, from http://www.ia-labs.com/ViewVideoGameMarket.aspx?ID=7

Lanningham-Foster, L., Jensen, T.B., Foster, R.C., Redmond, A.B., Walker, B.A., Heinz, D., & Levine, J.A. (2006). Energy expenditure of sedentary screen time compared with active screen time for children. *Pediatrics, 118*(6), 2535.

Lehrer, J. (2006). How the Nintendo Wii will get you emotionally invested in video games. *Seedmagazine.com. Brain & Behavior,* Nov 16. Retrieved April 30, 2008, from http://www.seedmagazine.com/news/2006/11/a_console_to_make_you_wiip.php

Long, J., & Sanderson, I. (2001). The social benefits of sport. Where is the proof? In C. Gratton & I. Henry (Ed.), *Sport in the city: The role of sport in economic and social regeneration* (pp. 187-203). London, UK: Routledge.

Maddison, R., Mhurchu, C.N., Jull, A., Jiang, Y., Prapavessis, H., & Rodgers, A. (2007). Energy expended playing video console games: An opportunity to increase children's physical activity? *Pediatric Exercise Science, 19*(3), 334-343.

Marentakis, G., & Brewster, S. (2006). Effects of feedback, mobility and index of difficulty on deictic spatial audio target acquisition in the horizontal plane. In *Proceedings of CHI 2006* (pp. 359-368). Quebec, Canada: ACM Press.

Merleau-Ponty, M. (2002). *Phenomenology of perception: An introduction*. (2nd Ed.). USA: Routledge.

Mokka S., Väätänen A., & Välkkynen, P. (2003). Fitness computer games with a bodily user interface. In *Proceedings of ICEC 2003* (pp. 1-3). Pittsburgh: ACM Press.

Morris, L., Sallybanks, J., & Willis, K. (2003). Sport, physical activity and antisocial behaviour in youth. *Australian Institute of Criminology Research and Public Policy Series, 249*, 1-6.

Mueller, F., Agamanolis, S., & Picard, R. (2003). Exertion Interfaces: Sports over a distance for social bonding and fun. In *Proceedings of CHI 2003*. Fort Lauderdale, USA: ACM Press.

Mueller, F., O'Brien, S., & Thorogood, A. (2007). Jogging over a Distance. In *Extended Abstracts CHI 2007*. California, USA: ACM Press.

Nintendo. (2008a). *Wii Sports*. Retrieved April 30, 2008, from http://www.nintendo.com/games/detail/1OTtO06SP7M52gi5m8pD6CnahbW8CzxE.

Nintendo. (2008b). *Wii Fit*. Retrieved April 30, 2008, from http://www.nintendo.com/wiifit

O'Brien, S., Mueller, F., & Thorogood, A. (2007). Jogging the Distance. In *Proceedings of CHI 2007.* California, USA: ACM Press.

Oliver, N., & Flores-Mangas, F. (2006). MPTrain: A mobile, music and physiology-based personal trainer. In *Proceedings of HCI-Mobile 2006* (pp. 21-28). USA: ACM Press.

Pate, R.R., Pratt, M., Blair, S.N., Haskell, W.L., Macera, C.A., Bouchard, C., Buchner, D., Ettinger, W., Heath, G.W., King, A.C., Kriska, A., Leon, A.S., Marcus, B.H., Morris, J.,

Paffenbarger, R.S., Patrick, K., Pollock, M.L., Rippe, J.M., Sallis, J., & Wilmore, J.H. (1995). Physical activity and public health: A recommendation from the Centers for Disease Control and Prevention and the American College of Sports Medicine. *Journal of the American Medical Association, 273*(5), 402–407.

Peak Training System Sensor Kit. (2008). *Sensors.* Retrieved April 30, 2008, from http://www.riderunrow.com/products_sensors.htm

Pedersen, J.B. (2002). Are professional gamers different? - Survey on online gaming. *Game Research.* Retrieved April 30, 2008, from www.gameresearch.com/art_pro_gamers.asp

Porcari, P., Foster, C., Dehart-Beverly, M., Shafer, N., Recalde, P., & Voelker, S. (2002). *Prescribing exercise using the talk test.* Retrieved April 30, 2008, from http://www.fitnessmanagement.com/FM/tmpl/genPage.asp?p=/information/articles/library/cardio/talk0801.html

Putnam, R. (2000). *Bowling alone.* New York, USA: Touchstone, Simon & Schuster.

Snider, M. (2008). Designer Miyamoto makes video games pulse with life. *USA Today.* Retrieved April 30, 2008, from http://news.yahoo.com/s/usatoday/20080515/tc_usatoday/designermiyamotomakesvideogamespulsewithlife

Toscos, T., Faber, A., Shunying, A., & Mona Praful, G. (2006). Chick Clique: Persuasive technology to motivate teenage girls to exercise. In *Extended Abstracts CHI 2006* (pp. 1873-1878). Quebec, Canada: ACM Press.

Wankel, L., & Bonnie, G. (1990). The psychological and social benefits of sport and physical activity. *Journal of Leisure Research, 22*(2), 167-182.

Winograd, T., & Flores, F. (1987). Understanding computers and cognition: A new foundation for design. Norwood, NJ: Addison-Wesley Professional.

Wulf, V., Moritz, E.F., Henneke, C., Al-Zubaidi, K., & Stevens, G. (2004). Computer supported collaborative sports: Creating social spaces filled with sports activities. In *Proceedings of the Third International Conference on Entertainment Computing ICEC 2004* (pp. 80-89). Heidelberg: Springer LNCS.

# Chapter X
# Double Play:
## How Video Games Mediate Physical Performance for Elite Athletes

**Lauren Silberman**
*Massachusetts Institute of Technology, USA*

## ABSTRACT

*Just at the moment when gaming has achieved broad cultural acceptance, a new way of using commercial sport video games is emerging, which adds a new perspective on the educational and social value games may offer for learning. This research calls attention to how elite athletes are currently using commercial video games for training purposes and the potential the games may afford for all elite athletes who can play as their "second-self."*

## INTRODUCTION

In March of 2005, a University of Wisconsin's Men's basketball player, with whom I was performing a naturalistic observation, announced to me that he better understood the offense of his opponent, the Minnesota Gophers, by playing a video game. On further explanation, Jake (name changed to protect player's privacy) reported that before the scheduled Saturday game, he had played "virtually" as one of the Gophers against his own team, in the commercial video game Electronic Arts' (EA) *NCAA® Basketball 2005.* Without any coaches, Jake had "whizzed" through a simulation he thought was very close to the real game he later played.

For a form of entertainment, the video game had an impressive ability to help Jake understand a component of the game. Indeed, there is a long list of research that is exploring the educational qualities and good learning principles inherent in good commercial video games (Gee, 2003; Squire, 2007). But how did such a simple form of entertainment help an elite athlete better understand his physical opponent? This is where my journey investigating how and why athletes are using video games for physical training begins.

It is hoped that this paper will help those trying to grapple with how to teach someone a simple component to build higher-level intelligence. Anyone who has ever contemplated how a commercial video game can be used for learning will find something resonant in the research and anecdotes this paper offers from athletes using their sport game counterpart. For teachers trying to understand ways to teach students, in an American educational system where "Every Child Gets Left Behind" (only 52% of students in the United States 50 largest cities complete high school with a diploma); the video game may someday be widely accepted as a multimodal learning tool that can adapt to every student's need. It is my hope that by presenting results from a survey and observational research with the Division I NCAA® University of Wisconsin Men's soccer team, and anecdotes from interviews with other elite athletes, that we can then begin to understand how athletes are using games and the potential they may hold as learning tools.

## BACKGROUND

While the emphasis in classical game theory has been on what constitutes play, sport video games deliberately blur the lines between a game and an experience, and between places of play (the "magic circle") and places of everyday life (Salen & Zimmerman, 2003). The understanding of sport video games as immersive and realistic enough to be an effective simulation, runs in direct opposition to the theorization of play as open-ended, although kept intact by the acknowledgement that these simulations give players numerous choices and are fun. The use of video games for athletic training purposes supports the notion and model of game based learning environments (Squire, 2007).

The term "edutainment" has been suggested for video games that have an educational purpose as well as providing entertainment. But this term is still inadequate to define sport video games. As Resnick (1998) explains, when people think about "education" and "entertainment" they tend to think of them as services that someone else provides. Studios, directors and actors provide an individual with entertainment; schools and teachers provide education. Edutainment companies try to provide both. Sport video games are primarily thought to be strictly a form of entertainment. However, they potentially afford athletes a designed experience that provides an opportunity for usage as a tool for learning, while still being a form of entertainment.

The line is now blurred between entertainment and education, with athletes such as NASCAR® drivers and professional coaches starting to use commercial-off-the-shelf sport games as tools for training (Rosewater, 2004). As professional racecar driver, Carl Edwards explains, "A video game helps you get the rhythm down – helps you find a place where speed is made up and speed is lost." Whenever he has to drive a track he regularly has trouble with – like Martinsville in Virginia or Bristol in Tennessee – he will spend a couple of hours in his trailer with the game.

Joe Paterno, the 82-year-old coach at Penn State is a great example of a coach using games with athletes. Every season he is reported to give incoming players a cartridge with the new seasons plays on it. Joe Paterno is known for his ability to teach and adapt to younger players. This level of adaptation and a willingness to teach is especially significant when one considers that Joe Paterno does not even use a computer.

While sports video game players are undoubtedly having fun while learning, they are far from passive recipients of either entertainment or educational services. In fact, a great deal of the training potential of these games derives from the fact that users play as both coach and player, making team decisions and executing individual player moves, play-after-play. In this way, they gain opportunities in the game, such as the ability to determine the plays, which they might not experience as a player on the field. Similarly, they get to try out different moves while simulating play and are able see the same play from different angles, in a perspective they would never experience in reality. Sport video games also provide the opportunity to perform many moves one could not perform physically. As a consequence, the athlete's understanding of the sport becomes both broader and deeper.

Also of note is how much time even elite athletes are devoting to playing these games. The Wisconsin men's soccer team averaged 4.8 hours playing video games during the soccer season and 6.7 hours playing video games during the off-season. It is particularly noteworthy that when they are playing video games these many hours, they are typically playing sport video games. The soccer players reported sport games to be their favorite genre by 70%, followed by shooter (12%) and strategy (9%) games. In particular, the FIFA™ franchise is the game consistently played.

An important advantage enjoyed by collegiate or professional users of the sports video game is that they are able to play avatars that simulate actual members of their own team; that is, avatars with the same statistics as their real life counterpart on the player's team. As Diego (name changed to protect player privacy), the University of Wisconsin soccer player tells me, "I don't know any athletes that are in the game who don't usually play as themselves." Not only is the athlete playing with a virtual double of his real team, he can play against an opposing team with virtual characteristics that match those of his real life opponents. This enhances the learning experience as well as the fun. If the athlete knows he is going up against a particular team, he can play in a game against a team with equal statistics, and he can play it alongside statistical replicas of his real team members, in a stadium with weather conditions that replicate the real physical games. These commercial off-the-shelf sports video games are thus able to teach players as they are having fun, in large part because of the enculturation and social team play that naturally happens when highly skilled athletic teams play their sport's virtual counterparts.

The highest learning seen in observations occurred with elite athletes who played the video games with their actual teammates between physical training sessions. These successful athletes were those who could play (or as I would argue, train) using avatars that were close simulations of them.

Significantly, EA Games upload these athletes' physical characteristics in the game, because they are members of certain college or professional sport teams. EA invites all professional basketball players that just entered the draft to participate in EA draft camp where EA does motion-capture tracking on the players so their physical movements are more accurately reflected in the game.

The sophistication of the training potential of video games for collegiate and professional football players aligns with data from my recent research concerning the University of Wisconsin men's soccer team. One of the most significant findings suggests that soccer video games were successful in helping the athletes have a better understanding of various styles of play.

Most of the Men's soccer team plays together on a weekly basis. It is no surprise that the majority of players said they play together. Almost all of the players believed that soccer video games may help them familiarize themselves with professional soccer team statistics. In addition, most of the team believed that games helped them have a starting point to discuss various professional level statistics.

But the video games offered no directions for how players should approach and play these games, nor did the games suggest to athletes how the concepts seen in the game related to the physical sport of soccer. Similarly, the video games did nothing to suggest what users may do with the knowledge in the virtual game to extend their interest beyond the game, such as using the embedded knowledge to learn more about professional soccer leagues. Although the games are successful at their core–for encouraging conceptual models for other types of play–their design is ineffective at directly helping a player to access all the learning that can be gained from the game.

Given the delta in the use of sport video games between the learning going on and the learning that is achievable, it is helpful to situate this discussion in a framework of cognitive apprenticeship that embeds learning in activity and makes deliberate use of the social and physical context. This will allow us to understand that it is not the game itself that teaches, but rather in what context one uses it. This understanding will aid those who wish to better apply the game as a training tool. How knowledge is situated is in part a product of the activity, context, and culture in which it is developed and used (Brown, Collins & Duguid, 1989).

The sport video game is designed to be a fun activity: key characteristics that make it fun, are its realism and the opportunity it provides to take on roles one cannot in real life. This safe environment or "sandbox" (Gee, 2003), where learners are put into a situation that feels like the real thing, but with risks and dangers greatly mitigated, means players can learn well and still feel a sense of authenticity and accomplishment. Sport games help to teach and motivate players while allowing them to make mistakes and experiment.

While many games have a sense of conflict, sport games allow players to design the conflict, because they choose which teams and players battle each other. In practice sessions or during games, even professional athletes cannot always choose the plays; be both coach and player; take certain risks; stall the action to review mistakes; look at the field from different angels; or try out numerous plays. But in a video game, the player can stop and replay the action; attempt another approach; try passing to one player or another, or on one side or another; take out a player or put him back in on the same play; and then watch the effects of these decisions on the outcome. All of this can be done in the video game, especially if the player is using an avatar that closely tracks his own attributes, those of his team members; or those of real opposing teams, in a game environment that simulates the stadium, weather conditions, and actual playbooks of the teams. The realistic simulations and increased controls of the videogame are ideally suited to allow the user to experience both fun and learning.

## HOW DOES VIRTUAL REALITY ENHANCE ENGAGEMENT?

A defender on the University of Wisconsin Men's soccer team spoke to me about being in the game and having control over its environment. As he explained, "I'm here playing so I'm inside of the virtual environment. I'm also watching things that happen, which makes it more like an image on television." Even though the player understands there are some similarities between video games and television, it is also very clear that the player is not simply gazing at a screen when playing a video game, as they would a television set. Rather, they are immersed in a designed environment that they are controlling.

As Bolter and Grushin (2000) argue, in our present state of media (or IT) history, one medium is always experienced, used, and defined by its relation to other media. They define a medium as one that "remediates." It is that which appropriates the techniques, forms and social significance of other media, and attempts to rival or refashion them in the name of the real. In their view, a medium in our culture can never operate in isolation, because it must enter into relationships of respect and rivalry with other media. While video games are like television in that they both communicate ideas, with video games the player is not passive: the player designs their experience. Jenkins (2006) has described this type of practice as belonging to a "participatory" culture where anyone, regardless of skill level, is welcome to contribute to a variety of expressive and community-based practices, or to even start their own. In essence, participation is positioned as the opposite of more passive types of media engagement, such as watching TV or other manifestations of mass media.

## FANTASY SPORTS

About 18 million Americans participated in sports fantasy games in 2005 (Janoff, 2005). The premise of fantasy sports is to select a group of real players for a fictional team, then use the real-life statistics of those players to determine which team in a fantasy league is doing the best. Also interesting is that 57% of fantasy football players watch at least four NFL® games every week. The average player visits the host site seven times per week and another site for information approximately five times a week. While in the past users would just watch the game, now they are no longer spectators: they are virtual participants.

In sport video games and on highlight reels from ESPN (the leading cable sports broadcaster, reaching more than 97 million US homes), we see that what was once "reality" on ESPN relies on video game technology as a tool to imagine game scenarios. The video game signifies the re-playing of what is on ESPN. There are even simulated ESPN television commentators appearing in the video games, reporting on plays as they occur. ESPN in turn expropriated from the videogame the behind-the-quarterback perspective and multiple camera angles, and now uses them in its television broadcast, which is referenced once again in the videogame's simulated ESPN broadcast. Examining how the relationship between television and video games has evolved through the years helps to contextualize the object of the game itself, and to explore the "new" way we watch sports and virtually play them today.

What was originally created as a form of entertainment, the sport video game is now helping us to watch and enjoy the original medium. ESPN and other sport program reels are racing to replicate the next generation graphics and new angles that sport video games are creating. The Wii™ camera mode in *Wii™ Sports Tennis* is now in an older ESPN mode, with the camera situated on the side-line for certain aspects of the game, while also giving players the evolved aerial and action shots. On ESPN sidelines, we can see commentators "talking" directly from the game sidelines. The transfer was first from the video game angle to the ESPN angle. Now we have the video game mimicking things that the ESPN highlight reels used to distinguish themselves, such as branded microphones.

Some of the symbiotic features of how sport video games and televised sports reporting has historically evolved are as follows:

1. Physical sport played on fields
2. Radio broadcast of sport with commentators
3. TV broadcast of sport with commentators
4. Simulation of the sport in the video game
5. TV commentator first seen in video games – John Madden
6. TV incorporates video game camera features (e.g. quarterback camera angle) and also video game animations to present real game highlights and replays
7. Replay and highlight features are adopted from TV into video games. Replay features are now options at any point in the play of many sport video games. A player can choose when he wants a replay and of what play, and can view the replay from multiple camera angles.
8. ESPN and EA team up to offer live sport statistics while playing video games online.
9. In-game fantasy statistics can now be seen scrolling at the bottom of TV screens dur-

ing games. Live television statistics can also be seen on tickers at the bottom of fantasy sport sites. Fantasy sports consumers often have one eye on their TV and one eye on the computer screen, monitoring the progress of their teams or players while games are in progress.

The sport game player also gets the benefit of learning how his game would play in the media, because he hears the simulated ESPN commentator in the videogame report on a play as it happens. On the real field, the player or coach does not hear or see the reportage until after the game, long after most television viewers. The simulated immediate feedback may help improve play. To a limited extent, game announcers offer training reinforcement to intelligent or non-intelligent analysis of the game. For example, if a player makes a failed fourth down conversion when it made no sense to attempt it, the commentators will generally add something like, "I really have no idea what the coach was thinking on that play."

Not only does one get to play every position on the field and decide what camera angle to play (aerial, action on field/in car, or side-line view), but also sport video games now allow the player to take on the role of television director and producer. Users can replay any point of action on the field and watch it instantly, or save it to view later. Through these choices, the users can create their own individual story or environment. The virtual environment and the player need one another to function.

## SPORT GAMES ARE RICH LEARNING ENVIRONMENTS

Understanding why the game player and the game need one another to function is a key element in understanding why these games provide a rich space for learning. Sport video games have great potential for leading players to enhance physical performance and professional athletic roles, while having fun. That simulation is only an approximation of reality, but it does not make it less useful than reality as a place to learn athletic skills. In fact, for learning sports, video games as a learning tool have some unique advantages when used as an adjunct to physical play: they are a safer place than the field to make mistakes; they can be more easily and frequently accessed; they provide for individualized learning; they allow the user to play more roles and see more of the field; and they provide the opportunity for experimentation, because the player controls more variables.

The Wisconsin Men's soccer players believed that watching video performances was the most helpful among categories of activities for strengthening on-field performance (see Figure 1).

*Figure 1. What helps on field performance*

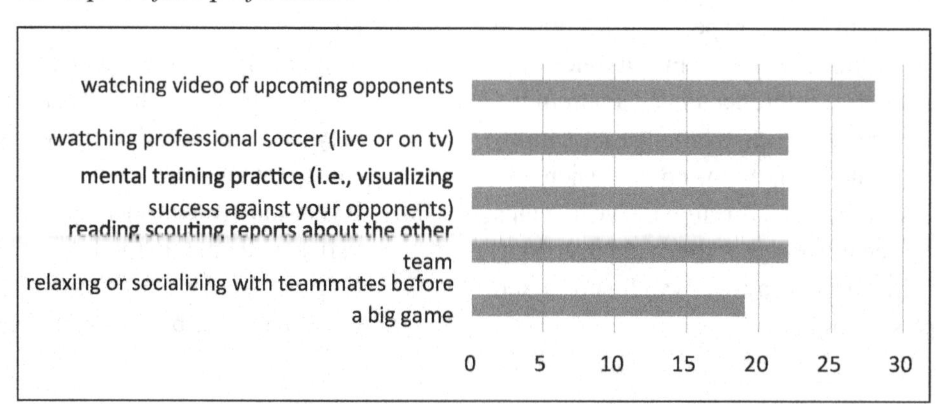

Gaming is a natural step from the use of watching video or taped performances. It takes viewing the video recording of a performance one step further, in that the player can create specific scenarios using virtual representations of the other team.

Significantly, if someone is playing the video game as himself, he also has a unique opportunity to gain self-awareness. This is because when using the video game, he is both playing and observing himself playing. My goal, therefore, is to illustrate why we should create a mental imagery model in which athletes can use the video game for overall learning.

## REALISM OF SPORT VIDEO GAMES

Although there is little empirical evidence for the effectiveness of simulation training, it may help athletes by counteracting the tendency for unusual or unexpected stimuli to distract them (Moran, 2004). The closer the training is to the real-life situation, the more beneficial it should be. Unlike the practice session, the video game can replicate crowd noise and other characteristics of physical games. Tiger Woods' father, Earl Woods, would intentionally drop clubs or cough to train Tiger to block out gallery noises while swinging (Andrisani, 2002). The video game also allows the athlete to practice with referees. This is important for mental training, because during an actual match, an athlete may become emotionally affected by changes in the behavior of officials. Bull, Albinson & Shambrook (1996) recommend "poor official" training. The designers at Electronic Arts have taken into account home team advantage and other location based factors in its game designs (Silberman, 2005). For example, the range of weather choices for a particular location in its college and professional football games vary, based on an average of actual weather statistics for that location. Similarly, the size of the field, the noise level and size of the audience are adjusted in the game based on real statistics for the sports arena selected.

## ASSIGNED ROLES TO A PARTICIPATORY CULTURE

Creative vision is the essence of game design. One's vision for the game is not another's vision. The future of athletes using video games for training purposes will be how one applies their vision to video games. The measure of success with this type of training is the extent to which it becomes a social phenomenon. A hit movie is a topic of culture, in that we all discuss it and can recite lines from it. Spectator sports are a social activity that most everyone can play. It is common for the public to talk about the big hits and long runs from championship games, with more fan talk happening at a local level. However, the video games now allow teammates to create matches and talk among themselves during the game (live or through live-chat). These scenarios among teammates generate banter previously unheard in the locker room. Professional players will often times make it a point of contention to beat each other in the video game. Further, interviews with professionals reveal that they want better statistics in the video game, which motivates them to become better on the field. As Michael Strahan, Defensive End for the NFL® New York Giants, describes the motivation,

*Back in 1993, I was playing the game (Madden) and I sucked. And I'm not talking about sucking at playing the game. My character in Madden sucked. I was the worst guy on the team...but it made me want to work harder. I seriously wanted to get better in real life so I wasn't so bad in Madden.*

Participation is not just happening within the domain of the sports field. This participation is extending outside of the game-space. As Washington Redskins player, defensive end, Erasmus James described it to me, "how your player is and how you play in *Madden* is a bragging right. We all play together on Xbox Live® and things really heat up."

## APPROPRIATION: IDENTIFICATION AS A PROFESSIONAL

A key component to these sport video games is the ability to adopt any persona, including one that closely mimics a professional athlete. Erikson's (1950) notion of adolescence as a "psychosocial moratorium," a period of "time out" in which young people can experiment with different potential identities and engage with risks of various kinds, seems appropriately applied to how players experiment with and enact a professional athlete's identity in sport video games.

Gee (2003) points out all deep learning involves a committed engagement of the learner's self. Games manage this by creating interesting virtual identities that players can juxtapose with their mundane real-world selves. Just as a good high school science class encourages students to imagine themselves as laboratory scientists conducting experiments, these games put the user in the role of a sort of demi-god player and professional coach. Fans also have the opportunity to play against simulated elite athletes after winning national tournaments against other game-players.

When given the opportunity, most of the professional athletes I spoke with play as themselves. EA and ESPN have both picked up on this trend and have *Madden* tournaments where professionals play against each other as their own virtual avatars, live on television. That professional athletes choose themselves as avatars may simply be a matter of ego, but it also affords them an interesting learning opportunity they might otherwise miss. A lot of sports training is designed to improve an individual's kinesthetic awareness. While many of us may be blithely unaware of our physical posture and muscle engagement, an athlete must be aware of how he is moving in order to improve performance. How helpful it is then, to watch ourselves in order to better see what we are doing, or perhaps, who we are. As humorously described by *The Sims*™ player, Chuck Klosterman (2004), in *Sex, Drugs, and Cocoa Puffs*:

*Who Am I? Or (Perhaps More Accurately) Who Else Could BE ME?*

*I would build a character who looked just like me, and I would name him "Chuck Klosterman." I would design his home exactly like my own, and I would have him do all the things I do every day. Perhaps I unconsciously assumed I would learn something about myself through this process... Maybe it was just the desire to watch myself live. Pundits like to claim that a game like The Sims™ taps into the human preoccupation with voyeurism, but it's really the complete opposite. I don't care about peeping into anyone else's keyhole; I only want to see into Chuck's. I designed my digital self as accurately as possible: pasty skin, thick glasses, uncommitted haircut, ill-fated trousers. Outlining my character's personality traits was a little more complicated, because nobody (myself included) truly knows how they act.* (p. 16)

As an identical twin, I know firsthand the benefits of being able to see someone just like me struggle with the same activities I am trying to learn. Even as babies, identical twins have been known to benefit from seeing their likeness perform the same task (Schoffeleers, 1991). They can learn from each other's success and mistakes, and because they have a point of comparison, twins may develop greater self-awareness (Schoffeleers, 1991). There are many examples of athletic twins, from the Bryans and the Barbers, to the Ferraro and Hamm brothers. Some credit to their success may be attributable to their ability to read each other closely on the field and to have a partner of like skills with whom to practice. Research also demonstrates that the common genes shared by identical twins contribute to their close social relations and, in turn facilitate practice, performance and success (Segal, 2006).

But in learning a sport, identical twins also enjoy a clear advantage in being able to observe their identical physical selves tackle it. One of an identical twin pair of star skiers on the Colby team explained to a reporter that, as a result of this, a twin can profit from the other's technical information or lessons learned.

When an athlete chooses to play a sport video game as himself, the avatar becomes his own identical twin. By watching his twin, he becomes much more aware of himself kinesthetically. The user can also test out and see how his twin handles a variety of changes. In actual practices, coaches will encourage players to try out using their non-dominant hand or foot, or to put on extra weights while trying to hit the ball. They want the players to know the difference in feel when exerting various strengths. These are the types of options that can be tried by the player on his identical avatar. This would hopefully provide the athlete with confidence if they have trouble with skills that come less naturally.

## IDENTITY ACHIEVEMENT

It is the establishment of a clear sense of identity, or "who I am,“ that allows the individual to experience psychological well-being and feelings of self-worth, usually after exploration and introspection about life experiences (Marcia, 1994). This skill requires long-term development and seems especially important for athletes, because it involves resistance to conformity and sub-cultural pressure based on the controlling nature of elite sport. The more self-awareness an athlete can gain, the more able he may be to focus his attention on goals, achievement and performance. It seems likely that the athlete's ability to see himself clearly and form a stronger identity would be enhanced by observation of his virtual self on the field, performing under various situations and pressures.

## MENTAL IMAGERY AND SEEING AND COGNITION

Because the technique of imagery has been shown to effectively enhance athletic performance, research is beginning to examine the effectiveness of imagery packaged in different ways using a variety of strategies or models of intervention. The research to date points toward the effectiveness of modeling mental imagery practice by watching videos. In an imagery training program designed to improve golf putting, performance was enhanced more by mental practice using audiotapes and videotapes, than by mental practice using written scripts that were read by the golfers (Smith & Holmes, 2004).

Philosophical counseling (Raabe, 2001), a fairly new approach to mental imagery training, focuses on helping individuals come to a better philosophical understanding of themselves in relation to their personal context. Since "seeing" is a critical part of cognition, video games enable the user to visualize himself contextually. In this sense, sport video games are particularly interesting for the way they lay interpretive scripts over the action and visually present the play from a variety of perspectives.

## WHOLE SYSTEM PERSPECTIVE VS. PARTIAL TASK SIMULATORS

Athletes, especially professionals, are trained to maintain an acute awareness of everything that is happening around them at any given moment. Just as chess players see patterns on the board instead of just individual pieces, highly skilled athletes exercise a gestalt view of the action on the field, despite being limited to one perspective during play.

When I stop video game users during play and ask them to turn to me and tell me what is happening, they may first tell me exactly where they are in the game, as they are still thinking about the exact moment of play. They may also say something like, "I just passed the ball to my teammate" or they may tell me about their own individual role, such as, "I'm the striker." But if we keep talking, they will always point out the score, where the ball is on the field, where the referee is standing, or how the offense or defense is shifting in response to what is happening. They are aware of the entire field, or whole system perspective.

It is the same with skilled users of multiplayer games like *Battlefield*™. Experienced players of this game exercise a system perspective - responding dynamically to their teammates' actions, locations, classes and so on, while inexperienced players adopt the role of the one man army, focusing only on what matters in the context of their individual experiences on the battlefield.

Recognizing the transferability of these skills to the real battlefield, the military now uses commercial-off-the-shelf games such as *America's Army*® to help soldiers train for battle. The similarities between football and infantry combat are not hard to find. A commanding general from the Army Infantry School in Fort Benning, Georgia illustrates the connection this way:

*In football, as in infantry combat, a player must be aware of both the wider situation on the field, and the area immediately surrounding him. The situation changes rapidly and the enemy is always adapting his tactics. Physical injuries abound in both places. Football is as close to fighting a war as one can come without guns and explosives.*

Given such similarities, it should come as no surprise that video game simulations would also be useful to train athletes.

## CONCLUSION

There are a growing number of athletes who are already reporting enhanced physical performance from their virtual play. By recognizing the value of sport video games as training tools, we can help all athletes fully realize the potential of these simulations for training. This paper helps us understand the didactic characteristics of these games and their link to improved performance. In so doing, this discussion may encourage further acceptance of the sport video game as a training tool and game design that reflects this purpose.

## REFERENCES

Andrisani, J. (2002). *Think like Tiger: An analysis of Tiger Woods' mental game*. New York: Penguin Putnam.

Bolter, J.D., & Grushin, R. (2000). *Remediation-Understanding new mMedia*. USA: MIT Press Paperback.

Brown, J.B., Collins, A., & Duguid, P. (1989). *Situated cognition and the culture of learning*. USA: Educational Researcher.

Bull, S.J., Albinson, J.G., & Shambrook, C.J. (1996). *The mental game plan: Getting psyched for sport*. Eastbourne, UK: Sports Dynamics.

Erickson, E.H. (1950). *Childhood and society*. New York: Norton.

Gee, J.P. (2003). *What video games have to teach us about learning and literacy*. New York: Palgrave/Macmillan.

Janoff, B. (2005, April 11). Marketers of the next generation. *Brandweek, 46*(15), 38.

Klosterman, C. (2004). *Sex, drugs, and cocoa puffs: A low culture manifesto*. New York: Scribner.

Marcia, J.E. (1994). Ego identity and object relations. In J.M. Masling & R.F. Bornstein (Eds.), *Empirical perspectives on object relations theory* (pp.59-93). Washington, DC: American Psychological Association.

Moran, A.P. (2004). *Sport and exercise psychology: A critical introduction*. Hove, UK: Routledge.

Raabe, P.B. (2001). *Philosophical counseling: Theory and practice*. Westport, CT: Praeger.

Resnick, M. (1998). Learning in school and out. *Educational Researcher, 16*(9), 13-20.

Rosewater, A. (2004). Hey, I'm practicing here, not playing: Video games so precise they help many drivers. *USA Today, Aug 8*, 6f.

Salen, K., & Zimmerman, E. (2003). This is not a game: Play in cultural environments. Retrieved September 15, 2008, from http://www.gamesconference.org/digra2003/2003/index.php?Abstracts/Salen%2C+et+al.

Schoffeleers, M. (1991). *Twins and unilateral figures in Central and Southern Africa: Symmetry and asymmetry in the symbolization of the sacred.* Journal of Religion in Africa.

Segal, N.L. (2006). *Twinsburg, Ohio; Twinsburg Research Institute Twin Study Summaries; The Outside World.*

Silberman, L.B. (2005). *Athletes' use of video games to mediate their play: College students' use of sport video games.* Paper delivered at the 2005 Seminar Series, Caladonian University School of Computing and Mathematical Sciences, Glasgow, Scotland.

Smith, D., & Holmes, P. (2004). The effect of imagery modality on golf putting performance. *Journal of Sport and Exercise Psychology, 26*, 385-395.

Squire, K. (2007). *Open-ended video games: A model for developing learning for the interactive age*. USA: MIT.

# Chapter XI
# A League of Our Own: Empowerment of Sport Consumers Through Fantasy Sports Participation

**Donald P. Roy**
*Middle Tennessee State University, USA*

**Benjamin D. Goss**
*Missouri State University, USA*

## ABSTRACT

*The explosion of fantasy sports and the dearth of research about it create a need for investigation in this relatively new form of sport spectatorship. This chapter proposes a conceptual framework for marketers to utilize in their examinations of influences on the consumption of fantasy sports by postmodern sports fans. The framework is based on literature from psychology, sociology, sport management/marketing, general management/marketing, and consumer behavior. It leads to the proposition that fantasy sports consumption is impacted by the interplay of psychological characteristics internal to consumers, social interactions, and external influences controlled by fantasy sports marketers.*

## INTRODUCTION

While estimations concerning the exact size and scope of the fantasy sports industry vary, what remains unchallenged is its status as a maturing industry connected to large portions of leisure time and activity, escalating numbers of participants, and increasing marketplace fertility (Janoff, 2005). Janoff (2005) and Fisher (2007) estimated fantasy sports to be a $2 billion industry in 2005, including website fees, game add-on features, videogames, and so on. Less well known are the underlying dynamics of the unique market of fantasy players and the major transformations occurring within the fantasy sports business (Russo & Walker, 2006). The explosion of popularity in fantasy sports participation and the dearth of research about it create a need for investigation in this relatively new form of sport spectatorship (Davis & Duncan, 2006), not to mention new challenges for marketers, media companies, and others within the sport industry who wish to capitalize on the fantasy sports audience (Russo & Walker, 2006).

The purpose of this chapter is to propose a conceptual framework for marketers to utilize in their examinations of influences on the consumption of fantasy sports by postmodern sports fans. The framework is based on literature from psychology, sociology, sport management/marketing, general management/marketing, and consumer behavior.

## BACKGROUND

### History

Fantasy sports leagues, first known as rotisserie leagues, were started using the sport of baseball in the United States during the early 1980's by American journalists Glen Waggoner and Daniel Okrent (Hu, 2003). Okrent, a former editor of *The New York Times*, concocted the basic rules of the game on a plane trip in 1987, then passed them on to friends at a Manhattan restaurant called LaRotisserie, from whence the game's original moniker was derived (Bernhard & Eade, 2005; Walker, 2007). Much like modern versions of fantasy sports, rotisserie owners drafted original teams from a pool of active players, tracked season-long statistics, and declared winners based on statistical performances (Bernhard & Eade, 2005; Diamond, 2004; Walker, 2007). The advent of high-speed computers and the Internet revolutionized statistical calculations for these rotisserie leagues, which now operate competitively on roster management elements such as fixed spending allowances or round-robin drafts, trades, waivers, and lineup changes ("Fantasy Baseball", n.d.).

According to Bernhard and Eade (2005), fantasy sports have two universal fundamental foundations. First, scoring is formulaically computed and winners are determined using the current season statistics of a sport's players. Naturally, these scoring formulas can vary widely and may be customized from league to league. Second, fantasy leagues begin their seasons with drafts in which players are chosen to fill the rosters of the fantasy teams.

Beyond these fundamental foundations, Bernhard and Eade (2005) note that other typical fantasy sport league characteristics include:

- A commissioner, who manages the league by establishing its rules and resolving disputes
- Team names, which are often unique creations of team owners or variants of actual sport franchise names
- The season, in which points are scored and from which winners are determined through head-to-head competition and/or rankings through accumulated points
- Season conclusion, in which a league winner is determined through head-to-head competition and/or accumulated points.

Recent movement from subscription models to widely available free fantasy games began when these games were offered by online providers and portals – such as ESPN.com, NFL.com, FoxSports.com, and AOL.com – to capitalize on robust online ad markets and compete with Yahoo!'s long-standing practice of offering free games (Russo & Walker, 2006).

Because of their widespread popularity and subsequent abilities to generate considerable sums of revenue, proprietary rights to various elements involved in fantasy sports games have undergone much recent consideration and even litigation. In 2005, Major League Baseball Advanced Media (MLBAM) entered a five-year, $50 million deal with the MLB® Players Association for its members' interactive rights. St. Louis-based CDM Fantasy Sports sued MLBAM after failing to obtain a fantasy sports license from MLBAM, asserting that the use of raw MLB® player names and statistics is public domain (Fisher, 2006a). Though MLBAM contended that commercial use of such data without a license violated its rights, the court ruled in favor of CDM, saying that names and playing records of MLB® players as used in fantasy games are not copyrightable. The verdict will likely cause a restructuring of the MLBAM-MLBPA deal, as well as sublicenses to Yahoo!, ESPN, CBSSportsLine.com, ProTrade, and Fox-Sports.com, which are each worth approximately $2 million per year (Fisher, 2006b).

In addition to the litigation it has generated, the fantasy sports industry has also spawned two trade associations. The Fantasy Sports Trade Association (FSTA) was started in 1999. It provides a forum for interaction and networking between hundreds of existing and emerging companies in the fantasy sports industry, including entrepreneurs, corporations, investors, advertisers and sponsors (Fantasy Sports Trade Association, n.d.; Fisher, 2006c). In 2006, the Fantasy Sports Association (FSA) was started as the fulfillment of a goal by the National Football League (NFL®) Players Association. Joined by all major national fantasy sports licensees, including Yahoo!, ProTrade and CBSSportsLine, the FSA immediately proclaimed its desire to become the fantasy sports industry's leading voice, though it also indicated that it had no desire to directly compete with the FSTA (Fisher, 2006c).

Fantasy sports competition now exists for nearly every major sport, including football, baseball, basketball, hockey, soccer, stock car racing and golf, as well as variations of competition for several sports as discussed later in this chapter (Birch, 2004). For the month of April 2006, seven major providers, including AOL Sports, ESPN.com, The Sporting News, Fox Sports, Sports Illustrated, Yahoo! Sports, and CBS SportsLine averaged 8.307 million unique visitors who spent an average of 28 minutes and 38 seconds on their websites (see Table 1). These providers bolstered their fantasy football offerings in anticipation of a considerable increase in 2006 fantasy football participants and to make sites as attractive as possible to participants (Fisher, 2006d).

*Table 1. Major Fantasy Sports (FS) Providers, April 2006*

| Fantasy Football Provider | Number of Unique Website Visitors (millions) | Time Spent on Website (Minutes:Seconds) | Major Attraction, Addition, or Renovation |
|---|---|---|---|
| AOL Sports | 5.44 | 16:46 | FS: critical component that scores well among hardcore fans |
| ESPN.com | 14.5 | 42:03 | King of sports Internet landscape relies heavily on FS |
| The Sporting News | .669 | 20:57 | promotes paid FS leagues to engaged, high-income base of sports fans through heavy community aspect |
| Fox Sports | 10.47 | 22:25 | Technical malfunctions marred its 2005 fantasy football season |
| Sports Illustrated | 6.49 | 17:03 | Enhanced all FS offerings; made "Fantasy Plus" section a larger, permanent component of magazine; added fantasy football column by Peter King |
| Yahoo! Sports | 11.35 | 45:30 | Early establishment as FS powerhouse placed it far ahead of late-arriving competition |
| CBS SportsLine | 9.23 | 33:56 | Deep base in FS |

*Source: Fisher, 2006d*

## Appeal

The rapidly increasing appeal of fantasy sports has been attributed to several major factors by mainstream and sport industry media. Perhaps the most widely acknowledged influential factor involves the amount of direct control that the operation of a fantasy sports team allows ordinary individuals. It affords them chances to live vicariously as franchise owners/general managers/coaches of their own team(s) by drafting teams and structuring starting rosters of athletes who, in a sense, play for them (Birch, 2004; Davis & Duncan, 2006). The popularity of fantasy sports participation is also attributed to the construct of competitive experiences among friends and/or with strangers, including match-ups of sport expertise and knowledge between individuals (Birch, 2004; Davis & Duncan, 2006). Another factor widely supposed to drive the growth of fantasy sports participation is the possibility of extrinsic rewards, which may involve provider- or sponsor-offered cash or other prizes for winners, not to mention wagering among participants in various forms and varying amounts (Birch, 2004; Thompson, 2007).

## Demographics

*Market Sizes.* From 2003 to 2006, an estimated 15-18 million people per year engaged in some form of fantasy sports participation (Birch, 2004; Fisher, 2006c, 2007; Janoff, 2005). In football, widely regarded as the most popular fantasy sport, other estimates include 9 million unique fantasy football users in October 2005, as well as an estimated 10 million fantasy football players for all of 2005, including offline players (Russo & Walker, 2006); this market was estimated to grow to more than 11 million for 2006 (Fisher, 2006e).

*Spending.* An August 2006 estimate indicated that approximately 12 million people spent more than $1.5 billion on fantasy sports per year, with some players spending $265 a year as the industry grows 7-10% annually. Major League Baseball® alone generated about $2 million in fantasy licenses in 2005 (Fisher, 2006f).

*Participant Profile.* Combined data from several sources reveals that the typical profile for a fantasy sports participant is that of a young, well educated, White, relatively affluent man (Levy, 2005; Weekley, 2004). According to Eric Bader, senior vice president of MediaVest Digital, "They've got a high household income, most are married, educated, big spenders. And they spend a lot on fantasy sports" (Klaassen, 2006, p. 4). In 2006, the average age of players was 36 years; these players, who spent 3.8 hours per week managing teams, possessed undergraduate degrees and received an annual income of $89,566 (Klaassen, 2006). Interestingly, only 49% of those who play fantasy football describe themselves as die-hard sports fans (Weekley, 2004). Data concerning the average fantasy football user revealed that s/he played 2.1 teams in 2005, spent 5.2 hours per week online managing teams, and visited fantasy game sites seven times per week (Russo & Walker, 2006). Also, nearly 70% of fantasy sports fans checked or managed their fantasy sports teams while at work (Janoff, 2005). Estimates of stock car racing fantasy sports participation indicate that its player base is approximately one-eighth the size of fantasy football's player base, and 20-40% of that for baseball (Fisher, 2007).

*Growth.* Numerous sources predict sizeable future growth for the fantasy sports industry. The total number of fantasy players is expected to reach 30 million by the end of the decade (Fisher, 2006c), with a prediction of more than 6 million fantasy baseball players alone for 2006, up from 5 million in 2005 (Fisher, 2006f). More than 30% of 2005 fantasy football players were first-timers; NFL.com indicated that nearly 50% of its 2005 fantasy audience consisted of first-timers (Russo & Walker, 2006). Fantasy sports has become such a booming phenomenon that at least four websites have been set up to serve as independent fantasy sports arbitrators, which examine disputes between and among players for a case-by-case fee (Thompson, 2007).

*Gender.* Though described by Thompson (2007) as "an outlet for misdirected testosterone, fueled by ego and trash talk" (p. A1), demographic growth segments in fantasy sports may not be limited to males only. Numerous studies indicate that several different media (including fantasy sports) offer men opportunities to emphasize masculine ideals in sport by providing viewing pleasures for male spectators (Duncan & Brummett, 1989); emphasizing empowerment through sports knowledge (Duncan & Brummett, 1989; Gantz & Wenner, 1995; Kennedy, 2000); and reinforcement of masculinity through hypermasculine sport fanship choice (Messner, Dunbar & Hunt, 2000; Sargent, Zillman & Weaver, 1998; Sullivan, 1991). However, Weekley (2004) indicated that 25% of fantasy football players in 2004 were women.

Another source indicated a major upswing in female fantasy sports participation related to a sports property long known to boast a female fan base of approximately 40%: the National Association of Stock Car Auto Racing (NASCAR®). According to an April 2007 study released by the Fantasy Sports Association and developed in conjunction with Interactive Sports Marketing, 33% of the more than 1.2 million people who play stock car racing fantasy games are females: far higher than the female participation rates of any other fantasy sports games and much larger than had been assumed industry-wide (Fisher, 2007). According to industry experts, these findings showed considerable demographic expansion for the fantasy sports industry, since female participation in other fantasy sports hovers in single-digit or low double-digit percentages (Fisher, 2007). Of additional importance were data indicating that stock car racing fantasy participants, who spent an average of 4.49 hours per week playing and researching the sport online, were almost as engaged time-wise as fantasy football players (5.05 hours per week) and fantasy baseball players (4.97 hours per week) (Fisher, 2007).

## Marketing

With the increased growth in fantasy sports participation, marketing dynamics surrounding the industry have heightened organizational restructuring and reformulation of marketing initiatives related to it (Fisher, 2006f). NBC's September 2006 unveiling of its fully redesigned and rebranded version of NBCSports.com featured a heavy emphasis on fantasy sports. NBC also purchased Allstar Stats Inc., which powers a weekly fantasy contest centered on its Sunday Night Football broadcast (Fisher, 2006g). Current fantasy sports sponsorships regularly command seven figures (Klaassen, 2006).

According to David Katz, Vice President for Entertainment and Sports Programming for Yahoo!, advertisers may currently fear a close association with user-generated content, because of the lessened guarantees of quality controls and variable standards of practice. However, Katz stated that advertising will likely evolve to accept user-generated content in more meaningful ways: "Part of our job is to show them [advertisers] that it is a safe environment and to create as many safe environments as we can for advertisers. That to me is a big leap that is going to help us move our industry forward in terms of getting advertisers to support the whole industry" (Oliver, 2005, p. 10).

Another way in which the fantasy sports boom has permanently impacted marketing tied to event consumption patterns in sport, lies within the subsequent diminished fan focus on the success of teams and the increased notoriety of individual performances (Birch, 2004). This heightened focus on individual performances has spurred interest even in the most mundane games, thereby increasing the value of sponsorships. This is because of the decreased likelihood of waning fan interest due to competitive factors, such as one-sided games, less-than-exciting match-ups, or the lack of seasonal success by a fan's favorite team.

However, perhaps the largest seismic shift in the fantasy sports marketing landscape has occurred offline with the advent of numerous fantasy season preview magazines, advertising across all forms of media, and numerous promotional offers to join leagues (Fisher, 2006e). Greg Ambrosius, editor of *Fantasy Sports* magazine and president of the Fantasy Sports Trade Association, said that more than 25 magazines devoted to fantasy football were published in 2005, with sales for more successful titles ranging from 150,000 to 400,000.

CBSSportsLine.com General Manager Steve Snyder said, "From a marketing perspective, putting a magazine into people's hands, and draft boards up on the walls—that all makes up the [fantasy sports] experience and its very important to us" (King, 2005, p. 20). For example, in addition to its other revamped online offerings, NBC Sports also plans a weekly online fantasy football show entitled "NBC Fantasy Fix" (Fisher, 2006g). Such an increased emphasis on fantasy team draft parties; the introduction of new, simpler games; and the sharp rise in TV programming devoted to fantasy football (Fisher, 2006e) has inevitably led to expanded marketing initiatives through promotions.

## Promotions

Most major marketing promotional initiatives tied to fantasy sports have involved fantasy baseball and fantasy football.

*Baseball.* On February 16, 2006, MLBAM held an expanded fantasy Opening Day marketing event that featured video content on MLB.com and celebrity drafting tips. CBSSportsline.com, ESPN and Yahoo! each used a series of event marketing initiatives, title sponsorships, or celebrity spokespersons to promote their fantasy baseball games in 2006. This included linking Mitchum®, a grooming products brand, to a presenting sponsorship for CBSSportsline.com's launch of its fantasy baseball (Fisher, 2006f). ESPN signed EarthLink® as a presenting sponsor of its fantasy baseball draft activities and combined that sponsorship with appearances by ESPN talent in Chicago, New York, and Atlanta, as well as a fantasy draft special on ESPN's "Baseball Tonight" show (Fisher, 2006f).

*Football.* Each of the National Football League's (NFL®) TV partners produced some sort of special fantasy draft TV program in August 2006. ESPN initiated a half-hour, football-season-long television show devoted exclusively to fantasy sports during the 2006 NFL® season (Fisher, 2006e). The NFL® staged Fantasy Draft Week beginning August 28, 2006, as a series of bar/restaurant events held in conjunction with major beer sponsor Coors. CBSSportsLine.com teamed with Morton's Steakhouses, and FoxSports.com teamed with Buffalo Wild Wings to hold similar events, while Yahoo! held a pre-draft press event in Manhattan with celebrity NFL® quarterbacks Eli Manning and Matt Leinart (Fisher, 2006e).

The explosive growth in fantasy sports participation, and utilization of fantasy sports as a marketing vehicle by sponsors and media properties, prompts questions about why people invest time, money and emotions to participate. A more complete view of fantasy sports consumption is warranted, given the little inquiry about the forces that influence one to become a fantasy sports player.

# A CONCEPTUAL FRAMEWORK OF FANTASY SPORTS CONSUMPTION

A person's decision to play fantasy sports games is influenced by several variables. As is the case with any purchase decision, consumption of fantasy sports can be driven by a combination of internal and external influences. A review of relevant literature leads to the conceptual framework illustrated in Figure 1. The framework demonstrates that fantasy sports consumption is impacted by the interplay of *psychological characteristics* internal to consumers, as well as *social interactions* and *marketer-controlled influences* that are external to consumers, but can affect consumption decisions. These three factors are discussed in the following sections.

## Psychological Influences

One set of influences on fantasy sports consumption resides within individual consumers. These variables, labeled psychological influences, are individual characteristics of consumers that could affect one's decision to participate in fantasy sports. Psychological influences identified as having a potential impact on the fantasy sports consumption decision are: the ability to exert *control*; the desire to *escape* from reality; and the feelings of *achievement* experienced when success is attained in competitive play.

*Control.* A strong psychological influence on fantasy sports consumption is the feeling of control created through participation in fantasy sports games. Owning a team in a fantasy sports league allows a person to participate vicariously in professional sports (Bernhard & Eade, 2005). Such decisions as drafting players for a team, acquiring players from other fantasy teams via trades, and claiming players from free agent pools provide fantasy team owners opportunities to experience a measure of the careers of certain sport management professionals. Exerting control as a fantasy sports team owner may be viewed as a way for men to demonstrate their masculinity (Davis & Duncan, 2006). Development and application of sports knowledge is highly correlated with the control experienced by fantasy team owners. The

*Figure 1. Influences on fantasy sports consumption*

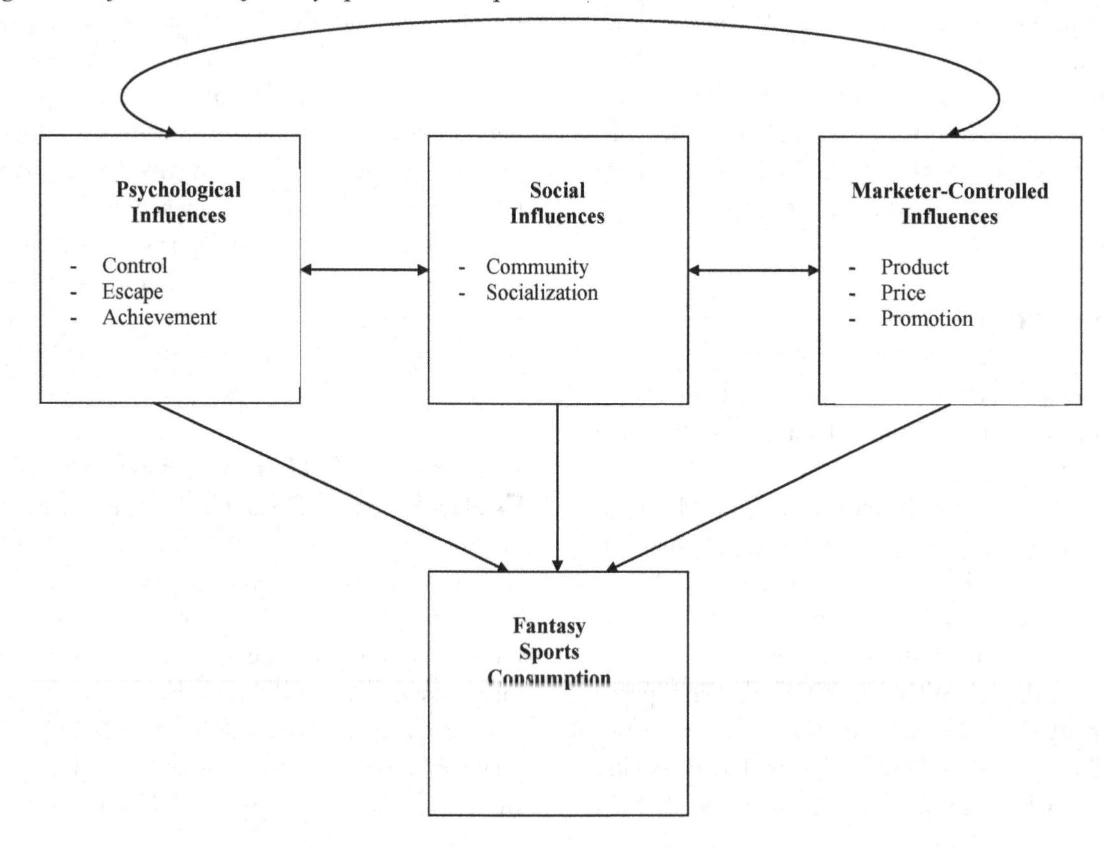

more knowledge one has about players' statistics, injuries, or other variables that can impact the performance of players on one's fantasy team roster, the stronger the feelings of control and confidence in decision-making become for fantasy team owners. For example, fantasy football players with more in-depth knowledge about NFL® players and events that could affect players' performances, may experience a stronger sense of exerting control, have greater confidence in selecting players in their league's draft, as well as their weekly line-ups than fantasy players with less knowledge.

Fantasy game marketers recognize consumers' desires to experience feelings of control as team owners. Advertising campaigns for fantasy sports games have utilized messages of individuals exerting control as appeals to participate (Shipman, 2001). Similarly, marketers of support products for fantasy game players often appeal to the desire to feel in control in their marketing efforts. For example, the website of fantasy baseball information provider Big Dawg Baseball promotes its 2007 draft kit products by using such copy as "... start preparing to dominate your fantasy baseball league" and "Get a huge leg-up on the rest of your league and sign up today with Big Dawg Baseball" (Big Dawg Baseball, 2007).

*Escape.* Sports offer people an opportunity to forget their worries and briefly escape into another world (Wann, Melnick, Russell & Pease, 2001). Escape is central to the notion of consumption as a form of play (Holt, 1995). The escape motive cuts across all demographic segments and is not limited to particular age groups, income levels, or occupation categories. Regarding fantasy sports, consumers who feel the need to get away from daily routines or relieve stress, regardless of their demographic characteristics, can find refuge in the simulated world of professional sports executives. The get-away-from-it-all ability provides fantasy sports players with an outlet to release tensions and engage in play with others, while immersing themselves in a particular sport. Escape attained via fantasy sports participation is congruent with the motives of hedonic consumption, which is driven by consumers' need to engage in experiences filled with fun and pleasure (Holbrook & Hirschman, 1982).

Based on their premise of allowing one to use fantastic imagery to assume the role of a sports team owner, fantasy sports games are ideal vehicles for enabling consumers to fulfill their desire to escape. Fantasy team owners become involved in professional sports in the only way most of them could ever possibly experience: assuming the role as a sports team executive in a simulated league in which performance is based on statistics of actual professional players.

Research into feelings experienced in fantasy game play and online gaming points to the importance of escape among consumers of these games. Escape from the everyday world is an important part of consumption experiences for many fantasy game players, as are preparation rituals conducted before engaging in game play, such as devising game strategies (Martin, 2004). Another way fantasy sports provide escape is by offering consumers a means to expend and recharge energy (Kim, Park, Kim, Moon & Chun, 2002). In addition to providing a getaway outlet, players benefit from fantasy sports games by strengthening analytical skills and problem-solving abilities (Bernhard & Eade, 2005). Thus, the escapes that consumers experience when engaged in fantasy sports may be more than mere diversions from reality; they may become means of sharpening intellectual skills that can benefit consumers upon return to reality.

*Achievement.* Another internal motive that influences fantasy sports consumption is achievement. According to Heckhausen (1967), achievement motivation is "the striving to increase, or keep as high as possible, one's own capability in all activities in which a standard of excellence is thought to apply and where the execution of such activities can, therefore, either succeed or fail" (p. 4). The need for achievement can be met through the competitive nature of fantasy sports in which the goal is to outperform other fantasy teams in certain statistical categories. While casual consumers of fantasy sports are unlikely to be drawn to fantasy games because of competition, most highly involved fantasy game players are thought to engage in fantasy sports because of their desires to engage in competitive activity. This need to achieve success and outperform other league players motivates players to invest time and money in acquiring and analyzing information that could improve their fantasy teams' performances.

Fantasy sports meet consumers' need for competition through frequent competitive experiences, such as weekly match-ups against other team owners in a fantasy football league (Davis & Duncan, 2006). Fantasy game marketers recognize that some consumers can be motivated to play due to opportunities to compete and achieve success through victories over other fantasy teams and even winning league championships. Advertising strategies for some fantasy sports games have been based on the suggestion that successful players (i.e. fantasy league champions) can enjoy public recognition of their sports knowledge (Shipman, 2001).

## Social Influences

Fantasy sports can be viewed as a form of consumption that is based on playful interaction with other people. Holt (1995) described "consumption as play" as a type of consumption practice in which a desire exists to interact with fellow consumers. Two forms of playing identified by Holt are *communing* and *socializing*. These forms of play appear to be influential forces on a consumer's decision to participate in fantasy sports and are discussed to follow.

*Community.* Fantasy sports games provide a forum for people with shared interests to interact with one another. Fantasy game players who play in public leagues are joining a group of people with whom they have had no prior interactions. In a public league, the group of team owners brought together to compete have only one known shared interest when the league is formed: a supposed interest in the sport upon which the fantasy game is based. Fantasy game tools such as live online drafts, message boards, and e-mails between team owners facilitate community building among fantasy consumers. Davis and Duncan (2006) posit that fantasy sports provide a particular form of community building: male bonding. They point to the fact that the large majority of fantasy sports consumers are males, often creating a "men's club" that provides a platform for reaffirming masculinity.

Participation in fantasy sports not only provides a means of communing with other players in one's league, but it enables communing through shaping one's identity. Just as people may define themselves in part through identification with a sports team or athlete (Wann et al., 2001), fantasy sports participation fosters identity development through identification with fantasy sports play in general (e.g. the ability to identify one's self as a fantasy football player). Also, sport identification can be enhanced through fantasy sports consumption. For example, fantasy baseball consumers may develop a stronger identification with Major League Baseball® as a result of closely monitoring game scores, league standings, players' statistics, and other information used to make decisions in fantasy game play. Identification with fantasy sports in general, or a particular sport through fantasy games, can become an element in one's identity and can aid in gaining acceptance among communities with similar interests.

*Socialization.* A strong influence on the decision to consume fantasy sports is the opportunity to socialize with family, friends, co-workers, or other people within previously established social networks. Fantasy sports participants can create private leagues, which allow those players to control the league's other participants. Whereas public league players are brought together through shared interest in the sport, private leagues are formed by dual influences of prior social interaction and sport interest. Social connections can even outweigh sport interest in some private leagues. For example, participation in a private fantasy football league created for a group of former college classmates may be motivated more by social affiliation (i.e. connecting with old friends) than members' interest in football.

Researchers investigating consumer behavior of sport and gaming have identified socialization as a prime motive for consumption. Desire for group affiliation and spending time with family are key reasons why people consume sport (Wann et al., 2001). The rapid evolution of fantasy sports as a mass-market product has enabled more consumers to meet these needs through sport. A negative stereotype of fantasy sports players that emerged in its early years was one of a group of dedicated players who were viewed by Bernhard and Eade (2005) as "an enthusiastic but socially disconnected cult" (p. 32). This stigma has been removed, largely through easy access to games via the Internet, making fantasy games a more common, attractive vehicle for social interaction.

## Marketer-Controlled Influences

In addition to psychological and social factors that influence consumers' choices to consume fantasy sports, marketers of fantasy games can impact consumer behavior through their selection of marketing mix elements. Potential for influencing consumers via the marketing mix is greatest through *product*, *price*, and *promotion* decisions.

*Product.* The product element of fantasy sports games includes game *branding*, game *formats*, and design of the *user experience*. The branding strategy used by a game marketer can affect consumers' perceptions about the quality and credibility of a game. For example, the primary fantasy game offered by the NFL® on its website during the 2006 season was "NFL® Fantasy 2006", with a shadow endorsement "Powered by CBSSportsLine.com." Football fans who had never played fantasy football previously may have considered fantasy consumption through their knowledge of the NFL® brand. Conversely, the NFL® is able to promote its association with CBSSportsLine, which is recognized by regular fantasy players as a well established, leading fantasy games provider. Similarly, Major League Baseball®'s fantasy games are co-branded with *The Sporting News*, a media outlet historically recognized for its coverage of baseball. These co-branded game offerings can enhance the perceived value of the games among baseball-savvy consumers.

Offering multiple types of fantasy game formats for a given sport can enhance the product offering of fantasy games. In the case of fantasy baseball games offered by Major League Baseball®/*The Sporting News*, players can choose to play games using the classic roto-scoring system (in which fantasy teams earn points based on their league rank in statistical categories), or a head-to-head scoring system (pairs of teams compete in terms of their players' statistical performances). Another game format offered by MLB® is a salary cap game in which team owners select a team with players who have variable salary values; owners must draft rosters whose total salaries do not exceed the salary cap. This format is noteworthy because MLB® does not have a salary cap in reality. Other variations of fantasy baseball games are available too, giving consumers many options for finding a game format that appeals to them.

A third element of the fantasy game product that can influence consumption of fantasy sports is the design of the user experience. The advent of online fantasy sports games predicates the importance of the user experience design, which becomes the epicenter of fantasy sports play. Just as sportscapes (physical environment in which sport is contested) influence consumers' patronage of spectator sports (Wakefield & Blodgett, 1994, 1996; Wakefield, Blodgett & Sloan, 1996), the environment in which fantasy sports players interact with competitors and make game decisions can influence consumption decisions and re-patronage intentions.

Since the development of a fantasy game's physical environment can be influential in creating a positive user experience, game marketers use a variety of tools to enhance the experience. Many fantasy game offerings enable team owners to select a team logo and/or color scheme that is represented graphically on a football helmet or baseball jersey, creating tangibility for an owner's fantasy team. Another element that enhances the product offering is message boards that allow team owners within an individual league to interact with one another by posting messages about player trades, strategy or rules questions, and even engaging in various forms of banter with opponents. Message boards not only serve as a means of information exchange, but they provide a channel for facilitation of meeting players' communing and socializing needs too.

Another way that the user experience design can influence fantasy sports consumption is the availability of statistics. Fantasy team owners with high involvement are particularly interested in accessing player statistics to make better decisions about roster changes and monitor performance of their team, as well as others in their league. Fantasy game marketers recognize that differing needs exist among fantasy game players. For example, Yahoo!'s 2007 fantasy football game offering had basic NFL® player statistics available for free, but it also offers other features for high involvement fantasy team owners, such as in-depth NFL® player and team reports and analysis that can be purchased for a one-time fee of $19.99.

*Price.* The opportunity for consumers to engage in fantasy sports consumption has been impacted greatly by pricing tactics employed by fantasy game marketers. Widespread adoption of fantasy sports play has occurred largely through free fantasy games offered by mainstream media companies such as ESPN and Yahoo!, as well as free games offered on professional sports leagues' websites. MLB®, NFL®, National Basketball Association (NBA®), National Hockey League (NHL®), and NASCAR® all offer free fantasy games on their league websites. Free game offerings are important for attracting new fantasy sports consumers, enabling them to engage in product trial at no cost.

In addition to free games that appeal to casual and price-conscious fantasy sports consumers, fantasy game marketers typically also offer fantasy games for which consumers must pay to participate. Consumers who participate in these pay fantasy leagues often have higher involvement and are more experienced fantasy sports players. Many pay fantasy leagues offer cash or other prizes for league winners, with team owners subsidizing the rewards through their payments to play in the league. Pricing of pay fantasy sports products is often structured to encourage more involvement and consumption. For example, ESPN's pricing for its pay fantasy football product in 2006 was $29.95 for one team, $49.95 for three teams, and $69.95 for five teams (ESPN.com, 2006). The pricing structure of fantasy sports games enables marketers to appeal to various market segments in terms of product usage rate, with free games for light-to-moderate users and pay games for heavy users.

*Promotion.* Consumption of fantasy sports can be influenced through promotion programs, particularly the use of incentives to encourage or reward game play. In addition to the aforementioned pricing incentives, fantasy game marketers use sales promotion methods intended to persuade consumers to play fantasy sports games. For example, in 2006 CBSSportsLine.com offered a $20 discount and a premium item (a free cap) for consumers to purchase Fantasy Basketball Commissioner: a fee-based fantasy basketball league product.

Rewarding performance is another incentive used to influence fantasy sports consumption.

The types and value of incentives vary depending largely on whether the league is a free league or pay league. Some free leagues do not offer any rewards for their champions. Other free leagues and some pay leagues offer prizes of nominal value, such as a t-shirt or banner denoting the recipient as a fantasy league champion. While low in financial value, these rewards can be a source of pride for recipients. Cash rewards are typically restricted to pay fantasy games and prize amounts vary: often these are correlated with the cost required to play in a pay league. For example, CBSSportsLine.com offers four different levels of pay fantasy football leagues, ranging in cost from $14.95 to $499.99. Prizes awarded to the league champions for the games range from $200 to $3,500 (CBSSportsLine, 2006). Offering different levels of risk and reward through different price and prize amounts allow fantasy game marketers to segment the fantasy player market and appeal to a broad group of consumers.

Another element of fantasy game marketers' promotional mix is the use of interactive marketing tactics, particularly e-mail marketing. As consumers sign up to play free or pay fantasy games, their contact information typically is fed into a game provider's database. This information can be used to influence customer loyalty to a fantasy game provider by cross-selling other fantasy sports games. For example, consumers who sign up to play a free fantasy baseball game on *The Sporting News* web site might receive e-mails containing information for fantasy games in other sports (e.g. NFL®, NASCAR®, or Professional Golfers Association [PGA®]). Such cross-selling tactics could increase visits to the game providers' website beyond the time periods of the initial fantasy games, thus increasing the value of the sites to advertisers that support the free fantasy games.

## FUTURE TRENDS

### Transition From Linear Experience to Engaged Community

According to Chris Russo, president of CR Media Ventures, "The universe of participants is getting both broader and deeper" (Fisher, 2006e, p. 7). This expansion of the fantasy sports universe is likely due to a demonstrated existence of fantasy sports participation as a pleasurable social experience (Davis & Duncan, 2006). As users increasingly desire connections to true online communities with social networking for quality levels of engagement, fantasy sports providers must ensure that associative opportunities become critical components of the entire user experience (Fisher, 2006d). Those providers who are able to offer users such community-based connectivity through enhanced participative experiences will generate greater percentages of Internet page views and wireless data consumption; this in turn will provide more viable, safe platforms for marketers and their initiatives (Fisher, 2006e; Oliver, 2005).

### Dynamics of Transition

Russo and Walker (2006) identify two chief dynamics that will benchmark successful transition of the fantasy sports industry from the linear experience model to the engaged community model.

First, movement will occur away from subscription models to ad-supported models that position fantasy sports games as multidimensional marketing vehicles. Strong engagement metrics (e.g. page views, visits, time spent on the website) and large numbers of new players in 2005 suggest strong future growth for ad-supported models, and a more limited niche for subscription-based products/services (Russo & Walker, 2006). Approximately 63% of fantasy football users recalled seeing an ad from one of the five major fantasy advertisers, and those who recalled fantasy advertising had a much better impression (15-20% better) than users with no recall/association (Russo & Walker, 2006).

Second, further development and evolution of a wide range of offline fantasy sports products and services to compliment online games must be generated. The increased marketing potential inherent in such offline activity surrounding fantasy sports must be embraced with greater creativity by marketers. According to Russo and Walker (2006), sponsors are presently included in online game headers and banners, but site providers rarely create custom interactive features that incorporate the sponsors, nor do most sponsors effectively leverage fantasy football in offline marketing. Klasseen (2006) identified the premise that game providers have historically had so little trouble selling fantasy sports advertising packages, little incentive existed to move past basic banner ads and sponsor buttons. However, Klasseen (2006) notes movement toward the design of more creative marketing packages in 2006 to drive sponsor initiatives.

## Keys to Unlocking Potential of Engaged Fantasy Sports Communities

Given the estimated number of players and monetary volume of the industry, the stability and viability of fantasy sports as a marketing platform is no longer questionable. Tremendous potential to leverage fantasy football as a marketing property apparently exists, but marketers and fantasy game providers must work together to develop more creative, high-impact programs, promotions, and executions in both online and offline elements. This development must be tempered by the realization that such engaged fantasy sports communities exist at the intersection of television viewing and fantasy play (Russo & Walker, 2006). The pending transformation must involve brand marketers embracing fantasy sports, as well as fantasy providers developing new products and services that extend fantasy sports beyond merely online games (Russo & Walker, 2006). According to Clay Walker, chairman of the Fantasy Sports Association and senior vice president of Players Inc. (the commercial arm of the NFL® Players Association), advertisers have not yet determined how to maximize their investments in fantasy sports (Fisher, 2006c). Such maximization, Fisher (2006c) says, will not occur until leagues and/or providers have "significantly better, richer data to bring to market" (p. 6). Conclusively, as with all social-network advertising, marketers must tread lightly so as not to disrupt the consumer experience (Klaassen, 2006).

Fueled by the discovery of fantasy stock car racing's considerable female base, the apparently increasing numbers of women make fantasy sports a fertile ground for marketing investment. Marketers should explore initiatives that encourage women who do not currently participate to sample fantasy sports participation. Plus, they must also drive those who already participate to more involved levels of the engaged community. This will require designing marketing campaigns that offer incentives, encouragement and acceptance.

Creation and dissemination of meaningful statistical information will be important to attract fantasy participants who crave and attempt to utilize even the smallest fragment of statistical data to their advantages, much as general managers and coaches do in actual sport franchises. According to Dick Glover, NASCAR®'s vice president of broadcasting and new media, the growth of stock car racing's overall and female fantasy participation rates can be directly attributed to specific efforts by NASCAR® in conjunction with Stats Inc. to create and disseminate more statistical information and concrete performance data (Fisher, 2007).

Another important implication, particularly for fantasy sports providers, is the emergence and growth of games/contests that apply fantasy sports game principles to Hollywood awards; reality television shows; cover appearances on national lifestyle magazines; and other entertainment elements (Fisher, 2007). Internet portals may find these to be particularly salient streams of new revenue if developed well.

## CONCLUSION

Ultimately for marketers, the major implication of the growth in the fantasy sports industry is that its evolution is another unambiguous signal of the emergence of an increasingly empowered consumer, in this case, the empowered sports fan. This empowerment of postmodern sports fans stems greatly from two growing elements of the sports universe: fantasy sports and sports talk radio (Fisher, 2006d). However, unlike sports talk radio, the fantasy sports industry affords its participants unparalleled opportunities for bonding (Davis & Duncan, 2006). Such opportunities have further spurred empowerment. Fantasy sports providers, through their expanded game offerings and marketing/promotional initiatives, have begun to accommodate the burgeoning fantasy sports market, as its movement tracks from online elements to offline elements that create movement from an individualized *linear experience* to participation in an *engaged community.*

## REFERENCES

Bernhard, B.J., & Eade, V.H. (2005). Gambling in a fantasy world: An exploratory study of rotisserie baseball games. *UNLV Gaming Research & Review Journal, 9*(1), 29-42.

Big Dawg Baseball. (2007). 2007 fantasy projections now available! Retrieved February 5, 2007, from http://www.bigdawgbaseball.com/?gclid=CIOQpcP4l4oCFRsZVAod_3S3nA

Birch, D. (2004, August 25). Fantasyland. *Modesto Bee*, F1.

CBSSportsLine.com. (2006). CBSSportsLine.com fantasy rules. Retrieved October 21, 2006, from http://football.sportsline.com/splash/football/spln/single/rules

Davis, N.W., & Duncan, M.C. (2006). Sports knowledge is power: Reinforcing masculine privilege through fantasy sport league participation. *Journal of Sport & Social Issues, 30*(3), 244-264.

Diamond, D. (2004). Rotisserie baseball—what is it, and why should I play? Retrieved August 4, 2004, from http://www.kcmets.com/RotoWorld

Duncan, M.C., & Brummett, B. (1989). Types and sources of spectating pleasure in televised sports. *Journal of Sport & Social Issues, 6*, 195-211.

ESPN.com. (2006). *Rules – legal restrictions*. Retrieved October 21, 2006, from http://sports.espn.go.com/fantasy/football/ffl/story?page=fflruleslegal

Fantasy Baseball. (n.d.). Retrieved October 2, 2006, from http://www.answers.com/main

Fantasy Sports Trade Association. (n.d.). *Welcome to the official site of the FSTA*. Retrieved October 24, 2006, from http://www.fsta.org

Fisher, E. (2006a, May 8). Lawsuit over fantasy stats set for trial. *Sports Business Journal, 6.*

Fisher, E. (2006b, August 14). Fantasy ruling could force MLBAM to revisit deal with union. *Sports Business Journal, 7.*

Fisher, E. (2006c, May 8). New group aims to be voice of fantasy sports. *Sports Business Journal,* 6.

Fisher, E. (2006d, June 12). MySpace race: Fan participation drives change on Web. *Sports Business Journal, 1.*

Fisher, E. (2006e, July 31). Fantasy football gets early jump on season. *Sports Business Journal*, 7.

Fisher, E. (2006f, February 20). New stat for fantasy: More than 6M players. *Sports Business Journal, 5.*

Fisher, E. (2006g, September 4). NBC Sports will roll out redesigned site on Tuesday. *Sports Business Journal, 44.*

Fisher, E. (2007, April 30). Women fuel NASCAR's fantasy growth. *Sports Business Journal, 1.*

Gantz, W., & Wenner, L.A. (1995). Fanship and the television sports viewing experience. *Sociology of Sport Journal, 12*, 56-74.

Heckhausen, H. (1967). *The anatomy of achievement motivation*. New York: Academic Press.

Holbrook, M.B., & Hirschman, E.C. (1982). The experiential aspects of consumption: Consumer fantasies, feelings, and fun. *Journal of Consumer Research, 9*(2), 132-140.

Holt, D.B. (1995). How consumers consume: A typology of consumption practices. *Journal of Consumer Research, 22*(1), 1-16.

Hu, J. (2003). *Sites see big season for fantasy sports*. Retrieved October 5, 2006, from http://news.com/2102-1026_3-5061351.html

Janoff, B. (2005, April 11). Marketers of the next generation. *Brandweek, 46*(15), 38.

Kennedy, E. (2000). You talk a good game. *Men and Masculinities, 3*(1), 57-84.

Kim, K.H., Park, J.Y., Kim, D.Y., Moon, H.I., & Chun, H.C. (2002). E-lifestyle and motives to use online games. *Irish Marketing Review, 15*(2), 71-77.

King, B. (2005, November 14). Magazines cater to growing industry. *Sports Business Journal*, 20.

Klaassen, A. (2006, August 7). That's real money—$1.5B—pouring into made-up leagues. *Advertising Age, 77*(32), 4.

Levy, D. (2005). *Sports fanship habitus: An investigation of the active consumption of sport, its effects and social implications through the lives of fantasy sports enthusiasts*. Unpublished doctoral dissertation, University of Connecticut.

Martin, B.A. (2004). Using the imagination: Consumer evoking and thematizing of the fantastic imaginary. *Journal of Consumer Research, 31*(1), 136-149.

Messner, M.A., Dunbar, M., & Hunt, D. (2000). The televised sports manhood formula. *Journal of Sport & Social Issues, 24*(4), 380-394.

Oliver, J. (2005, November 21). Sports media & technology conference. *Sports Business Journal*, 10.

Russo, C., & Walker, C. (2006, May 8). Fantasy sports growth hinges on marketing, offline efforts. *Sports Business Journal, 21.*

Sargent, S. L., Zillman, D., & Weaver, J.B. (1998). The gender gap in the enjoyment of televised sports. *Journal of Sport & Social Issues*, 22(1), 46-64.

Shipman, F.M. (2001). Blending the real and virtual: Activity and spectatorship in fantasy sports. *Proceedings of the Conference on Digital Arts and Culture*. Retrieved October 3, 2006, from http://www.csdl.tamu.edu/~shipman/papers/dac01.pdf

Sullivan, D.B. (1991). Commentary and viewer perception of player hostility: Adding punch to televised sports. *Journal of Broadcasting & Electronic Media, 35*(4), 487-504.

Thompson, A. (2007, March 10). In fantasy land, sports judges hear imaginary cases. *The Wall Street Journal*, A1.

Wakefield, K.L., & Blodgett, J.G. (1994). The importance of servicescapes in leisure service settings. *Journal of Services Marketing, 8*(3), 66-76.

Wakefield, K.L., & Blodgett, J.G. (1996). The effect of the servicescape on customers' behavioral intentions in leisure service settings. *Journal of Services Marketing, 10*(6), 45-61.

Wakefield, K.L., Blodgett, J.G., & Sloan, H.J. (1996). Measurement and management of the sportscape. *Journal of Sport Management, 10*(1), 15-31.

Walker, S. (2007). *Fantasyland: A sportswriter's obsessive bid to win the world's most ruthless fantasy baseball league.* New York: Penguin Group.

Wann, D.L., Melnick, M.J., Russell, G.W., & Pease, D.G. (2001). *Sports fans: The psychology and social impact of spectators.* New York: Routledge.

Weekley, D. (2004, September 7). Fantasy football numbers on the rise. *Charleston Gazette*, 3B.

# Chapter XII
# Computational and Robotic Pool

**Jean-Pierre Dussault**
*University of Sherbrooke, Canada*

**Will Leckie**
*Nortel, Canada*

**Michael Greenspan**
*Queen's University, Canada*

**Marc Godard**
*Queen's University, Canada*

**Jean-François Landry**
*University of Sherbrooke, Canada*

**Joseph Lam**
*Queen's University, Canada*

## ABSTRACT

*We introduce pool and its variants, and describe the challenges of computationally simulating the game to create a robot capable of selecting and executing shots on a real table. A proficient pool player performs accurate shots and recovers good position with reasonable alternatives to play following each shot, which requires planning. The computational and robotic simulation of a high level player requires vision, calibration, and accurate robot positioning, as well as an ability to precisely anticipate the table arrangement through simulation of the planned shots. Such a system also requires strategic planning in order to recreate the human's ability to clear table after table. Ultimately, the challenges associated with the creation of a pool robot will promote new ways of using existing AI methods and provide, if successful, a training tool for players wanting to improve their game.*

## INTRODUCTION

The interest in computational pool is motivated by a variety of factors. The mechanics of pool has long served as a focus of interest for physicists, an interest which has extended naturally to the realm of computer simulation. Realistic and efficient simulators have led to the proliferation of computer pool games. These games include significant computer graphics components, often including human avatar competitors with unique personalities, as well as an element of rudimentary artificial intelligence to simulate strategic play. In addition to the interesting challenges of computational pool, the interest in robotic pool stems from the requirement of dealing with a physical system. The issues of robotic pool include the identification and accurate localization of balls through computer vision techniques, and accurate calibration and positioning of the robotic mechanism. Robotic pool systems are geared toward serving as advanced training platforms to assist and improve human play, and also hold the promise of competing directly against proficient humans, which is in some sense the ultimate challenge.

Though in the context of this chapter the primary objective is the creation of a robotic pool player, we wish to extend our research beyond the scope of robotics. We hope that by researching the best way of making a perfect player we can develop new AI approaches, and possibly contribute to other problems of that nature. Coincidently, a sub-goal is also to come up with a system good enough so that a human player can use it for training. Although the problem we wish to solve is deterministic in nature, subject to the laws of physics, it is so easily influenced by many small factors that it can actually be seen as stochastic. This makes it very hard to create a perfect player, because even if he never misses, the outcome of the game is never pre-defined.

In this chapter we will explore the technical challenges of computational and robotic pool. Some recent work on artificial intelligence methods for strategic play is presented. The graphics elements of computational pool are not covered in this chapter, as they are considered to be similar to other computer games, and not worth discussing uniquely. We also present a review of previous work in robotic pool, as well as a detailed description of the Deep Green® robotic pool system.

For simplicity, and somewhat loosely, we use the term *pool* here to encompass all cue-sports, and we limit ourselves to those variants which involve pockets. The basic elements common to all of these games are as follows:

- **Cue:** The stick used to strike the balls.
- **Cue Ball:** The solid white ball which is always struck first with the cue.
- **Object Balls:** The various balls on the table which need to be sunk in pockets.
- **Table:** The flat playing surface. The table size varies, depending on the game played, but is always rectangular, being twice as long as is wide. In most games, the table has a pocket in each of the four corners, and one at the centers of each long side. The table is covered in a textured felt (or baize) material, usually of green color, to add a frictional damping effect to the shots.
- **Rail:** A rubberized edge running along the inner boundary of the table, to accommodate rebounds following the collision of a ball.
- **Table State:** the position of all balls at rest on the table, ready for the next shot.
- **Shot:** The basic element of play, where the player strikes the cue ball with the cue tip, and the cue ball then collides with object balls. Depending upon the game variant and table state, a shot may be considered "legal" or "illegal" (i.e. "foul"). Different classes of shot type include:
    - **Direct Shot:** A shot where the cue ball hits an object ball, which then reaches a pocket.
    - **Bank Shot:** A shot where the cue ball hits an object ball, which then rebounds on a rail before reaching a pocket.
    - **Kick Shot:** The cue ball first rebounds off of a rail, and then hits an object ball which reaches a pocket.
    - **Combination Shot:** The cue ball impacts an object ball, which then collides with another object ball. The second object ball then reaches a pocket.
- **Spin:** Also known as *English*, spin is purposefully imparted to the cue ball during the shot using a variety of techniques to enact a desired effect. Spin can be communicated from the cue ball to the object ball, and can greatly affect the outcome of the shot, especially with respect to the resulting table state. The three types of spin are:

- **Draw:** Also known as back spin, draw results from striking the cue ball below center. One of the effects of draw is to shorten, or possibly reverse, the path of the cue ball following a collision with an object ball.
- **Follow:** Also known as top spin, follow results from striking the cue ball above center. An affect of follow is to lengthen the path of the cue ball following a collision.
- **Side Spin:** This occurs when the cue ball is struck either left or right of center. Side spin is used mostly to alter the path of the cue ball as it rebounds off of a rail.

## Rules of 8 Ball

The game of 8-ball is likely the most popular of all pool variants, partly due to the simplicity of its rules, which makes it an attractive entry game for the novice. The objective is to be the first player to sink the 8-ball, after having sunk all seven balls in one's assigned group of either solids (ball one to seven) or stripes (ball 9 to 15). Sinking the 8-ball prior to sinking all other balls in one's group results in a loss of game.

These basic rules are often all that a beginning player needs to know to enjoy a game, but beyond these, there are a number of more specific situations that can occur that have been codified. In North America, one of the most commonly used rules in tournament play are those of the Billiard Congress of America (*BCA*) (BCA, 2008). The largest point of difference between the BCA rules and common casual play is that of ball nomination. Ball nomination is the process of declaring which object ball is intended to be sunk in which pocket, prior to placing a shot. Pocketing an unnominated ball within one's group is not considered to be a foul, but does result in loss of turn. Only one ball need be nominated at each turn: pocketing two balls in one turn, only one of which was nominated, results in continuation of the turn.

There are a number of situations where fouls can occur in the BCA rules. If the cue ball first makes contact with a ball within the opponent's group, then this is a foul. Similarly, if the cue ball is pocketed, or if no balls (cue or object) touch a rail on a given shot, then a foul results. Pocketing an opponent's ball is not a foul, and pocketing an opponents ball while pocketing one's own nominated ball results in continuation of the turn.

A foul is particularly advantageous for one's opponent, as it leads to a condition called *ball-in-hand*, whereby the opponent can place the cue ball anywhere on the table prior to her next shot. This is a distinct advantage, and serves to limit a player from purposefully committing a foul to place the opponent in a difficult situation. Similar to ball-in-hand is *ball-behind-line*, which in the BCA rules occurs only when a foul is committed on a break, e.g. if no object ball is pocketed, and fewer than four object balls touch a rail. In ball-behind-line, the opponent places the cue ball anywhere behind the head string, and must shoot down toward the foot of the table.

There are a number of other rules which, for the sake of brevity will not be detailed here.

# COMPUTATIONAL POOL

The mechanics of pool has long been of interest to researchers of applied physics, due to the nearly pure Newtonian interactions of the colliding balls, and the observable conservation laws. The earliest and most famous investigator of the physics of pool was Coriolis (Coriolis, 1835), who is more commonly known for his work on fluid dynamics. More recent efforts include the works of Marlow (Marlow, 1995), Petit (Petit, 2004), Shepard (Shepard, 1997), and currently Alciatore (Alciatore, 2004), who have aimed to successively refine the simplifying assumptions of the basic model, and to impart an understanding of the underlying physical concepts to those who play the game. The problematic of billiards was even used by the philosopher David Humes to illustrate the principle of cause and effect. He pointed out how a ball seemingly going straight at another ball will not always finish at the same precise position on the table after a collision, thus showing how the effect is actually distinct from the cause.

An understanding of the physical model led to executable simulators, typically based on numerical integration. These physics simulators were then coupled with computer graphics technology, which led to the development of a number of consumer-level computer pool games. These games are used both as training aids, mostly to facilitate shot planning, and also as entertaining diversions in their own right. Today, computer pool games are available for both PC and proprietary gaming consoles. Some games are Web-enabled and allow remote opponents to compete on a common graphical interface over the internet.

Computational pool strategy continues to be a subject of academic interest in its own right. Approaches have included fuzzy logic (Chua et al., 2002), (Chua et al., 2007), (Chua et al., 2005) grey decision making (Lin et al., 2004), optimization-based methods (Dussault and Landry, 2006a), (Dussault and Landry, 2006b), evolutionary approaches (Alian et al., 2004a), Monte-Carlo approaches (Leckie and Greenspan, 2006a), and look ahead through game tree expansion (Smith, 2006).

## Physical Model

To computationally simulate a game of pool, a mathematical model is needed based on the physics of rolling and colliding balls. We begin by introducing the basic physical equations used by both the simulation engine, and the AI algorithms. We summarize the model describing a ball rolling on a table, and aspects of the collisions. We do not attempt to provide a thorough treatment of the physics equations, but rather give enough flavor for the reader to appreciate the level of detail that is involved, and that accuracy in pool depends upon the details.

## Rolling Balls

The movement of a ball may be split into two phases: the sliding phase, and the (pure) rolling phase. Actually, when the ball stops, it may still spin around its vertical axis. This stationary spinning could be considered to be a third spinning phase, during which the ball's position is constant.

Both sliding and rolling phases are described with a quadratic parametric equation, the friction coefficient being much lower in the rolling than in the sliding phase. Even if no "English" is used, after a collision, balls are in sliding phases, and exhibit spinning behavior.

Formally, let the table be in the $x$ - $y$ plane, and let the vector $e_3 = (0\ 0\ 1)'$ point in the table's vertical direction. The sliding movement is characterized by a non-vanishing relative velocity of the contact point of the ball with the cloth, i.e. $0 \neq v_r(t) = v(t) + Re_3 \times \omega(t)$. We may assume that the frictional force is constant and given by the relation $f = -mg\mu_s \hat{v}_r(0)$. We denote by $I$ the inertia tensor for the sphere, $I = 2mR^2/5$, where $m$ is the ball's mass, $R$ its radius, and $g$ is the gravitational constant. Also, $\mu_s$ is the sliding friction coefficient, ~ 0.2.

We may now deduce the equation for the linear and angular velocities:

$$v(t) = v(0) - g\mu_s \hat{v}_r(0)t \quad (1)$$

$$\omega(t) = \omega(0) - \frac{Re_3 \times f}{I}t. \quad (2)$$

Similarly, the linear and angular positions are given by:

$$p(t) = p(0) + v(0)t - \frac{g\mu_s \hat{v}_r}{2}t^2 \quad (3)$$

$$a(t) = a(0) + \omega(0)t - \frac{Re_3 \times f}{I}t^2. \quad (4)$$

Using the aforementioned developments, we may deduce that the relative speed will become null at a time

$$\tau_r = \frac{2}{7g\mu_s} v_r(0), \tag{5}$$

after which the ball will be in a pure rolling motion. In rolling motion, the force is $f = -mg\mu_r \hat{v}_r$ and the friction coefficient $\mu_r \sim 0.01$. Of course the friction coefficients will vary depending on the table on which the game is played, which is why professional players often take practice shots to evaluate these factors and "calibrate" their own intuitive model of the frictional coefficients. The coefficients we mention are loosely based on Marlow's calculations (Marlow, 1995) and (Sénéchal, 1999). Since side spin does not influence a ball's trajectory, we may consider the two first components of $\omega$, so that the angular quantities depend on the linear ones:

$$\begin{aligned} \omega(t) &= \frac{1}{R} e_3 \times v(t) \\ v(t) &= v(\tau_r) - g\mu_r \hat{v}_r (t - \tau_r), \\ p(t) &= p(\tau_r) + v(\tau_r)(t - \tau_r) - \frac{g\mu_r \hat{v}_r}{2}(t - \tau_r)^2 \end{aligned} \tag{8}$$

and

$$a(t) = a(\tau_r) + \frac{1}{R} e_3 \times \left(v(\tau_r)(t - \tau_r) - \frac{g\mu_r \hat{v}_r}{2}(t - \tau_r)^2\right). \tag{9}$$

Moreover, we deduce that the ball will stop at time $\tau_f = v(\tau_r)\, g\mu_r$.

For most shots, the cue stick is held horizontally, and thus does not impart and curvilinear motion to the cue ball. Therefore, although parametrized quadratically, the motion is linear. The common belief that side spin imparts a curvilinear motion on the ball is not true; for a curved shot, one needs to use a **"massé"** technique, holding the queue high, and hitting the cue ball with a significant vertical angle. Similarly, a shot hit directly in the center of the cue ball results in a sliding motion for the cue ball. In order to achieve a pure rolling motion, one has to hit the cue ball above the center, at a height of exactly $\frac{7}{10}$ of the ball's diameter.

## Collision Models

Of course, an important aspect of pool is understanding the ball—ball collisions. Ball—rail collisions are also important, as are cue tip—cue ball impacts. We assume that all balls share the same mass, which may not be the case on commercial coin operated tables since the cue ball is sometimes larger and heavier in order to be identified by the ball return mechanism.

**Ball—Ball Collisions.** In most situations, it should be enough to describe the collision as if it were perfectly elastic, and frictionless. However, there is a slight deflection (or "throw") induced by the small amount of friction that exists between balls, as well as the fact the the collision is not actually perfectly elastic. On long shots, if unaccounted for, this throw may be enough to miss a shot, and so it must be an explicit component of a computational model.

**Frictionless and Elastic Ball—Ball Collisions.** The ideal collision corresponds to the "90 degree" rule: after impact, the object ball and the cue ball will move orthogonally, with the cue ball in the tangent direction of both balls, and the object ball in the orthogonal direction, on the line joining the centers. As hinted before, angular velocity is important. Immediately after the collision, both balls maintain their spin. If the object ball was at rest, it starts sliding, but soon convert its motion energy into a pure rolling state. If the cue ball was in a pure rolling state, then it slides along a parabolic trajectory, to reach its pure rolling state again.

For example, we may obtain the velocity of the cue ball after impact by the simple formulæ:

$$\begin{aligned} \bar{v}_{0x} &= \frac{v(\bar{t})_x\delta_y - v(\bar{t})_y\delta_x}{\delta_x^2 + \delta_y^2}\delta_y \\ \bar{v}_{0y} &= -\frac{v(\bar{t})_x\delta_y - v(\bar{t})_y\delta_x}{\delta_x^2 + \delta_y^2}\delta_x \end{aligned} \quad (10)$$

where $\delta$ is the direction of the line through the center of the balls at the collision, $v(\bar{t})$ the speed of the cue ball at the impact time $\bar{t}$, and the object ball is at rest.

**Imperfect Elastic Ball—Ball Collision With Friction.** In practise, the balls do not exhibit pure elastic frictionless collisions. The elasticity does not influence the object ball direction, but does influence its speed. The instantaneous rebound direction of the cue ball is also affected, reducing slightly the 90° angle of the collision. Friction further reduces the 90° angle, this time by affecting the object ball's instantaneous direction. In (Alciatore, 2004, TP A.5), computations for a 30° cut angle with a 0.06 friction coefficient and 0.94 coefficient of restitution reduces the 90° rule to as low as 83.259°.

**Ball—Rail Collisions.** In a sense, since the rail is fixed, the collision between a ball and a fixed object is simpler than collision between two balls, both of which are in motion. However, the rail is far from perfectly elastic, and is deformable. This deformability is difficult to model, and is typically abstracted in a collision model.

**Cue tip—Cue Ball Collisions.** The interaction of the cue with the cue ball is typically modeled as a point collision. In practise, the cue tip itself has a shape which is curved, and in fact mimics the curvature of the local point of contact with the ball. The mechanical properties of the cue tip are also quite involved. The leather that is used has the correct combination of elasticity and stiffness to result in a satisfying transfer of energy from the cue to the ball. Chalk is also added to increase the frictional coefficient between the cue tip and the ball. Most current computational physical models of pool ignore these factors, although it is understood that adding these components could increase the accuracy of a model.

For human players, the weight of the cue is an important characteristic. In (Petit, 2004), the foreword explicitly mention that the book explains (among other things) "Why one must use a light-weight cue, and hold it lightly in order to succeed in a draw shot". Such analysis involves the way cue speed is transfered into ball velocity when its tip hits the ball. Similar analysis will have to be completed for robot-actuated cues. Will the robot request to change cue for certain shots?

## Poolfiz® Simulation Engine

The aforementioned mathematical model has been used within an asynchronous simulation model: time is not discretized, only relevant event times are computed. The simple equations allow the computation in closed form of the exact time when a given ball changes its motion state, e.g. stops sliding, or comes to rest (Leckie and Greenspan, 2005). Similarly, collision times may be obtained by solving a quartic polynomial equation (Leckie and Greenspan, 2006b). The resulting simulation engine is efficient and accurate, especially as it does not use numerical integration and therefore does not have to use any discretization of time.

We did not present the cue stick–ball collision model. The weight of the cue, type of leather, and player's technique all impart a significant effect to the actual shot. However, this aspect is so far specific to human players, and we may make an abstraction of those considerations in assuming that an impact is imparted to the cue ball, regardless of the type of cue and other details important for human players.

The interface to both the simulation engine, and to the Deep Green robot, is reduced to five shot parameters which we now describe.

- $a$ and $b$ represent the horizontal and vertical offset of the cue tip from the center of the cue ball;

- $\theta$ the elevation angle of the cue stick;
- $\varphi$ the aiming direction angle;
- $V$ the speed of the cue stick.

Given legal and physically possible values for the five shot parameters, Poolfiz provides detailed information on the simulated ball dynamics. The sequence of all events until the table state is again motionless is available. This has been used to act as a table in virtual competitions in the two first editions of the computer pool Olympiads. Also, the computer players may rely on this information to predict their shot's outcomes, and ultimately build their game strategy on this information.

## AI Player Styles and Strengths

Excellent human players shoot with accuracy and planning. Depending on the specific variant of pool, accuracy may be more important than planning, or the opposite can also be true. Therefore, when devising a computer player, both aspects should be addressed.

Some comments are in order now to balance the accuracy and planning aspect of players. Using the Poolfiz engine, we ran a few experiment *without noise*. Each player was then perfectly accurate. Nevertheless, the stronger planners would very seldom reach table states where no shot is possible while weak planners would get stucked much more often. Thus, an hypothetical perfectly accurate robot will benefit from planning. As observed in (Leckie and Greenspan, 2006b), the advantage of a deeper search tree was magnified for players with greater shooting precision.

### Shot Difficulty

Early publications in robotic pool were concerned with evaluating the easiest shot available on a given table state (Chua et al., 2007, Alian and Shouraki, 2004, Alian et al., 2004a, Chua et al., 2005). While this is not sufficient to build a strong player, it is nevertheless necessary for any player to be able to estimate the difficulty of a proposed shot. Actually, a player will not always select the easiest shot, but when he departs from this, it will be because planning considerations outweigh the risk of attempting a more difficult shot.

No definitive technique has proved yet its superiority in this concern. A mathematical approach models the geometry of the shot, and estimate the sensitivity with respect to small errors in the execution. A computer science–statistical approach precomputes a look up table of several situations, and relies on this table to estimate the shot difficulties during actual play. An important complication arrives when taking into consideration bank, kick or combination shots. For difficulty estimation purposes, most approaches have converted those complicated shots to sequences of simpler ones.

### Control of the Cue Ball

Another crucial aspect of the game is the ability to predict, even control where the cue ball will stop after the shot is finished. Leckie and Greenspan describe two paradigms for billiards shot generation in (Leckie and Greenspan, 2006a): shot discovery and shot specification.

Human players always rely on the shot specification principle: they see the table state, plan to sink (at least) the next ball, and perform the shot in order to bring the cue ball in a good position for the subsequent shot.

The AI player has the possibility to randomly simulate several spin–strength combinations, and retain the one yielding the most favourable situation. This has been called *shot discovery*.

### Planning Ahead

The first attempts in computer pool competitions were concerned with the game of 8–ball, a game in which using 1- or 2-ply look ahead played impressively well. Statistical experiments revealed that a deeper level of look ahead is worthwhile provided that the shot accuracy is relatively high.

In more challenging games such as straight pool, winning streaks consist of sinking as many as 150 consecutive balls. Similarly, in snooker, a perfect game consists of sinking a red ball followed by the black ball, and so on fifteen consecutive times, ending up the frame by sinking the colored balls without missing. Professional players reputedly plan the whole table ahead, which amounts to a 15 or more ply search tree if done naively.

### Defensive Play

Once a situation is reached where a player expects no success, it is possible for him to call a defensive shot, trying to leave his opponent with an poor table state.

## Competitions: Computational Pool at the International Computer Olympiad

A computational pool tournament was held in two consecutive years during the yearly International Computer Olympiad (Greenspan, 2005; 2006). One difference between pool and board games is that the state of the table (i.e. the position of the balls) is a continuous rather than a discrete domain. There exist a truly infinite number of table states and games that can be played, rather than the huge but finite number that exists in chess. The practical effect of this when two computer systems compete is that, whereas in chess and other discrete board games it is common for an operator to communicate the opponent's move to a computer with a few keystrokes, in pool it is impossible to rely upon a human operator to input the state of the table between shots. Each ball in its resting state is described by it's $(x,y)$ coordinates, which are represented as double precision floats. If we approximated the ball position to just the micron level of precision, this would require the input of a possible 256 keystrokes per shot, which would be tedious, error prone and impractical.

The remedy for this was to design a client-server based match management software system called *QMin*®. In this system a competition involves 3 processes. The QMin server acts as the table and referee, and each player connects to the server through a QMin client. Identical copies of the Poolfiz simulator reside on the server and with both clients, and the table state and shot information is passed between the clients and server using a defined communication protocol.

At the beginning of a game, the breaking player computes a cue ball location and shot parameters, and communicates them to the server. Each commanded shot comprises 5 parameters (2 angles, 2 offsets, and a velocity), and is executed by the server using its physics simulator. Prior to executing a shot, the server checks certain conditions to determine if the shot is physically possible (e.g. the cue ball is on the table) and legal (e.g. the cue ball is behind the head string on a break). After executing a shot, the server determines the shot outcome (e.g. legal potted so continue shooting, foul so opponent's shot, etc.) and passes control of the table to the appropriate player. Once the 8 ball is potted, the server declares the winner of the game and tabulates the points earned, and proceeds to rack the next game in an N-game match.

In addition to completely automating the interface between players, another benefit of the QMin client-server architecture is that it allows the application of noise to each shot. In human play, only the most skilful can perfectly execute an intended shot. In computational pool, with a full quantitative knowledge of the table state and underlying physics model, it is quite possible to plan shots that are either humanly impossible or far too risky to seriously contemplate. To compensate for this omniscience, which would lead to unrealistic and uninspiring shots, the server has been designed to add an amount of random noise to each commanded shot. For each shot specification received by the server from a client, random zero-mean Gaussian noise is added to each of the 5 shot parameters prior to the server executing the shot. The noise model was calibrated so that roughly one potted ball would be missed per player per table, which was believed to be a level of skill similar to that of an average professional player. The main purpose of this noise was to make the game more realistic and challenging, and the majority of the competitors explicitly included this noise model in their systems to determine more reliable shots.

A final benefit of the QMin client-server architecture is that it is internet enabled, which allowed the players to execute on remote (and potentially exotic) hardware. There are a number of other useful features of the QMin system, such as its ability to log all games to a database for subsequent review. There were also tools developed to query the database and visualize the games using the graphics simulator embedded within poofiz.

## The Competitors

There were five programs competing in the most recent tournament: PickPocket® (Smith, 2006); PoolMaster® (Dussault and Landry, 2006b;Landry and Dussault, 2007); Snooze®; SkyNet® (Leckie and Greenspan, 2006b; and Elix®. The tournament was organized as a round robin, with each competitor playing every other competitor in matches of 50 games. There was also a time limit of 10 minutes per player per game. Each competitor played 200 games, for a total of 500 games and a maximum of 83 hours of play.

The five competing programs encompassed four distinct approaches to the game. The first approach was inspired by classical board game solution methods and involved the expansion and search of a game tree. Both PickPocket and SkyNet applied this approach, although these programs were developed completely independently.

The basic idea is to generate a series of potential shots based upon a consideration of the current table state (i.e. the position of the balls on the table). Each of these potential shots is executed in simulation, resulting in a set of potential *depth 1* table states. The process is then applied to these depth 1 table states, resulting in depth 2 table states, and can be further repeated until either: a maximum depth is reached; there are no further desirable shots; a time budget is exceeded; or all balls are cleared. The decision of which shot to place is therefore not only based upon the current table state, but also upon advantageous positioning of the cue ball for future shots.

The existence of noise is an aspect of pool which differentiates it from traditional board games. When a human executes a shot, the result is often quite different from what was intended, due to inaccuracies in the human control of the cue (as well as perceptual inaccuracies). With noise, it is necessary for the shot planning methods to look ahead to future table states and choose shots strategically, so that a successful current shot would open up a number of possibilities for future shots. Both PickPocket and SkyNet searched the tree to depth 2.

All of the five programs included some notion of evaluating the probability of success of a potential shot by considering the effects of noise, and this led to one of the significant differences between PickPocket and SkyNet. SkyNet determined the probability of success of a potential shot by randomly sampling the five-dimensional shot parameter space around the nominal shot parameters. For a given potential shot, the known noise distributions for each shot parameter was used to generate a random set of shots that follows this distribution, and the percentage of successful outcomes was used as an indication of the probability of success of the shot. This was done online, and required simulating a number (tens or hundreds) of shot outcomes for each potential shot, which could be time consuming. PickPocket took a similar approach, but rather than calculating each shot online, a series of lookup tables were generated offline that effectively characterized the shot parameter space for a significant subset of conditions. While the resultant tables tended to be large (tens or hundreds of MBytes), once generated they could be accessed efficiently, and the time saved could then be applied elsewhere.

The PoolMaster program followed the second approach known as *position play*, wherein the emphasis is on how to position the cue ball advantageously following the current a shot. Unlike PickPocket and SkyNet, who generate many potential shots by randomly varying the shot parameters and evaluating the utility of each resulting table state, PoolMaster instead explicitly selects the desired resulting position of the cue ball. The desired cue ball position is determined based upon a function which analyzes the current table state and constructs level sets of the shot difficulty function. Once the cue position has been determined, a numerical optimization process is then invoked to determine the shot parameters that will position the cue ball at or close to this position. PoolMaster did not consider any table states following the current shot, so it can be equated to a depth 1 search.

The Elix program characterizes the third approach, which is based upon a set of heuristics that mimic human play. Elix's developer, Marc Godard, is an experienced pool player, and has encoded many elements of human strategic play into his program. The main structure for shot selection performs a depth 2 search on the most favourable four shots on the table (two straight shots and two bank shots). There are also a number of specific heuristics which are applied to particular circumstances (e.g. ball-in-hand, ball-behind-line, safety).

The fourth and final approach was that of Snooze. Snooze was a late entry to the competition, and was developed over a period of only a few months, as opposed to the year or greater development time for the other four programs. Snooze did not use any look ahead (i.e., it was a depth 1 search), and selected the next shot by considering the set of possible shots from the current table state. Straight shots, bank shots, kick shots, and combination shots were all evaluated, but no combination of these were considered. From this set of potential shots, the shot parameters were varied uniformly over a range for each shot, and an evaluation function was applied to assess the outcome. Due to the short development time, only a rudimentary strategy was applied to the ball-in-hand condition, which always placed the cue ball on the head spot, and easily cost the program a few games.

The tournament results are listed in Table 1. The scores for each competitor were the sum of the number of games won for all matches. PickPocket took first place decisively winning 148 of the 200 games played (74%). Curiously, despite the large number of games played, both SkyNet and Elix tied for second place with exactly 105 wins each. It should be noted that, despite being in the same lab, the approaches were different, and there was no code shared by the developers of these two programs.

## Observations

It has been shown empirically that there is an inverse relationship between the level of noise and the optimal search depth (Leckie and Greenspan, 2006b). If the noise level is too small then most

*Table 1. Tournament results*

| | Pocket | SkyNet | Elix | PoolMaster | Snooze | Total Won | % Won | Rank |
|---|---|---|---|---|---|---|---|---|
| PickPocket | - | 34 | 38 | 37 | 39 | 148 | 74.0 | 1 |
| SkyNet | 16 | - | 24 | 28 | 37 | 105 | 52.5 | 2 |
| Elix | 12 | 26 | - | 31 | 36 | 105 | 52.5 | 3 |
| PoolMaster | 13 | 22 | 19 | - | 22 | 76 | 38.0 | 4 |
| Snooze | 11 | 13 | 14 | 28 | - | 66 | 33.0 | 5 |

shots will succeed, regardless of their level of difficulty, and planning the current shot based upon future table states does not tend to improve the level of play (although it does consume cycles). As the noise level increases, it becomes necessary to consider future table states so that the current shot will always set up a subsequent shot that has a good chance of succeeding. If the noise level becomes too large, then the table state following a shot is largely indeterminate, and planning future shots once again becomes less useful. The same effect seems to apply to human play, which may be one reason why novice and occasional players tend not to appreciate the strategic aspects of the game.

One aspect of the game which was handled much more effectively in the latest than in the previous tournament was the safety shot. In 8 Ball, a player maintains control of the table as long as a shot is successful, i.e., the called ball is sunk in the called pocket. There can be situations, however, where a player would like to relinquish control of the table to the opponent even if the current shot is successful. An example is when there exists a cluster of balls (i.e. a set of balls which are touching) which need to be broken up after the current shot. It is usually difficult to predict the outcome of breaking up a cluster, and the expectation is that whoever breaks the cluster will not pot a ball leaving the next shot with many new opportunities to the opponent. It is therefore desirable to call a safety on the shot prior to a table state that contains only clusters.

Another situation where a safety may be called is when Player A has only a low probability of succeeding with the current shot. Rather than attempting to sink the ball and missing, possibly leaving Player B with an advantage, the alternative is for Player A to call a safety and focus on placing the cue ball in a position where Player B has no clear shot. In the ideal case, Player A will place the cue ball such that Player B will foul (e.g., the cue ball will contact Player A's colour group first) leaving Player A with ball-in-hand, a distinct advantage.

Both PickPocket and SkyNet handled safety shots in a similar way. If the best shot resulting from the tree search had a probability of success that was below a specified threshold, then a safety procedure would be invoked. The basis of the safety procedure is to randomly alter the shot parameters and choose a shot that results in a difficult table state for the opponent.

## ROBOTIC POOL

Whereas computational pool aims to simulate the game of pool, the further objective of robotic pool is to actuate the results of such a simulation. This robotic actuation preferably takes place on a standard table, and may be in the form of either a main-versus-machine system, or a purely machine-versus-machine system, to evaluate the effectiveness of different strategies and components.

Pool is a game that requires a high degree of positioning accuracy, in five degrees-of-freedom (DOFs). While standard robotic systems are precise and repeatable, they tend to lack absolute positioning accuracy. A major challenge of robotic pool is therefore to construct a system that has sufficient accuracy to position the cue correctly and execute the desired shot.

There are two approaches to achieve an accurate robotic system, the first being the design of the mechanism itself. Most robotic pool systems have been based on a ceiling-mounted gantry platform, with a 2 DOF spherical wrist to facilitate aiming. It is possible to achieve fine positioning accuracy using gantry-style mechanisms. Coordinate Measurement Machines, for example, can achieve ~ 25 micron accuracy over similar workspaces. Such accuracy comes at a cost, however, as these machines tend to be heavy and brittle, and are unlikely to absorb the significant impacts required to shoot a ball without requiring extensive recalibration.

The second approach to achieving fine positioning accuracy demands less from the primary gantry device, and instead relies on external camera sensors to improve accuracy. This has been the more common approach, and permits the use of standard and relatively inexpensive gantry mechanisms. In all cases, a camera is mounted on the ceiling aimed down at the table, and in some systems there are additional cameras mounted on or near the end-effector. The cameras do require calibration, some of which is standard, and some of which is the subject of academic enquiry.

## Previous Work

The first effort to develop a robotic pool playing system was "The Snooker Machine" from the University of Bristol (Chang, 1994). This system was developed from the late 1980's though to the mid 1990's, and was based upon an inverted 6 DOF articulated manipulator mounted on a 3 DOF gantry system. The system employed a ceiling-mounted camera, as well as an end-effector mounted camera. The workspace of the system was reasonably large, enveloping a small quarter-sized snooker table. There was a non-trivial amount of interest in the system at the time, which culminated in a televised man-versus-machine event on the Discovery Channel in the U.K.

There was very little effort at further developing pool robotics until the mid 2000's. when in 2004, Alian et al. from Sharif University in Iran published details about "Roboshark"(Alian et al., 2004b),(Alian and Shouraki, 2004),(Alian et al., 2004a). This system was based upon a custom constructed gantry system positioned above a toy (~ 1/8 sized) pool table. An unnamed system was also at the Multimedia University in Malaysia (Cheng et al., 2004; Chua et al., 2005;Chua et al., 2007; Lin et al., 2004). Like the Iranian effort, this system was also based upon a custom constructed gantry robot and a toy pool table. Finally, there is the Deep Green system from Queen's University (Greenspan et al., 2008). This is the only such system that is known to still be in existence, and is described in detail in the following section.

In addition to the aforementioned efforts, there have been a number of projects into pool automation that, while they fall short of full robotic actuation, involve a number of similar elements. In (Denman et al., 2003), methods where explored to apply computer vision techniques to automatically analyze video streams of televised pool competitions to extract information about player shot selection. at the University of Aalborg have developed the "Automatic Pool Trainer"(Larsen et al., 2002). This system involves an overhead video camera to identify the table state, and a steerable laser to inscribe potential shots directly onto the table surface. A similar concept, called "Mixed Reality Pool", is currently being pursued at Pace (Hammond, 2007).

## Deep Green

Deep Green is a robotic pool system that is currently under development at Queen's University, in Kingston, Canada. The goal of the project is to design a system that is capable of competing with proficient human player. Figure 1 shows the major hardware components of Deep Green, which include: a ceiling-mounted gantry robot, a cue end-effector, a ceiling-mounted camera called the Global Vision System (GVS), a wrist-mounted camera called the Local Vision System (LVS), and a standard 4' × 8' pool table. The major software components include: robot control, robot and vision system calibration, vision algorithms for ball identification and localization, pool physics simulation, and shot planning and strategic play routines.

### Gantry Mechanism

Deep Green is based on a seven degree-of-freedom (DOF) industrial gantry robotic system. The gantry itself provides three linear DOFs, translational movement in the direction of the x, y , and z axes through joints 1, 2, and 3. Attached to the gantry is a three DOF spherical wrist, which

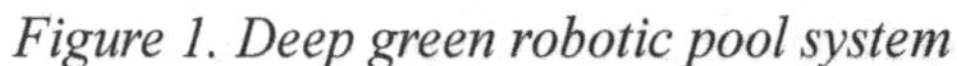

*Figure 1. Deep green robotic pool system*

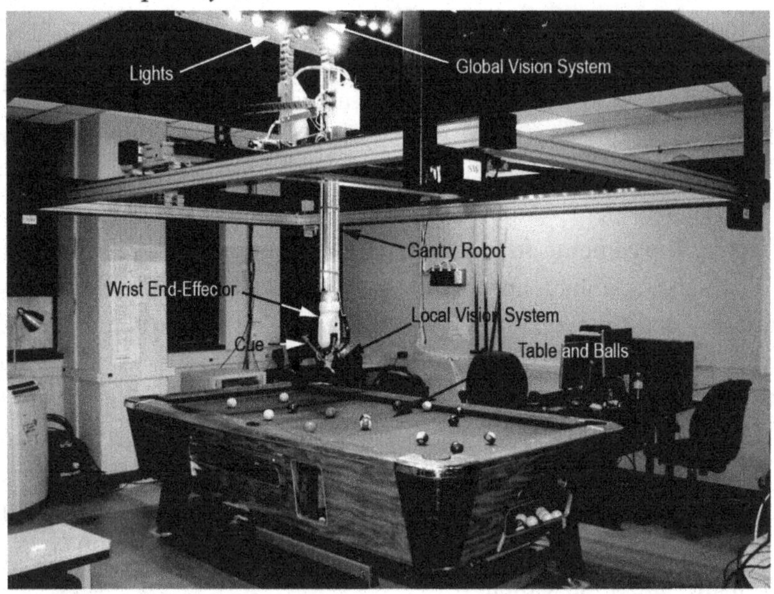


M. Greenspan, J. Lam, W. Leckie, M. Godard, I. Zaidi, K. Anderson, D. Dupuis, and S. Jordan, "Toward a Competitive Pool Playing Robot", /IEEE Computer Magazine,/ vol. 41(1):46-53, Jan. 2008. © [2008 IEEE]

*Figure 2. Deep green end-effector*

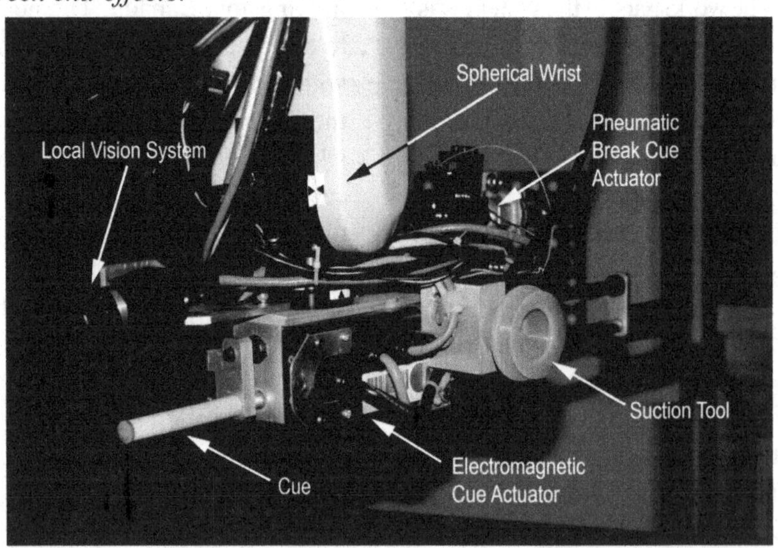


M. Greenspan, J. Lam, W. Leckie, M. Godard, I. Zaidi, K. Anderson, D. Dupuis, and S. Jordan, "Toward a Competitive Pool Playing Robot", /IEEE Computer Magazine,/ vol. 41(1):46-53, Jan. 2008. © [2008 IEEE]

provides rotational movement through joints 4, 5, and 6. Joint 7 is a one DOF cue end-effector, which is an electrically-powered linear actuator that can be finely controlled to strike the cue ball with any desired velocity up to a maximum of ~ 3 m/s. The end-effector and all peripheral devices are illustrated in Figure 2. In addition to the electric cue actuation, there is also an alternative pneumatically-powered cue that is used only for power breaks, and strikes with a velocity of 12 m/s. Also mounted to the side of the end-effector is a pneumatic suction cup powered by a vacuum generator. The suction cup can pick and place balls at precise locations on the table, which is necessary for ball-in-hand and ball-behind-line situations, and can also be used for automatic racking at the beginning of a game. The robot and all peripheral devices are connected to a single PC workstation through serial interfaces.

As described before, pool is a sport that demands high accuracy. Unfortunately, large, rugged and relatively inexpensive gantry robots like the one upon which Deep Green is based, while very precise, are not terribly accurate. These devices typically suffer from sources of error which lead to nonlinearities between the joint encoder readings and the Cartesian frame of the robot. An alternative would be to use a robot with fine-grain accuracy over the desired workplace, such as a Coordinate Measurement Machine, which can have an accuracy of $\sim 15\mu m$ over a similar working volume. This high degree of accuracy comes at a cost, however, and such a device would be expensive, delicate, and unlikely to maintain accuracy while absorbing the impacts required when placing shots. To achieve the high accuracy that pool demands, the approach taken with Deep Green is to rely on computer vision and pool-specific visual servoing techniques to improve the positioning accuracy of the robot.

## Vision System

The are two major vision systems for Deep Green, the GVS and the LVS. The GVS (Global Vision System) is the primary sensor of the system. It is mounted on the ceiling above the pool table, and offers a 2-D birds-eye view of the table. The primary functions of the GVS are ball localization and identification. For localization, the positions of the individual balls are determined within the table coordinate reference frame through a combination of background subtraction, blob detection, circle fitting, and homographic rectification. For identification, the unique number of each ball (required in 8-ball for ball nomination) is determined by analysing the color content of each ball using color indexing methods. The GVS camera is a off-the-shelf high resolution camera, an 8.2 megapixel Canon Digital Rebel. The Canon was selected for its high resolution, excellent lens, and the existence of public domain software that allows for remote image acquisition.

In contrast to the GVS, the LVS (Local Vision System) offers a much closer look at the scene. The purpose of the LVS is to improve upon robot positioning accuracy using local landmarks. The LVS camera is mounted at the end-effector of the robot, with the optical axis pointing in the direction of the cue. The camera is a Point Grey Flea2 digital video camera with a resolution of 1024 by 768 pixels. An eye-in-hand visual-servoing algorithm using the LVS was designed to line up a perfect straight shot. The algorithm relies on observing three important landmarks within the LVS image: the cue ball center, the cue ball radius, and the ideal shot trajectory that passes through the object ball and cue ball centers.

Both the LVS and GVS cameras are calibrated to determine their intrinsic parameters, such as focal length and radial distortion parameters. The GVS is further calibrated to determine the homography that relates the GVS camera retinal plane to the metric coordinate reference frame of the table (Lam et al., 2006). The robot is also calibrated, using a combination of GVS and an additional end-effector mounted camera, to relate its encoder readings directly to table frame coordinates (Long et al., 2004). Finally, the kinematic parameters that relate the optical frame of the LVS with the frame centered at the end-effector is determined (Lam and Greenspan, 2007), as are the characteristics that indicate an ideal straight shot from the vantage of the LVS. It has been shown that the use of the LVS improves the shooting accuracy of the system by a factor of three (Lam and Greenspan, 2008).

### Software

While the Deep Green hardware elements are significant and prominent, they have wherever possible been based on standard off-the-shelf components. The bulk of the effort has been devoted to the development of software methods to calibrate and control the hardware, and interpret the sensor data. The software functionality includes low level robot control, calibration processes, computer vision and visual-servoing algorithms, pool physics simulation, with some computer graphics elements, and artificial intelligence for game strategy.

The majority of code is written in C/C++ on a linux platform. The robot control programming is the lowest level of software, providing basic operations to the robot such as move, home, stop, and shoot. The calibration processes are extremely important, and comprise the majority of the codebase and effort, along with the computer vision methods described in the previous section.

The interaction between different two competing artificial intelligence gaming systems (AIs) is implemented in a client-server architecture. Since all components are independent modules (i.e. physics simulator, server, AI module), the infrastructure of the system can be maintained with less effort when organized as a client-server system. The server saves the current table state and is the platform that runs the simulation. The AIs are the clients that log on to the server, which provides information to the clients about the current table state following each shot, as well as various control information about the game (e.g. which AI has control of the table, etc.). The clients are asked by the server to provide the parameters for each shot, and the target ball that it is aiming for (i.e. ball nomination). The system architecture allows the AIs to play on either a simulated table (using the simulation engine) or on the actual pool table (using Deep Green). When playing on the actual table, rather than calculating the table state resulting from a specified shot through simulation, the GVS is used instead to acquire and update the table state before and after the robot places the shot. The described configuration allows the system to be operated in two different modes: either a player can choose to play against an AI, or two AIs compete against each other sharing the same hardware, possibly remotely.

## FUTURE TRENDS

While the research field of computational and robotic pool is still in its relative infancy, enough has been demonstrated to warrant the high expectations expressed by those involved. The AI aspects of computational pool have been slightly more developed than the other robotic aspects, thanks to the accelerating effects of the computer pool tournaments at the Olympiads. The next few Olympiads, aiming to widen the game variants (snooker, 9 ball, straight pool), will stimulate the player developers. We expect that a super player may emerge by combining the strong points of the competing players at the first two Olympiads, merging the search trees principles, clever ad hoc analysis, and optimization for fine tuning precise position.

As the robot gets more accurate and reliable, the AIs will meet the additional challenge to deal with *real* noise levels. Olympiads to compare AIs on a real robot (like Deep Green) should be held in some very near future. As the robot improves, the need to apply it to a more precise table using new balls will trigger further improvements to both the AIs and the simulation engine. According to (Petit, 2004, p. 94), worn or dirty balls may increase the ball–ball friction coefficient from ~ 3% to as much as 20%.

An important feature will be the ability for the system (robot and simulation engine) to self calibrate. In imitation of professional players, the system will try several shots, and calibrate the various friction and elasticity constants so that the simulated shots and the observed ones are as close as possible.

## CONCLUSION

We have presented the first steps of a new research field, computational and robotic pool. We have shown that the AI aspect of the game has still a lot room left for improvement, but that present research show promising results; enough for this research to be very interesting and challenging. We have also shown that the robotic aspect of the research, though a work in progress, evolves very quickly. It comprises many challenges in various fields, which in return promotes very interesting collaborations, and development of new methods and techniques.

The challenges of this field of research are real and important. Human champions are incredibly accurate, their performances being possible by their intuitive understanding of very delicate physics phenomena, their planning abilities, and of course the tremendous physical precision in the execution of their shots. Therefore, one could be sceptical that a robot could eventually compete at this level.

A few decades ago, bets were that the computer would — or would not — eventually beat human champions in chess; we now know that humans are not strong enough to challenge computers in that realm. Although a considerable amount of work remains to be done, we believe that in some relatively near future, no human will be able to compete with a well designed robotic system in any variant of pool.

## REFERENCES


Alciatore, D.G. (2004). *The Illustrated Principles of Pool and Billiards.* Sterling Publishing.

Alian, M., Lucas, C., & Shouraki, S. (2004a). Evolving game strategies for pool player robot. In *4th WSEAS Intl. Conf. on Sim., Mod. and Opt.*

Alian, M., & Shouraki, S. (2004). A fuzzy pool player robot with learning ability. In *WSEAS Trans. on Electronic, 1,* 422–425.

Alian, M.E., Shouraki, S., Shalmani, M., Karimian, P., & Sabzmeydani, P. (2004b). Roboshark: A gantry pool player robot. In *ISR 2004: 35th Intl. Sym. Rob.*

BCA (2008). Billiard congress of america. Retrieved from http://www.bca-pool.com.

Chang, S.W.S. (1994). *Automating Skills Using a Robot Snooker Player.* PhD thesis, Bristol University.

Cheng, B., Li, J., & Yang, J. (2004). Design of the neural-fuzzy compensator for a billiard robot. In *IEEE Intl. Conf. Networking, Sensing & Control,* (pp. 909–913).

Chua, S., Wong, E., Tan, A.W., & Koo, V. (2002). Decision algorithm for pool using fuzzy system. In *iCAiET 2002: Intl. Conf. AI in Eng. & Tech.,* (pp. 370–375).

Chua, S.C., Wong, E.K., & Koo, V.C. (2005). *Intelligent Pool Decision System Using Zero-Order Sugeno Fuzzy System, 44.* Springer Netherlands.

Chua, S.C., Wong, E.K., & Koo, V.C. (2007). Performance evaluation of fuzzy-based decision system for pool. *Appl. Soft Comput.,* 7(1), 411–424.

Coriolis, G.-G. (1835). *Théorie mathématique des effets du jeu de billard.* J. Gabay, Paris, France.

Denman, H., Rea, N., & Kokaram, A. (2003). Content-based analysis for video from snooker broadcasts. *Computer Vision and Image Understanding, 92*(2/3), 176–195.

Dussault, J.-P., & Landry, J.-F. (2006a). Optimization of a billiard player—position play. In H.J. van den Herik, S. chin Hsu, T. sheng Hsu, and H.H.L.M., Donkers (Eds.), *ACG, 4250 of Lecture Notes in Computer Science,* (pp. 263–272). Springer.

Dussault, J.-P., & Landry, J.-F. (2006b). Optimization of a billiard player—tactical play. In

*Proceedings of the Computer and Games Conference, Italy, May.* Springer Verlag Heidelberg Germany, lncs series. CG06.

Greenspan, M. (2005). Uofa wins the pool tournament. *Intl. Comp. Gaming Ass. Journal, 28*(3), 191–193.

Greenspan, M. (2006). Pickpocket wins pool tournament. *Intl. Comp. Gaming Ass. Journal, 29*(3), 153–156.

Greenspan, M., Lam, J., Leckie, W., Godard, M., Zaidi, I., Anderson, K., Dupuis, D., & Jordan, S. (2008). Toward a competitive pool playing robot. *IEEE Computer Magazine, 41*(1), 46–53.

Hammond, B. (2007). *A computer vision tangible user interface for mixed reality billiards.* Master's thesis, Pace University.

Hickok (2008). *Billiard—history.* Retrieved from http://www.hickoksports.com/history/billiard.shtml.

Lam, J., & Greenspan, M. (2007). An iterative algebraic approach to tcf matrix estimation. In *IEEE/RSJ 2007 Intl. Conf. Intell. Rob. Sys.*, (pp. 3848–3853).

Lam, J., & Greenspan, M. (2008). Eye-in-hand visual servoing for accurate shooting in pool robotics. In *5th Can. Conf. Comp. Rob. Vis.*

Lam, J., Long, F., Roth, G., & Greenspan, M. (2006). Determining shot accuracy of a robotic pool system. In *CRV 2006: 3rd Can. Conf. Comp. Rob. Vis.*

Landry, J.-F., & Dussault, J.-P. (2007). AI optimization of a billiard player. *Journal of Intelligent and Robotic Systems, 50*(4), 399–417.

Larsen, L., Jensen, M., & Vodzi, W. (2002). Multi modal user interaction in an automatic pool trainer. In *ICMI 2002: 4th IEEE Intl. Conf. Multimodal Interfaces*, (pp. 361–366).

Leckie, W. and Greenspan, M. (2005). Pool physics simulation by event prediction 1: Motion transitions. *Intl. Comp. Gaming Ass. Journal*, 28(4):214–222.

Leckie, W. and Greenspan, M. (2006a). Monte carlo methods in pool strategy game. In *Advances in Computer Games 11.* To appear.

Leckie, W., & Greenspan, M. (2006b). Pool physics simulation by event prediction 2: Collisions. *Intl. Comp. Gaming Ass. Journal, 29*(1), 24–31.

Lin, Z. M., Yang, J., & Yang, C. (2004). Grey decision-making for a billiard robot. In *IEEE Intl. Conf. Systems, Man and Cybernetics.*

Long, F., Herland, J., Tessier, M.-C., Naulls, D., Roth, A., Roth, G., & Greenspan, M. (2004). Robotic pool: An experiment in automatic potting. In *IROS 2004: IEEE/RSJ Intl. Conf. Intell. Rob. Sys.*, (pp. 361–366).

Marlow, W.C. (1995). *The physics of pocket billiards.* MAST, Palm Beach Gardens, Florida.

Petit, R. (2004). *Billard: Théorie du jeu.* Chiron Éditeur.

Sénéchal, D. (1999). *Mouvement d'une boule de billard entre les collisions.* Unpublished manuscript.

Shamos, M. (1995). *A brief history of the noble game of billiards.* http://www.bca-pool.com/aboutus/history/start.shtml.

Shepard, R. (1997). *Amateur physics for the amateur pool player.* self published.

Smith, M. (2006). Running the table: An ai for computer billiards. In *AAAI 2006: The 21st Nat. Conf. on AI.*

# Section II
# Business Applications

# Chapter XIII
# A Framework for the Adoption of the Internet in Local Sporting Bodies:
## A Local Sporting Association Example

**Scott Bingley**
*Victoria University, Australia*

**Stephen Burgess**
*Victoria University, Australia*

## ABSTRACT

*Sport plays a major part in the Australian psyche with millions of people participating every year. However organised sport at the local or social level in Australia relies on volunteers to support the needs of associations and their participating clubs. There is evidence that Internet applications are being adopted within associations and clubs for administration purposes (such as committee members using email to communicate with each other, or use of the Internet to record match results and calculate player performance statistics online). However, how are these being adopted, what are they being used for and what is the effect of the adoption on the associations and their volunteers? Using the Rogers' (2005) innovation-decision process as a basis, this chapter describes the development of a framework that traces the adoption of an Internet application from initial knowledge of the application, through the decision to adopt and eventual confirmation of the usefulness of the application by continuance or discontinuance of its use. As local sporting clubs and associations are part of a larger group known as community based organisations and are predominantly run by volunteers, literature related to Internet application use by these groups is used to inform the framework. Lastly, an actual example of the adoption of an online statistics program in a local sporting association is mapped onto the framework, to show it may be applied in a practical situation.*

## INTRODUCTION

Many people participate actively in sporting clubs based in local communities throughout Australia. Participation in sporting activities, specifically in local sporting clubs, play an important role in Australia with a significant proportion of the population being involved. These participants are supported by a group of volunteers that provide structure to their activities through the provision of club administration functions, such as player registration, fund raising activities and so forth. Recently, there is evidence to suggest that Internet technologies are being used in various ways to support these activities. What is not known is the effect of these adoptions on local sporting associations, their clubs and their volunteers. The primary aim of this chapter is to develop a framework that identifies the factors that determine how the Internet is being adopted within local sporting associations and clubs and to determine the factors of the adoption that effect local sporting clubs and their volunteers from initial knowledge of the application, through to its adoption and use and eventual decision whether or not to continue with the adoption. The applicability of the framework is then assessed by mapping an actual adoption of an Internet application (by a local cricket association in Australia) onto it.

## SETTING THE SCENE: ISSUES AND PROBLEMS

Sport plays a major part of the Australian psyche. It provides benefits for the community and individuals, such as improved health, social networking, and self esteem improvement. Local sporting clubs play an important role in Australia. For instance, in the 2005/6 season, from a population of around 20 million people, there were almost 550,000 participants playing cricket in Australia (Cricket Australia, 2006). Even a less high-profile sport, such as field hockey, had nearly 140,000 participants (Hockey Australia, 2006). Sport plays a significant role amongst Australian youth, with 10% of boys (aged 5-14 years) participating in the game of cricket, behind outdoor soccer (20%), swimming (13%), and Australian Rules football (13%) (Australian Bureau of Statistics, 2003). Belonging to such a group can bring a sense of community, which is "a feeling the members have of belonging, a feeling that members matter to one another and the group and a shared faith that members' needs will be met through their commitment together" (McMillan & Chavis, 1986, p. 9). Pretty, Andrewes and Collett (1994) suggest that there is a link between sense of community and individual well-being. In many countries around the world, the introduction of sporting programs, activities and events to the community is largely reliant on volunteers to invest their time and energy (Cuskelly, 1995).

Local sporting clubs are part of the larger group known as community based organisations (CBOs). CBOs as a sector rely heavily on volunteers to support their activities. In the case of local sporting clubs, this reliance is usually on their members (both playing and non-playing), who typically perform a number of administrative and other support activities on a volunteer basis to ensure their clubs remain operational.

With their involvement in local sporting clubs over a number of years, the authors have observed the introduction of Internet technologies into many member activities. These activities range from the use of email to improve communications between committee members, to the introduction of online systems to handle match scores and statistics related to player performance – the latter eliminating repeated data entry and saving countless labour hours. The adoption of these different applications of Internet technologies are sometimes driven from the 'top' (that is, from the club or even cricket association level) and imposed upon members in the clubs. However, in some instances the adoption may have been driven from the 'bottom', via 'technology savvy' members keen to apply the technology as part of their duties.

However, the introduction of any new technology has an impact – and whilst improved communication and increased efficiencies might seem desirable outcomes – there can be negative impacts on those that feel marginalised from the technology, or for those volunteers for whom the task of recording player statistics may have been a desirable use of their time. It is in this context that this research is being conducted.

As the reader may imagine, there is not a great deal of literature available addressing the adoption and use of Internet applications in local sporting bodies. As local sporting bodies are small organisations, literature related to the use of information and communications technologies (ICTs) in business and in CBOs is used to identify some of the relevant issues to inform the development of the framework. In addition, literature related to the use of ICTs by volunteers will also be used to inform the framework. These are discussed in the next few sections.

## Information and Communication Technologies in Business

ICTs are increasingly considered to be one of the main contributors to high economic growth rates at national levels (Ciarli & Rabellotti, 2007). During the 1990s, the widespread adoption of ICT through firms and households was deemed to be one of the core explanations for the high growth rate of the economy throughout the United States and some countries in Europe (Bassanini & Scarpetta, 2002). Information is the basic requirement for business creation, growth and survival, and ICTs are capable of easing information gaps in the business sector (Diaz, 1977). Of course, two of the inherent characteristics of the Internet (as an ICT) are that it can facilitate communication and also allow the transfer of vast amounts of information (Turban, Leidner, McLean & Wetherbe, 2006, p. 730). ICTs can also be a catalyst for change within strategic, structure, operations and management sectors of an organisation (Carr, 2001). These capabilities, according to Wreden (1997), support the following business objectives: improved productivity, reduced costs, improved decision making, enhanced customer relations, and the development of new strategic applications.

The primary aim of this study is to determine what Internet applications are being adopted within local sporting clubs and to determine the adoption factors that affect them and their members. ***Do these business benefits of ICT use (that is, improved productivity and so forth) translate to locate sporting clubs when they adopt Internet applications?***

## Community Based Organisations

A community is a group of people with common interests, concerns and functions (Bellah et al., 1985; Brookfield, 1983; Hamilton & Cunningham, 1989; Hiemstra, 1993; Roberts, 1979). Community based organisations are "non-profit organisations that operate in urban neighbourhoods to benefit neighbourhood residents and address their concerns. CBOs typically serve a relatively small geographic area, tend to have a small full time staff and often depend on neighbourhood volunteers for programme delivery" (Kellogg, 1999, p. 447). Local sporting clubs are typically CBOs that serve a relatively small geographic area. There is a lag in the amount of research conducted in the CBO sector when compared to more commercial sectors, particularly in relation to their use of ICTs (Hall & Banting, 2002). ICTs are not necessarily used in CBOs to provide commercial benefits, as may be the case in businesses. Their use is generally aimed at improving a CBO's ability to share ideas and information to meet social needs (MacKay, Parent & Gemino, 2004).

CBOs face a number of barriers to the successful use of ICTs and the Internet. Typically, these barriers are a lack of ICT expertise, lack of time to exploit ICTs and lack of capital to fund them adequately (Karanasios et al., 2006). ***Do these barriers apply in local sporting clubs when they adopt Internet applications?***

## Volunteers

Volunteering Australia (2005) define volunteering as "an activity which takes place through not for profit organisations or project". Volunteering is an activity in which individuals donate time and effort for the benefit of others (Wilson, 2000). Volunteering 'gives back' to the community in two ways: by the time and effort the volunteer donates to the community, and by having volunteers engaged in activities to reduce the risk of them descending into antisocial behaviour (Eley & Kirk, 2002). The Australian Bureau of Statistics (2001) indicates that the greatest volunteer involvements were in the areas of community/welfare (26%), sport/recreation (21%) and religion (19%).

Many organisations that rely on volunteers face large demands from their patrons and this is complicated with the very limited supply of volunteers (Edwards & Watts, 1983).

A report issued by the Australian Government, *Giving Australia* (2005) found that over six million Australians over the age of 18 years were active volunteers, and that the two most common reasons for volunteering are to give to the community (47%) and personal satisfaction (43%) (Australian Bureau of Statistics, 2001).

Volunteers are drawn to sporting committees in order to "make a contribution", "establish social relations", and/or "achieve recognition" (Doherty & Carron, 2003, p. 130). The motivation to volunteer also differs greatly from one person to another over time. For instance, different age groups can vary in relation to their motives (Eley & Kirk, 2002). In 2004 there were 1.5 million persons (which is nearly 10% of all persons aged 15 and over) involved in at least one *non-playing* role in organised sport and physical activity. Almost a third of these were involved in two or more non-playing roles, with 0.6 million involved as a coach, instructor or teacher; the same number as a committee member or administrator; and 0.3 million as a referee or umpire (Australian Bureau of Statistics, 2006).

Coleman (2002) reports that volunteer managers fulfilled multiple roles. Team selection (99%) and coaching (74%) were two of the major roles undertaken by volunteer managers. Junior sports require even greater support from adult volunteers than senior sport, as parents and adult club members typically volunteer for committee and coaching roles, and also manage the logistics of transporting participants to and from training, events and games. The use of the Internet technologies by local sporting clubs and their volunteers is increasing – it is important that the factors that affect its adoption and its impacts are understood.

Volunteer boards or executive committees typically administer non-profit amateur sporting clubs and organisations. They contain elected, appointed, invited or self-selected members who are responsible for the operations of the organisation (Doherty, Patterson & Van Bussel, 2004). At a local, or community level, non-profit amateur sports clubs rely more or less exclusively on volunteer administrators (Doherty & Carron, 2003). Members of these committees are, as Shibli, Taylor, Nichols, Gratton and Kokolakakis (1999) label them, "*systematic volunteers*" who "*have a clearly defined role and are required to make a regular commitment to the operation of the club*" (p. 10). ***Are these roles influenced by or affected by the adoption of Internet applications?***

We know that ICTs can also have negative effects on individuals in the business arena. These effects include: information overload, dehumanisation and other psychological impacts, and impacts on health and safety (Turban et al., 2006). Very few studies have investigated the role of volunteer use of ICTs (Boyle et al., 1993; Madon, 1999; Morgon, 1995). Although there is research examining Internet adoption in SMEs, this cannot necessarily be generalised to small CBOs and their volunteers. Therefore, more specific research is needed to investigate implications of the significant differences between the use of ICTs by volunteers and that in businesses (MacKay, Parent & Gemino, 2004). Daniel, Wilson and Myers (2002) found that the volunteer sector has amongst the lowest levels of e-commerce adoption in the UK. Many organisations that rely on volunteers do not see the perceived need to integrate these technologies in their ongoing business due to risk, cost, time, and lack of understanding of the full potential of the e-commerce and its strategic benefits. Therefore, the perceived potential of e-commerce is much lower in the volunteer sector (Saxton & Game, 2001). ***Do these limitations effect the adoption of Internet applications in local sporting clubs?***

## IDENTIFYING A SOLUTION: A FRAMEWORK FOR THE ADOPTION OF INTERNET APPLICATIONS IN LOCAL SPORTING CLUBS

In the previous sections, the discussion of the factors that may affect the adoption of Internet applications in local sporting clubs identified a number of research questions:

- Do the benefits of using ICTs, and in particular Internet applications, in business translate to local sporting clubs? In particular, do they apply to the typical roles played by volunteer members of these clubs, such as committee member, administrator, selector and coach?
- Do the barriers typically faced by CBOs in relation to their use of ICTs, and in particular Internet applications, translate to local sporting clubs? Do the factors affecting low adoption rates of e-commerce by CBOs translate to the adoption of Internet applications by local sporting clubs?
- Are the roles that volunteers play in local sporting clubs affected by the adoption of Internet applications?
- Do the negative impacts associated with the use of ICTs, and in particular Internet applications, translate to local sporting clubs and particularly their members?

We would like to add one more question that we feel may influence the adoption of certain Internet applications in sporting clubs:

- Does the *nature* of the sport being played (or more specifically, its record keeping requirements) make a local sporting body more or less likely to adopt Internet applications?

We believe that some sports, especially sports that rely heavily on statistics as part of their record keeping (such as baseball and cricket) might be more inclined to take advantage of the benefits that Internet applications can offer.

The purpose of this chapter is to introduce the framework we will be using as the basis for the study. So far we have identified a series of factors relating to adoption of ICTs in large and small businesses and CBOs that may affect the adoption of Internet applications in local sporting clubs. In addition, we have also identified a series of factors related to ICT adoption that affect employees of businesses and volunteers in CBOs that may relate to the adoption and use of ICTs by members of local sporting clubs. We have added the 'type of sport' factor for reasons mentioned earlier.

To assist us in developing a suitable framework to examine our research questions we have also used the innovation-decision process (Rogers, 2005) to help inform its initial version, which can be viewed in Table 1.

The Rogers' (2005) innovation-decision process "is the process through which an individual (or other decision making unit) passes from gaining initial knowledge of an innovation, to forming an attitude towards the innovation, to making a decision to adopt or reject, to implementation of the new idea, and to confirmation of this decision".

Rogers' innovation-decision process is a component of his well-known *diffusion of innovations* theory, originally published in 1962. The most recent (the fifth and final) edition of the book reflects updates to the theory over previous editions. Rogers actually notes in the most recent edition that the Internet has "spread more rapidly than any other technological innovation in the history of humankind... the term *digital divide* indicates those individuals who are advantaged versus those who are relatively disadvantaged by the Internet... such interactive communications technologies may be changing the diffusion process in certain fundamental ways, such as by removing, or at least greatly diminishing, the role of spatial distance in who talks to whom about a new idea" (2005, p. xix). Thus, he refers to the Internet as an innovation in its own right and as a means by which information about innovations can be disseminated.

Rogers' (2005) innovation-decision process provides the initial, general framework to which we can add the factors we have already identified. The framework not only provides the initial factors to be considered that affect such adoptions, it also provides a series of stages which encompasses the innovation process as a component of the study. These are:

- **Knowledge:** When a decision maker is made aware of an innovation.
- **Persuasion:** When a decision maker forms an attitude towards an innovation. In this instance it encompasses comparing the relative advantages and disadvantages of using Internet applications (how does the innovation compare with what is already happening?). This assessment includes consideration of level of *compatibility* (with other technologies and processes); level of *complexity* (how difficult is it to understand its use?); level of *trialability* (is it possible to trial the innovation?) and level of *observability* (is it possible to observe it in use?).
- **Decision:** When a decision maker engages in activities that lead to either choosing the innovation or rejecting it.
- **Implementation:** When a decision maker puts the new innovation in place.
- **Confirmation:** When a decision making unit wants reinforcement about the decision made to use the innovation, in order to decide whether to continue its use or not.

The innovation-decision process has been used elsewhere as a basis for framing research into the adoption of Internet technologies. Kendall et al. (2001) used the process to scope their research into the adoption of e-commerce by Singapore small and medium sized enterprises. Fogelgren-Pedersen (2005) used the knowledge, persuasion and decision stages of the innovation-decision process as a theoretical lens for examining the adoption of 'third generation' mobile technologies in Denmark. Li and Lindner (2006) used the innovation-decision process as the theoretical basis for examining the behaviour of faculty at a Chinese university in relation to the adoption of web-based distance education.

One of the changes over time in the various incarnations of Rogers' work is that he recognised the influence that factors *other than* the characteristics on an innovation (such as the socio-technical factors we have identified earlier) can play in the adoption of the innovation. It is anticipated that examination of these factors at organisational (club and association) and individual (member) levels will identify influences in both 'directions' related to pressure to adopt Internet technologies and tensions (especially on individuals) related to the use of the technologies.

The research aims to trace the adoption of Internet applications from initial knowledge of the application, through the decision to adopt, and confirmation by continuance or discontinuance of their use. Referring again to Table 1, Rogers' stage model is represented in the centre row, with factors relating to the organisational level (local sporting association and clubs) in the upper row, and factors relating to the individual (member/volunteer) level in the lower row. We believe that factors such as the limitations faced by CBOs, the type of sport and volunteer uses of ICTs will initially influence adoption at the persuasion stage. Rogers also recognises that adoption decisions can be driven from the individual or organisation level, as we mentioned earlier in the chapter. In the implementation stage, we are interested in what applications are adopted within local sporting clubs. At the organisational level of the confirmation stage, we are interested if the benefits of ICT use recognised in businesses are reflected in local sporting club Internet application use. At the individual level, we are interested predominantly in whether the negative effects of ICT use that can occur for employees in organisations, or volunteers in CBOs, are replicated where members use ICTs for local sporting club duties.

## AN APPLICATION OF THE FRAMEWORK

We will now use cricket clubs in a typical Australian local sporting association as an example of how the framework may be applied in a practical setting. Internet technologies have become increasingly used in the activities of club committee members. These activities range from the use of email to improve communications between committee members, to the introduction of commercial online systems to handle match scores and statistics related to player performance – the latter eliminating repeated data entry and saving countless hours of time. The adoption of these technologies may be driven from the 'top' (that is, the club or even cricket association level) and imposed upon volunteers in the club – or in some instances, the adoption may have been driven from the 'bottom', via 'technology savvy' volunteers keen to apply the technology as part of their duties. However, the introduction of any new technology has an impact – and whilst improved communication and increased efficiencies might seem desirable outcomes – there can be negative impacts on those that feel marginalised from the technology, or for those volunteers for whom the

*Table 1. Roger's nodel of the innovation-decision process combined with the relevant adoption factors to form our initial framework*

| Level | Rogers' Stage of the Innovation-Decision Process | | | | |
|---|---|---|---|---|---|
| | Knowledge | Persuasion | Decision | Implementation | Confirmation |
| Organisation (Local Sporting Association & Clubs) | | -CBO Use of ICTs<br>-Type of Sport | -Top Down Influences | -What applications are adopted? | -Impacts of ICTs |
| **Rogers' Stage Model (Features/ Characteristics)** | -Awareness | -Relative advantage<br>-Compatibility<br>-Complexity<br>-Trialability<br>-Observability | -Accept<br>-Reject | -Possible Tensions<br>-Uncertainty? | -Reinforcement for decision<br>-Dissonance<br>-Discontinuance |
| **Individual (Club Member/ volunteer)** | | -Volunteer use of ICT | -Bottom up Influences<br>-ICT Champion | | -Employee use of ICT<br>-Volunteer use of ICT |

task of recording player statistics may have been a desirable use of their time. We will discuss the adoption of an Internet application within a cricket association in Australia - an association-wide online package to store match results and player statistics.

At the highest level, cricket is predominantly played in Commonwealth countries (such as Australia, New Zealand, England, Pakistan, India and South Africa). Cricket has some similarities with baseball, as the winning team scores the most runs and there are a set number of innings, each involving a team batting and a team on the field (with positions reversed at the conclusion of an innings). At the top level, a game of cricket can be played over five days (a 'test match'). However, local cricket clubs predominantly play shorter versions that are played over two days or even a few hours. A local cricket match in Australia will typically take place over two Saturday afternoons, with each team usually having one innings on each day.

In order to represent how the framework might be applied in this context, a representation of an actual situation involving one of the authors and a local cricket association is presented and described in the form of a case study. This perhaps could be described as *action research*, since the author was involved in the situation being described (as outlined to follow). However, one of the differences here is that, at the time, the author did not realise the situation would be presented as a piece of research, so it is probably more appropriate to describe it as *historical research* (as per Johanson, 2002), where the author takes on the role of observer, minor participant and reporter!

The situation involved a cricket association that operates in the suburbs of an Australian city, a few years after the turn of the millennium. The association had operated for over 50 years, having been merged from two other local cricket associations. Over recent years, the association has been comprised of 20-30 clubs, incorporating some 30-40 senior teams (in five or six grades) and some 40-50 junior teams (over four age groups). So, allowing for some duplication of roles and taking into account administrative, officiating and other duties, at any one time during the cricket season the activities of the association would involve approximately 1,000 people in its activities.

For almost all of its existence, records secretaries in individual clubs and for the association recorded the results of matches and player statistics in a similar manner. As matches were played, two scorebooks were updated on a ball-by-ball basis (manually) by a scorer for each club in a scorebook. At the end of an innings of a batting side, the totals of player runs and bowler performances were added together by the scorers and verified by umpires. As mentioned, a match is generally made up of one or two innings from each side. Entering details in a scorebook whilst the match was under way may seem like quite a simple task, however in many instances a regular scorer was not available and it would be up to two players (often both from the batting team) to update the scorebook. The scorers (who varied in level of scoring 'ability') would rotate as batsmen were dismissed, or just to have a rest. In other words – it could be quite a task to 'balance' the two scorebooks as scorers could be easily distracted or may just make mistakes - but at the end of the day it was usually managed in some manner. Despite the introduction of computers to many other facets of society, and quite sophisticated cricket scoring software now being used at the professional levels of cricket, for the average local sporting cricket club this process remains virtually the same as it has for decades.

In the particular cricket association we are referring to, the match results and individual player statistics from the scorebooks were transcribed by each club onto a single sheet, which was then forwarded to the association. It was then common practice for the association records secretary and the club record secretary (for each club) to separately update the player statistics for the purposes of end-of-season awards and, in some instances, career statistics. Thus, it was not uncommon for individual player statistics to actually be recorded on a number of occasions:

- **Twice:** On the original scorebook (once on the scorebook of each team).
- **Twice:** On the match summary sheet provided to the association by each participating club.
- **Once:** By the association records secretary for end-of-season association awards (copied from the match summary sheets).
- **Once:** By each records secretary when recording their own statistics for end-of-season club awards (copied from the scorebook or the match summary sheet).

Six times in total!

This also meant that there was some duplication in the calculation of player statistics for end-of-season awards as these were calculated at both club and association level. In some instances, these were calculated using informally developed applications (such as spreadsheets) and in a few instances dedicated cricket statistics software would be used. However, the results would still need to be separately entered into these systems.

In addition to this, each team had to supply a list of players that participated in a game to the association registrar – whose job it was to record the number of games that each player played in different grades for the purposes of determining qualifications for finals series (which are similar to playoff rounds in other sports) and also for monitoring player movements between different grades of competition.

A few years into the millennium, the association was approached by a company that offered an online service related to storing cricket match results and player statistics. A number of members in the association were already aware of this service, as they had heard about it from other associations that were using it and some had even seen it in action.

What would (supposedly) happen is that the scorebooks would still be filled out in the same manner, but the match and player details could then be entered *directly online* by one team (usually the team designated as the *home* team). The *away* team could then access the scores online after this and verify that the results were entered correctly (there were procedures in place to deal with disputes, which could also occur in the old system if the match summary sheets did not agree). However, the main difference is the number of times that scores would be entered:

- **Twice:** On the original scorebook (once on the scorebook of each team, as before).
- **Once:** When the match scores and player statistics were entered by the home club.

As mentioned, in addition to this there was also the added task that the away team had to verify the scores.

However, there were some other major improvements:

- Player performances for club and association awards would be automatically calculated 'online'.
- The number of games played by each player in different grades could also be easily determined.

Thus, the role of the association records secretary and the registrar would be simplified a great deal (less manual entry and far less calculation time), as would the role of the club record secretaries.

It seemed almost too good to be true. At that stage, it was.

One problem, as perceived by the association, was the amount being charged for the service. At the time, the association was asked to pay $50 (Australian) per team to use the service. Of course, this would have to be passed on directly to the clubs. Now, cricket is quite an expensive game to play. All of the clubs were non-professional and covered their expenses by charging annual player membership fees, weekly fees to participate, and by running social events. A club with (say) four teams would therefore be asked to find an extra $200 a season.

On top of this, it was not a situation where some clubs could participate and some could not. It was either 'all in' or 'none in'.

Note that the decision in this instance was probably much harder for a cricket association than it would have been for a business in a similar situation. Any system that was able to offer a business this number of labour hours saved would almost certainly have to be adopted. However, in the case of an association and clubs, the labour devoted to manually entering and checking all of the data on multiple occasions occurred using *volunteer* labour – which of course cost the association and clubs *nothing* in relation to the direct outlay of money. Thus, this system could be (and was) viewed as adding cost, but bringing in no extra revenue.

The business offering the online service was invited to give a presentation to the executive committee of the association. This is where one of the authors was involved – as an invitee to the meeting to offer an opinion as to what the association should do. The business presented a simulation of the service to a monthly association club delegate's meeting. The demonstration went quite well for them. A number of questions were asked relating to how the service operated, the changes in processes needed by the association and its clubs, the security of the system, and so forth. Having asked a number of questions during the presentation about the systems and having seen it in operation in another association, the author suggested at the time that it seemed like a sound system and that it would be a good idea to adopt it. However, the committee members again reiterated their concern at the costs that had to be passed on to member clubs.

There was another challenge facing the committee in their decision. Whilst all of the clubs had at least one member that had a computer and an Internet connection, it was not necessarily the records secretary. If a decision was made to adopt the system, then it would be incumbent on each club to ensure that they had the facilities to enter their match data online – another potential expense. Of course, the majority of training in relation to using this technology would come from the usual sources in community based organisations – club members, family members and friends – all volunteers. In the case of the association committee, the association made a decision that it would finance the Internet connection for any of its executive committee that required such a connection to operate the system for the association.

Before we move on to discuss the actual decision that was made there is one other factor that should be considered that makes the adoption of this system different for a local club than for a business. In many instances, the committee members of clubs and the association had been filling out the various forms of match reports and performing their calculations for years, sometimes decades. These jobs became part of their lives during cricket season. There was a real chance that removal of these staple tasks might have negative social effects on some of these people.

After the meeting where the demonstration occurred, the author lost touch with the association for a few months. The next contact, with a member of the association's executive committee, occurred just before the commencement of the new season. The author was informed that, just as the association was to make its decision about whether to adopt the system, a new alternative became available – offered by the State peak body for cricket. To provide some perspective, cricket in Australia at a national level is managed by one peak body, Cricket Australia. At the next level down, there are a number of States which each

*Table 2. Application of Initial Framework to the adoption of an online cricket statistics cricket program*

| Level | Rogers' Stage of the Innovation-Decision Process | | | | |
|---|---|---|---|---|---|
| | Knowledge | Persuasion | Decision | Implementation | Confirmation |
| **Organisation (Local Sporting Association & Clubs)** | 'EXISTING' SYSTEM<br>-Association contacted by business offering system<br>-Other committee members had seen the system in action elsewhere<br>'PEAK BODY' SYSTEM<br>- association contacted by Peak body | **-CBO Use of ICTs**<br>- Chance for ICTs to reduce duplication of effort and save time (BOTH systems)<br>- Inexpensive alternative (PEAK BODY system)<br>-Type of Sport<br>- Cricket is a sport based on statistics – highly suited to this use (BOTH systems) | **-Top Down Influences**<br>- After decision was made by association, all clubs had to 'opt in'<br>- Author used as external 'consultant' [unpaid!] | **- What applications are adopted?**<br>- PEAK BODY system was adopted<br>- Most of the tensions occurred as clubs tried to position themselves (PCs; ICT expertise) for the adoption of the system | **-Impacts of ICTs**<br>-'teething' problems lead to many early errors and inconsistencies |
| Rogers' Stage Model (Features/ Characteristics) | -Awareness | **-Relative advantage:** savings in relation to labour hours and duplication of effort (BOTH systems)<br>**-Compatibility:** it was recognised that there would need to be a change in processes; although many association executives were older than 50, there was an acceptance this was inevitable; some clubs would need new technologies for records secretaries (BOTH systems)<br>**-Complexity:** it was recognised that there would be some difficulty in implementing such a system and that this would differ across clubs depending upon how technology *savvy* they were (BOTH systems)<br>**-Trialability:** the demonstration allowed limited Trialability of the EXISTING system<br>- the PEAK BODY system was not trialled<br>**-Observability:** the EXISTING system had been in operation and its use had been observed<br>- the PEAK BODY system had not been observed | **OPTIONS:**<br>-Accept<br>- adopt EXISTING system<br>- adopt PEAK system<br>**-Reject**<br>- continue 'as is' | **-Possible Tensions**<br>**-Uncertainty?** | **-Reinforcement for decision**<br>**-Dissonance**<br>**-Discontinuance**<br>- in the end, there was enough evidence to suggest that it was a good idea to actually **adopt** a new system, but that perhaps the incorrect option had been chosen<br>- the association decided to move to the EXISTING system in subsequent years. |
| **Individual (Club Members)** | | **-Volunteer use of ICT**<br>- where to source ICT expertise if the system was adopted? | **-Bottom up Influences**<br>**-ICT Champion**<br>- Apart from some member clubs making up the members of the association there was little 'bottom-up' influence | - Possible tensions related to loss of traditional duties and apprehension about use of technology | **-Employee use of ICT**<br>**-Volunteer use of ICT**<br>- Possible problems related to the use of technology as experienced by employees and volunteers – feelings of displacement and fear of using technology |

have their own peak body. The cricket associations sit (sometimes as part of regional divisions) within these peak bodies. The alternative offered was that the State peak body were building their own system, with much (but not all) functionality of the existing system, which they would offer to their associations for free. This set off 'alarm bells' immediately for the author, as it was quite close to the start of cricket season and the new system was untested. However, at this stage the decision had already been made to adopt the alternative system for the association as their major concern was dealt with - they did not have to ask clubs to outlay any extra money to participate (apart from any technology needed to connect to the system).

Before going on to describe what actually happened as a result of the adoption, it might be useful at this stage to show how this situation might be represented in the proposed framework by examining Rogers' innovation-decision stages (refer Table 2). As mentioned earlier, *Knowledge* of the existence of the existing system was gained through various means. Knowledge about the peak body system was provided after contact by the peak body.

*Persuasion* occurred after this. The *relative advantages* in relation to existing processes have already been discussed. Note that in the 'organisation' row of the table it is noted that the innovation could provide benefits similar to those delivered by ICTs in other CBOs and also small businesses – in this instance benefits related to the saving of labour hours and reductions in duplication of effort. It is also noted that cricket is the type of (statistics based) sport that appears suited to this type of application. Levels of *compatibility* and *complexity* were similar for both systems.

However, the levels of *trialability* and *observability* were different between the two systems, with the alternative (peak body) system not having been previously tested in real-life conditions. This might provide the first hint as to what was going to happen afterwards. In relation to the 'individual' row of the table, the main concern was where required levels of expertise might be sourced if they were not available within clubs. This is a situation faced by many CBOs and their volunteers.

The *Decision* to adopt came down to three options:

- *Accept* the innovation and adopt one of the two possible alternatives
- *Reject* the innovation and continue with existing processes.

As is pointed out in the *Decision* column of Table 2, the decision was made by the association committee members on behalf of all of the clubs. Although these members were also individual club members, there was no influence from them as ICT champions, or from a perspective that it would benefit their own club. In the end it was decided that the relative advantages of adoption outweighed those of non-adoption and thus a decision was made to adopt *a* system – but which one? It is probably of no surprise to the reader to find out that the fact the system was offered for no outlay by the peak body outweighed the fact that it had not been trialled before. Subsequent discussions with the committee members suggested that there was perhaps a naivety towards what the system was capable of doing, and that an experienced ICT expert might have suggested that adopting an untested system at any time was a tremendous risk!

So, the peak body system was adopted and *implemented*. What followed were a number of weeks of confusion, as there were numerous problems after its inception that needed to be fixed as they occurred. For instance, some of the teams in the association found that the system had drawn them to play against teams in *other* associations (that had also adopted the system). There was also some confusion in relation to players with identical names and also when they transferred from one club to another mid season.

Some clubs had to adjust (in relation to setting up technology and Internet connections), but most club records secretaries were able to use the technology already in their own home, with some coming to special arrangements regarding funding of Internet connections. In Figure 2 we have noted the possibility of individual members feeling displaced or threatened by the introduction of the new innovation, but this was not assessed at the time (or since). Again, one of the limitations of historical research is that the complete picture is not always available to the investigator, so the situation can only be interpreted on the basis of what is known to have happened at the time (Johanson, 2002).

By the *confirmation* stage, the vast majority of the problems with the peak body system had been corrected, but the association had been affected by the problems so much that it switched to the other system for the next cricket season (and subsequent seasons). Interestingly, the peak cricket body still operates its cheaper system (now fully operational, but less functional) and also works with associations that have adopted the other system to provide links between the peak body website (which has information about all associations and clubs) and their statistics. Thus, associations can choose between the less functional, cheaper system and the more expensive system, and know they will be supported by the peak body.

## Suitability of the Framework: The Future

It is interesting to consider whether the stages of Rogers' innovation-decision process actually exist in reality. Rogers himself (2005, p. 195) suggests that a definitive answer to this question is impossible to provide. However, we believe that the stages do provide a useful means to *classify* the events that have lead to the eventual adoption (or non-adoption) of a particular innovation. In our instance, the idea of splitting the activities into these five stages does provide a structured means to classify the events leading to the adoption of the peak body software, and to highlight the factors in the innovation-decision process that lead to what history has suggested was a less-than-optimal solution (in the short-term anyway). We believe that the additions that we have made to the innovation-decision framework help to highlight the role that sporting bodies (such as clubs and associations) play, as opposed to the roles that individuals (club members and volunteers) play in the adoption of Internet technologies. In the case of the example we have provided, many of the influences, drivers and effects occurred at the club and association level. The innovation adoption was certainly driven from the 'top'. Experience tells us that other recent innovations, such as the use of email by club and association members, have predominantly been driven at the *individual* level (or, from the 'bottom') by club members that can see the benefits of using it (*ICT champions*). In this case, the representation of the adoption of this innovation would show much more detail in the *individual* (lower) row of the framework.

The framework therefore helps us to map the separate processes involved in the adoption of these technologies more easily than if we had just used the five stages of the innovation-decision process by themselves. In the case of the example we have presented, an awareness of the various stages of the innovation-decision process has highlighted the shortcomings made in the initial decision of the cricket association. Increased awareness of how adoption decisions are made and the consequences of these decisions can help to better inform future decisions in this arena.

We are intending to extend this study to test and refine the framework by collecting cases related to the use of Internet technologies by different sporting bodies and in different countries (Australia, New Zealand, the US and the UK). At the time of writing, data collection has commenced in three of these countries.

## CONCLUSION

Local sport has a major, positive impact in our society and relies heavily on the huge number of volunteers who devote many hours of their time to the activities that allow it to occur. In this chapter we outlined the development of a framework, specifically for the adoption of Internet applications by local sporting associations, clubs and their members. The framework has used Rogers' (2005) innovation-decision process as its base, but added in factors that specially relate to the adoption of ICTs at the organisational (community-based organisation) level and individual (volunteer and employee) level. An example (the adoption of an online match results and player statistics application in a local sporting association) illustrates how the framework may be applied in a practical situation to represent the adoption of Internet technologies and, more importantly, how it may be able to identify lessons that can help to inform improved decision making in local sporting bodies in relation to these adoptions.

## REFERENCES

Australian Bureau of Statistics. (2001). *Voluntary work Australia,* (Catalog No. 4441.0). Canberra: ABS.

Australian Bureau of Statistics. (2003). *General social survey. Summary results Australia,* (Catalog No. 4159.0). Canberra: ABS.

Australian Bureau of Statistics. (2003). *Sport and recreation: A statistical overview,* (Catalog No. 4156.0). Canberra: ABS.

Australian Bureau of Statistics. (2006). *Sport and recreation: A statistical overview, Australia, 2006 Edition 2,* (Catalog No. 4156.0). Canberra: ABS.

Australian Government. (2005). *Giving Australia: Research on philanthropy in Australia.* Canberra: ACOSS.

Bassanini, A., & Scarpetta, S. (2002). Growth, technological change, and ICT diffusion:

Recent evidence from OECD countries. *Oxford Review of Economic Policy,* (pp. 324-344).

Bellah, R., Madsen, R., Sullivan, W., Swidler, A., & Tipton, S. (1985). *Habits of heart: Individualism and commitment in American life.* New York: Harper and Row.

Bocij, P., Chaffey, D., Greasley, A., & Hickie, S. (2006). *Business information systems.* (3rd ed.). Harlow: Pearson Education Limited.

Boyle, A., Macleod, M., Slevin, A., Sobecka, N., & Burton, P. (1993). The use of information technology in the voluntary sector. *International Journal of Information Management,* (pp. 94-112).

Brookfield, R. (1983). *Adult learners, adult education and the community.* New York: Teachers Collage Press.

Carr, N. (2001). *The digital enterprise.* Boston: Harvard Business School Press.

Charity Commission for England and Wales. (2002). *Giving confidence in charities: Annual report 2001–2002.* London.

Ciarli, R., & Rabellotti, R. (2007). ICT in industrial districts: An empirical analysis on adoption, use and impact. *Industry and Innovation,* (pp. 277-303).

Clary, E., & Snyder, M. (1991). *A functional analysis and prosocial behaviour: The case of volunteerism.* Newbury Park: Sage.

Coleman, R. (2002). Characteristics of volunteering in UK sport: Lessons from cricket. *Managing Leisure,* (pp. 220-238).

Cricket Australia. (2006). *Annual report 2005-2006.* Melbourne: Cricket Australia.

Cuskelly, G. (1995). The influence of committee functioning on the organisational commitment of volunteer administrators in sport. *Journal of Sport Behaviour,* (pp. 254-270).

Daniel, E., Wilson, H., & Myers, A. (2002). Adoption of e-commerce by SMEs in the UK:

Towards a stage model. *International Small Business Journal,* (pp. 253-270).

Daveri, R. (2002). *The new economy in Europe (1992–2001).* Working Paper Series, WP No. 213 IGIER, Universita Bocconi, Milan.

Dawley, D., Stephens, R., & Stephens, B. (2005). Dimenstionality of organisational commitment in volunteer workers: Chamber of Commerce board members and role fulfilment. *Journal of Vocational Behaviour,* (pp. 511-525).

Diaz, J. (1977). *Communication and rural development.* Paris: United Nations Educational, Scientific and Cultural Organisation (UNESCO).

Doherty, A., & Carron, A. (2003). Cohesion in volunteer sport executive committees. *Journal of Sport Management,* (pp. 116-141).

Doherty, A., Patterson, M., & Van Bussel, M. (2004). What do we expect? An examination of perceived committee norms in non-profit sport organisations. *Sports Management Review,* (pp. 109-132).

Eley, D., & Kirk, D. (2002). Developing citizenship through sport: The impact of a sport based volunteer programme on young sport leaders. In D. Eley & D. Kirk (Eds.), *Sport, education and society* (pp. 151-166). London: Routledge.

Edwards, P., & Watts, A. (1983). Volunteerism and human service organizations: Trends and prospects. *Journal of Applied Social Sciences,* (pp. 225-245).

Fogelgren Pedersen, A. (2005). The Mobile Internet: The pioneering users' adoption decisions. In *Proceedings of the 38th Hawaii International Conference on Systems Sciences.* Hawaii.

Hall, M., & Banting, K. (2002). *The nonprofit sector in Canada: An introduction.* Working Paper, School of Policy Studies, Queen's University.

Hamilton, E., & Cunningham, P. (1989). Community-based adult education. In S. Merriam & P. Cunningham (Eds.), *Handbook of adult and continuing education* (pp. 439-450). San Francisco: Jossey-Bass.

Hiemstra, R. (1993). *The educative community.* (3rd ed.). Syracuse: Syracuse University Adult Education Publications.

Hockey Australia. (2006). *2006 Hockey Australia Annual Report.* Melbourne: Hockey Australia.

Johanson, G. (2002). Historical research. In K. Williamson (Ed.), *Research methods for students, academics and professionals – Information management and systems.* Wagga Wagga: Centre for Information Studies, Charles Sturt University.

Karanasios, S., Sellitto, C., Burgess, S., Johanson, G., Schauder, D., & Denison, T. (2006). The role of the Internet in building capacity: Small businesses and community based organisations in Australia. In *Proceedings of the 7th Working for E-Business Conference.*

Kellogg, W. (1999). Community-based organisations and neighbourhood environmental problem solving: A framework for adoption of information technologies. *Journal of Environmental Planning,* (pp. 445-469).

Kendall, J., Tung, L., Chua, K.H., Ng, C.H.D., & Tan, S.M. (2001). Receptivity of Singapore's SMEs to electronic commerce adoption. *Journal of Strategic Information Systems, 10*(3), 223-242.

Li, Y., & Lindner, J.R. (2005). Faculty adoption behaviour about web-based distance education: A case study from China Agricultural University. *British Journal of Educational Technology, 38*(1), 83-94.

MacKay, N., Parent, M., & Gemino, A. (2004). A model of electronic commerce adoption by small voluntary organisations. *European Journal of Information Systems,* (pp. 147–159).

Madon, S. (1999). International NGOs: Networking, information flows and learning. *Journal of Strategic Information Systems,* (pp. 251-261).

McMillan, D., & Chavis, D. (1986). Sense of community: A definition and theory. *American Journal of Community Psychology,* (pp. 6-23).

Montazemi, A. (1988). Factors affecting information satisfaction in the context of the small business environment. *MIS Quarterly,* (pp. 239-256).

Morgon, G. (1995). ITEM: A strategic approach to information systems in voluntary organisations. *Journal of Strategic Information Systems,* (pp. 225-237).

Pearce, J. (1993). *Volunteers: The organisational behaviour of unpaid workers.* London: Routledge.

Pretty, G., Andrewes, L., & Collett, C. (1994). Exploring adolescent's sense of community and its relationship to loneliness. *Journal of Community Psychology,* (pp. 346-357).

Raymond, L. (1985). An empirical study of management information systems sophistication in small business. *MIS Quarterly,* 37-52.

Roberts, H. (1979). *Community development: Learning and action.* Toronto: University of Toronto Press.

Rogers, E. (2005). *Diffusion of innovations.* (5th ed.). New York: The Free Press.

Saxton, J., & Game, S. (2001). *Virtual promise: Are charities making the most of the Internet revolution.* London: Third Sector.

Shibli, S., Taylor, P., Nichols, T., Gratton, C., & Kokolakakis, T. (1999). The characteristics of volunteers in UK sports clubs. *European Journal for Sport Management,* (pp.10-27).

Suraya, R. (2005). Internet diffusion and e-business opportunities amongst Malaysian travel agencies. In *Proceedings of the Hawaii International Conference on Business* (pp. 1-13). Honolulu.

Turban, E., Leidner, D., McLean, E., & Wetherbe, J. (2006). *Information technology and management.* Hoboken: John Wiley & Sons.

Volunteering Australia. (2005). *Definitions and principles of volunteering.* Retrieved May 28, 2007, from http://www.volunteeringaustralia.org.

Wilson, J. (2000). Volunteering. *Annual Review of Sociology,* (pp. 215-240).

Wreden, N. (1997). Business boosting technologies. *Beyond Computing, 27.*

# Chapter XIV
# Online Questionnaires and Interviews as a Successful Tool to Explore Foreign Sports Fandom

**Anthony K. Kerr**
*University of Technology, Sydney (UTS), Australia*

## ABSTRACT

*Globalisation and advances in communications technology have greatly expanded the potential marketplace for professional teams, especially for those sports internationally popular. As a result, many team brands profit from millions of satellite fans, or supporters, worldwide. However, the reasons satellite supporters identify with their team remain largely unexplored. Therefore, this chapter describes three studies designed to examine the team identification of these supporters and highlights how mixed methods can be successfully employed online to engage with distant sports fans.*

## INTRODUCTION

Globalisation and advances in communications technology have greatly expanded the potential marketplace for professional teams, especially for those sports internationally popular. Moreover, professional team sport, like other forms of popular culture, has become a marketable commodity subject to the logic of the marketplace. As a result, sports teams are commodified, courtesy of the 'Faustian pact' that exists between mass media and professional sport, and tailored to appeal to indirect consumers, either through television or, increasingly, the Internet. Due to this global exposure, Spanish football club, Barcelona F.C., is believed to have 70 million supporters worldwide, while English rival, the Liverpool F.C., claims to have 28 million fans (Rice-Oxley, 2007). Indeed, global powerhouses such as Real Madrid and Manchester United could have as many as 350 million supporters (Henriksen, 2004).

The increased importance of these foreign consumers, especially with respect to the contribution they could make to a team's brand equity, led Kerr and Gladden (2008) to coin a new term, the 'satellite fan', to encompass these individuals. However, given that the strength of their loyalty is unknown, Kerr (2008; in press) suggested, it was perhaps more prudent to refer to these individuals as 'satellite supporters'. The objectives of this chapter, therefore, are threefold: to describe the evolution of a series of studies designed to understand the team identification of satellite supporters; to briefly discuss some key findings from these studies; and to highlight how online research methods can successfully contribute to sports fan research, especially as researchers seek to engage with distant participants.

## BACKGROUND

In a competitive sports marketplace, satellite supporters represent significant revenue streams. For instance, 20 percent of merchandise sold through the National Basketball Association's (NBA) official website are to overseas fans (Eisenberg, 2003); Japanese baseball fans were expected to spend $USD500 million on New York Yankees tickets and souvenirs in support of their countryman, Hideki Matsui (Whiting, 2003); and Real Madrid now earns 60 percent of merchandise revenue from international markets, up from 10 percent only five years ago (Jones, Parkes, & Houlihan, 2006). Furthermore, the greater the degree of identification between an individual and their chosen team, or "the extent to which a fan feels psychologically connected" (Wann, Melnick, Russell, & Pease, 2001, p. 3), the more likely they are to attend their team's games (Fink, Trail, & Anderson, 2002; Fisher, 1998; Wann & Branscombe, 1993); monitor their team in the media (Fisher, 1998; James & Trail, 2005); purchase team merchandise (Fisher & Wakefield, 1998; Greenwood, 2001; James & Trail, 2005); and to both recognise (Gwinner & Swanson, 2003; Lascu, Giese, Toolan, Guehring, & Mercer, 1995) and purchase (Gwinner & Swanson, 2003; Madrigal, 2000, 2004) products from team sponsors.

The complex nature of sports fandom also has implications for the choice of suitable methods. While studies on the sports fan have traditionally favoured quantitative methods, there is now greater acceptance of a qualitative approach. Furthermore, mixed methods, Jones (1997b) claimed, can provide "a fuller understanding of the sports fan". Although the use of mixed methods to explore sports fandom is becoming increasingly popular, its adoption is still relatively limited. In addition, although there have been attempts to examine sports fans through the Internet, using for instance online surveys (Nash, 2000), online message boards (End, 2001; Lewis, 2001; Mitrano, 1999), and online 'interviews' (Heinonen, 2002; Mitrano, 1999; Silk & Chumley, 2004), it is still a relatively recent approach.

## ISSUES, CONTROVERSIES, PROBLEMS

Although satellite supporters might prove to be the future lifeblood of a sports franchise, and despite the increased attention paid to team identification (notably, Greenwood, Kanters, & Casper, 2006; Gwinner & Swanson, 2003; Jacobson, 2003; Jones, 1998; Kolbe & James, 2000; Sutton, McDonald, Milne, & Cimperman, 1997), the reasons these foreign consumers identify with their chosen sports teams remain largely unexplored. Indeed, Richardson and O'Dwyer (2003) claimed, research on factors that contribute to a fan's original choice of sports team would be a welcome addition to the field.

The series of studies presented in this chapter therefore seeks to address the imbalance. This chapter reports the results of research on the team identification of satellite supporters. Furthermore, it seeks to validate the use of online mixed methods to explore, and better understand, the team identification and behaviour of foreign sports consumers.

## SOLUTIONS AND RECOMMENDATIONS

The adopted research methodology was a case study approach using mixed methods. The case study is an appropriate methodology when a holistic, in-depth exploration of a phenomenon is required, and when the individual viewpoints of participants are important (Orum, Feagin, & Sjoberg, 1991; Tellis, 1997). Furthermore, it has been increasingly adopted as a vehicle to examine sports fans (see, for instance, Heinonen, 2002; Jacobson, 2003; Jones, 1998; Nash, 2000; Reimer, 2004), and even supporters of foreign-based teams (Nash, 2000; Reimer, 2004). Ultimately, the use of the case study approach is especially beneficial when literature on the issue is poor or scarce (Jacobson, 2003). Indeed, literature regarding the team identification of foreign supporters has been almost non-existent.

The English Premier League's Liverpool F.C. was an appropriate subject for the case study. The Merseyside club has a global fan base, possesses a rich history, and is arguably one of the world's premier team brands. In addition, access to enthusiastic research participants was possible given the number of online fan organisations dedicated to the club. These issues were critical to the study's success for, as Stake (1995) explained, it is beneficial that cases are chosen which "are easy to get to and hospitable to our inquiry" (p. 4). Sports fan research has begun, as earlier explained, to use the Internet to engage with distant participants, and so this technology enabled access to the satellite supporter irrespective of location.

As the first stage in this quantitative research, pilot studies were developed and conducted so as to ensure the validity of the survey instrument. The pilot study, according to van Teijlingen and Hundley (2001), is a mini-version of a full-scale study and allows for specific pre-testing of a particular research instrument, and is "a crucial element of a good study design" (p. 1). Indeed, testing the adequacy of a questionnaire, Moser and Kalton (1971) claimed, is "probably the most valuable function of the pilot survey" (p. 49).

The success of the pilot studies resulted in the development of a suitable questionnaire; however, since this approach used mixed methods, it also included qualitative data from semi-structured interviews. The adoption of mixed methods allowed for the triangulation of data gathered from satellite supporters of the Liverpool F.C. and enhanced the overall validity of the research. In addition, semi-structured interviews have been increasingly used in sports fan research (for instance, Derbaix, Decrop, & Cabossart, 2002; Dolance, 2005; Tapp & Clowes, 2000), and can often be more insightful than other methods (Yin, 2003). Consequently, a number of online interviews were conducted until both sufficiency and saturation of information had been satisfied. These interviews were conducted via instant messaging technology through an online chat room established for this express purpose. Indeed, given advances in technology, Fontana and Frey (2000) suggested, it was "only a matter of time" before qualitative researchers began to conduct 'virtual interviews' (p. 667).

## THE USE OF ONLINE QUESTIONNAIRES

### Overview

Research into sports team identification has consistently relied upon survey research (for example, Jacobson, 2003; Jones, 1997a, 1998; Kolbe & James, 2000; Wann, Tucker, & Schrader, 1996). However, Alreck and Settle (2004) explained, the use of the Internet in survey research has become increasingly popular, largely due to greater online access and acceptance, increased availability of broadband connections, and the time and cost savings when compared to more traditional survey approaches. Furthermore, there appear to be key advantages to such an approach. Computer-driven surveys can give respondents a greater feeling of anonymity, thus leading to more honest responses, and participants often seem to provide more thought when completing online surveys. They might also prove ideal for "studying respondents in remote locations all over the world" (Bailey, 1994, p. 205). Despite Alreck and Settle's (2004) concerns that online surveys were not always appropriate, the decision to adopt an online questionnaire was logical since participants could reside anywhere in the world, and cost and access were important considerations. Furthermore, Hoyle, Harris and Judd (2002) claimed, the existence of online communities make it possible to target very specific and difficult to reach populations. As such, a self-administered Internet questionnaire was selected as a cost-effective vehicle to explore satellite supporters and their fandom.

Before the final questionnaire was administered, however, online pilot studies were developed and conducted so as to ensure the validity of the final research instrument. The pilot study, Beebe (2007) argued, is important in "yielding data to assess cost, feasibility, methodology, and data analysis for future studies" (p. 213), and helps researchers "identify design flaws, develop data collection and analysis plans, and gain experience with participants" (p. 213). Two pilot studies were therefore conducted: an exploratory, and a team-specific, study.

## The Exploratory Pilot Study (AFANA)

Similar to Wann, Tucker and Schrader's (1996) 'exploratory examination' of team identification, an online questionnaire was designed to solicit open-ended responses from a community of satellite supporters. A detailed description of this study and its findings can be found in Kerr (in press). At this initial stage it was important that the results were not limited to any one sports team so the Australian Football Association of North America (AFANA) was chosen as a suitable candidate. The association's U.S. members were satellite supporters of the Australian Football League (AFL) and management was enthusiastic about the project.

### Participants

Sixty-three AFANA members responded to the online questionnaire, however, 13 participants failed to fully complete the survey and were therefore removed from the sample. Furthermore, closer examination revealed that 11 participants were either Australian or had an immediate family member from Australia. Although it came as little surprise to find that many AFANA members were in fact Australian and, as a consequence, supported an AFL team, the primary purpose of the study was to explore the team identification of satellite supporters. As a result, these individuals, perhaps better classified as 'expatriate fans', were excluded from the study and their presence reflected the challenge of locating a suitably-large number of satellite supporters. Consequently, the final sample comprised 39 individuals (34 male, 5 female) whereby nearly two-thirds of respondents were aged between 26-45 years.

### Materials and Procedure

Respondents followed a link on the official AFANA website, www.afana.com, which led them to the survey hosted by www.surveymonkey.com, an independent third-party which specialises in the creation and management of online surveys. The questionnaire was available for a period of two weeks and contained six sections.

The first section introduced the researcher and the purpose of the study, while the second asked participants to provide a range of demographic items such as age, gender and zipcode. The third section asked participants to name their favourite AFL club and the reasons why they supported that sports team. These questions were open-ended and allowed individuals to express themselves in their own words for, as Lavrakas (2004) claimed, this facilitates "richer, more detailed responses" (p. 903). The section concluded by asking whether they had a connection to Australia or the host city of their favourite AFL team. Participants in the final sections were asked to rate the importance of certain factors upon their initial team support and how often each season they engaged in specific sports consumption behaviour. Both these questions employed Likert-scales.

## Discussion

Respondents were asked why they originally supported their favourite AFL team. A number of logical themes emerged from the questionnaire and were mostly consistent with those antecedents found in the sports fan literature, notably, the role of socialisation agents, team performance, and other team-related factors. Furthermore, the data provided a solid cross-section of team support, as only fans of two of the 16 AFL teams, the Melbourne Demons and Port Adelaide Power, were not represented in the study.

Five popular themes emerged for their initial team support: early media exposure (33.3 percent of cases), logo design and/or name (20.5 percent), team-specific appeal (20.5 percent), personal connection to Australia (12.8 percent), and the presence of a particular player(s) (12.8 percent). These themes accounted for nearly two-thirds of total responses. Team success and reputation and/or tradition ranked sixth and seventh respectively, although given that successful teams provide an opportunity for fans to 'bask in reflected glory' or BIRG, as Cialdini, Borden, Thorne, Walker, Freeman and Sloan (1976) explained, it was a surprise that these did not feature more prominently in the sample. For instance, it has been shown that successful sports teams often have greater support than those which perennially struggle (Branscombe & Wann, 1991; End, Dietz-Uhler, Harrick, & Jacquemotte, 2002; Mahony, Howard, & Madrigal, 2000; Wann et al., 1996). The later Likert-scaled question confirmed the importance of many of these factors, especially reputation and/or tradition and the presence of a particular player(s).

A key socialisation agent, geographic proximity, was clearly absent for these supporters. The majority of AFANA's members reside in the United States, thousands of kilometres from their favourite Australian team, and so it appears that the media acts instead as an agent of socialisation for many of these individuals. Approximately a third of subjects, double any other response, indicated that they originally supported their favourite AFL team because they were exposed to the product at an early age through the media. Indeed, the role of the media in this process has received some support in the fan literature (Jacobson, 2003; Mahony, Nakazawa, Funk, James, & Gladden, 2002; Sutton et al., 1997; Wann et al., 1996). International media arrangements ensure the foreign sports product is available to the satellite supporter, without which foreign teams are unlikely to develop significant brand awareness in foreign markets.

Piloting questionnaires is especially important when the research instrument is self-administered (Lancaster, Dodd, & Williamson, 2004). As a result, this preliminary study was critical in the development of a team-specific pilot survey to further explore this phenomenon. The subsequent online study examined satellite supporters of AFC Ajax of Amsterdam.

## The Team-Specific Pilot Study (Ajax USA)

Data analysis from the initial pilot study confirmed the research validity, or that the questions accurately measured what they were designed to do, namely, which factors were, and continue to be, important in the team identification of the satellite supporter. The majority of respondents clearly understood the questions and often provided extensive accounts of their fandom. The open-ended questionnaire successfully served as an 'exploratory examination' and certain factors surfaced as potentially important upon their team identification.

Likert-scaled items were introduced to rank the importance of these factors and therefore replaced the original study's open-ended design. Further questions were generated from the initial pilot and additional factors identified through the literature as potentially important. This was consistent with Bailey (1994), who explained that researchers can begin with open-ended questions and, after analysis, construct closed-ended categories from the earlier data. This combination of items served to provide a more detailed account of this phenomenon.

Ajax USA, an online supporter group for fans of the Dutch club, AFC Ajax of Amsterdam, was chosen as a suitable candidate for the second study. As the organisation explained: "Our club is open to all true fans of Ajax and the beautiful game. We hope Ajax USA helps you to support your team and to make friendships with other Ajax fans around the world" (www.ajaxusa.com). Ajax F.C. has been European champion four times and, along with Feyenoord and PSV, consistently dominates the Dutch national football league, the Eredivisie. Furthermore, in an effort to expand its brand to foreign markets, Ajax FC also established a South African franchise, Ajax Cape Town (Browne, 1999). Ajax USA has more than 1,000 registered members, acts as a forum for satellite supporters of the team, and management agreed to promote the study.

In addition, access to Ajax F.C. fans allowed this survey to be tailored to an individual team, in contrast with the original pilot and, in particular, a European football team. This was consistent with the choice to study satellite supporters of the Liverpool F.C. and allowed the instrument to be easily adapted. Reds' fans were not chosen for this revised pilot survey for two reasons: their participation was critical for later data collection and so I did not wish to exhaust this resource at this early stage and, secondly, some argue, pilot participants should not be included in the main study (Lancaster et al., 2004; van Teijlingen & Hundley, 2001).

## Participants

One hundred and eighty seven Ajax USA members responded to the online questionnaire, however, 34 participants failed to fully complete the survey and were therefore removed from the sample. Furthermore, closer examination revealed that 25 participants were either Dutch or had an immediate family member from the Netherlands. For instance, Jon, an American fan, explained, "I have a Dutch ancestry ... so I choose to follow Ajax from Amsterdam, the Capital of the Netherlands" so, similar to the earlier study, these 'expatriate fans' were excluded. Although most subjects resided in the U.S. (63 individuals), the sample represented more than 40 countries. In addition, while almost a third of participants classed themselves as American, almost 50 other nationalities were represented. The final sample comprised 128 individuals (all male) and nearly 73 percent of respondents were aged between 18-35 years.

## Materials and Procedure

Respondents followed a link on the Ajax USA website, www.ajaxusa.com, which led them to the survey hosted by www.surveymonkey.com. The questionnaire was available for a period of two weeks and contained five sections. The first section introduced the researcher and the purpose of the study, while the second asked participants to provide a range of demographic items such as age, gender, country of residence and nationality. The third section asked participants to rank the importance of certain factors upon their identification with the Amsterdam club, while the fourth asked them to do the same for their current support. These factors were consistent with a review of the literature and the initial pilot, and questions used five Likert-scale items where response options ranged from 'not important' to 'extremely important'. Finally, fans were asked whether they would ever stop supporting the Ajax F.C. and which activities they pursued to express their fandom.

Analysis of the data supported the value of a web-based research instrument and confirmed that an online questionnaire can be a valuable research tool to explore a cross-section of satellite supporters. The team-specific pilot study generated an enthusiastic response from participants and encouraged confidence in a larger global questionnaire, in this instance, focussed on satellite supporters of the English Premier League's Liverpool F.C.

## The Global Fan Study (Liverpool FC)

Both the exploratory and team pilot studies confirmed the appropriateness of the research instrument. Respondents appeared to understand the online questionnaire and it accurately accounted for their original identification with their favourite professional sports team. However, one question was modified for the final Liverpool F.C. questionnaire. The question regarding reputation and tradition was split into three separate questions: the Merseyside club's history of success, history of ethical behaviour, and style of play. This was done for both their original and continued support of the Merseyside club. Although reputation and tradition encompasses all of these elements, this modification allowed a greater understanding of which elements were believed to be most important. The two pilot studies had generated an extremely-high completion rate and enthusiastic responses from participants. Furthermore, they proved extremely valuable and encouraged confidence in a larger global fan questionnaire. As earlier explained, this research adopted a case study approach using mixed methods, questionnaires and semi-structured interviews, in an online environment. However, justification for the selection of the Liverpool F.C. as a suitable subject remains to be made.

## Case Selection

Veal (2005) emphasised the key importance of case selection in case study research. However, certain criteria likely dictate the selection of appropriate case(s). For instance, it is highly unlikely that professional sports teams are equally able to market themselves worldwide, due largely to the international popularity of their sport. Indeed, the globalisation of a sports brand, Richelieu and Pons (2006) claimed, is limited to the sport's worldwide acceptance and popularity. While it might be intriguing to examine foreign fans of the National Rugby League's (NRL) South Sydney Rabbitohs or National Hockey League's (NHL) Toronto Maple Leafs, the relatively limited reach, and participation levels, of these sports severely limits the international potential of their brands. Given the successful globalisation of football (Giulianotti, 2002), and the awareness of teams such as Manchester United and Real Madrid, it is understandable why football teams are better situated to become global brands, and hence a more appropriate choice of case for this research.

Europe is the focus of world football. Its major leagues, the English Premier League, Spanish la Liga, Italian Serie A, and German Bundesliga, are televised worldwide, and attract the best international players in the world (Gieske & Forato, 2004). Indeed, the UEFA Champions League ranks second only to the quadrennial FIFA World Cup in terms of football importance. The European leagues alone have an estimated fan base of 130 million and, in the case of the English and Spanish leagues, international fans outnumber their domestic counterparts (*The Brand Champions*

*League: Europe's Most Valuable Football Clubs*, 2005). Brand awareness is critical to the creation of brand equity, and international media exposure ensures the product is available to satellite supporters. It is thus a necessary condition for their identification with these football teams, and so case selection is limited to these leagues.

There is a strong relationship between a team's brand value, according to brand consultant *FutureBrand,* and the revenue generated by their brand. The more supporters a club has, they argued, the greater the potential to increase revenue and hence brand value. Therefore, the one dimension common to premier brands was their ability to promote themselves overseas and thereby attract fans from all over the world (Gieske & Forato, 2004). Furthermore, elite teams often feature prominently in their respective title races and therefore benefit from European competition, either UEFA Cup or, preferably, Champions League. This provides substantial prize money and enables them to showcase their brand to an audience of billions.

The value of Europe's top football brands reflects the emergence, and increased importance, of the satellite supporter. A number of consultants (*BrandFinance*™, *FutureBrand* and *Deloitte*) produce annual reports that rank the top European football teams based on financial performance and brand valuations. It is no coincidence that many of the same teams consistently feature in these reports. Thirteen clubs, in particular, have appeared in all nine editions of the *Deloitte Money Football League*™. A shortlist of potential case candidates therefore emerged and is arranged in alphabetical order according to their league (Table 1).

Five of these clubs, notably Manchester United, Real Madrid, Juventus, Bayern Munich and AC Milan are what *Deloitte*™ terms, the 'Money League elite', due to their financial performance over the past decade. They are "truly global clubs with significant and increasing international support to add to their millions of domestic fans" (Jones et al., 2006, p. 23). However, a number of others are fighting to join this elite company. These 'second-tier' clubs are of particular interest in this research, for they have the potential to develop their foreign fan base, increase revenue streams and, ultimately, emerge as premier team brands. Professional sports teams that consistently receive international media exposure are likely to possess, or have the potential to develop, a solid foreign fan base. Media exposure allows teams to develop brand awareness abroad and is more likely for teams that participate in top-flight European leagues and the highly-televised UEFA competitions.

The Liverpool F.C., and its 28 million registered fans worldwide (Rice-Oxley, 2007), present, as Stake (2001) suggested, an ideal 'opportunity to learn' about this particular phenomenon, the team identification of satellite supporters. It is one of football's top brands, largely in part to its sizable fan base, and has a rich history of both success and tragedy. There is significant evidence for the satellite supporter phenomenon and the Merseyside club: the existence of 150 international supporter branches and published journal articles on foreign Liverpool F.C. supporter groups. The English club

*Table 1. Case study candidates*

| Premier League | La Liga | Serie A | Bundesliga |
|---|---|---|---|
| Arsenal<br>Chelsea<br>Liverpool<br>Manchester United<br>Newcastle<br>Tottenham Hotspur | Barcelona<br>Real Madrid | AC Milan<br>AS Roma<br>Inter Milan<br>Juventus<br>SS Lazio | Bayern Munich<br>Schalke 04 |

is an ideal candidate due to the aforementioned criteria, the availability of secondary data, and the existence, accessibility and cooperation of supporter organisations dedicated to the Merseyside club. As Stake (1995) explained, it is beneficial if cases are chosen which "are easy to get to and hospitable to our inquiry" (p. 4).

## The Liverpool F.C. Story

The Liverpool F.C. is the most successful club in the history of English football. They have won the League Championship 18 times, compared to their nearest rival, Manchester United, who has won the title on 17 occasions. However, their most recent title came in 1990 before the introduction of the English Premier League. They are also the most successful English representative in European competition: they have won five European Cup (or Champions League) trophies, the last being their memorable 3-2 triumph over Italian rivals, A.C. Milan. A complete list of their major honours can be seen in Table 2.

In February 2007, the Liverpool F.C. became yet another English club to be run by overseas owners. Announcing they were "ready to restore the glory days back to Anfield" (Wood, 2007), Americans, George Gillett and Tom Hicks, were welcomed, in contrast to Malcolm Glazer's takeover of Manchester United. The club has a global fan base of between 18 (Gage, 2006) and 100 million fans (Estridge, 2007). Although the brand appears ready to tackle the world, Liverpool F.C. came from humble beginnings.

## Online Supporter Branches

Satellite supporters, by virtue of their geographic location, are indirect consumers of the sports product, and therefore are not as accessible to researchers as more 'traditional' local team fans. Furthermore, indirect consumers, as Wann et

*Table 2. Major Liverpool F.C. honours*

| Year | Honour | Manager |
|---|---|---|
| 1901/1906 | League Champions | Tom Watson |
| 1922/1923 | League Champions | David Ashworth |
| 1947 | League Champions | George Kay |
| 1964/1966/1973<br>1965/1974<br>1973 | League Champions<br>F.A. Cup<br>UEFA Cup | Bill Shankly |
| 1976/1977/1979/1980/1982/1983<br>1976<br>1977/1978/1981<br>1981/1982/1983 | League Champions<br>UEFA Cup<br>European Cup<br>League Cup | Bob Paisley |
| 1984<br>1984<br>1984 | League Champions<br>European Cup<br>League Cup | Joe Fagan |
| 1986/1988/1990<br>1986/1989 | League Champions<br>F.A. Cup | Kenny Dalglish |
| 1992 | F.A. Cup | Graeme Souness |
| 1995 | League Cup | Roy Evans |
| 2001<br>2001<br>2001/2003 | F.A. Cup<br>UEFA Cup<br>League Cup | Gerard Houllier |
| 2005<br>2006 | European Cup<br>F.A. Cup | Rafael Benitez |

al. (2001) explained, witness the sports contest through some form of mass media. While fans of the Liverpool F.C. are abundant in the north of England, they proved more difficult to locate in large numbers abroad. This final global fan study was therefore conducted with the cooperation of online supporter organisations dedicated to the Merseyside club, specifically those affiliated with the Liverpool F.C.'s Association of International Branches. These branches are categorised according to geography: England, Europe, Ireland, Northern Ireland, Scotland, Wales, and the rest of the world. Although there were a handful of supporter branches in Australia (Queensland, Adelaide, Perth and Melbourne), those overseas had more members, hence the potential for a larger sample size. For instance, the Scandinavian Branch boasts more than 25,000 members. Moreover, the participation of international supporter branches enhanced the diversity of data collected. Many of these international branches also maintain their own websites, so it was logical that the satellite supporter would use the Internet to enhance their fan experience, hence the Internet's value to engage with members. Indeed, the Internet, as seen earlier, has been increasingly used to engage with distant participants.

The management of 54 international branches were emailed regarding their participation in the study, however, since many email addresses were incorrect, or no longer in use, emails were often unable to be delivered. Despite this, the opportunity to participate was warmly welcomed by a number of Liverpool F.C. supporter clubs worldwide (Table 3). It was decided to focus exclusively on those fan clubs outside Great Britain since their members were more likely to be satellite supporters rather than expatriate fans. In addition, two online fan forums (Soccer Fans Network and YNWA) also offered to promote the questionnaire. An expatriate fan, in this instance, would be an English Liverpool F.C. fan that lived abroad. These individuals were not of interest to this research, as they were simply local fans who happened to live overseas. This study instead sought to determine why satellite supporters identify with the Liverpool F.C. in the absence of a geographic connection.

## Participants

More than 1,500 (1,515) Liverpool F.C. fans from 37 countries responded to the online questionnaire, however, 340 participants failed to fully complete the survey and were therefore removed from the sample. Furthermore, closer examination revealed that 47 participants were best classed as expatriate fans and excluded from the study. Therefore, the final sample comprised 1,128 individuals (1087 male, 41 female) and nearly three-quarters of respondents were aged between 26-35 years, perhaps reflecting the younger generation's acceptance, and use, of technology and the Internet.

*Table 3. Participating online LFC supporter branches*

| Branch | Location | Web Address |
|---|---|---|
| The Calgary Branch of the Liverpool FC Supporters Club | Calgary, Canada | www.lfccalgary.com |
| Indonesia's Official Liverpool FC Supporters Club | Indonesia | www.big-reds.org |
| Liverpool FC New York Supporters' Club | New York, U.S.A. | www.lfcny.org |
| Liverpool FC Supporters Club Scandinavian Branch | Norway | www.liverpool.no |
| Melbourne Liverpool Supporters Association | Melbourne, Australia | www.liverpoolfc.com.au |
| Official Canadian Supporters Club of Liverpool FC | Canada | www.liverpoolfc.ca |
| United States Supporters of Liverpool F.C. | U.S.A. | www.uslfc.com |

### Materials and Procedure

An explanatory paragraph was written which detailed the study's goals and a link to the online survey hosted by www.surveymonkey.com. Management of the supporter branches promoted the online survey in one, or all, of the following methods: a) posted the paragraph and the survey link in their fan forums, b) emailed the paragraph and link to registered members, or c) wrote a story about the research and posted the link on their official website. The questionnaire was available for a period of five weeks and its format was almost identical to the previous team-specific pilot study. Therefore, the online questionnaire contained five sections, utilised closed- and open-ended items, and Likert scales. The final section thanked participants and asked if they would be interested in further discussion about their support of the Liverpool F.C. This contact information, and consent, was important for the qualitative phase of this research. Indeed, 40 percent of participants provided either an email address or contact telephone number such was their willingness to discuss their team fandom.

Analysis of the data further supported the value of a web-based research instrument. Although respondents were directed to the questionnaire through online supporter organisations dedicated to the Merseyside team, 37 nationalities were represented, and the participation of more than 1,500 satellite supporters was extremely welcome. The online questionnaire is therefore a valuable research tool to explore satellite supporters, irrespective of location.

## THE USE OF ONLINE INTERVIEWS

### Overview

The interview can take many forms and range from structured to unstructured (Minichiello, Aroni, Timewell, & Alexander, 1995). Between these two types lies the semi-structured interview. Semi-structured interviews have been increasingly used to understand sports fans worldwide (for instance, Derbaix et al., 2002; Dolance, 2005; Tapp & Clowes, 2000), and due to the qualitative nature of this method, more unstructured interviews can provide a greater breadth of data than other types (Fontana & Frey, 2000). In this research, the interview was conducted in the virtual environment, a process foreshadowed by Fontana and Frey (2000) who suggested that it was "only a matter of time" before qualitative researchers used the Internet to conduct 'virtual interviews' (p. 667).

Although the Internet's value as a research tool has previously been discussed in this chapter, namely its use as a cost-effective vehicle to access key individuals, Crichton and Kinash (2003) explained that the online interview has other major strengths. For instance, participants are likely to take more care with their responses, since they can 'take back their words' before posting them rather than commit misspoken comments to print. The instantaneous nature of online communication also keeps the conversation spontaneous and unrehearsed, while the absence of visual, bodily cues, and the fixed nature of the printed word, they claimed, allows participants to remain oriented to the other's intentions, while the absence of such cues means participants do not second-guess the expectations of the 'other' (Crichton & Kinash, 2003). The use of these interviews was advantageous in this research since it provided cost-effective access to foreign Liverpool F.C. supporters scattered worldwide.

### Participants

The final sample comprised 30 satellite supporters from eight nations. Nationalities represented were, in order of the number of respondents, Norway, USA, Sweden, Indonesia, Ireland, Canada, Finland, and the Faroe Islands. Participant ages ranged from 18 to 65 years and 10 percent of the sample were female.

## Materials and Procedure

An interview guide was prepared after analysis of the questionnaire data. Qualitative interviews provided an invaluable opportunity to speak with satellite supporters and explore their identification with the Liverpool F.C. in greater detail. These qualitative interviews were conducted via instant messaging technology with the assistance of Online Institute LLC, an independent third-party which specialises in the creation and management of online chat rooms and forums. The interactive nature of the technology allowed the 'conversations' to take place in real-time, be instantaneous, and provided an instant, and accurate transcription of each interview. This proved invaluable for archival purposes and data analysis. A three-month subscription to the service was purchased whereby the company hosted the fan forum (http://www.liverpoolfc.olicentral.com/) for that period. Each participant was provided a username and password to the forum and at the scheduled time both parties logged on to the website. Interviews were one hour duration and allowed an exploration of their identification with the Liverpool F.C. and the extent of their team loyalty.

Respondents in the online questionnaire had earlier provided an email address and willingness to engage in future discussions about their fandom. As a result, 12 of these earlier subjects were chosen at random for an initial round of online interviews which were held during a three-week window in late 2007. A number of factors were found and reinforced those from the questionnaire data. After this initial round proved workable, a second round of interviews was undertaken until a sufficient number had been completed. This number was derived according to two criteria: sufficiency and saturation of information (Seidman, 1998). Sufficiency concerns the requirement that the sample has sufficient numbers to reflect the range of participants that comprise the population, while saturation is that point whereby additional interviews do not unearth new information. When 30 interviews had been conducted, the researcher was content that both these criteria had been fulfilled.

## CONCLUSION

The sports landscape, Euchner (1993) argued, has changed dramatically in recent years. Besides increased commodification, he claimed, sport has 'delocalised' or, more precisely, has become "less dependent on attachment to a specific place" (p. 26). Sport, and other industries, therefore has to adapt to the challenges presented by globalisation. Although globalisation and advances in communications technology have greatly enhanced the popularity, and bottom line, of many professional teams, there has been little research on the foreign consumer, or satellite supporter. In a competitive global marketplace, there is a need to understand who these consumers are, and why they support a foreign-based sports team. Unlike their domestic counterparts, satellite supporters cheer for their adopted team from far-flung locations and, as a result, access to these individuals is a challenge. The emergence of online supporter organisations and products designed to conduct research online does appear to somewhat ease these concerns. Regardless of the challenges these individuals present, if sports teams wish to best exploit international opportunities, such efforts need to be made.

The complex nature of sports fandom has implications for the choice of suitable methods and, as earlier explained, while studies have traditionally favoured quantitative methods there is now greater acceptance of a qualitative approach. In addition, mixed methods can provide "a fuller understanding of the sports fan" (Jones, 1997b). This chapter therefore explored foreign fandom through mixed methods (questionnaires and semi-structured interviews) while the Internet proved an ideal vehicle to engage with distant sports fans. Although locating, and securing the cooperation of, large numbers of suitable respondents remains a challenge, these preliminary studies generated valuable insights.

In the absence of geographic proximity, these online studies reinforced the importance of the media as a key socialisation agent for the satellite supporter. Indeed, the Internet has served as "a vehicle for socialisation at a distance" (Assael, Pope, Brennan, & Voges, 2007, p. 337). The ability to watch the team play due to media coverage was instrumental in all three studies and media coverage allowed them to better support their teams. For instance, more than two-thirds of respondents visited their team's official website, or unofficial supporter websites, at least once per week. Indeed, some indicated they habitually visit these websites and "can't go a day without checking out news from the club". Furthermore, 83 percent of the Ajax F.C. supporters watched televised club games at least once per season while 86 percent of the Liverpool F.C. supporters said they did so at least once per week. This disparity is likely due to the phenomenal media exposure given to the English Premier League, a league broadcast to more than 600 million households in nearly 200 nations (Huggins, 2005), while the Dutch national football league, the Eredivisie, is not one of the top-four European leagues.

Successful sports teams often attract great support due to the possibility to BIRG. Consistent with this, support for team success or a positive team reputation and/or tradition featured prominently in the creation of supporter team identification. In addition, the presence of a particular player(s) was consistently reinforced in these studies. Finally, although one might expect satellite supporters to be more fickle and their support for foreign-based teams less stable, this was not the case. Nearly 85 percent of these supporters expressed loyalty to their adopted AFL team, 87 percent to the Ajax F.C. and nearly 96 percent to the Merseyside club.

Although the use of mixed methods to explore sports fandom is becoming increasingly popular, its adoption is still relatively limited. Furthermore, although there have been attempts to examine sports fans through the Internet, using online surveys (Nash, 2000), online message boards (End, 2001; Lewis, 2001; Mitrano, 1999), and online 'interviews' (Heinonen, 2002; Mitrano, 1999; Silk & Chumley, 2004), it is still a relatively recent approach. This chapter examined the team identification of satellite supporters, and pilot tests, questionnaires, and subsequent interviews were successfully administered with the cooperation of online supporter clubs worldwide. Recent technological advances have allowed access to satellite supporters irrespective of their geographic location. This chapter therefore highlights how mixed methods can be successfully employed online to explore fandom and allow future investigators to engage with distant research participants.

## REFERENCES

Alreck, P.L., & Settle, R.B. (2004). *The Survey Research Handbook* (Third ed.). Boston, MA: McGraw-Hill Irwin.

Assael, Pope, N., Brennan, L., & Voges, K. (2007). *Consumer Behaviour.* Brisbane: John Wiley and Sons.

Bailey, K.D. (1994). *Methods of Social Research* (Fourth ed.). New York, NY: The Free Press.

Beebe, L.H. (2007). What Can We Learn From Pilot Studies? *Perspectives in Psychiatric Care, 43*(4), 213-218.

*The Brand Champions League: Europe's Most Valuable Football Clubs.* (2005). London: Brand Finance.

Branscombe, N. R., & Wann, D. L. (1991). The Positive Social and Self Concept Consequences of Sports Team Identification. *Journal of Sport and Social Issues, 15*(2), 115-127.

Browne, K. (1999). The branding of soccer. *Finance Week*(January 29), 26.

Cialdini, R.B., Borden, R.J., Thorne, A., Walker, M.R., Freeman, S., & Sloan, L.R. (1976). Basking in Reflected Glory: Three (Football) Field Studies. *Journal of Personality and Social Psychology, 34*(3), 366-375.

Crichton, S., & Kinash, S. (2003). Virtual Ethnography: Interactive Interviewing Online as Method [Electronic Version]. *Canadian Journal of Learning and Technology, 29.* Retrieved July 13, 2006 from http://www.cjlt.ca/content/vol29.2/cjlt29-2_art-5.html.

Derbaix, C., Decrop, A., & Cabossart, O. (2002). Colors and Scarves: The Symbolic Consumption of Material Possessions by Soccer Fans. *Advances in Consumer Research, 29*, 511-518.

Dolance, S. (2005). "A Whole Stadium Full": Lesbian Community at Women's National Basketball Association Games. *The Journal of Sex Research, 42*(1), 74-83.

Eisenberg, D. (2003, March 9). *The NBA's Global Game Plan.* Retrieved March 16, 2007, from http://www.time.com/time/magazine/article/0,9171,430855,00.html

End, C.M. (2001). An Examination of NFL Fans' Computer Mediated BIRGing. *Journal of Sport Behavior, 24*(2), 162-181.

End, C.M., Dietz-Uhler, B., Harrick, E.A., & Jacquemotte, L. (2002). Identifying With Winners: A Reexamination of Sport Fans' Tendency to BIRG. *Journal of Applied Social Psychology, 32*(5), 1017-1030.

Estridge, H.L. (2007, May 11). Rangers owner talks sports business. Retrieved September 11, 2007, from http://dallas.bizjournals.com/dallas/stories/2007/05/07/daily47.html

Euchner, C.C. (1993). *Playing the Field: Why Sports Teams Move and Cities Fight to Keep Them.* Baltimore, MD: Johns Hopkins University Press.

Fink, J.S., Trail, G.T., & Anderson, D.F. (2002). An Examination of Team Identification: Which Motives are Most Salient to its Existence? *International Sports Journal, Summer*, (pp. 195-207).

Fisher, R.J. (1998). Group-Derived Consumption: The Role of Similarity and Attractiveness in Identification with a Favorite Sports Team. *Advances in Consumer Research, 25*, 283-288.

Fisher, R.J., & Wakefield, K. (1998). Factors Leading to Group Identification: A Field Study of Winners and Losers. *Psychology & Marketing, 15*(1), 23-40.

Fontana, A., & Frey, J.H. (2000). The Interview: From Structured Questions to Negotiated Text. In N.K. Denzin & Y.S. Lincoln (Eds.), *Handbook of Qualitative Research* (pp. 645-672). Thousand Oaks, CA: SAGE Publications.

Gage, J. (2006, April 17). Winner With Losses. Retrieved May 16, 2007, from http://www.forbes.com/free_forbes/2006/0417/084.html

Gieske, C., & Forato, M. (2004). *The most valuable football brands in Europe*. London: FutureBrand.

Giulianotti, R. (2002). Soccer Goes Glocal. *Foreign Policy, 131*(July-August), 82-83.

Greenwood, P.B. (2001). *Sport Fan Team Identification in a Professional Expansion Setting.* Unpublished Masters, North Carolina State University, Raleigh, NC.

Greenwood, P.B., Kanters, M. A., & Casper, J. M. (2006). Sport Fan Team Identification Formation in Mid-Level Professional Sport. *European Sport Management Quarterly, 6*(3), 253-265.

Gwinner, K., & Swanson, S.R. (2003). A model of fan identification: antecedents and sponsorship outcomes. *Journal of Services Marketing, 17*(3), 275-294.

Heinonen, H. (2002). Finnish Soccer Supporters Away from Home: A Case Study of Finnish National Team Fans at a World Cup Qualifying Match in Liverpool, England. *Soccer and Society, 3*(3), 26-50.

Henriksen, F. (2004, September 13). Lure of United dims as Real reel in fans. Retrieved August 7, 2006, from http://www.realmadrid.dk/news/article/default.asp?newsid=5194

Hoyle, R.H., Harris, M.J., & Judd, C.M. (2002). *Research Methods in Social Relations*. Fort Worth, TX: Wadsworth.

Huggins, T. (2005, May 2). Chelsea seek to paint world blue. Retrieved November 18, 2005, from http://www.rediff.com/sports/2005/may/02foot1.htm

Jacobson, B.P. (2003). *Rooting for Laundry: An Examination of the Creation and Maintenance of a Sport Fan Identity.* Unpublished Doctoral dissertation, University of Connecticut.

James, J.D., & Trail, G.T. (2005). The Relationship Between Team Identification and Sport Consumption Intentions. *International Sports Journal, Winter*, (pp. 1-10).

Jones, D., Parkes, R., & Houlihan, A. (2006). *Football Money League: Changing of the Guard.* Manchester: Sports Business Group at Deloitte.

Jones, I. (1997a). A Further Examination of the Factors Influencing Current Identification with a Sports Team, A Response to Wann, et al. (1996). *Perceptual and Motor Skills, 85*, 257-258.

Jones, I. (1997b, December). Mixing Qualitative and Quantitative Methods in Sports Fan Research. Retrieved June 20, 2007, from http://www.nova.edu/ssss/QR/QR3-4/jones.html

Jones, I. (1998). *Football Fandom: Football Fan Identity and Identification at Luton Town Football Club.* Unpublished Doctoral dissertation, University of Luton, Luton.

Kerr, A.K. (2008). Team Identification and Satellite Supporters: The Potential Value of Brand Equity Frameworks [Electronic Version]. *Papers from the 6th Annual Sport Marketing Association Conference*, 48-66. Retrieved July 19 from http://www.usq.edu.au/sma08/conference/callforpapers/default.htm

Kerr, A.K. (in press). Australian Football Goes For Goal: The Team Identification of American A.F.L. Sports Fans. *Football Studies, 11*(1).

Kerr, A.K., & Gladden, J.M. (2008). Extending the understanding of professional team brand equity to the global marketplace. *International Journal of Sport Management and Marketing, 3*(1/2), 58-77.

Kolbe, R.H., & James, J.D. (2000). An Identification and Examination of Influences That Shape the Creation of a Professional Team Fan. *International Journal of Sports Marketing & Sponsorship, February/March*, (pp. 23-37).

Lancaster, G.A., Dodd, S., & Williamson, P. R. (2004). Design and analysis of pilot studies: recommendations for good practice. *Journal of Evaluation in Clinical Practice, 10*(2), 307-312.

Lascu, D., Giese, T.D., Toolan, C., Guehring, B., & Mercer, J. (1995). Sport Involvement: A Relevant Individual Difference Factor in Spectator Sports. *Sport Marketing Quarterly, 4*(4), 41-46.

Lavrakas, P.J. (2004). Questionnaire. In M. S. Lewis-Beck, A. Bryman & T. Futing Liao (Eds.), *The SAGE Encyclopedia of Social Science Research Methods* (pp. 902-903). Thousand Oaks, CA: SAGE Publications.

Lewis, M. (2001). Franchise Relocation and Fan Allegiance. *Journal of Sport & Social Issues, 25*(1), 6-19.

Madrigal, R. (2000). The Influence of Social Alliances with Sports Teams on Intentions to Purchase Corporate Sponsors' Products. *Journal of Advertising, 29*(4), 13-24.

Madrigal, R. (2004). A Review of Team Identification and Its Influence on Consumers' Responses Toward Corporate Sponsors. In L.R. Kahle & C. Riley (Eds.), *Sports Marketing and the psychology of marketing communication* (pp. 241-255). Mahwah, NJ: Lawrence Erlbaum Associates.

Mahony, D.F., Howard, D.R., & Madrigal, R. (2000). BIRGing and CORFing Behaviors by Sport Spectators: High Self-Monitors Versus Low Self-Monitors. *International Sports Journal, Winter*, (pp. 87-106).

Mahony, D.F., Nakazawa, M., Funk, D.C., James, J.D., & Gladden, J.M. (2002). Motivational Factors Influencing the Behaviour of J. League Spectators. *Sport Management Review, 5*, 1-24.

Minichiello, V., Aroni, R., Timewell, E., & Alexander, L. (1995). *In-depth Interviewing: Principles, Techniques, Analysis.* South Melbourne: Longman.

Mitrano, J.R. (1999). The "Sudden Death" of Hockey in Hartford: Sports Fans and Franchise Relocation. *Sociology of Sport Journal, 16*, 134-154.

Moser, C., & Kalton, G. (1971). *Survey Methods in Social Investigation* (Second ed.). London: Heinemann Educational Books.

Nash, R. (2000). Globalised Football Fandom: Scandinavian Liverpool FC Supporters. *Football Studies, 3*(2), 5-23.

Orum, A.M., Feagin, J.R., & Sjoberg, G. (1991). The Nature of the Case Study. In J.R. Feagin, A.M. Orum & G. Sjoberg (Eds.), *A Case for the Case Study* (pp. 1-26). Chapel Hill, NC: The University of North Carolina Press.

Reimer, B. (2004). For the love of England. Scandinavian football supporters, Manchester United and British popular culture. In D.L. Andrews (Ed.), *Manchester United. A Thematic Study* (pp. 265-277). Abingdon, England: Routledge.

Rice-Oxley, M. (2007, June 8). English fans pool case to buy their own soccer team. Retrieved June 11, 2007, from http://www.csmonitor.com/2007/0608/p01s03-woeu.html

Richardson, B., & O'Dwyer, E. (2003). Football Supporters and Football Team Brands: A Study in Consumer Brand Loyalty. *Irish Marketing Review, 16*(1), 43-53.

Richelieu, A., & Pons, F. (2006). Toronto Maple Leafs vs Football Club Barcelona: how two legendary sports teams built their brand equity. *International Journal of Sports Marketing & Sponsorship, May*, (pp. 231-250).

Seidman, I.E. (1998). *Interviewing as Qualitative Research: A Guide for Researchers in Education and Social Sciences.* New York, NY: Teachers College Press.

Silk, M., & Chumley, E. (2004). Memphis United? Diaspora, s(t)imulated spaces and global consumption economies. In D.L. Andrews (Ed.), *Manchester United: A Thematic Study* (pp. 249-264). Abingdon: Routledge.

Stake, R.E. (1995). *The Art of Case Study Research.* Thousand Oaks, CA: Sage Publications.

Stake, R.E. (2001). Case Studies. In N.K. Denzin & Y.S. Lincoln (Eds.), *Handbook of Qualitative Research* (pp. 435-454). Thousand Oaks, CA: Sage Publications.

Sutton, W.A., McDonald, M.A., Milne, G.R., & Cimperman, J. (1997). Creating and Fostering Fan Identification in Professional Sports. *Sport Marketing Quarterly, 6*(1), 15-22.

Tapp, A., & Clowes, J. (2000). From "carefree casuals" to "professional wanderers". Segmentation possibilities for football supporters. *European Journal of Marketing, 36*(11/12), 1248-1269.

Tellis, W. (1997, September). Application of a Case Study Methodology. Retrieved June 2,

2007, from http://www.nova.edu.ssss/QR/QR3-3/tellis2.html

van Teijlingen, E., & Hundley, V. (2001). The importance of pilot studies. *Social Research Update*(35), 1-4.

Veal, A.J. (2005). *Business Research Methods: A Managerial Approach* (Second ed.). Frenchs Forest, Sydney: Pearson Addison Wesley.

Wann, D.L., & Branscombe, N.R. (1993). Sports Fans: Measuring Degree of Identification with Their Team. *International Journal of Sport Psychology, 24*, 1-17.

Wann, D.L., Melnick, M.J., Russell, G.W., & Pease, D.G. (2001). *Sport Fans. The Psychology and Social Impact of Spectators*. New York, NY: Routledge.

Wann, D.L., Tucker, K.B., & Schrader, M.P. (1996). An Exploratory Examination of the Factors Influencing the Origination, Continuation, and Cessation of Identification with Sports Teams. *Perceptual and Motor Skills, 82*, 995-1001.

Whiting, R. (2003, April 21). Hideki Matsui. Godzilla vs. the Americans. Retrieved November 18, 2005, from http://www.time.com/time/asia/2003/heroes/hideki_matsui.html

Wood, C. (2007, February 7). Americans Kop Liverpool deal. Retrieved September 3, 2007, from http://www.dailymail.co.uk/pages/live/articles/sport/football.html?in_article_id=434446&in_page_id=1779&in_a_source=&ct=5

Yin, R.K. (2003). *Case Study Research: Design and Methods* (Third ed.). Thousand Oaks, CA: SAGE Publications.

# Chapter XV
# Virtual Digital Olympic Museum

**Gaoqi He**
*Zhejiang University, China & East China University of Science and Technology, China*

**Zhigeng Pan**
*Zhejiang University, China*

**Weimin Pan**
*Zhejiang University, China*

**Jianfeng Liu**
*Zhejiang University, China*

## ABSTRACT

*Virtual reality and the Olympic Games Museum are used to create a virtual digital Olympic museum (VDOM). This is available solely through the medium of digital technology. VDOM extends in a comprehensive way the main functionalities of the traditional physical Olympic museum. Thus virtual reality technologies and the Olympic Games motivate the researchers. Three characteristics, namely sports, humans, and entertainment/ education, are crucial for the development of the VDOM. In developing the VDOM, the four major concerns were data storage and retrieval oriented to the Olympics task; modeling and rendering of the digital museum; the virtual demonstration of sports with virtual humans; and virtual reality based sports simulation. Appropriate solutions to these problems are proposed. Finally, a prototype of the VDOM using these technologies is demonstrated. This validates the efficiency of the proposed methods.*

## INTRODUCTION

Physical museums have long been one of the most important channels for transmission of knowledge and culture. People who visit physical museums browse directly the collections behind the glass boxes through text and picture, and are educating themselves. However, several limitations still exist for this form of physical demonstration, especially the constraints of time, space and interaction channels. With the aid of computer technologies and other advanced information technologies, the digital museum has been developed as one efficient and promising alternative.

The Olympic Games have a long history, which among other objectives is to lead to international friendship and understanding through competition in the sports stadium. This is commemorated in the physical Olympic museum in Lausanne, Switzerland and its largely conventional website (www.olympic.museum.org). It is attractive to demonstrate the history, culture, highlights and sports gymnasium of the Olympic Games in the VDOM. Therefore, the two entities, virtual reality technologies and the Olympic Games, motivate research into a virtual digital Olympic museum.

Conventionally, a digital museum is defined briefly as the digitalization of all the demonstrations in a physical museum using extant digitizing technologies. These digitizing technologies include, but are not limited to, panorama photography, videography, 3D scan, model reconstruction, audio recorder and network transmission. Thus, users are able to browse all the demonstrations anytime and anywhere. Collections of rare artifacts start to come into sight of the general public. And more importantly, the mode of people browsing the digital museum is greatly changed from seeing what they are given, to seeing what they choose to see. Moreover, the digital museum facilitates research, comprehensive communication and academic research all around the world.

Recently, there are two interesting research fields and applications that have come together with regard to the digital museum. One is a digital museum in the virtual environment via a network. The other is the Olympic Games in a digital museum format. Due to the former, virtual reality is increasingly used. Virtual reality technology is one of the most advanced human-computer-interaction techniques. It has arrived with the rapid growth of computer graphics, sensor, communication and hardware. Virtual reality provides users with more sense of immersion, interactivity and a greater ability to use their own imagination.

Using virtual reality, the VDOM intends to supply users with a remote 3D information service; a variety of simulations of participation in different sports; demonstrations of Olympics-related artistic works; the reappearance of previous Olympic Games; a walk-through of the Olympic gymnasium and so on. Browsers can visit the virtual Olympic museum with the help of an intelligent guide, or freely walkthrough the virtual environment by controlling one avatar. Accordingly, this chapter will focus on the design of the VDOM, its relationship to virtual reality, the use of digital networks and related technologies. This is all to improve users' experience in the VDOM, where the physical Olympics museum in Lausanne is impossible to travel to for most.

## BACKGROUND

Among the functions of a digital museum, the most basic are the presentation and preservation of cultural artifacts and artistic works. As large scale sports, presentation of the Olympic Games requires dynamic elements, rather than static ones. However, digital museums have their own distinct characteristics beyond these. Consequently, the following three characteristics should be considered when engaging in research and development of the VDOM. One is sport itself, portraying the Olympic spirit of faster, higher and stronger. Fitness is not only one fashion, but is necessary for the health of people in modern society under great pressure. The second is human, the entity taking part in sports. How to realistically characterize human beings and human emotion in sports are challenging issues. The third is to engage the audience through entertainment and education, the users' ultimate purpose in visiting either the physical or digital Olympic Museum. Correspondingly, researchers of a networked VDOM touch upon multiple areas of research and applications, including object modeling and real-time rendering, multimedia data storage and retrieval, virtual human motion control, and virtual reality sports simulation.

A large amount of literature has been published about the related technologies for digital museums in the aforementioned fields (Gray, 1999, 2001; Usaka, 1996). However, limitations still exist in current research. Static demonstrations still contribute a bigger proportion to digital museums. It is not surprising then that virtual reality has not been made full use of. And the current digital Olympic museum lacks virtual simulation of sports and humans. Poor interaction reduces the users' interest to browse the system. Also, low rendering quality in both speed and photorealistic greatly influence the usability of the system.

Consequently, the following section focuses on the aforementioned issues and tries to find efficient solutions. Major concerns include:

- Digital-museum oriented data storage and retrieval
- Modeling and rendering of the digital museum
- Virtual demonstration of sports and virtual humans
- Virtual reality based sports simulation.

## DIGITAL-MUSEUM ORIENTED DATA STORAGE AND RETRIEVAL

### Data Storage, Compression and Transmission

There are a large number of data that need to be dealt with in VDOM, including images, audio, video and the geometry model. For example, storage space required is commonly reached to several tens megabit for a segment of 5 minutes of sports video or a model for a virtual gymnasium. In spite of modern mass storage devices and high speed networks, such large scale data have to be specifically treated in the practicable applications.

Data compression is the most efficient way to reduce data storage requirements. With regard to image, JPEG2000 format supports not only compression for area of interest and lossless compression, but also progressive transmission. With progress transmission, coarse data is first transmitted, and then gradually, progressively more detailed data are transmitted to improve image quality. As a consequence, the load of each communication process is significantly reduced. With regard to video data, MPEG4 is currently the most suitable encoding method. The MPEG4 format has the following distinct strengths: content based interaction, high compression rate and common capability of access.

There are some efficient solutions for geometry model storage and transmission. One is geometry simplicity and compression. The other is progress transmission. The original model is expressed with multiple levels of detail format. The client each time receives a relatively small model for rendering.

Client based cache technologies are efficient to reduce storage needs. One form is only saving the most frequently used resource on the client side. The other is avoiding to download all the rather large resources one time. Stream media based video transmission is one typical example of the latter, with the added burden on continuous network bandwidth.

### Content Based Retrieval

A content based retrieval technique breaks through the limitations of traditional text based retrieving. It directly extracts semantic features from media data and its context, through which the index is constructed and the retrieval is completed. Retrieved objects are no longer limited to common text, but also 2D image and even 3D geometry models. Precise, objective and quick retrieving methods are important to digital-museum oriented data. Similarity matching of retrieval objects is one important means of a content based retrieval system. Gray image is frequently used for 2D data, and moreover, other formats of image data can be easily transformed into gray image. Multi-scale analysis is integrated with generic Fourier descriptor for robustness against noise during gray image retrieval (Zhang et al., 2005). With regard to 3D geometry model similarity, a visual projection similarity based algorithm is proposed for robustness against rotation transform, noise and model degeneracy (Zhang & Yang, 2006).

## Multi-Scale Generic Fourier Descriptor for Gray Image Retrieval

A large number of gray images are created and available for VDOM, since more and more gray images and digitizing tools are developed for different applications. Contour existing in the images indicates the content information of the whole image.

Fourier descriptor (FD) is usually used to analyze the similarity of image contour. However, it is not easy to extract the contour of one image, and image contour cannot represent the shape information in an image completely. So, generic Fourier descriptor (GFD) is proposed to process the whole image, with the advantages of scale and rotation invariance. However, GFD is sensitive to the noise and other changes of an image. Thus, multi-scale generic Fourier descriptor (MGFD) is proposed to get corresponding results in different scales. MGFD is defined as the following:

$$MGFD^{\alpha} = \left\{ \frac{|PF^{\alpha}(0,1)|}{|PF^{\alpha}(0,0)|}, \frac{|PF^{\alpha}(0,2)|}{|PF^{\alpha}(0,0)|}, \ldots, \frac{|PF^{\alpha}(k,t)|}{|PF^{\alpha}(0,0)|} \right\} \quad (1)$$

Which

$$C_{\alpha}(b,\delta) = \frac{k}{\sqrt{|\alpha|}} \int_{R} \int_{0}^{2\pi} f(r,\theta)\varphi\left(\frac{k-b}{\alpha}, \frac{\theta-\delta}{\alpha}\right) dk d\theta \quad (2)$$

$$PF^{\alpha}(\rho,\varphi) = \sum_{b}\sum_{\delta} C_{a}(b,\delta)\exp\left[j2\pi\left(\frac{b\rho}{N} + \frac{\delta\varphi}{M}\right)\right] \quad (3)$$

In the aforementioned definition, $b$ and $\delta$ are translation parameters. Parameter $\alpha$ is scale. $\rho = 0,1...,N-1$, $\phi = 0,1...,M-1$, $k$ is the proportional coefficient.

From the definition, multi-scale analysis is applied firstly to the original image, then the FD to each scale. Moreover, GFD can be considered as one special case of MGFD.

After the GFD analysis, the similarity distance of the original image is obtained for different scales. In a practical gray image retrieval system, appropriate weight is added to each distance, which embodies both whole and detail similarity.

To validate the performance of GFD, a serial of Lena images are used and different operations are applied on these images. In the experiment, Gauss or uniform noise is added into the original image, and the original images are rotated and scaled, respectively. The experimental results show that MGFD is not only capable of multi-scale analysis, but is robust against noise and other changes of image. It also is invariant to rotation and scale. So it can be used in gray image retrieval.

## Visual Projection Similarity for 3D Models Matching

The 3D geometry model is one of the most important and common formats of target object. Compared with grey image, the 3D model contains more geometry details, which makes the users' interaction with objects of interest possible. 3D geometry models are widely used in computer animation, CAD & CAM, as well as daily life, like virtual objects in the file head and advertisements, or architecture of real buildings.

For the content based 3D model retrieval, how to define and get the similarity among 3D models is essential. Current research (Cui &

Shi, 2004) comprise contour based, topology based and visual based similarity algorithms. The visual projection similarity algorithm is one of the visual based similarity algorithms, which has the advantages of simplicity, efficiency and robustness. The complete algorithm is made up of the following four steps.

Firstly, viewpoints are uniformly placed around the model. The major purpose of this step is to get sufficient visual images with the least viewpoints and minimum computation complexity.

Secondly, the 3D model is re-sampled to increase the vertices density. Then the processed vertices are orthogonal projected to the plane.

Thirdly, features of the orthogonal projected image are characterized by Zernike descriptor and Reeb graph. Reeb graph expresses well the topology of the visual image, while Zernike can effectively express the distribution of luminance.

Finally, similarity distance is computed for the 3D model. After all the features of projected images from each viewpoint are obtained, the minimum similarity distance between two projected images is used as the similarity distance of 3D models.

The experimental results indicate that this visual-based approach is robust against rotation transform, noise, model degeneracy, and achieves fairly good performance for 3D model retrieval.

According to the research in the fields of data storage, transmission and retrieval, a large number of objects obtained in VDOM can be efficiently processed and used for the upcoming purpose, like modeling and rendering.

## MODELING AND RENDERING OF THE DIGITAL MUSEUM

Representative objects should be selected to characterize the digital Olympic museum. Referring to the physical Lausanne Olympic museum, architecture of the digital museum is easily constructed. All types of torches in all previous Olympic Games are able to be modeled from several images. All the final images, video and models will be present for the visitors' enjoyment through advanced rendering technologies. Modeling the obtained multimedia resources and rendering them are therefore essential to implement this VDOM.

### Geometry and Image Based Modeling and Rendering

Research into geometry based modeling and rendering have a long history (Hoppe et al., 1992), and many off the shelf can be used directly for VDOM.

Level of detail (LOD) is able to accelerate graphics rendering. In complex virtual scenes, many objects are very small, or far away from the viewpoint. If such objects are represented with multiple levels of detail, they can be rendered in due time with a 'rough' model without loss of overall visual quality. Creating, choosing and switching level of detail for an original model are the three key steps for the technology of LOD.

The purpose of creating level of detail is to simplify the geometry description of an input model, and keep the appearance as much as possible. Frequently used methods are vertices clustering, face clustering and vertices removal. Vertices clustering means all the vertices lying in one square grid will merge into one vertex. Face clustering is reached through merging all the almost coplanar and neighbor faces into one face. Vertices removal technology is to remove some vertices on the input grid and then fill the hole using triangulation.

Measures for choosing an appropriate level of detail can be based on distance of the object from viewpoint, or the projected area of the object. Sometimes, several levels are chosen simultaneously for one mixed level.

A switch between two adjacent levels can cause a vision jump. One acceptable way is to produce geometry deformation from one level to the next level, which needs to define the corresponding geometry relationship between adjacent levels.

Geometry based modeling and rendering used in creating realistic scenes consumes large amounts of computation resources and storage space. An alternative is the technology of image based modeling and rendering (IBMR), which generates different views of an environment from a set of pre-acquired images. According to how much the geometry information is used, IBR can be classified into three categories: explicit usage, implicit usage and no usage. The best advantage of IBMR is that the rendering workload is proportional to the number of pixels that need to be rendered in the final scene.

Accordingly, geometry based and image based modeling and rendering are two important methods in the field of computer graphics. They have obvious merits and demerits respectively. In VDOM, it is necessary to choose one of these two technologies, or mix them according to the usage context and evaluated users' requirements. When the strategy of modeling and rendering is confirmed, then specific technologies are considered to get better rendering quality.

## Photorealistic and Real-Time Rendering

Photorealistic and real-time rendering are the two ultimate purposes of computer graphics. The scene and objects in VDOM are always very large and complex. It will take a long time to finish the rendering task, especially in a network environment. Moreover, end users easily discover the visual artifact in the rendering result. So photorealistic and real-time rendering are crucial to the VDOM system and should be carefully considered.

Shade provides the relative position information for the target object, which enhances the hierarchical perception of a scene. So shade is one important means for photorealistic rendering. However, computation of shade is generally complex, which will take a long time to render. With the aid of GPU, shade also can be generated in a large scene. Firstly, scene and light source are transformed into perspective space through the camera matrix. Then taking the position of light source as viewpoint, standard shade of scene is generated after scene rendered on unit cube. Finally, percentage-closer filtering and fuzzy techniques are used to improve the final result. GPU offloads large computation from CPU, which accelerates the graphics rendering.

Sports are always dynamic and quick with high speed. However, an audience cannot capture all the interested detail at a time. The audience aspires to take part in game playing directly, and watches the game at an arbitrary point in the scene from any direction and mode. But limitations are caused by fixed and limited number of cameras. Video based rendering is an efficient way to satisfy both speed and interaction requirements. Video streams are recorded simultaneously from several cameras located around the playground. Then the rendering scene at an arbitrary viewpoint is automatically generated in real time through interpolation of video based rendering. This method supplies the user with a fresh and interesting way to watch a sport game.

Collision detection has been of major interest in interactive computer graphics. Numerous approaches have been developed to detect interfering objects in various applications such as computer animation, robotics and games. Fast and precise collision detection helps to improve the realistic feeling of human-computer-interaction virtual environments. In VDOM, collision detection existed in various conditions, such as team competition and the performance of opening ceremonies. While collision detection for rigid bodies is well-investigated, collision detection for deformable objects introduces additional challenging problems. In cloth simulation of sports, self-collisions of cloth as well as the interaction of cloth with animated avatars need to be handled. Collision detection algorithms that are appropriate for deformable and animated objects are required.

Efficient approaches for collision detection among deformable objects include bounding volume hierarchies, distance fields, and spatial partitioning. Further, image-space techniques and stochastic methods are frequently considered. Two-pass rendering calculates the potential collision set, which can solve the issue of collision detection for the complex scene including many deformable and fragile objects.

## VIRTUAL DEMONSTRATION OF SPORTS AND VIRTUAL HUMANS

Sports cannot be separated from humans in the virtual environment. Relative technologies of virtual humans comprise virtual human modeling, motion control, action animation and virtual crowd simulation. There are four types of virtual human modeling: (1) interaction based human body modeling, (2) 3D measurement based human body reconstruction, (3) image-based human body reconstruction and (4) human body deformation. Motion control is of considerable interest, which addresses such issues as how to create virtual humans' motion, how to control the motion, and how to store and edit motion data. Current research directions comprise motion capture, editing, synthesis, retargeting and trajectory.

### Footprint Sampling Based Motion Editing

Motion capture now is one of the most promising techniques for character animation. A large quantity of realistic motion data is captured through an optical or magnetic motion capture system. However, the usability of raw captured data is quite limited, as they can only be used for an avatar with a similar skeleton configuration to perform captured actions in a specific virtual environment. Motion editing tools provide animators with the ability to manipulate the prerecorded motion data. The key issue in motion editing algorithms is modeled as space-time constraint. Editing systems should provide a friendly interface to enable animators to express their editing demand and map them back into space-time constraints. The space-time constraint of biped locomotion is the stance foot constraint, for the stance foot cannot slide on the ground. The location and orientation of the footprint model the stance foot constraint very well, and are also a friendly interface for animators to edit the biped animation. So footprint sampling based motion editing algorithm for the human biped locomotion captured by motion capture device is presented here (Xu, Pan & Zhang, 2003).

Our algorithm adopts footprints to describe the space-time constraints, which should be satisfied during biped locomotion, and footprints are also used as an interface to enable the user to control the space-time constraints directly. A real-time Inverse Kinematics (IK) solver is adapted to compute the configuration of the human body and then motion displacement mapping is constructed with hierarchical B-spline. In order to facilitate IK solver, we propose a sampling based scheme to generate root trajectory. Hermit interpolation is then used to generate the whole root trajectory. System architecture is shown as Figure 1.

The proposed algorithm is tested on running. The length of captured running data is about 120 frames. Figure 2 is the original running data. In Figure 3, the plane running is edited to running on a stair.

This scheme provides a speedup to root trajectory generation. The performance of our algorithm is further enhanced by the real-time IK solver, which computes directly the displacement angles as solution.

### Virtual Human Crowds Simulation and Group Calisthenics Rehearsal

The Olympic games attracts audiences of hundreds of thousands in the physical world and hundreds of millions in electronic form. Many sports are

*Figure 1. System architecture*

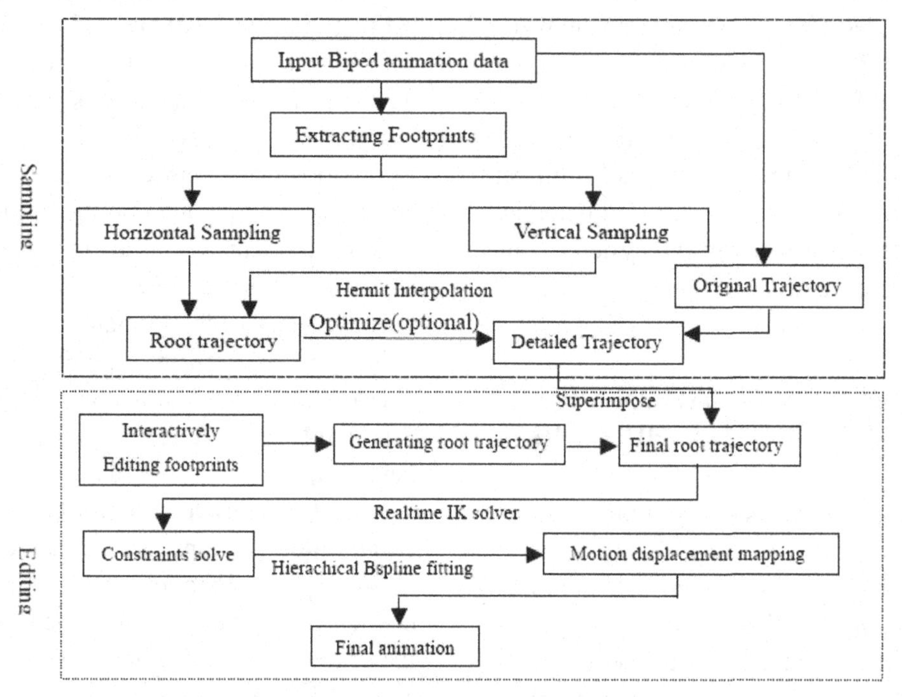


*Figure 2. Original data (running straight on flat ground)*

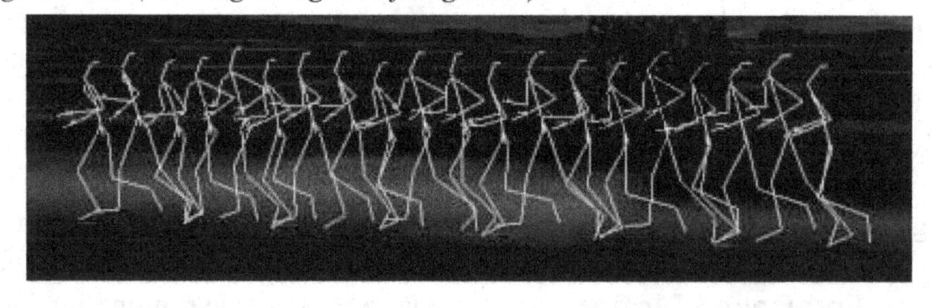

*Figure 3. Adapted to running on stairs*

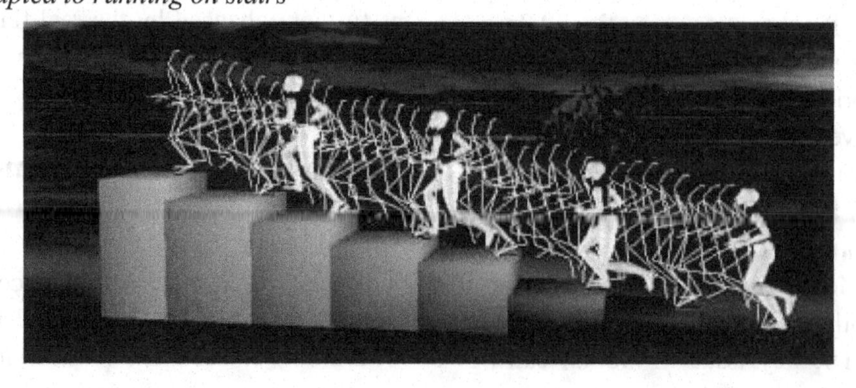

team activities. Virtual human crowd simulation is necessary to provide visitors a more real experience in VDOM.

An intelligent agent has the primitive action set in one specific field, through which more complex action animation is generated and the purpose is reached. Moreover, the crowd, group and individual form a 3-level structure in the virtual environment and have a distinct motion pattern. Research of action animation is extended in three aspects: action design, hierarchical structure and behavior decision mechanism. Virtual human crowd simulation involves some key technologies, like procedural animation, kinematics, motion capture, collision detection and real rendering. Virtual arrangement and rehearsal of group calisthenics is present as one example of crowd simulation (Ji et al., 2004).

Group calisthenics is a comprehensive collective performance item, an integration of sports and arts involving many participants. The prototype system consists of two parts: design system of formation or patterns, and a virtual rehearsal system. The former has the functionalities of formation editor and dot matrix generator. The latter comprises such components as virtual environment set, task planning, path planning, collision detection, events processing and crowd simulation. Experiments of square and circle group calisthenics are demonstrated as Figure 4.

## VIRTUAL REALITY BASED SPORTS SIMULATION

Sports simulation is one kind of systems engineering science. Through technologies of computer simulation, it reproduces the teaching experience of teachers, training intention of coaches, administrators' organization scheme, and athletes' training process. Sports simulation helps to interpret, analyze, predict, organize and evaluate a sports system. Virtual reality is one advanced human-computer-interaction technique, which provides users with a greater sense of immersion, interactivity and imagination. Correspondingly, virtual reality based sports simulation satisfies both athlete and coach, as well as other participants, and is widely used in the field of sports.

### Main Functionalities and Key Technologies

Virtual reality based sports simulation systems generally have the following functionalities:

- Constructing of a virtual training scene
- Capturing motion data
- Collecting physiology, biochemistry and psychology data
- Reproducing action
- Illustrating and analyzing training effects.

Correspondingly, key technologies of sports simulation include physics and physiology based modeling, virtual human animation, motion data capture, real-time rendering, and natural human-computer interaction.

### Typical Prototypes

Now VDOM is extended from the traditional Olympic Games to other types of sports and functions. Some mass sports are also integrated into VDOM. Several virtual reality based sports simulation systems have been successfully implemented, like Easybowling (Pan & Xu, 2003) and virtual network marathon (VNM). They both integrate body exercise into game playing.

***Easybowling.*** Players throw a real bowling ball on a 2-meter-long track, and then the EasyBowling system uses a PC Camera to detect the speed and direction of the bowl. After the motion parameters are computed, the movement of the bowling ball and its collision with pins are simulated in real-time, and the result is displayed on a large display

*Figure 4. Prototype of Easybowling (left) and collision simulation (right)*

screen. Since it only requires a small space, this real game mode machine can also be a family game machine. Figure 4 shows the prototype of Easybowling.

***Virtual network marathon.*** Users do physical exercise and competition in a distributed virtual environment. When users run on the elliptical machine, their body motions are captured by sensors. They can observe their actions and current environment in real-time, rendered through an output screen, and control their actions by multimodal interactions or interactions with other users by network. There are three game modes: training, exercising and competition. Also, users are able to choose the appropriate distance to run. Figure 5 shows the exercise modes of VNM (walkthrough in the museum).

*Figure 5. Walkthrough in the museum*

*Figure 6. Interface of VDOM*

## PROTOTYPE OF VDOM

All the aforementioned techniques can be integrated into the construction of VDOM. Figure 6 shows a snapshot of our VDOM. Visitors can enter the system, browse the digital Olympic museum and communicate with others.

## FUTURE TRENDS

For the future trends, at least three fields are currently being explored and are worthy of attention.

One is the web based digital Olympic museum. Although parts of current prototypes are running through networks, network bandwidth is still bottlenecked and only a limited number of users are supported. Yet, the number of web users increases quickly. Research of the Web based digital Olympic museum includes investigation of distributed architecture and user-oriented human-machine interaction.

Another is the handhold devices oriented digital Olympic museum. Now the handhold device is prevalent with the development of embedded devices and wireless communication. But, handhold devices are small screen, low computation and battery powered, while wireless communication has the disadvantage of low transmission speed. All these issues should be solved in this interesting field.

The last field is high photorealistic rendering. Users are caring for visual enjoyment more and more. However, advanced computer graphics and image technologies need to be exploited and improved.

## CONCLUSION

Digital sports is a promising research field with broad research directions. It will be acknowledged in the near future. With the internationalization of the Olympic Games as its research purpose, this chapter concentrates on relative technologies and application of VDOM. Sports, humans, and entertainment/ education are described as the main three features of VDOM. Correspondingly, solutions are proposed for the four major concerns, including digital-museum oriented data storage and retrieval; modeling and rendering of the digital museum; virtual demonstration of sports and virtual humans; and virtual reality based sports simulation. The prototype of VDOM is running well, thereby validating the present work in this field.

## ACKNOWLEDGMENT

This project is co-supported by Key NSFC (grant no: 60533080) and 863 project in China (grant no: 2006AA01Z303), Doctoral Fund of Ministry of Education of China (grant no: 200802511025), and Science Foundation for The Excellent Youth Scholars of ECUST (grant no:YH0157124). Many people contribute to this project, including Mr. Zhongfu Li, Mr. Zi Li and Mr. Yuhu Ma for modeling of the museum. The system is implemented on the game engine called Lucid, which is provided by Prof. Gino Yu in the Hong Kong Polytechnic University. The retrieval part is implemented by Prof. Gangshan Wu and his students in Nanjing University.

## REFERENCES

Cui, C., & Shi, J. (2004). Analysis of feature extraction in 3D model retrieval. *Journal of Computer-aided Design & Computer Graphics, 16*(7), 882-889.

Gray, G. (1999). Discussion on future directions of global digital museum: Museum and GUI. In S. Sugita, J.K. Hong, J. Reeve & G. Gay (Ed.), *Senri Ethnological Report, Global Digital Museum for Museum Education on the Internet*. Osaka, Japan: National Museum of Ethnology.

Gray, G. (2001). Co-construction of digital museums. *Spectra, 28*(1), 12-14.

Hoppe, H., DeRose, T., Duchamp, T., McDonald, J., & Stuetzle, W. (1992). Surface reconstruction from unorganized points. *ACM SIGGRAPH Computer Graphics, 26*(2), 71-78.

Ji, Q., Pan, Z., et al. (2004). Virtual arrangement and rehearsal of group calisthenics. *Journal of Computer-aided Design & Computer Graphics, 16*(9), 1185-1190.

Pan, Z.G., & Xu, W.W. (2003). Easybowling: A small bowling machine based on virtual simulation. *Computers & Graphics, 27*(2), 231-238.

Usaka, T. (1996). *A large scaled multi-user virtual environment system for digital museums*. Master's thesis, Division of Science, Graduate School of University of Tokyo, Japan.

Xu, W., Pan, Z., & Zhang, M. (2003). Footprint sampling based motion editing. *International Journal of Image and Graphics, 3*(2), 311-324.

Zhang, Z., Pan, Z., et al. (2005). Multiscale generic fourier descriptor for gray image retrieval. *Journal of Image and Graphics, 10*(5), 611-615.

Zhang, Z., & Yang, B. (2006). Visual projection similarity for matching similarity among 3D models. *Journal of Computer-Aided Design & Computer Graphics, 18*(7), 1049-1053.

# Chapter XVI
# The Market Structure and Characteristics of Electronic Games

**Kerri-Ann L. Kuhn**
*Queensland University of Technology, Australia*

## ABSTRACT

*A multi-billion dollar industry, electronic games have been experiencing strong and rapid growth in recent times. The world of games is not only exciting due to the magnificent growth of the industry however, but due to a host of other factors. This chapter explores electronic games, providing an analysis of the industry, key motivators for game play, the game medium and academic research concerning the effects of play. It also reviews the emerging relationship games share with sport, recognizing that they can replicate sports, facilitate sports participation and be played as a sport. These are complex relationships that have not yet been comprehensively studied. The current chapter serves to draw academic attention to the area and presents ideas for future research.*

## INTRODUCTION

A multi-billion dollar industry, electronic games have been experiencing strong and rapid growth in recent times. A review of the industry and key trends shaping the market shows they have grown to become a key segment in the entertainment category, drawing an increasing number of players from a broader demographic. Despite their magnificent growth however, this remains a medium that is still little understood. The characteristics of games have not been formally developed in the literature, nor do we have conclusive evidence of their influence, or the effects of play generally. This is particularly true in the context of sport. The objective of the current chapter is to review the market structure and characteristics of electronic games, with particular emphasis given to their relationship with sport. Such a review is a worthy pursuit, considering the significant economic, social and cultural impact of electronic games (Castronova, 2005).

## INDUSTRY BACKGROUND

The global entertainment and media industry is predicted to grow 6.6% annually to 2012, to then be valued at US $2.2 trillion (PricewaterhouseCoopers, 2008). Within that industry grouping, the gaming market alone is expected to grow 10.3% annually to be worth $68.4 billion by 2012, up from $41.9 billion in 2007 (Caron, 2008; PricewaterhouseCoopers, 2008). Digital technology advances are driving growth, facilitating product innovation and the development of new hardware and software. Worldwide, video game and interactive entertainment industry revenue is forecast to reach $57 billion in 2009, including revenue from video game hardware and software; dedicated portable system hardware and software; PC games; as well as online PC and console games (DFC Intelligence, 2008). Overall, the industry has become more structured and stable: revenues for manufacturers are increasing, software sales are rising and a more diverse audience playing. These issues are examined next.

### Game Platforms and Software

Broadly, electronic games can be divided into five categories: console/ handheld, wireless, online, PC and arcade. Leading industry bodies such as the Entertainment Software Association (ESA) and PricewaterhouseCoopers define the games market as being made up of these groupings. It must be recognized however, that it is becoming increasingly difficult to distinguish between the game formats. For example, console-based games are often played using console systems connected to a television, but hardware connectivity has allowed for their extension to related handheld devices (such as Sony PlayStation®Portable and Nintendo Game Boy® or the current DS™). Technological advances also mean they can be played online (the new consoles and handhelds allow for internet connection for multiplayer gaming), and sometimes PC versions are released. Past console games can also now be played on mobile phones. Despite the convergence that is apparent, the nature of the games, the way they are played, and the types of gamers differ across the platforms, so it is necessary to analyze each independently.

Consoles remain the dominant platform for play. The electronic games market is controlled by three major console manufacturers who account for 80% of all game revenue (Taub, 2004): Sony Computer Entertainment which launched its PlayStation® in 1994, Microsoft Corporation which launched Xbox® in 2001 and Nintendo Company which launched the Nintendo Entertainment System (NES®) in 1985. The latest generation consoles released in 2005-2006 include Sony PlayStation®3, Microsoft Xbox 360™ and Nintendo Wii™. All three are direct competitors, but each of the manufacturers targets a slightly different market: Sony targets the 16 to 24 year old demographic, Microsoft's focus is on 18 to 34 year olds, while Nintendo has traditionally concentrated on a younger audience (8 to 18 years) (Schilling, 2003), though this has changed with the advent of the Wii™, which is targeted to more of a mass market.

Sony is the largest and most successful manufacturer, dominating with a market share of 69% in 2005, followed by Microsoft with 16% and Nintendo with 15% (Edwards & Greene, 2005). It also has the strongest customer base; by October 2007, Sony had sold more than 120 million PlayStation®2 units worldwide and more than a billion software units from its library of over 1500 titles (Sony, 2007). However, although PlayStation®2 is the best-selling gaming platform ever released, sales for PlayStation®3 have fallen below expectations, with only 13 million units sold worldwide by May 2008, compared to 25 million Wiis™ and 19 million Xbox 360s™ (Reisinger, 2008). This is significant, because unlike other media, each of the manufacturers operates a proprietary system, so the first-mover advantage and mass acceptance of the product are critical success factors (Gallagher & Park, 2002).

Growth in the video games industry follows a cyclical pattern, with alternating revenue curves for hardware and software. The manufacturers derive the bulk of their money from software sales (Grossman, 2005; Schilling, 2003). The console units are sold below cost in order to create a large installed user base, which provides the critical mass to generate required profits. This also attracts game developers, who derive a royalty from each software unit sold (Schilling, 2003; Williams, 2002). A strong network of publishers and developers represents the console segment's key strength and a major barrier to entry in the industry. There is much speculation concerning Sony's future in the games market (see, for example, Brightman, 2007), though it remains to be seen who will win the latest console war. There is however a lot at stake: console games are expected to dominate overall sales in the games industry, growing 6.9% annually from $24.9 billion in 2007 to $34.7 billion in 2012 (Caron, 2008).

Handhelds are considered a sub-sector of the console market (ESA, 2006a; PricewaterhouseCoopers, 2007). In fact, the term "video game" is often used to refer to both consoles and handhelds. Nintendo dominates the portable gaming sector, offering several handheld systems including the Game Boy® Advance, DS™ and extensions of the two such as the Game Boy® Micro and Nintendo DS™ Lite. The other major handheld device is the Sony PlayStation®Portable (PSP®). Consoles and handhelds tend to offer the same games and are similar in terms of the way they are played, with consistency between the controls. Like their console counterparts, handhelds are also sophisticated, offering advanced graphics, hi-fi sound and internet capabilities. A key difference, however, is that of convenience. A major advantage of handheld systems is that since they are portable, they can be played almost anywhere. DFC Intelligence (2007) estimates worldwide revenue from portable game systems exceeded $10 billion in 2007, making handhelds one of the fastest growing categories.

Wireless gaming (also called mobile gaming) represents a distinct type of gaming, separate from the handheld category. In this instance, play occurs using a mobile phone or PDA. Wireless games tend to be rudimentary, easy to learn games (casual mini-games Hyman, 2005), which involve minimal plot or character development and relatively basic animation. They lack the visual and auditory elements typical of console games and tend to be more simplistic in terms of their narrative, music and artwork. It is predicted that wireless gaming will remain a smaller category, though worldwide revenue may increase 16.1% to $4.5 billion in 2008, and potentially reach $13.5 billion in 2012 (Caron, 2008; Gartner, 2008; Hartig, 2006).

Online games have become the fastest growing segment of online entertainment and represent one of the top entertainment destinations on the internet (Kretchmer, 2004; PQ Media, 2006). Online game play is markedly different from that for consoles, offering the key advantage of multiplayer gaming. The games are also vastly different. Online games may range from simplistic stand-alone games that are free, to those that are part of elaborate online communities, which require players to subscribe and pay a fee (for a complete discussion, see Moore, 2006). *Second Life*® is perhaps the most sophisticated example. It allows users to escape their everyday lives and inhabit a new reality where their alter egos can function (for a complete discussion, see Dower, 2007). Massively multiplayer online games (MMOGs) in particular are popular, with market value in Europe and the United States (U.S.) reaching $1 billion for the first time in 2006 (Reportlinker, 2007). Online game subscriptions are still in their infancy.

Computer game sales continue to be surpassed by those for console-based games: over 36 million units were sold in the U.S. in 2007, compared to more than 231 million for video games (ESA, 2008). This represented an 8% drop in computer game sales from 2006, but more than a 14.5% increase for video games (ESA, 2008). Worldwide, overall PC gaming is a $10.7 billion industry (GamersCircle, 2008). Retail sales account for only 30%, but online PC gaming is worth almost double at $4.8 billion, driven largely by the Asian market (GamersCircle, 2008). Computer game sales are expected to continue to decline with the release of the next generation consoles, though online gaming should continue to grow. Online games are forecast to reach $14.4 billion in 2012 (Caron, 2008).

Arcade gaming represents the final category. Less data is available concerning basic industry metrics in the coin-operated game community, though it is known that arcade gaming has been in decline. The U.S. market has declined from 1.4 million coin-operated video games in the early 1980s to 327 000 in 2006, and revenue has decreased from $7.3 billion in 1982 to just $866 million in 2004 (Hurley, 2008). Today, a successful arcade game will only sell around 4000 to 6000 units globally (Hurley, 2008). As a result, many of the manufacturers in the arcade sector such as Capcom, Konami, Namco and Sega have consolidated, closed some operations and changed focus. However, although the number of arcades in the U.S. dropped from 10 000 in the early 1980s to 2500 in 2003, this number has grown to 3500 in 2008 (Hurley, 2008). The recent growth is attributable to a new focus on providing technology and an experience that cannot be duplicated in the home. The introduction of novel machines has sparked renewed interest in arcade gaming (see Libaw, 2000), as has the shift to online tournaments for coin-operated games (see Williams, 2005). Growth potential may be restricted though, as a result of competition from other online gaming activities, and the repackaging of many arcade games for console use.

## Game Players

A key driver for electronic games overall is the expanding demographic of players, allowing the industry to achieve strong performance by delivering content that appeals to a broader audience (Paterson, 2003). The increased maturity and diversity of this audience is demonstrated by the variety of available game titles and genres, as well as the rise in games rated "Mature" (RocSearch, 2004).

Electronic games represent one-third of the toy industry in the U.S. (Paterson, 2003), but they are no longer a form of in-home entertainment just for children. Approximately 50% to 60% of all Americans over the age of six now play (ESA, 2003), with an average age of 35 years (ESA, 2008). In fact, 35% of the most frequent gamers are between 18 and 35 (ESA, 2006b). It is estimated 112 million people aged 13 years or older participate in some form of electronic gaming (Shields, 2005). Worldwide, more than 300 million people play electronic games (JWT Intelligence, 2006), with almost 217 million accessing online gaming sites specifically (comScore, 2007).

With regards to computer and video games, approximately 65% of American households play (ESA, 2008; NPD Group, 2007a). The penetration of consoles (estimated at 41%) is not as large as that for televisions, but around 45.7 million U.S. households have one or more game systems, up from 38.6 million in 2004 (Nielsen, 2007). Console gamers tend to be predominantly males in their 20s, though it is estimated 40% of women now play (ESA, 2008). In the online environment, young adult males are the key target for MMOGs (Griffiths, Davies & Chappell, 2003), though social games are becoming increasingly popular amongst women. Players of *The Sims™ Online*, for example, are roughly 60% female (Gaither, 2003). Females (particularly teenage girls) are also more likely to play wireless games on mobile phones (Taylor, 2005).

Broadly, game players can be segmented into two groups: the hard core/ avid gamers (estimated to make up around 15% of all gamers) and the more casual users (McCandless, 2003; Williams, 2002). Game usage time is increasing amongst both groups. Americans on average spent 64 hours each playing electronic games in 2002, almost double that of five years earlier (Delaney, 2004). They now play an average of 3 hours every week, or 156 hours per year (Taylor, 2005). Increased usage is diverting consumer leisure time and dollars away from other entertainment options like television, movies and music (ESA, 2004a). Males aged 18 to 34 years, for example, watched 12% less prime-time television in 2004 than in 2003 (Devaney, 2005). It is estimated they now spend 12 hours a week watching television, compared to 14 hours playing games (Foege, 2005). Game usage is expected to continue to increase in coming years, with growth in consumption expected to outpace other forms of media. Overall, electronic games have truly transformed into a type of mass market entertainment.

## KEY MOTIVATORS FOR GAME PLAY

There are many factors which explain why people play electronic games. According to Youn and Lee (2003), people play to escape reality, engage in competition, relieve boredom, ease stress and overall to have fun. Similarly, Jansz and Martens (2005) claim that the key reasons why people play are for competition, leisure/ pleasure and social aspects. In a study conducted by the Entertainment Software Association (ESA, 2004b), the most frequent gamers cited a number of reasons for playing: 81% played games because they are fun, they are challenging (72%), they are an interactive social experience that can be shared with family and friends (42%), and because games provide a lot of entertainment value for the money (36%). Three factors are consistently identified as motivators: challenge/ competition, enjoyment and social factors.

### Challenge and Competition

First, games are intentionally challenging. This is consistently identified as the most important aspect of good game design (see, for example, Sweetser & Wyeth, 2005) and serves as a key motivator for people to play (Kim, 1995; Kim, Park, Kim, Hak & Chun, 2002). The challenge stems from the fact that, in a game, players are engaged in a competition whereby they must battle an opponent (whether that be another player or the machine) to achieve a valued outcome: winning. Game play is therefore a goal directed activity. Certain skills are required if an individual is to succeed and achieve their goal of winning the competition.

Electronic games have high range whereby they offer many possibilities for action at any given time (Steuer, 1992). In this mediated environment, users require a high level of skills and concentration if they are to successfully carry out many activities simultaneously. Playing a game is an active, controlled exercise, which demands not only visual attention, but also motor actions (Grodal, 2000). In other words, both mental and physical effort is necessary. Studies have demonstrated how intense this activity is for players, finding a number of effects from competitive game play including increased blood pressure and heart rate (Harrison et al., 2001; Turner & Carroll, 1985). It can also increase mental rotation (De Lisi & Wolford, 2002); improve eye-hand coordination and reaction time (Griffith, Voloschin, Gibb & Bailey, 1983); as well as alter visual skills by increasing visual attention and spatial visualization (Dorval & Pepin, 1986; Green & Bavelier, 2003).

Researchers have warned about the confounding potential of individual differences in experience with computerized media (Jih & Reeves, 1992), as level of experience has been shown to affect attitudes toward computer programs (Kieras & Polson, 1985; Vincente, Hayes & Williges, 1987). When an individual's skills are strong in an interactive environment they have high self-confidence and, as a result, more favorable attitudes toward the medium (Gardner, Dukes & Discenza, 1993). Skill level may also generate emotional responses (De Rivera, 1984; Izard, 1977). Overall, competence enhances a user's experience in a mediated environment (Howe & Sharkey, 1998), which would seem to be particularly pertinent in the context of a competitive exercise like game play.

## Enjoyment

Enjoyment is at the heart of game playing and serves as another key motivator for the activity. Sweetser and Wyeth (2005) argue that player enjoyment is the single most important goal for games. This is supported by research conducted by the ESA (2004b). To be enjoyable, Sweetser and Wyeth (2005) contend a game must require a user's concentration; it should be challenging and test a player's skills; provide players a sense of control; present clear goals and feedback; involve/immerse players; and support, as well as create, opportunities for social interaction.

Electronic games are designed to engage players in an inherently enjoyable form of entertainment. Studies have linked game play to physiological arousal and affective experience (see, for example, Anderson & Bushman, 2001; Grigorovici, 2003; Koepp et al., 1998), with players found to report strong emotions being experienced both during and after play (Molesworth, 2006). Psychological responses stemming from interactivity, such as presence, have also been found to be arousing in a game context (Grigorovici & Constantin, 2004).

Games have become immersive and complex virtual worlds, which allow for not only the engagement of a player's physical activity, but also their imagination, thoughts and feelings. They involve sight, sound, touch and feel, and can influence emotions, giving players the ability to laugh at, talk about, and connect with them (Brown & Cairns, 2004; Sweetser & Johnson, 2004). Electronic games have been likened to art, because they can be a personal experience that excites emotions, with the potential to convey feelings and experiences more strongly than other forms of media (Kap, 2004).

## Social Aspects

Finally, people often play electronic games in groups (ESA, 2004a; NPD Group, 2007a). The opportunity to share a social experience with others is a notable motivating factor. Lazzaro (2004) suggests that social interaction represents not only an important reason why some people play, but it is another element for the enjoyment of games.

Around 59% of American gamers report playing games with others in-person, up from 51% in 2006 (ESA, 2008). Parents in particular often play with their children, because they see game play as a fun activity for the entire family and an opportunity to socialize (ESA, 2008). Heavier gamers are more inclined to play alone than casual users, but both enjoy playing games as a family, group or party activity, and value gaming as a way to bring their families closer together (NPD Group, 2007a). Social motives are especially strong amongst networked gaming participants, whether that is players at players at Local Area Network (LAN) events (Jansz & Martens, 2005) or online gamers (Hsu & Lu, 2004). In fact, social motives are so important, Williams (2006) hypothesizes that the human need for social contact has driven the growth of networked gaming.

# GAME MEDIUM CHARACTERISTICS AND PSYCHOLOGICAL RESPONSES

The successful use of a computer system is not a product of the design specifications, but rather of the interaction between a human user and the system (Siekpe & Hernandez, 2006). System characteristics however, work to determine the psychological experience of the user (Steuer, 1992). Each medium carries its own meanings and connotations, so it is necessary to understand the characteristics of a medium and how people interact with the technology. Understanding games and the way players experience this media environment therefore represents an important research priority. To date, the characteristics of games have not been formally developed in the literature. Some attempts have been made to do this in the online environment (see, for example, Sweetser & Wyeth, 2005), but exactly what makes electronic games different from traditional media and the corresponding impact of these features on players' psychological responses remains unclear. Literature concerning the internet offers some insights.

One of the most important dimensions that differentiates new media from traditional media is the level of realism they provide (Coyle & Thorson, 2001). This is the result of two key features: interactivity and vividness. Interactivity and vividness represent distinct characteristics of the environment, though some authors do not distinguish between the two (see, for example, Steuer, 1992). Fortin and Dholakia (2005) highlight this is not necessarily accurate, since certain forms of communication can be high or low on either dimension (take for example television, which is highly vivid but non-interactive). With respect to games however, advances in technology have enabled the creation of environments that are both interactive and vivid.

## Interactivity

Interactivity has been defined as "the degree to which two or more communication parties can act on each other, on the communication medium, and on the message, and the degree to which such influences are synchronized" (Liu & Shrum, 2002, p. 54). It is a function of the different ways content in a mediated environment can be manipulated by users in real time, the speed of this manipulation, and mapping (the similarity between the controls and manipulation in the mediated environment and those in a real environment) (Steuer, 1992). Unlike traditional media, which are passive and one-way in nature, interactive media provide users greater control, allowing them to receive and alter information and images. In other words, an individual can interact with and act on the medium, as well as the message. Further, interactive media facilitate communication between two parties. Console games are based on the traditional model of mass communication following a process of "one-to-many," but online games facilitate many-to-many communication.

The effects of interactivity remain unclear due to inconsistencies in the literature as to how it should be defined and operationalized (for a complete discussion, see McMillan, 2002). Plus, its effects appear to be dependent on the situation and the user, including their goals, attention, level of involvement and experience (Balabanis & Reynolds, 2001; Bezjian-Avery, Calder & Iacobucci, 1998; Hoffman & Novak, 1996; Liu & Shrum, 2002). Some researchers have found interactivity attracts attention and positively influences user attitudes (Cho & Leckenby, 1999; Wu, 1999). Others have found that interactivity imposes additional information processing demands on individuals (Ariely, 2000), because managing it requires significant cognitive effort (Sicilia, Ruiz & Munuera, 2005).

## Vividness

Vividness represents the second key attribute that distinguishes electronic games from traditional media. Vividness refers to the richness of a mediated environment and the way that environment presents information that engages the senses (Steuer, 1992). It is increased by media tools such as video, audio, animation and 3D imaging, which are capable of enhancing the richness of the experience (Coyle & Thorson, 2001). The effects of vividness have been studied in the context of investigations concerning distinctiveness: a stimulus characteristic relative to other stimuli presented, or to a previously defined set of stimuli (Beattie & Mitchell, 1985).

A game is a competitive, dense-with-messages environment, which offers rich and animated imagery. Games are high in both breadth (they activate a number of sensory channels simultaneously) and depth (each sensory channel is strongly influenced) (Steuer, 1992). With the presentation of intense auditory and visual stimuli, they can facilitate sensory immersion.

## Presence

Interactivity and vividness represent characteristics of the computer-mediated environment, which are capable of facilitating unique psychological responses from users. First, both are important determinants of presence (also called telepresence) (Coyle & Thorson, 2001; Steuer, 1992). Presence represents an individual's mediated perception of an environment so that it is imagined as real (Lombard & Ditton, 1997). Induced by vividness, interactivity and focused attention, presence provides a feeling of being present in an environment and enables an individual to experience psychological states, such as virtual experience (Hoffman & Novak, 1996; Li, Daugherty & Biocca, 2001). It has been suggested that playing a game can enhance a sense of presence (Lessiter, Freeman, Keogh & Davidoff, 2001).

## Flow

Presence is closely related to, and actually enhanced by, flow. Flow is a process of optimal experience accompanied by a loss of self-consciousness, which occurs as a result of a seamless sequence of responses facilitated by machine interactivity (Hoffman & Novak, 1996; Novak, Hoffman & Yung, 2000). It generally occurs in structured activities where action follows action (Privette, 1983), as is the case in a game. Bambauer (2006) and Sweetser and Wyeth (2005) argue that flow is relevant in a game context. For example, sticky web sites which attract and hold visitors' interest, such as gaming sites (O'Guinn, Allen & Semenik, 2006) can promote a flow state (Hamman, 2000). There are several antecedents however, that must be present for an individual to experience flow: the interaction must be fun and enjoyable (Csikszentmihalyi & LeFevre, 1989); the user must focus their attention on the interaction; and they must perceive a balance between this challenge and their skills (Hoffman & Novak, 1996).

With flow comes more positive subjective experiences, including positive affect (feelings) and satisfaction (Csikszentmihalyi, 1977). In the case of computer-mediated environments, higher degrees of pleasure and involvement during computer interactions have been shown to lead to concurrent subjective perceptions of positive mood and emotion (the two components of affect) (Starbuck & Webster, 1991; Webster & Martocchio, 1992). In a game context, flow has been found to positively influence attitude toward a game (Bambauer, 2006). However, it can be distracting and has been linked to over-involvement, which can lead to mental and physical fatigue (Csikszentmihalyi, 1977; Hoffman & Novak, 1996; Novak et al., 2000).

## Involvement

Finally, the characteristics of computer-mediated environments can increase involvement. Johnson et al. (2006) demonstrate a positive relationship between interactivity and involvement, highlighting that the more interactive the communication, the more involved a user becomes. It is this involvement that can help facilitate a state of flow and may represent the essence of experiencing presence (Schubert, Friedmann & Regenbrecht, 2001). Gamers commonly report deep involvement with games (Johnson & Wiles, 2003), whereby they can become so absorbed by the activity, that they feel removed from their immediate environment (Molesworth, 2006). With deep involvement players can become less aware of their surroundings and the passing of time (Brown & Cairns, 2004). This immersion is a psychological state and an outcome of the medium's ability to focus users' attention (Witmer & Singer, 1998).

By their very nature, games demand and automatically elicit attention. Immersion, absorption and engagement are considered key for game effectiveness (Sweetser & Wyeth, 2005). Current literature on usability and user experience presents many heuristics for game design and evaluation to achieve these outcomes (in terms of interface design, mechanics and game play). It has been recommended that games should: be detailed to capture and maintain a user's attention (Pagulayan, Keeker, Wixon, Romero & Fuller, 2003); progressively increase a player's workload (Lazzaro & Keeker, 2004); gradually increase in difficulty to be challenging (Desurvire, Caplan & Toth, 2004; Juul, 2004); minimize distractions from major game tasks; and maximize the amount of screen taken up with game action (Johnson & Wiles, 2003). In order to facilitate immersion in the environment, they should also: offer mastery of the interface and control system (Adams, 2005); actively involve players with different options so they can shape the environment (Gee, 2004); make the interface invisible (Federoff, 2002); and use audio and narrative that will draw players in (Sweetser & Johnson, 2004).

Grigorovici and Constantin (2004) suggest the structural features of a virtual environment in 3D gaming increase affective engagement with the environment. Others highlight that features such as interactivity and vividness encourage cognitive involvement (Li et al., 2002), which may lead to higher satisfaction (Dellaert & Kahn, 1999; Judge, Bono & Locke, 2000), information processing (Liu & Shrum, 2002), and enhanced learning (Zimmerman, 2000).

### Summary

It is apparent that electronic games possess key characteristics, namely interactivity and vividness, which distinguish them from traditional media. In fact, some authors consider interactivity to be the most important characteristic of games (see, for example, Grodal, 2000; Vorderer, 2000). Interactivity and vividness can facilitate several psychological responses, including presence, flow and involvement. There is a growing body of work which has examined the influence of these outcomes on users, but mostly in the online environment (see, for example, Chou & Ting, 2003; Wan & Chiou, 2006). The majority of academic work has concentrated not on the characteristics of games per se, but rather the influence of game play.

## A REVIEW OF ACADEMIC RESEARCH CONCERNING THE EFFECTS OF GAME PLAY

Some academic attention has been given in the social sciences to the impact of electronic games on gamers pertaining to four key areas: violence (for a review, see Gentile, 2005; Gentile & Stone, 2005), social development (see Rauterberg, 2004), learning (for a review, see Dipietro, Ferdig, Boyer & Black, 2007; Mitchell & Savill-Smith, 2004), and physical well-being (Vandewater, Shim & Caplovitz, 2004). These studies have largely been motivated by concerns over the influence of games on children: an active gaming segment particularly in the online environment. Online, games appear on over 55% of web sites aimed at children and teens (Neuborne, 2001). It is estimated that approximately 64% of children aged 5 to 14 years who access the internet do so to play games (DeBell & Chapman, 2003) and spend an average of 30 minutes on each gaming site (Pereira, 2004). Their game usage time is also increasing (NPD Group, 2007b). Business, government, consumer advocacy groups, civil libertarians and parents have raised concerns over the use of games by children, but academic research has produced inconclusive evidence of the effects on players (for a complete discussion, see Ferguson, 2007).

Game violence is one area which has attracted particular attention. Concerns about violent content have increased, with suggestions the impact of games is intensified because of their active component and heightened reality (Calvert & Tan, 1994; Chambers & Ascione, 1987). Some research has found children, adolescents and young adults tend to become desensitized to real violence (Carnagey, Anderson & Bushman, 2007) and more aggressive following exposure to violent electronic games (Anderson & Bushman, 2001; Dominick, 1984). Other studies however, indicate this is not the case (Graybill, Strawniak, Hunter & O'Leary, 1987; Winkel, Novak & Hopson, 1987). It is claimed that for most people, playing video games will not cause severe psychopathology or major adjustment problems (Gibb, Bailey, Lambirth & Wilson, 1983; Kestenbaum & Weinstein, 1985), but there is evidence of negative outcomes for subgroups of players. Ellis (1984), for example, found a positive relationship between video arcade involvement and deviant behavior, but only for a minority of children whose involvement in arcades was associated with weak parental control. Funk and Buchman (1996) also found that spending more time playing video games was associated with lower self-esteem for adolescent girls. Overall, there is insufficient causal research to confirm the impact of playing violent games, and inadequate research to identify high-risk players or game playing habits.

The impact of game play on social development is another topic which has received academic attention. In this area too, there is disagreement concerning effects. Some psychologists have expressed concern that electronic games can negatively impact the social skills of children, who may use games as a substitute for relationships and become socially isolated (Dominick, 1984). In the case of violent games, some researchers have found children demonstrate less pro-social behavior (see, for example, Chambers & Ascione, 1987), but others have shown games do not impact social introversion (see Kestenbaum & Weinstein, 1985). It has even been suggested that video game play is a positive social activity that can actually enhance social interaction, as well as motor, intellectual and affective development (Gros, 2003; Snider, 2003). Shimai, Masuda and Kishimoto (1990) found that video gamers performed better socially in some areas than did non-gamers, while McClure and Mears (1984) showed that frequent players were more outgoing and less accomplishment-oriented than those who played less often. Contradicting these findings, Sakamoto (1994) found no relationship between video games and social behavior.

The impact of electronic games on learning has also been explored. Gros (2003) suggests that games can be used as learning tools, as they have the ability to reach, motivate and fully involve learners. It has been claimed that through play, students can develop abilities and strategies, as well as acquire digital literacy (Gros, 2003; Robertson & Good, 2005; Soloway, 1991). Some research has found video games can help teach perceptual and motor skills (Jones, Kennedy & Bittner, 1981; Kennedy, Bittner & Jones, 1981), as well as leadership (Comer, 2001). They can also alter visual attention processing (Bavelier & Green, 2003), promote inductive reasoning, and facilitate observational learning (Funk & Buchman, 1996). For children in particular, they may have a more powerful influence than the classroom (Gee, 2003; Rosas et al., 2003). It has even been recommended that electronic games be used as an adjunct for skill training in laparoscopic surgery (Rosser et al., 2007). There are some suggestions however, that electronic games can have a negative impact on learning (Randel, Morris, Wetzel & Whitehill, 1992; Stoll, 1999). There is also little agreement in the literature as to the theoretical underpinnings of why games should be used, or how they should be designed to support learning.

Finally, the relationship between game play and physical health has been a focus in recent research studies. Obesity, particularly among children, has become a major health problem in many countries, and games have been identified by some as a contributing environmental factor associated with this epidemic. A series of studies have identified a positive relationship between time spent playing electronic games and obesity (Gordon-Larsen, Adair & Popkin, 2002; Stettler, Signer & Suter, 2004; Vandewater et al., 2004). At the same time however, there are those who argue that games can be used to promote health consciousness and encourage physical activity (Baranowski et al., 2003; Brown, 2006; Dorman, 1997). Indeed, Mhurchu, Maddison, Jiang, Jull and Prapavessis (2008) found that regularly playing active video games (also called physical games or exer-games) might have a positive effect on children's overall physical activity levels. The most "active" electronic games can be found in the sports genre. The relationship between games and sport is examined next.

## ELECTRONIC GAMES AND THEIR RELATIONSHIP WITH SPORT

Sport is one of most commonly purchased video game genres (RocSearch, 2004). In 2007, the majority of console players purchased action games (22.3%), followed by sports (14.1%), shooter (12.1%) and then racing (8.3%) (ESA, 2008). Children under 12 years in particular are a key target, making up 25% of sports video game purchases in 2007 (Hein, 2008). Online, action and sports games represent one of the most frequently played genres (ESA, 2008).

Several motivators have been identified for playing sports games, including social interaction, fantasy, competition, entertainment, pastime, knowledge application and interest in sport (Kim & Ross, 2006). Interestingly, many of the same factors that motivate sports video game use also serve as motivators for sports participation. Gould and Petlichkoff (1988) identified six primary motivations for youth sport participation, which include: improving skills, having fun, being with friends, experiencing thrills and excitement, achieving success through competition and developing fitness. The same three motivators identified earlier for general game use are evident when it comes to sports video games specifically, and sports participation generally. First, competition is a key reason why people play both games (particularly sports games) and sports (Jansz & Martens, 2005). Second, having fun is a key goal: just as level of enjoyment is a mediating factor when choosing to allocate time to active leisure pursuits such as sport (Dishman et al., 2005), so too is it important for games. Finally, social aspects are key. Jansz and Martens (2005) liken shared gaming to football, suggesting both offer similar opportunities for bonding and the creation of social and emotional ties.

Traditionally, video games and sports have been viewed as unrelated activities. In fact, they are often seen as competitors, battling one another for an individual's leisure time. Their relationship however, has become more apparent. Aside from motivational factors, electronic games share a profound relationship with sport in three main ways.

### Electronic Games and the Replication of Sport

The first relationship games have with sport is through their replication: electronic games can replicate sports and sporting events. As an example, Nintendo, in conjunction with Sega, recently released a game based on the largest sporting event in the world: *Mario & Sonic at the Olympic Games*™. This is the official video game of the 2008 Beijing Olympics, licensed by the International Olympic Committee. There are also many game titles that are based on pro sports such as football, basketball and hockey.

Compelling content is one factor that drives the games market. Marketing and production costs for games are now comparable to feature films (Adams, 2006), yet product lifecycles are highly variable. The keys for publishers are strong marketing and distribution networks, as well as hit titles (Williams, 2002). For this reason, a strong convergence is becoming apparent between games and other entertainment categories. This includes sport. Licensed brands are growing in importance, because they have mass market appeal, and their easily recognizable titles modeled after successful franchises gives them greater sales potential (DFC Intelligence, 2004; NPD Group, 2004). Electronic Arts is the most successful and largest game publisher, responsible for such titles as *Madden NFL, FIFA Soccer, NBA Live* and others in the EA Sports™ series.

The aforementioned games represent simulations of sports entertainment. Creative and technological advances are allowing developers to deliver more detailed and realistic environments, which is creating more exciting consumer experiences (Lienert, 2004). Perhaps one of the most sophisticated categories is motor racing. There are many game titles which replicate motorsport events, from *Nascar® SimRacing™*, to *World Rally Championship* and *V8 Supercars*. These games are so realistic that professional race drivers often use them to prepare for track events (Rosewater, 2004). In the case of *V8 Supercars*, professional drivers have labeled it "the most immersive racing experience ever" (Games Universe, 2004).

*V8 Supercars* is the official racing game licensed by V8 Supercars Australia™. It is based on the V8 Australian Touring Car Championship, which is Australia's largest, professional motor sport series, now telecast to 400 million homes in over 70 countries (McKay, 2004). It attracts more than 1.7 million spectators per annum (V8 Supercars, 2007), making motor racing the third highest attended sport in Australia (ABS, 2007). In the game, the motor sport teams, locations, imagery, events, circuits, drivers and sponsors appear naturally, almost as they do in the real series. Authentic sound recording has been used for all the vehicles, the car mechanics and handling are modeled on actual brands, and collisions are based on the system used by the crash test industry (Product Review, 2004). *V8 Supercars 3™* also allows for customizable car set-ups to provide a true-to-life driving experience.

Technology provides game developers the means to not only deliver virtual realities for target groups of users, but also game worlds that can be custom designed by them (Pennington, 2001). Through their active participation, players can impact a game and co-create its content, because interactivity gives them more control over messages received and the ability to customize these according to their needs (Liu & Shrum, 2002; Roehm & Haugtvedt, 1999). Players can therefore truly be part of the sports action. They may perform the role of coach or player and construct their own sports teams, make decisions about players, execute plays and so on: activities that, for most people, are not possible in reality. In this way, sports games allow users to build on, maintain, or experience dreams, and realize fantasies (Shapiro & McDonald, 1992), plus each time they play, a gamer can act out different scenarios and receive different experiences (Frasca, 2003; Gee, 2004). New levels of sophistication will be established in the next ten years and graphics quality will improve, as the technology and artistry of games continues.

## Electronic Games and the Facilitation of Sports Participation

The second relationship games have with sport is through the facilitation of sports participation: electronic games provide players the opportunity to participate in sports. For example, in a console racing game players can select, accessorize, tune and drive production cars under realistic conditions. They may even use after-market accessories, such as a steering wheel and pedals, to gain a more realistic driving experience. However, activities are carried out by a game character, under the direction of a player who selects various options in the game. It is therefore the mechanical device that performs most of the work. Arcade machines on the other hand often require greater physical involvement. Arcade games like *Daytona USA®*, *Super GT* and *Alpine Racer* require the player to sit or stand in a cab and perform certain physical moves, rather than simply push a button. For instance, in the case of many arcade racers, players must turn the steering wheel, move the gear stick and use the pedals to drive their on-screen vehicle.

In the aforementioned examples, playing a sport does not occur in a real world context; it is simulated in the medium. With the advent of the Nintendo Wii™ however, released in the U.S. in 2006, electronic games are now a medium that truly facilitates sports participation. This game console allows individuals to play imaginary sports games using a wireless handheld controller, whereby a player can physically simulate sports moves. For example, instead of pushing a button to swing a tennis racket, a player can swing the controller and hit the tennis ball. The controller contains solid-state accelerometers and gyroscopes that sense three-dimensional spatial movements (Bonis, 2007). A player must use this controller and physically interact with images on-screen in order to "play." Sony PlayStation®'s EyeToy™ digital camera offers similar functionality. The camera detects physical movement, allowing players to interact with the on-screen action in fitness, music and sports games by moving their hands, arms or legs (The Age, 2006). Likewise, Microsoft has an Xbox Live® Vision Camera for its Xbox 360™ console, which caters for motion-sensing games.

With hardware developments, a new generation of active electronic games has emerged, labeled "exer-games" (see Sinclair, Hingston & Masek, 2007). Take for example *Wii™ Sports Boxing*, in which players have to punch, dodge and block like a real boxer. Nintendo has also now introduced *Wii™ Fit* and the *Wii™ Balance Board*, which allow a player and their on-screen representation to interact with trainers in the practice of yoga, strength training and aerobics (Boehret, 2008). The system monitors and reports a users' weight loss, body mass index and balancing skills, providing tips for improvement. Although not technically a sport, the other major exer-game which has attracted attention is *Dance Dance Revolution®* (see, for example, White, Lehmann & Trent, 2007; Yang, Smith & Graham, 2008).

When both interactivity and vividness are high in a computer-mediated communication, and an individual is sufficiently stimulated by the artificial environment, their experience can become direct (Nicovich, 2005). With active games, players can gain a more true-to-life experience than in the case of previous formats. The technology is inspiring activity amongst a broader demographic, including older adults (Hrehocik, 2008) and the disabled (Pearson & Bailey, 2007), because unlike in a traditional game of sport, endurance and physical strength are not limiting factors. Preliminary studies show that even individuals suffering spinal injuries can experience aerobic training effects from playing (Fitzgerald et al., 2004; Widman, McDonald & Abresch, 2006). A new term is being used to reflect the use of the Wii™ sports games for physical therapy: "Wiihabilitation" (Tanner, 2008).

Games are often criticized for diverting consumer leisure time from more healthy pursuits (Williams, 2008), but it has been claimed that video games and exercise are no longer mutually exclusive (Baumann, 2007). The fitness industry is beginning to embrace the idea that playing the Wii™ and games such as tennis, bowling, golf and baseball provides a workout (Baumann, 2007). Indeed medical studies lend support.

Investigations of earlier games found they can have a metabolic and physiological impact, whereby games are capable of enhancing cardiovascular conditioning and increasing blood pressure (Baldaro et al., 2004; Wang & Perry, 2006). More active games however, can increase energy expenditure and heart rate to similar levels achieved from walking, jogging and skipping (Maddison et al., 2007). In a comparative study, Mellecker and McManus (2008) found that energy expenditure and heart rate was significantly higher for seated gaming than rest, but higher still in the case of active games. In fact, the mean heart rate from active game play was equal to that for vigorous exercise. Lanningham-Foster et al. (2006) found energy expenditure to be more than two times higher in the case of active games than traditional seated games. The cardiorespiratory measures for active game players can exceed the heart rate intensity necessary for developing and maintaining cardiorespiratory fitness, as recommended by the American College of Sports Medicine (Unnithan, Houser & Fernhall, 2006).

Game playing is so rigorous that new terms have been pioneered to diagnose patients suffering strains from playing. The term "Nintendinitis" was first coined by Brasington (1990) and has since been used to describe injuries stemming from the intensive use of recreational game technologies, especially those relating to tendons and ligaments (see, for example, Koh, 2000; Macgregor, 2000). Other terms include "Playstation thumb" (Karim, 2005) and more recently, "Wiiitis" (Bonis, 2007). In their article entitled *Wii Knee*, Robinson, Barron, Grainger and Venkatesh (2008) suggest that with the development of video games, many other types of sports injury typically associated with athletic activity could result from game play.

## Playing Electronic Games: A New Sport

Playing electronic games has also now become a professional sport in itself, labeled e-Sport. Essentially e-Sport is competitive video gaming, whether that is in the form of LAN events where gamers can link their PCs to play, or tournaments that are conducted in cyberspace. These physical and electronic events bring together gamers from across the world to compete against each other in real-time strategy, shooter and sport replications (soccer, auto racing, target shooting). Examples include the Electronic Sports World Cup™ (ESWC), the World Cyber Games (WCG), and the World e-Sports Games.

The ESWC is a major LAN gaming event. It attracts over 30 000 live spectators and several tens of millions who watch the coverage (ESWC, 2008). In the last four years, the event has grown from 380 to 800 gamers, and the prize money has increased from $156 000 to $400 000 (Game Guru, 2006). The WCG, on the other hand, is a global tournament where sport is conducted within the medium of cyberspace. It claims to have over a million participants from all over the world (WCG, 2008). The 2005 grand final attracted over 55 000 spectators and millions who watched online and television broadcasts (WCG, 2008). The WCG grand finals have been compared to the Olympic Games, because they share similar organizational and promotional strategies (Hutchins, 2006). Other competitions have been compared to an FA Cup Final (see, for example, Fifield, 2007). The magnitude of these events is astounding. They attract major sponsorship from corporations such as Samsung, Microsoft, NVIDIA and Intel, and pro-gamers (also called cyber-athletes) are treated like celebrities (c.f. Hutchins, 2006). Some fan clubs of top gamers have upwards of 700 000 members (Schiesel, 2006).

It is claimed that professional video gaming is now the world's fastest growing competitive sport (see, for example, Major League Gaming, 2006). In South Korea, which has more than 20 000 public PC gaming rooms (called PC bangs) that attract a million people per day, competitive gaming is one of the most televised sports (Schiesel, 2006). It is estimated that 10 million South Koreans regularly follow e-Sports (Schiesel, 2006). World game championships have been around since the 90s, but the internet has brought about the growth of more formal events and professional leagues, such as Major League Gaming, one of North America's biggest professional video game leagues. Many of these leagues were established with the mission to promote game competitions as a professional sport (see, for example, Cyberathlete Professional League, 2008). Interestingly, the activity does not appear to be labeled e-Sport due to the nature of the games that are played: the most popular are role-playing and strategy games, not those in the sports genre.

Aside from professional video gaming, or e-Sport, social online games in general have been labeled the electronic "sports" of the 21st century (Murphy, 2007). Social gaming involving the interaction of multiple players has existed since the late 70s, with the creation of the first Multi-User Dungeon (MUD) by Bartle and Trubshaw (1979). MUD essentially was an adventure game, which allowed many players to play at the same time over a network. Variations of MUD have continued to appear until today, with different terms now used to define them including Multi-User Virtual Environments and MU's. The 1980s saw the growth of arcades, which brought multiple players together in a physical space to compete. This is a form of asynchronous multiplay where players take turns (Bogost, 2004). The nature of multiplayer competition was changed however, with the introduction of persistent scoring on arcade machines. Gaming moved beyond one-on-one challenges to competitions that involved potentially all members of a community using the same machine, battling to overtake the player with the high score (Bogost, 2004).

The establishment of video gaming as a global phenomenon began with the introduction of the Nintendo Entertainment System (NES®) console in 1985 (Murphy, 2007). In the 1990s, console use grew rapidly, which encouraged competitions between multiple individuals in the home. At the same time, PC games gained popularity, starting initially with solitary games and then later, online games. Game technology advances, reduced costs and increased broadband penetration however, made MMOGs possible in the late 90s (Bogost, 2004). Online networked gaming grew from casual games of online cards or checkers, to these MMOGs, which enable synchronous play, where tens of thousands of individuals can play a game at the same time (Chan & Vorderer, 2006). In other words, individuals play simultaneously rather than in sequence. These games are becoming a powerful social force, and may be just as significant as other social activities for some groups, such as in the case of children and youth sports (Smoll & Smith, 2002). This, however, must be validated.

## AREAS FOR FUTURE RESEARCH

The key recommendation from the review performed in the current chapter is that further research is needed to understand the new and exciting world of electronic games. Given the size and growth of the industry, and the many issues and aspects associated with game play that are not yet understood, it is evident that this is an area worthy of academic attention. The opportunities for future study are almost endless. The following discussion serves not to present all the possibilities, but rather to stimulate ideas for potential work in the area.

First, research is needed concerning the personal context of game play. We need to better understand players, including who they are (e.g. gender, age and personality differences), why they play (i.e. motives), and how (i.e. gaming technologies and platforms used). This will provide further insights into the game playing experience. An individual's underlying motives and needs determine their game preferences (Klug & Schell, 2006), so research concerning players will also reveal more about what games are played. Future work should examine games across different platforms, including those played via console systems, handhelds, on mobiles, computers and in arcades. Durkin and Barber (2002) note that little research has documented the consequences or psychological correlates of playing different formats. Studies comparing offline and online gaming will make a particularly important contribution, especially since the motives for play (e.g. competition, enjoyment and social needs) are likely to be prioritized differently. The availability of blogs and newsgroups offers a rich data source and the opportunity for new methodological approaches to explore online games and their impact.

In addition to exploring what players do with games, it is necessary to understand what games do to them. A key weakness of many existing investigations is that they do not capture the complexity of the game playing experience. Much of the extant work conducted to date is focused on the short-term effects of exposure to games. However, electronic games have the ability to attract and hold interest, so people can spend long and intense periods of time playing, over hours, weeks and even months (Hamman, 2000). Users may play a game an average of 40 hours before tiring of it (Hein, 2004). The impact of extended play on psychological and physical responses across different gamer populations could prove interesting. Important areas for exploration have been identified in this chapter, including violence, social development, learning and physical well-being. Further, aside from the game playing experience and its influence, just as important is the need to study the characteristics of the game medium, the structural aspects of games in all their various forms, and influences on user response.

The situational context for play should also be studied, examining when and where games are used and how they fit into users' daily lives (see, for example, Sall & Grinter, 2007). This has some overlap with the social context for play and the need to understand with whom individuals share the gaming experience. The potential impact of shared viewing experiences and the role of coviewing behaviors in shaping game play outcomes contribute to the complexity of game playing and require research attention. This could include investigation of the impact of the group situation on moderating interactivity effects and the potential for flow, as well as other psychological responses. As part of this, there is an opportunity to study game playing culture, and its various sub-cultures such as multiplayer gaming, tournaments and battles.

Certainly the role of game playing in a wider social and cultural context also warrants attention (Mayra, 2007). As gaming spreads throughout society, there is a need to understand how it is transforming social systems. Important work has begun, which shows that gaming is having a real social, economic and cultural impact (see, for example, Castronova, 2005), but further research is necessary. Hutchins (2006) suggests we must look wider to contextualize game activities, events and indeed the whole industry within the evolving context of the Information Age. Researchers would be wise to continue careful monitoring of the games industry and its growing power.

Finally, in the domain of sports, there is much to be done. Games are shaping, and being shaped by, sports. Comparative studies of sports games based on true and fictional events should be conducted. Connectedness is a construct (Cohen, 1983) that awaits empirical testing in this context. Studies that examine a player's relationship with a sport, and the influence on their social and personal life may prove interesting. Individuals who are not only highly involved, but who define themselves by the sports they follow or play (and hence are true fans), may demonstrate a stronger commitment to related video gaming activities and could be more susceptible to their influence. Games allow for greater customization and involvement, allowing users to carry out desirable activities perhaps not possible in reality (such as playing the role of a sporting hero, racing a car and so on). The ability of a user to shape the game experience based on their preferences may have an as yet undetermined impact on them. Video game competitions represent a particularly exciting, and as yet untapped, avenue for study. The division between athlete and cyber-athlete appears to be getting smaller, but comparative studies will reveal by just how much. Further, exer-games are a new category that warrants attention.

The fact that all of the aforementioned areas are inter-related adds to the complexity of performing research concerning electronic games, but also reinforces the need to do so. Academics played an important role when previous media were introduced, in that they helped to understand the industry, bridged the gap between the public and the industry, and assisted with policy formation (Williams, 2002). Future research will require cross-disciplinary collaboration between academics and practitioners from such diverse fields as economics, marketing, law, politics, medicine, computer science, information technology, health sciences, communications and sport, among others.

## CONTRIBUTION OF THE CHAPTER AND CONCLUSION

Although a growing form of entertainment for an increasingly diverse audience, electronic games are a new medium, about which there is a fundamental lack of understanding. Information pertaining to the industry, games manufacturers, and players was presented in this chapter. That analysis facilitates understanding of market dynamics and trends concerning electronic games. Schilling (2003) and Williams (2002) appear to be the only other academic authors who have presented such a review. Schilling (2003) presents a commentary concerning the introduction and diffusion of technologies, using the games industry as an example, while Williams' (2002) focus is primarily on the structure and competition in the home video game industry. Likewise, even though industry reports and other practitioner papers are available, often they are incomplete, addressing only several core issues at the one time.

This chapter has also identified the key motives for game use and considered the characteristics that differentiate games from traditional media, as well as the associated user responses they facilitate. It contributes new knowledge to the games literature and provides a structured framework for comparison of games with different media types. This provides a foundation for further exploration of game medium characteristics as part of future research. Further, a review of the extant literature concerning the effects of game play has been presented. This review serves as a platform for future studies to investigate the influence of games.

Finally, the relationship between electronic games and sport has been examined. This chapter recognizes the multi-faceted relationship the two share, in terms of the ability of electronic games to replicate sport, facilitate sports participation, and be played as a sport. The evolution of games and expansion into this area has lessened the gap between sport in a physical space and that in an electronic space. The current chapter has summarized trends in this regard and highlighted opportunities for future study.

In conclusion, electronic games are exciting on a number of levels. First, the scale and scope of the industry is impressive. Second, the nature of the medium and its unique characteristics vis-à-vis more traditional media are of interest, as are the reasons why people play. Third, game playing can have a potentially strong influence on users, and in no case is this more true than in the case of sports. Finally, and perhaps most importantly, electronic games are exciting because of the research potential they offer.

## REFERENCES

ABS, Australian Bureau of Statistics. (2007). *Sports attendance, Australia, 2005-06,* (Catalog No. 4174.0). Canberra: ABS. Retrieved February 21, 2008, from http://abs.gov.au/AUSSTATS/abs@.nsf/Lookup/4174.0Main+Features12005-06?OpenDocument.

Adams, E. (2005). The Designer's Notebook: Bad game designer, no Twinkie! VI. *Gamasutra*. Retrieved August 8, 2007, from http://www.gamasutra.com/features/20050603/adams_01.shtml.

Adams, E. (2006). Cheer up! Video games are in great shape. *Gamasutra*. Retrieved August 8, 2007, from http://www.gamasutra.com/features/20060421/adams_01.shtml.

Anderson, C.A., & Bushman, B.J. (2001). Effects of violent video games on aggressive behavior, aggressive cognition, aggressive affect, physiological arousal, and prosocial behavior: A meta-analytic review of the scientific literature. *Psychological Science, 12*(5), 353-359.

Ariely, D. (2000). Controlling the information flow: Effects on consumers' decision making and preferences. *Journal of Consumer Research, 27*(2), 233-249.

Balabanis, G., & Reynolds, N.L. (2001). Consumer attitudes towards multi-channel retailers' web sites: The role of involvement, brand attitude, internet knowledge and visit duration. *Journal of Business Strategies, 18*(2), 105-132.

Baldaro, B., Tuozzi, G., Codispoti, M., Montebarocci, O., Barbagli, F., Trombini, E., & Rossi, N. (2004). Aggressive and non-violent videogames: Short-term psychological and cardiovascular effects on habitual players. *Stress and Health, 20*(4), 203-208.

Bambauer, S. (2006). Effects of brand placement in PC/ video games on the change of the attitude toward the advertised brand. In D. Grewal (Ed.), *Proceedings of the 2006 Summer Marketing Educators' Conference* (pp. 231-240). Chicago: American Marketing Association.

Baranowski, T., Baranowski, J., Cullen, K.W., Marsh, T., Islam, N., Zakeri, I., Honess-Morreale, L., & deMoor, C. (2003). Squire's Quest!: Dietary outcome evaluation of a multimedia game. *American Journal of Preventive Medicine, 24*(1), 52-61.

Bartle, R., & Trubshaw, R. (1979). *Multi-User Dungeon*. Colchester, UK: Essex University.

Baumann, M. (2007). Slimming down with Wii™ Sports. *Information Today, 24*(4), 47.

Bavelier, D., & Green, C. (2003). Action video game modifies visual selective attention. *Nature, 423*(6939), 534-537.

Beattie, A.E., & Mitchell, A.A. (1985). The relationship between advertising recall and persuasion: An experimental investigation. In L.F. Alwitt & A.A. Mitchell (Ed.), *Psychological processes and advertising effects* (pp. 129-155). Hillsdale, NJ: Lawrence Erlbaum Associates.

Bezjian-Avery, A., Calder, B., & Iacobucci, D. (1998). New media interactive advertising vs traditional advertising. *Journal of Advertising Research, 38*(4), 23-32.

Boehret, K. (2008). The Mossberg Solution: For Wii™ lovers, something worth sweating over. *Wall Street Journal, May 14*, D1.

Bogost, I. (2004). *Asynchronous multiplay: Futures for casual multiplayer experience*. Paper presented at the Other Players Conference on Multiplayer Phenomena, Copenhagen, Denmark.

Bonis, J. (2007). Acute Wiiitis. *New England Journal of Medicine, 356*(23), 2431.

Brasington, R. (1990). Nintendinitis. *New England Journal of Medicine, 322*(20), 1473-1474.

Brightman, J. (2007). WMS: PS3 to 'win' console war because of Blu-ray. *GameDaily*. Retrieved July 2, 2008, from http://www.gamedaily.com/articles/features/wms-ps3-to-win-console-war-because-of-blu-ray/70379/?biz=1.

Brown, D. (2006). Playing to win: Video games and the fight against obesity. *Journal of the American Dietetic Association, 106*(2), 188-189.

Brown, E., & Cairns, P. (2004). A grounded investigation of game immersion. In *Extended Abstracts of the 2004 Conference on Human Factors in Computing Systems* (pp. 1297-1300). New York: ACM Press.

Calvert, S.L., & Tan, S. (1994). Impact of virtual reality on young adults' physiological arousal and aggressive thoughts: Interaction versus observation. *Journal of Applied Developmental Psychology, 15*(1), 125-139.

Carnagey, N.L., Anderson, C.A., & Bushman, B.J. (2007). The effect of video game violence on physiological desensitization to real-life violence. *Journal of Experimental Social Psychology, 43*(3), 489-496.

Caron, F. (2008). Gaming expected to be a $68 billion business by 2012. *Ars Technica*. Retrieved July 2, 2008, from http://arstechnica.com/news.ars/post/20080618-gaming-expected-to-be-a-68-billion-business-by-2012.html.

Castronova, E. (2005). *Synthetic worlds: The business and culture of online games*. Chicago: University of Chicago Press.

Chambers, J.H., & Ascione, F.R. (1987). The effects of prosocial and aggressive videogames on children's donating and helping. *Journal of Genetic Psychology, 148*(4), 499-505.

Chan, E., & Vorderer, P. (2006). Massively multiplayer online games. In P. Vorderer & J. Bryant (Ed.), *Playing video games: Motives, responses and consequences* (pp. 77-90). USA: Routledge.

Cho, C.H., & Leckenby, J.D. (1999). Interactivity as a measure of advertising effectiveness: Antecedents and consequences of interactivity in web advertising. In M.S. Roberts (Ed.), *Proceedings of the 1999 Conference of the American Academy of Advertising* (pp. 162-179). Gainesville, FL: American Academy of Advertising.

Chou, T.J., & Ting, C.C. (2003). The role of flow experience in cyber-game addiction. *CyberPsychology & Behavior, 6*(6), 663-675.

Cohen, J.B. (1983). Involvement and you: 1000 great ideas. In R.P. Bagozzi & A.M. Tybout (Eds.), *Advances in consumer research* (vol. 10, pp. 325-328). Ann Arbor, MI: Association for Consumer Research.

Comer, D.R. (2001). Not just a Mickey Mouse exercise: Using Disney's The Lion King to teach leadership. *Journal of Management Education, 25*(4), 430-436.

comScore. (2007). Worldwide online gaming community reaches 217 million people. Retrieved July 20, 2008, from http://www.comscore.com/press/release.asp?press=1521.

Coyle, J.R., & Thorson, E. (2001). The effects of progressive levels of interactivity and vividness in Web marketing sites. *Journal of Advertising, 30*(3), 65-77.

Csikszentmihalyi, M. (1977). *Beyond boredom and anxiety*. San Francisco: Jossey-Bass.

Csikszentmihalyi, M., & LeFevre, J. (1989). Optimal experience in work and leisure. *Journal of Personality and Social Psychology, 56*(5), 815-822.

Cyberathlete Professional League. (2008). About us. Retrieved July 17, 2008, from http://www.thecpl.com/index.php?page_id=1090.

DeBell, M., & Chapman, C. (2003). *Computer and internet use by children and adolescents in 2001*. (NCES 2004-2014). Washington, DC: U.S.

Department of Education, National Center for Education Statistics.

Delaney, K.J. (2004). Space Invaders: Ads in videogames pose a new threat to media industry; Marketers pay for placement at the expense of TV; A small but growing area; The terrorist on the cellphone. *Wall Street Journal, Jul 28*, A.1.

De Lisi, R., & Wolford, J.L. (2002). Improving children's mental rotation accuracy with computer game playing. *Journal of Genetic Psychology, 163*(3), 272-282.

Dellaert, B.G.C., & Kahn, B.E. (1999). How tolerable is delay?: Consumers' evaluation of internet web sites after waiting. *Journal of Interactive Marketing, 13*(1), 41-54.

De Rivera, J. (1984). Development and the full range of emotional experience. In C. Malatesta & C. Izard (Eds.), *Emotion in adult development* (pp. 45-63). Beverly Hills, CA: Sage.

Desurvire, H., Caplan, M., & Toth, J.A. (2004). Using heuristics to evaluate the playability of games. In *Extended Abstracts of the 2004 Conference on Human Factors in Computing Systems* (pp. 1509-1512). New York: ACM Press.

Devaney, P. (2005). America: Advertisers find a new playmate in the computer games industry. *Marketing Week, 28*(33), 28-29.

DFC Intelligence. (2004). New DFC Intelligence reports look at growing convergence of video game, music and movie industries. Retrieved September 23, 2004, from http://www.dfcint.com/news/prmarch192004.html.

DFC Intelligence. (2007). DFC Intelligence estimates worldwide portable game market to exceed $10 billion in 2007. Retrieved August 5, 2008, from http://www.dfcint.com/wp/?p=74

DFC Intelligence. (2008). DFC Intelligence forecasts video game market to reach $57 billion in 2009. Retrieved August 5, 2008, from http://www.dfcint.com/wp/?p=222.

Dipietro, M., Ferdig, R.E., Boyer, J., & Black, E.W. (2007). Towards a framework for understanding electronic educational gaming. *Journal of Educational Multimedia and Hypermedia, 16*(3), 225-248.

Dishman, R.K., Motl, R.W., Saunders, R., Felton, G., Ward, D.S., Dowda, M., & Pate, R.R. (2005). Enjoyment mediates effects of a school-based physical-activity intervention. *Medicine and Science in Sports and Exercise, 37*(3), 478-487.

Dominick, J.R. (1984). Media effects on the young videogames, television violence, and aggression in teenagers. *Journal of Communication, 34*(2), 136-148.

Dorman, S. (1997). Video and computer games: Effect on children and implications for health education. *Journal of School Health, 67*(4), 133-138.

Dorval, M., & Pepin, M. (1986). Effect of playing a video game on a measure of spatial visualization. *Perceptual and Motor Skills, 62*(1), 159-162.

Dower, T. (2007). A parallel universe. *Marketing, Feb*, 16-19.

Durkin, K., & Barber, B. (2002). Not so doomed: Computer game play and positive adolescent development. *Applied Developmental Psychology, 23*(4), 373-392.

Edwards, C., & Greene, J. (2005). Who's got game now? *Business Week, May 16*(3933), 40.

Ellis, D. (1984). Video arcades, youth, and trouble. *Youth and Society, 16*(1), 47-65.

ESA, Entertainment Software Association. (2003). Game players are a more diverse gender, age and socio-economic group than ever, according to new poll. Retrieved September 23, 2004, from http://www.theesa.com/8_26_2003.html.

ESA. (2004a). Americans playing more games, watching less movies and television. Retrieved September 23, 2004, from http://www.theesa.com/5_12_2004.html.

ESA. (2004b). Games, parents and ratings. Retrieved September 23, 2004, from http://www.theesa.com/gamesandratings.html.

ESA. (2006a). Essential facts about the computer and video game industry. Retrieved July 4, 2006, from http://www.theesa.com/archives/files/Essential%20Facts%202006.pdf.

ESA. (2006b). Facts and research: Game player data. Retrieved January 12, 2008, from http://www.theesa.com/facts/gamer_data.php.

ESA. (2008). Essential facts about the computer and video game industry. Retrieved September 4, 2008, from http://www.theesa.com/facts/pdfs/ESA_EF_2008.pdf.

ESWC, Electronic Sports World Cup™. (2008). ESWC overview. Retrieved July 17, 2008, from http://www.eswc.com/info/overview_eswc.

Federoff, M. (2002). *Heuristics and usability guidelines for the creation and evaluation of fun in video games*. Unpublished thesis, Indiana University, Bloomington.

Ferguson, C.J. (2007). Evidence for publication bias in video game violence effects literature: A meta-analytic review. *Aggression and Violent Behavior, 12*(4), 470-482.

Fifield, A. (2007). Limbering up for an e-sport revolution Korean-style. *Financial Times, Sep 15*, 10.

Fitzgerald, S.G., Cooper, R.A., Thorman, T., Cooper, R., Guo, S., & Boninger, M.L. (2004). The GAME(Cycle) exercise system: Comparison with standard ergometry. *Journal of Spinal Cord Medicine, 27(5),* 453–459.

Foege, A. (2005). All the young dudes. *Mediaweek, 15*(31), 16-20.

Fortin, D.R., & Dholakia, R.R. (2005). Interactivity and vividness effects on social presence and involvement with a web-based advertisement. *Journal of Business Research, 58(3)*, 387-396.

Frasca, G. (2003). Simulation versus narrative: Introduction to ludology. In M.J.P. Wolf & B. Perron (Eds.), *The video game theory reader* (pp. 221-236). London: Routledge.

Funk, J.B., & Buchman, D.D. (1996). Playing violent video and computer games and adolescent self-concept. *Journal of Communication, 46*(2), 19-32.

Gaither, C. (2003). Battle for the sexes. *Boston Globe, Jan 13*. Retrieved May 20, 2008, from http://www.there.com/pressBostonGlobe_011303.html.

Gallagher, S., & Park, S.H. (2002). Innovation and competition in standard-based industries: A historical analysis of the U.S. home video game market. *IEEE Transactions on Engineering Management, 49*(1), 67-82.

Game Guru. (2006). Electronic Sports World Cup™ India Lan gaming tournament kicks off. Retrieved August 25, 2008, from http://www.gameguru.in/general/2006/19/electronic-sports-world-cup-india-lan-gaming-tournament-kicks-off/.

GamersCircle. (2008). PC gaming a $10.7 billion industry (Games Convention 2008). Retrieved October 5, 2008, from http://www.gamerscircle.net/2008/08/19/pc-gaming-a-107-billion-industry-games-convention-2008/.

Games Universe. (2004). V8 Supercars Race Driver. Retrieved September 24, 2004, from http://gamesuniverse.com.au/ProductPage.asp?ProductID=3131.

Gardner, D.G., Dukes, R.L., & Discenza, R. (1993). Computer use, self-confidence and attitudes: A causal analysis. *Computers in Human Behavior, 9*(4), 427-440.

Gartner. (2008). Gartner says worldwide mobile gaming revenue to surpass $4.5 billion in 2008. Retrieved August 5, 2008, from http://www.gartner.com/it/page.jsp?id=706407.

Gee, J.P. (2003). *What video games have to teach us about learning and literacy.* New York: Palgrave Macmillan.

Gee, J.P. (2004). Learning by design: Games as learning machines. *Gamasutra.* Retrieved August 8, 2007, from http://www.gamasutra.com/gdc2004/features/20040324/gee_01.shtml.

Gentile, D.A. (2005). Examining the effects of video games from a psychological perspective: Focus on violent games and a new synthesis. *National Institute on Media and the Family.* Retrieved August 13, 2007, from http://www.psychology.iastate.edu/faculty/dgentile/pdfs/Gentile_NIMF_Review%20_2005.pdf.

Gentile, D.A., & Stone, W. (2005). Violent video game effects on children and adolescents: A review of the literature. *Minerva Pediatrica, 57*(6), 337-358.

Gibb, G.D., Bailey, J.R., Lambirth, T.T., & Wilson, W.P. (1983). Personality differences between high and low electronic video game users. *Journal of Psychology, 114*(2), 159-165.

Gordon-Larsen, P., Adair, L.S., & Popkin, B.M. (2002). Ethnic differences in physical activity and inactivity patterns and overweight status. *Obesity Research, 10*(3), 141–149.

Gould, D., & Petlichkoff, L. (1988). Participation motivation and attrition in young athletes. In F.L. Smoll, R.A. Magill, & M.J. Ash (Eds.), *Children in sport* (3rd ed., pp. 161-178). Champaign, IL: Human Kinetics Books.

Graybill, D., Strawniak, M., Hunter, T., & O'Leary, M. (1987). Effects of playing versus observing violent versus non-violent video games on children's aggression. *Psychology, 24*(3), 1-8.

Green, C.S., & Bavelier, D. (2003). Action video game modifies visual selective attention. *Nature, 423*(6939), 534-537.

Griffith, J.L., Voloschin, P., Gibb, G.D., & Bailey, J.R. (1983). Differences in eye-hand motor coordination of video-game users and non-users. *Perceptual and Motor Skills, 57*(1), 155-158.

Griffiths, M.D., Davies, M.N.O., & Chappell, D. (2003). Breaking the stereotype: The case of online gaming. *CyberPsychology and Behavior, 6*(1), 81-91.

Grigorovici, D. (2003). Persuasive effects of presence in immersive virtual environments. In G. Riva, F. Davide & W.A. Ijsselsteijn (Eds.), *Being there: Concepts, effects and measurements of user presence in synthetic environments* (pp. 191-207). Amsterdam: IOS Press.

Grigorovici, D., & Constantin, C. (2004). Experiencing interactive advertising beyond rich media: Impacts of ad type and presence on brand effectiveness in 3D gaming immersive virtual environments. *Journal of Interactive Advertising, 5*(1). Retrieved August 24, 2005, from http://www.jiad.org/vol5/no1/grigorovici/index.htm.

Grodal, T. (2000). Video games and the pleasures of control. In D. Zillmann & P. Vorderer (Eds.), *Media entertainment: The psychology of its appeal* (pp. 197-214). Mahwah, NJ: Lawrence Erlbaum.

Gros, B. (2003). The impact of digital games in education. *First Monday Journal.* Retrieved September 24, 2004, from http://www.firstmonday.org/issues/issue8_7/xyzgros/index.html.

Grossman, L. (2005). Out of the X Box. *Time, 165*(21), 44-52.

Hamman, R. (2000). Forget the web ads: Online promos capture attention. *Marketing News, 34*(23), 28-29.

Harrison, L.K., Denning, S., Easton, H.L., Hall, J.C., Burns, V.E., Ring, C., & Carroll, D. (2001). The effects of competition and competitiveness on cardiovascular activity. *Psychophysiology, 38*(4), 601-606.

Hartig, K. (2006). Next-generation consoles drive global game revenues to more than $45 billion by 2010. *PricewaterhouseCoopers*. Retrieved September 5, 2007, from http://www.pwc.com/extweb/pwcpublications.nsf/docid/C987CEB2D179131F852572090083B4B3/$FILE/VideoGames_KH_ls.pdf.

Hein, K. (2004). Getting in the game. *Brandweek, 45*(7), 26-27.

Hein, K. (2008). EA opts not to keep it real with new 'Freestyle' brand. *Brandweek, 49*(18), 7.

Hoffman, D.L., & Novak, T.P. (1996). Marketing in hypermedia computer-mediated environments: Conceptual foundations. *Journal of Marketing, 60*(3), 50-69.

Howe, T., & Sharkey, P.M. (1998). Identifying likely successful users of virtual reality systems. *Presence, 7*(3), 308-316.

Hrehocik, M. (2008). ICAA defines 'active aging.' *Long-Term Living, 57*(4), 12-14.

Hsu, C.L., & Lu, H.P. (2004). Why do people play on-line games? An extended TAM with social influences and flow experience. *Information and Management, 41*(7), 853-868.

Hurley, O. (2008). Game on again for coin-operated arcade titles. *The Guardian, Feb 7.* Retrieved September 30, 2008, from http://www.guardian.co.uk/technology/2008/feb/07/games.it.

Hutchins, B. (2006). Computer gaming, media and e-sport. In V. Colic-Peisker, F. Tilbury & B. McNamara (Eds.), *Proceedings of the 2006 Australian Sociological Association Annual Conference* (pp. 1-9). Australia: University of Western Australia and Murdoch University.

Hyman, P. (2005). Sponsors go ape over advergames. *Hollywood Reporter.* Retrieved July 16, 2007, from http://www.hollywoodreporter.com/hr/search/article_display.jsp?vnu_content_id=1001523697.

Izard, C. (1977). *Human emotions.* New York: Plenum Press.

Jansz, J., & Martens, L. (2005). Gaming at a LAN event: The social context of playing video games. *New Media and Society, 7*(3), 333-355.

Jih, H., & Reeves, T.C. (1992). Mental models: A research focus for interactive learning systems. *Educational Technology Research and Development, 40*(3), 39-53.

Johnson, D., & Wiles, J. (2003). Effective affective user interface design in games. *Ergonomics, 46*(13/14), 1332-1345.

Johnson, G.J., Bruner II, G.C., & Kumar, A. (2006). Interactivity and its facets revisited. *Journal of Advertising, 35*(4), 35-52.

Jones, M.B., Kennedy, R.S., & Bittner, A.C. (1981). A video game for performance testing. *American Journal of Psychology, 94*(1), 143-152.

Judge, T.A., Bono, J.E., & Locke, E.A. (2000). Personality and job satisfaction: The mediating role of job characteristics. *Journal of Applied Psychology, 85*(2), 237-249.

Juul, J. (2004). Working with the player's repertoire. *International Journal of Intelligent Games and Simulation, 3*(1), 54-61.

JWT Intelligence. (2006). Gaming. Retrieved May 20, 2008, from https://016fd0d.netsolstores.com/index.asp?pageaction=viewprod&prodid=4.

Kap, J.T. (2004). But is it art? *PBS.* Retrieved September 23, 2004, from http://www.pbs.org/kcts/videogamerevolution/impact/art.html.

Karim, S.A. (2005). Playstation thumb – a new epidemic in children. *South African Medical Journal, 95*(6), 412.

Kennedy, R.S., Bittner, A.C., & Jones, M.B. (1981). Video game and conventional tracking. *Perceptual and Motor Skills, 53*(1), 310.

Kestenbaum, G.I., & Weinstein, L. (1985). Personality, psychopathology, and developmental issues in male adolescent video game use. *Journal of the American Academy of Child Psychiatry, 24*(3), 329-333.

Kieras, D., & Polson, P.G. (1985). An approach to the formal analysis of user complexity. *International Journal of Man-Machine Studies, 22*(4), 365-394.

Kim, K.H., Park, J.Y., Kim, D.Y., Hak, M., & Chun, H.C. (2002). E-Lifestyle and motives to use online games. *Irish Marketing Review, 15*(2), 71-77.

Kim, S.W. (1995). *The motive of a video game use and the types of enjoying of the youths*. Unpublished master's thesis, Han Yang University, Korea.

Kim, Y., & Ross, S. (2006). An exploration of motives in sport video gaming. *International Journal of Sport Marketing and Sponsorship, 8*(1), 34-46.

Klug, G.C., & Schell, J. (2006). Why people play games: An industry perspective. In P. Vorderer & J. Bryant (Ed.), *Playing video games: Motives, responses and consequences* (pp. 91-100). USA: Routledge.

Koepp, M.J., Gunn, R.N., Lawrence, A.D., Cunningham, V.J., Dagher, A., Jones, T., Brooks, D.J., Bench, C.J., & Grasby, P.M. (1998). Evidence for striatal dopamine release during a video game. *Nature, 393*(6682), 266-268.

Koh, T.H. (2000). Ulcerative "nintendinitis": A new kind of repetitive strain injury. *Medical Journal of Australia, 173*(11), 671.

Kretchmer, S.B. (2004). Advertainment: The evolution of product placement as a mass media marketing strategy. *Journal of Promotion Management, 10*(1/2), 37-54.

Lanningham-Foster, L., Jensen, T.B., Foster, R.C., Redmond, A.B., Walker, B.A., Heinz, D., & Levine, J.A. (2006). Energy expenditure of sedentary screen time compared with active screen time for children. *Pediatrics, 118*(6), 1831-1835.

Lazzaro, N. (2004). Why we play games: Four keys to more emotion without story. *XEODesign*. Retrieved August 8, 2007, from http://www.xeodesign.com/whyweplaygames/xeodesign_whyweplaygames.pdf.

Lazzaro, N., & Keeker, K. (2004). What's my method? A game show on games. In *Extended Abstracts of the 2004 Conference on Human Factors in Computing Systems* (pp. 1093-1094). New York: ACM Press.

Lessiter, J., Freeman, J., Keogh, E., & Davidoff, J.A. (2001). A cross-media presence questionnaire: The ITC-sense of presence inventory. *Presence, 10*(3), 282-297.

Li, H., Daugherty, T., & Biocca, F. (2001). Characteristics of virtual experience in electronic commerce: A protocol analysis. *Journal of Interactive Marketing, 15*(3), 13-30.

Li, H., Daugherty, T., & Biocca, F. (2002). Impact of 3-D advertising on product knowledge, brand attitude and purchase intention: The mediating role of presence. *Journal of Advertising*, 31(3), 43-57.

Libaw, O.Y. (2000). Dance machine: Quirky arcade game draws legions of fans. *ABC News*. Retrieved August 5, 2008, from http://www.ddrfreak.com/newpress/ABC%20News.htm.

Lienert, A. (2004). Video games open new path to market cars. *Detroit News*. Retrieved September 24, 2004, from http://www.detnews.com/2004/autosinsider/0402/15/b01-64356.htm.

Liu, Y., & Shrum, L.J. (2002). What is interactivity and is it always such a good thing? Implications of definition, person and situation for the influence of interactivity on advertising effectiveness. *Journal of Advertising, 31*(4), 53-64.

Lombard, M., & Ditton, T. (1997). At the heart of it all: The concept of presence. *Journal of Computer-Mediated Communication, 3*(2). Retrieved February 11, 2005, from http://jcmc.indiana.edu/vol3/issue2/lombard.html.

Macgregor, D.M. (2000). Nintendonitis? A case report of repetitive strain injury in a child as a result of playing computer games. *Scottish Medical Journal, 45*(5), 150.

Maddison, R., Mhurchu, C.N., Jull, A., Jiang, Y., Prapavessis, H., & Rodgers, A. (2007). Energy expended playing video console games: An opportunity to increase children's physical activity? *Pediatric Exercise Science, 19*(3), 334-343.

Major League Gaming. (2006). About Major League Gaming. Retrieved July 17, 2008, from http://www.mlgpro.com/about/index.

Mayra, F. (2007). The contextual game experience: On the socio-cultural contexts for meaning in digital play. In *Situated play: Proceedings of DiGRA 2007 Conference* (pp. 810-814). Tokyo: Digital Games Research Association.

McCandless, D. (2003). Just one more go.... *The Guardian, Apr 3*. Retrieved August 11, 2007, from http://www.guardian.co.uk/technology/2003/apr/03/onlinesupplement3.

McClure, R.F., & Mears, F.G. (1984). Video game players: Personality characteristics and demographic variables. *Psychological Reports, 55*(1), 271-276.

McKay, P. (2004). Polites takes aim to protect V8 Supercars. *Sydney Morning Herald*. Retrieved September 24, 2004, from http://www.smh.com.au/cgi-bin/common/popupPrintArticle.pl?path=/articles/2004/02/23/1077497513603.html.

McMillan, S.J. (2002). Exploring models of interactivity from multiple research traditions: Users, documents and systems. In L.A. Lievrouw & S. Livingstone (Eds.), *Handbook of new media: Social shaping and consequences of ICTs* (pp. 163-182). London: Sage.

Mellecker, R.R., & McManus, A.M. (2008). Energy expenditure and cardiovascular responses to seated and active gaming in children. *Archives of Pediatrics & Adolescent Medicine, 162*(9), 886-891.

Mhurchu, C.N., Maddison, R., Jiang, Y., Jull, A., & Prapavessis, H., (2008). Couch potatoes to jumping beans: A pilot study of the effect of active video games on physical activity in children. *International Journal of Behavioral Nutrition and Physical Activity, 5*(8). Retrieved August 30, 2008, from http://www.ijbnpa.org/content/5/1/8.

Mitchell, A., & Savill-Smith, C. (2004). *The use of computer and video games for learning: A review of the literature*. London: Learning and Skills Development Agency.

Molesworth, M. (2006). Real brands in imaginary worlds: Investigating players' experiences of brand placement in digital games. *Journal of Consumer Behavior, 5*(4), 355-366.

Moore, E.S. (2006). *It's child's play: Advergaming and the online marketing of food to children*. Menlo Park, CA: Kaiser Family Foundation.

Murphy, S.M. (2007). A social meaning framework for research on participation in social online games. *Journal of Media Psychology, 12*(3). Retrieved May 20, 2008, from http://www.calstatela.edu/faculty/sfischo/A_Social_Meaning_Framework_for_Online_Games.html.

Neuborne, E. (2001). For kids on the Web, it's an ad, ad, ad, ad, world; How to help yours see the sales pitches behind online games. *Business Week, Aug 13*(3725), 108.

Nicovich, S.G. (2005). The effect of involvement on ad judgment in a video game environment: The mediating role of presence. *Journal of Interactive Advertising, 6*(1). Retrieved August 9, 2007, from http://www.jiad.org/vol6/no1/nicovich/index.htm.

Nielsen. (2007). The state of the console. Retrieved August 6, 2008, from www.nielsenmedia.com/nc/nmr_static/docs/Nielsen_Report_State_Console_03507.pdf.

Novak, T.P., Hoffman, D.L., & Yung, Y.F. (2000). Modeling the flow construct in online environments: A structural modeling approach. *Marketing Science, 19*(1), 22-42.

NPD Group. (2004). The NPD Group reports on sales of licensed video game titles. Retrieved September 23, 2004, from http://www.npd.com/press/releases/press_040608.htm.

NPD Group. (2007a). Playing video games viewed as family/ group activity and stress reducer. Retrieved July 15, 2008, from http://www.npd.com/press/releases/press_071212.html.

NPD Group. (2007b). Amount of time kids spend playing video games is on the rise. Retrieved July 15, 2008, from http://www.npd.com/press/releases/press_071016a.html.

O'Guinn, T.C., Allen, C.T., & Semenik, R.J. (2006). *Advertising and integrated brand promotion.* (4th ed.). Mason: Thomson/South-Western.

Pagulayan, R., Keeker, K., Wixon, D., Romero, R., & Fuller, T. (2003). User-centered design in games. In J.A. Jacko & A. Sears (Eds.), *The human-computer interaction handbook: Fundamentals, evolving techniques and emerging applications* (pp. 883-905). Mahwah, NJ: Lawrence Erlbaum Associates.

Paterson, P.A. (2003). Synergy and expanding technology drive booming video game industry. *TD Monthly.* Retrieved September 23, 2004, from http://www.toydirectory.com/monthly/Aug2003/Games_Booming.asp.

Pearson, E., & Bailey, C. (2007). Evaluating the potential of the Nintendo Wii™ to support disabled students in education. In *ICT: Providing choices for learners and learning. Proceedings ascilite Singapore 2007* (pp. 833-836). Singapore: Center for Educational Development, Nanyang Technological University.

Pennington, R. (2001). Signs of marketing in virtual reality. *Journal of Interactive Advertising, 2*(1). Retrieved August 10, 2007, from http://www.jiad.org/vol2/no1/pennington/index.htm.

Pereira, J. (2004). Junk-food games; online arcades draw fire for immersing kids in ads; Ritz Bits Wrestling, anyone? *Wall Street Journal, May 3*, B1.

PQ Media. (2006). Alternative media research series II: Alternative advertising and marketing outlook 2006. Retrieved July 5, 2006, from http://www.pqmedia.com/execsummary/AlternativeAdvertisingMarketingOutlook2006-ExecutiveSummary.pdf.

PricewaterhouseCoopers. (2007). Industry previews: Video games. Retrieved September 5, 2007, from http://www.pwc.com/extweb/industry.nsf/docid/8CF0A9E084894A5A85256CE8006E19ED?opendocument&vendor=#video.

PricewaterhouseCoopers. (2008). Entertainment and media companies face a collaboration imperative for next five years, says PricewaterhouseCoopers annual outlook report. Retrieved August 5, 2008, from http://www.pwc.com/extweb/ncpressrelease.nsf/docid/6DD913426F4A05108525746B004C3C42.

Privette, G. (1983). Peak experience, peak performance and flow: A comparative analysis of positive human experience. *Journal of Personality and Social Psychology, 45*(6), 1361-1368.

Product Review. (2004). V8 Supercars Australia racing video game details. Retrieved September 23, 2004, from http://www.productreview.com.au/showitem.php?item_id=1075.

Randel, J.M., Morris, B.A., Wetzel, C.D., & Whitehill, B.V. (1992). The effectiveness of games for educational purposes: A review of recent research. *Simulation and Gaming, 23*(3), 261-276.

Rauterberg, M. (2004). Positive effects of entertainment technology on human behavior. In R. Jacquart (Ed.), *Building the information society* (pp. 51-58). Toulouse, France: Kluwer Academic Publishers.

Reisinger, D. (2008). Why the Xbox 360™ will win the console war. *cnet*. Retrieved July 31, 2008, from http://news.cnet.com/8301-13506_3-9939276-17.html.

Reportlinker. (2007). Western world MMOG market: 2006 review and forecasts to 2011. Retrieved July 31, 2008, from http://www.reportlinker.com/p046468/online-games.html.

Robertson, J., & Good, J. (2005). Story creation in virtual game worlds. *Communications of the ACM, 48*(1), 61-65.

Robinson, R.J., Barron, D.A., Grainger, A.J., & Venkatesh, R. (2008). Wii knee. *Emergency Radiology, 15*(4), 255-257.

RocSearch. (2004). Video game industry. Retrieved September 15, 2004, from http://www.rocsearch.com/pdf/Video%20Game%20Industry.pdf.

Roehm, H.A., & Haugtvedt, C.P. (1999). Understanding interactivity of cyberspace advertising. In D.W. Schumann & E. Thorson (Eds.), *Advertising and the World Wide Web* (pp. 27-39). Mahwah, NJ: Lawrence Erlbaum.

Rosas, R., Nussbaum, M., Cumsille, P., Marianov, V., Correa, M., Flores, P., Grau, V., Lagos, F., Lopez, X., Lopez, V., Rodriguez, P., & Salinas, M. (2003). Beyond Nintendo: Design and assessment of educational video games for first and second grade students. *Computers and Education, 40*(1), 71-94.

Rosewater, A. (2004). Hey, I'm practicing here, not playing: Video games so precise they help many drivers. *USA Today, Aug 8*, 6f.

Rosser, J.C., Lynch, P.J., Cuddihy, L., Gentile, D.A., Klonsky, J., & Merrell, R. (2007). The impact of video games on training surgeons in the 21st century. *Archives of Surgery, 142*(2), 181-186.

Sakamoto, A. (1994). Video game use and the development of sociocognitive abilities in children: Three surveys of elementary school students. *Journal of Applied Social Psychology, 24*(1), 21-42.

Sall, A., & Grinter, R.E. (2007). Let's get physical! In, out and around the gaming circle of physical gaming at home. *Computer Supported Cooperative Work, 16*(1-2), 199-229.

Schiesel, S. (2006). The land of the video geek. *New York Times, Oct 8*, 2.1.

Schilling, M.A. (2003). Technological leapfrogging: Lessons from the U.S. video game console industry. *California Management Review, 45*(3), 6-32.

Schubert, T., Friedmann, F., & Regenbrecht, H. (2001). The experience of presence: Factor analytic insights. *Presence, 10*(3), 266-281.

Shapiro, M.A., & McDonald, D.G. (1992). I'm not a real doctor, but I play one in virtual reality: Implications of virtual reality for judgments about reality. *Journal of Communication, 42*(4), 94-114.

Shields, M. (2005). Overload of game ads could defeat purpose. *Adweek, 46*(46), 9.

Shimai, S., Masuda, K., & Kishimoto, Y. (1990). Influences of TV games on physical and psychological development of Japanese kindergarten children. *Perceptual and Motor Skills, 70*(3), 771-776.

Sicilia, M., Ruiz, S., & Munuera, J.L. (2005). Effects of interactivity in a web site: The moderating effect of need for cognition. *Journal of Advertising, 34*(3), 31-45.

Siekpe, J.S., & Hernandez, M.D. (2006). The effect of system and individual characteristics on flow, and attitude formation toward advergames. In P. Rutsohn (Ed.), *Proceedings of the Annual Meeting of the Association of Collegiate Marketing Educators* (pp. 131-137). Oklahoma City, OK: Association of Collegiate Marketing Educators.

Sinclair, J., Hingston, P., & Masek, M. (2007). Considerations for the design of exergames. In *Proceedings of the 5th International Conference on Computer Graphics and Interactive Techniques in Australia and Southeast Asia* (pp. 289-295). Perth, Australia: ACM.

Smoll, F.L., & Smith, R.E. (2002). *Children and youth in sport: A biosychosocial perspective*. Madison, WI: Kendall Hunt.

Snider, M. (2003). Study surprise: Video games enhance college social life. *USA Today*. Retrieved September 23, 2004, from http://www.usatoday.com/tech/news/2003-07-06-games_x.htm.

Soloway, E. (1991). How the Nintendo Generation learns. *Association for Computing Machinery. Communications of the ACM, 34*(9), 23-27.

Sony. (2007). PlayStation®2 celebrates its seventh anniversary, more than 120 million consoles sold worldwide. Retrieved July 31, 2008, from http://www.us.playstation.com/News/PressReleases/431.

Starbuck, W.J., & Webster, J. (1991). When is play productive? *Accounting, Management and Information Technology, 1*(1), 71-90.

Stettler, N., Signer, T.M., & Suter, P.M. (2004). Electronic games and environmental factors associated with childhood obesity in Switzerland. *Obesity, 12*(6), 896-903.

Steuer, J. (1992). Defining virtual reality: Dimensions determining telepresence. *Journal of Communication, 42*(4), 73-93.

Stoll, C. (1999). *High tech heretic – Reflections of a Computer Contrarian*. New York: First Anchor Books.

Sweetser, P., & Johnson, D. (2004). Player-centered game environments: Assessing player opinions, experiences and issues. In M. Rauterberg (Ed.), *Entertainment computing – ICEC 2004: Third International Conference, LNCS 3166* (pp. 321-332). New York: Springer Verlag.

Sweetser, P., & Wyeth, P. (2005). GameFlow: A model for evaluating player enjoyment in games. *ACM Computers in Entertainment, 3*(3), 1-24.

Tanner, L. (2008). New form of physical therapy: Wii™ games. *LiveScience*. Retrieved August 19, 2008, from http://www.livescience.com/health/080209-ap-wii-therapy.html.

Taub, E.A. (2004). Video game makers play it safe. *International Herald Tribune*. Retrieved September 23, 2004, from http://www.iht.com/articles/540148.html.

Taylor, C. (2005). Who is playing games – and why. *Time, 165*(21), 52.

The Age. (2006). The next gaming wave. Retrieved August 19, 2008, from http://www.theage.com.au/news/games/the-next-gaming-wave/2006/11/28/1164476212099.html?page=fullpage.

Turner, J.R., & Carroll, D. (1985). Heart rate and oxygen consumption during mental arithmetic, a video game, and graded exercise: Further evidence of metabolically-exaggerated cardiac adjustments? *Psychophysiology, 22*(3), 261-267.

Unnithan, V.B., Houser, W., & Fernhall, B. (2006). Evaluation of the energy cost of playing a dance simulation video game in overweight and non-overweight children and adolescents. *International Journal of Sports Medicine, 27*(10), 804-809.

V8 Supercars. (2007). A success story - V8 Supercars Australia. Retrieved July 6, 2007, from http://www.v8supercar.com.au/content/about_avesco/the_v8_supercars_australia_success_story/?ind=M.

Vandewater, E.A., Shim, M.S., & Caplovitz, A.G. (2004). Linking obesity and activity level with children's television and video game use. *Journal of Adolescence, 27*(1), 71-85.

Vincente, K.J., Hayes, B.C., & Williges, R.C. (1987). Assaying and isolating individual differences in searching a hierarchical file system. *Human Factors, 29*(3), 349-359.

Vorderer, P. (2000). Interactive entertainment and beyond. In D. Zillmann & P. Vorderer (Eds.), *Media entertainment: The psychology of its appeal* (pp. 21-36). Mahwah, NJ: Lawrence Erlbaum.

Wan, C.S., & Chiou, W.B. (2006). Psychological motives and online games addiction: A test of flow theory and humanistic needs theory for Taiwanese adolescents. *CyberPsychology & Behavior, 9*(3), 317-324.

Wang, X., & Perry, A.C. (2006). Metabolic and physiologic responses to video game play in 7- to 10-year-old boys. *Archives of Pediatrics and Adolescent Medicine, 160*(4), 411-415.

WCG, World Cyber Games. (2008). About WCG: WCG concept. Retrieved July 17, 2008, from http://www.worldcybergames.com/6th/inside/WCGC/WCGC_structure.asp.

Webster, J., & Martocchio, J.J. (1992). Microcomputer playfulness: Development of a measure with workplace implications. *MIS Quarterly, 16*(2), 201-226.

White, M., Lehmann, H., & Trent, M. (2007). 31: Disco dance video game-based interventional study on childhood obesity. *Journal of Adolescent Health, 40*(2), S32.

Widman, L.M., McDonald, C.M., & Abresch, R.T. (2006). Effectiveness of an upper extremity exercise device integrated with computer gaming for aerobic training in adolescents with spinal cord dysfunction. *Journal of Spinal Cord Medicine, 29*(4), 363-370.

Williams, D. (2002). Structure and competition in the U.S. home video game industry. *International Journal on Media Management, 4*(1), 41-54.

Williams, D. (2006). Why game studies now? Gamers don't bowl alone. *Games and Culture, 1*(1), 13-16.

Williams, K. (2005). Manufacturers finally learn how to sell game networking to operators. *Vending Times, 45*(12). Retrieved August 5, 2008, from http://vendingtimes.com/ME2/dirmod.asp?sid=&nm=&type=Publishing&mod=Publications%3A%3AArticle&mid=8F3A7027421841978F18BE895F87F791&tier=4&id=EB40CAAFDA0643598C31564B24E0206A.

Williams, W. (2008). In electronic age, sports stores compete for youth. *The State Journal.* Retrieved September 8, 2008, from http://www.statejournal.com/story.cfm?func=viewstory&storyid=43197.

Winkel, M., Novak, D.M., & Hopson, H. (1987). Personality factors, subject gender, and the effects of aggressive video games on aggression in adolescents. *Journal of Research in Personality, 21*(2), 211-223.

Witmer, B.G., & Singer, M.J. (1998). Measuring presence in virtual environments: A presence questionnaire. *Presence, 7*(3), 225-240.

Wu, G. (1999). Perceived interactivity and attitude toward web sites. In M.S. Roberts (Ed.), *Proceedings of the 1999 Conference of the American Academy of Advertising* (pp. 254-262). Gainesville, FL: American Academy of Advertising.

Yang, S., Smith, B., & Graham, G. (2008). Healthy video gaming: Oxymoron or possibility? *Journal of Online Education, 4*(4). Retrieved September 30, 2008, from http://innovateonline.info/index.php?view=article&id=186&action=synopsis.

Youn, S., & Lee, M. (2003). Antecedents and consequences of attitude toward the advergame in commercial web sites. In L. Carlson (Ed.), *Proceedings of the 2003 Conference of the American Academy of Advertising* (p. 128). Pullman, WA: American Academy of Advertising.

Zimmerman, B.J. (2000). Self-efficacy: An essential motive to learn. *Contemporary Educational Psychology, 25*(1), 82-91.

# Chapter XVII
# Sport Video Game Sponsorships and In-Game Advertising

**Beth A. Cianfrone**
*Georgia State University, USA*

**James J. Zhang**
*University of Florida, USA*

## ABSTRACT

*This chapter introduces the new and unique sport promotional format of sport video game sponsorships and in-game advertising. Information on the growth of sport video games, unique features of this segment of the sport industry, and financial and technical value of in-game advertising and sponsorships are first introduced. Extensive discussions are made on the advantage of sport video games as a marketing tool and the importance of assessing the effectiveness of in-game advertising and sponsorships. The need to systematically understand consumer motivation and market demand for sport video games is highlighted. This chapter concludes with recognizing contemporary issues and recommended solutions.*

## INTRODUCTION

Sport video games are consistently a popular genre of video games. A part of their popularity is attributed to the visual likeness and authenticity of the games with traditional televised sports. In a merging of the classic business relationship between sports and advertising, corporations are advertising within sport video games as an avenue to reach sport fans, while the advertisements contribute to the realism of the game. This chapter introduces the new and unique sport promotional format of sport video game sponsorships and in-game advertising. The characteristics of sport video games (e.g., interactive, fantasy based, realistic and repetitive nature); the dynamics between the stakeholders (e.g., gaming publishers, the sport organizations, and in-game advertising firms); and the demographics of the target market (e.g., 18-34 year old males who have shifted their media consumption from television to video games) are discussed in relation to their impact on sport video game sponsorships. Various types of sponsorships and in-game advertising in sport video games, and their integration, are also explored. Contemporary research on the marketing effectiveness of sport video game sponsorships and in-game advertising, as well as consumer motivations associated with the demand of playing sport video games, is reviewed.

Specifically, the objectives of this chapter are to provide readers with information in the following areas:

1. The growing trend of sport video games.
2. The uniqueness of sport video game products.
3. The financial and functional values of in-game advertising and sponsorships.
4. Congruence of target markets between sport video games and sponsors.
5. Advantages associated with in-game advertising and sponsorships.
6. Measurement issues associated with studying the effectiveness of in-game advertising and sponsorships.
7. Basic concepts of motivational factors associated with the consumption of sport video games.
8. Contemporary issues, problems, and recommended solutions.

## SPORT VIDEO GAMES BACKGROUND

Sport video games have become a popular form of entertainment and a growing segment of the $24.5 billion worldwide gaming industry [Entertainment Software Association (ESA), 2007]. The sport video game genre is among the top video game sales in total business transactions each year and represented 17% of the total console video game units sold in 2006 (ESA, 2007). In the United States (U.S.), the best selling console video game in 2006 was an American football game, Electronic Arts (EA) Sports *Madden NFL 07* on the PlayStation2 format, with 1.8 million units sold. Another 825,000 units were sold in the Xbox 360 format ("Top Ten", 2007). In fact, during its first week of sales, *Madden NFL 07* grossed $100 million in retail sales ("First-Week", 2006). Generally speaking, sport video games are a growing segment of both the sport and entertainment industries.

Sport video games are popular for many reasons. Sport video games are an interactive, virtual media outlet through which people play a fantasy-based sport game. There are numerous sport video game titles; some mimic traditional organized sports, such as soccer, football, baseball, golf and tennis, while others appeal to fans of more untraditional sports, such as bass fishing, skateboarding and motocross racing. The games are sold for various platforms, and can be played on personal computers or with console gaming systems. A video game console is a computer system used exclusively for gaming purposes (Forster, 2005). The emergence of these interactive entertainment systems began with Magnavox in the 1970s, Atari in the early 1980s, and transgressed to Nintendo and Sega in the early 1990s (Forster, 2005). Within the past 10 years, Sony (PlayStation2 and PlayStation3), Microsoft (Xbox and Xbox 360), and Nintendo (GameCube and the Wii) have been the key console manufacturers (Forster, 2005). Handheld devices are also popular, such as the Sony PlayStationPortable (PSP) and the Nintendo DS. However, sport video games are sold more often as console platforms than computer or handheld platforms. Console systems are popular for sport video gaming due to a number of reasons. Consoles are developed completely for gaming usage, so they often work at faster processor speeds, allowing the athlete's game movements to be realistic. The nature of console gaming also allows the player to use a handheld controller to play the game, rather than a keyboard or joystick, while the game is displayed on a television, rather than a computer monitor.

This benefits gamers who enjoy watching their games on large screens. Console manufacturers are also keeping pace with technological advances. The Xbox 360 and PlayStation3 consoles have capabilities to utilize high definition gaming formats in conjunction with high definition televisions (HDTV). These two consoles also have online capabilities, allowing the owners to play gamers across the globe.

Some of the appeal of sport video games is the resemblance to televised sports and the technological control that is allowed in the games. Sport video games allow the gamer to act as a coach and/or player of the selected sport. The gamer has control over every aspect of the game, including the selection of teams/players to utilize while playing, starting lineups, uniforms, stadiums/arenas, weather conditions, play calling, athlete moves, and even celebration dances. The ability for people to "play" as their favorite player/coach/team has appeal to many sport fans (Kim & Ross, 2006). The experience to fulfill a fantasy is a unique characteristic of video games and has contributed to a high volume of sport video game consumption, as well as technological improvement by game publishers.

Sport video games act as a media source and can be likened to watching a televised sporting event, due to the games' real-life virtual viewing effects. Sport video game publishers pride themselves on the realistic nature of the games, utilizing the latest technology and real-life players to simulate the game for better effects. Most of the sport games resemble televised sports because they include an exact professional league's teams, uniforms, rosters, stadiums, players, and live commentators. Also, due to the inherent relationship between sport and sponsorships, brand logos or advertisements inserted within the virtual games create a more realistic game experience. The realistic features of sport video games help the gaming industry encompass all major professional and amateur sport organizations, individual professional athletes, and corporate sponsors (Fisher, 2006).

## Stakeholders

Sport video games can be very lifelike because of contracts and endorsements by the actual sport organizations themselves, which allow usage of league owned rights (e.g., logos, teams, players, stadiums). Professional sport organizations, such as the National Football League (NFL), Major League Baseball (MLB), National Hockey League (NHL), National Basketball Association (NBA), and Fédération Internationale de Football Association (FIFA) have licensing contracts with game publishers. Electronic Arts is a prominent game publisher; its sports segment, EA Sports, is the leader in sport video games with 70% of the market share (Adams, 2005). 2K Sports, owned by parent company Take-Two Interactive Software, is another major sport video game publisher. Publishers may have developers who create the games in-house, or may contract with other developers. The licensing contracts between sport organizations and game publishers allow the games to utilize the team logos, uniforms, and players of the leagues, adding credibility and realism to the video games. The exclusivity of a publishing company to create realistic games comes at a high price tag. As evidenced by the NFL and NFL Players Association's $300 million, five year contract with EA Sports, gaming is a lucrative licensing outlet for leagues and organizations (Lefton, 2004a). This contract made EA Sports the exclusive publisher of NFL football video games. In 2004, the NFL earned nearly 50% of its total licensing revenue from video game sales, suggesting sport video games are a worthwhile marketing tool for the league (Lefton, 2004a). In 2008, the parties involved extended the contract to 2012 (Thorsen, 2008). This exclusivity means 2K Sports and other publishers cannot associate with the likeness of NFL and must be creative if making football related sport video games. EA Sports has made the most of its exclusivity with numerous NFL games (e.g., *NFL Head Coach, NFL Tour, NFL Street* series, *Madden* series;

Thorsen, 2008). Similarly, Take-Two Interactive (2K Sports) has an exclusive contract to publish MLB games (Take-Two, 2005).

Many advertising corporations are getting involved with in-game advertising. IGA Worldwide, Massive, Engage Advertising, and Double Fusion are just a few of the corporations that develop in-game advertisements. They serve as the facilitators between game publishers and sponsoring/advertising companies. Massive was recently purchased by Microsoft, impacting in-game advertising in some popular sport video games. Five EA Sports games on the Xbox 360 or computer platform (*Madden NFL 08*, *NASCAR 09*, *NHL 08*, *Tiger Woods PGA TOUR 08*, and *SKATE*) have in-game advertising. Because these games are on the Xbox 360, which has online capabilities through Xbox Live, advertisements can be changed and customized to the consumer (Microsoft, 2007).

Numerous individual professional athletes also have sport video game contracts, endorsing games such as *Tiger Woods PGA TOUR* or *Tony Hawk's Pro Skater* games. These endorsements have created business relationships and ventures that continue to prosper. As a matter of fact, a critical area of the sport video game industry lies within game sponsorships, where corporations aim to advertise through this electronic medium. Considering that creating technologically enhanced games is costly for game publishers, sponsorships provide a relatively new financial resource for publishers to offset production costs and even generate profit. Related sport organizations often dictate the types of brands that are involved with in-game advertising.

## Growth of In-Game Advertising and Sponsorships

Video games are a growing outlet for sponsorships and advertisements. Game publishers have turned to in-game advertising to supplement rising costs. At the same time, many corporations have adopted sport video games as a marketing tool in an effort to reach sport video game consumers. Sponsoring and advertising within a sport video game offers companies a way to integrate and display their brands or products through a heavily consumed virtual medium. In 2005, about $56 million was spent on video game advertising. That figure will likely reach between $400 and $732 million by 2010 (Fisher, 2006; Howard, 2006). Some even predicted that in-game advertising spending would reach between $1.6 billion to $1.8 billion by 2010 (Shields, 2006). More recently, these numbers have been retracted to $971.3 million by 2011 (Shields, 2007). Although growth estimates have varied, the widespread utilization of sport video games as promotional outlets is imminent and certainly a marketing platform that is on the continued rise. Costs of in-game advertising vary and are evolving as the technology becomes more common and advanced. In 2004, EA Sports charged 10 cents per in-game advertisement (signage), multiplied by the number of games sold (Lefton, 2004b). Five companies paid a combined $1.5 million for in-game advertising in EA Sports *NASCAR 2005: Chase for the Cup* (Adams, 2004).

Today's fan consumes sport through many forms of media, including television, radio, print, Internet, and video games. The diversity of the sports media marketplace indicates that focusing marketing efforts on a repetitive and highly consumed media source, such as video games, may be beneficial. The advent of digital video recorders, such as TiVo which helps to skip commercials, has further prompted the need for creative marketing strategies to reach sport consumers (Fernando, 2004). Some corporations aiming to reach the 18-34 year old sport fans have even abandoned traditional television-based sponsorship and advertising in an effort to achieve more promising results through video games. As male gamers between 18-34 years old are playing sport video games more often than watching television, sponsors have delved into a different avenue of marketing (Della Maggiora,

2006). The rise in sport video game popularity has suggested a shift in marketing efforts from televised medium to the interactive sport video game medium (Della Maggiora, 2006).

## Target Market

To a great extent, increased interest in sport video game sponsorships can be attributed to the demographics of gamers and their consumption levels. The sport video game demographic is defined as the traditional sport demographic: 18-34 year old male consumers. Nearly 60% of men 18-34 years old own a video game system (Media Usage, 2006). The 18-34 year old male demographic is characterized by their impressionability in brand selection and large spending habits (King, 2006). This group's attention is highly coveted by television networks and their subsequent advertisers; meanwhile, this age group is also one of the major consumer segments targeted by sport organizations along with their advertisers or sponsors. Parks Associates found the 18-34 year old male demographic to be more open to in-game promotions than other age groups (Business Wire, 2006).

According to Radia and Harris (2006), the typical profile of a gamer is a 24 year old adult male who plays video games 5 days a week for 1.5 hours at a time. He plays *Madden Football* at the prime time of 8:00 p.m. to 1:00 a.m. on Monday through Friday evenings. In the meantime, prime time television viewing by this age group of audience is declining by an average of one hour per week in the U.S. (Radia & Harris, 2006). These contrasting numbers suggest a possible shift to online and console gaming from traditional television programs. It is apparent that the ways that young people consume media and sport are changing. In fact, Neilson reported more men aged 18-34 played console video games than watched major network television (Della Maggiora, 2006). This shift in electronic media consumption has some profound implications in that traditional televised promotions (commercials, product placement advertisements, and virtual advertisements) are not viewed by as many viewers as previously; instead, online and console gaming as new media outlets are receiving increased attention from gamers. Logically, it is beneficial for corporations to aim at the 18-34 year olds through sport video game sponsorships (Browne, 2006; Radia & Harris, 2006).

This shift in media consumption has led some marketers to find alternative promotional outlets to compensate for the loss in television viewership. The increase in sport video game consumption has created a new area of sponsorship for companies targeting the 18-34 year old male sport fan. Sport video game sponsorships and in-game advertising have quickly become a growing area of marketing, as companies look for a repetitive and highly consumed medium.

## SCOPE, ISSUES, CONTROVERSIES, PROBLEMS

In the U.S., approximately $6.4 billion is spent each year on sport sponsorships ("About us", 2008). Sport sponsorships are a common marketing tool. The partnership between sport organizations and sponsoring corporations is fulfilled through various promotional forms. A sponsoring corporation pays a fee to be associated with a sport, organization, or team in hopes of creating or enhancing an association between the consumer and the sponsoring brand (Mullin, Hardy & Sutton, 2007). The ultimate purpose of a sport sponsorship is to influence the behavior of sport consumers and generate sales (Crompton, 2004), which is usually accomplished via creating brand awareness, enhancing brand image, and associating the brand with positive feelings or emotions. The ability and effectiveness of achieving these sponsorship goals are often described in the advertising and sponsorship literature. Similar to traditional sponsorship, formulation of a sport video game sponsorship relies on the association between a sport organization and a sport video game publishing company.

## Types of In-Game Advertising and Sponsorships

Sport video game sponsorships can be fulfilled by virtually inserting static billboard signage with a corporation's brand logo, to more invasive approaches (McClellan, 2005). Brand logos of sponsoring companies have been inserted into specific segments of sport video games (e.g., the Halftime, Red Zone, and Player of the Game sponsorships in *NCAA Football*). Athletes and coaches through endorsement deals sponsor games (e.g., the *Tony Hawk's Pro Skater* series and *Madden NFL* series). The billboard signage insertions or game segment sponsorships are similar to virtual advertising on television, except for the fact that video games are all "virtual."

Billboard signage can be dynamic as well, changing depending on the segment of the game or the type of consumer. Some products are fully integrated into the game; for example, a golfer can select specific Nike clubs when playing *Tiger Woods PGA TOUR* (McClellan, 2005). One of the most extensive sponsorships is Burger King's usage of the "King" character in *Fight Club 2007*. Other types include on-screen venue signage, such as virtual logos, similar to the venue signage seen at a live or televised sporting event; commentary plugs such as a game announcer's fulfillment of sponsorships with pre-recorded messages that play in an appropriate game segment; and transition advertisements, which are the newest form of in-game advertising, where a logo or advertisement is displayed with a 4-5 second pause prior to reaching the next gaming level (Lefton, 2004b). The major benefit of all of the sport video game sponsorship types is that they cannot be skipped by the player. Overall, in-game advertisements are very prevalent within sport video games and are similar to sponsorships of televised sports.

## Advantages of In-Game Advertising and Sponsorships

The sport video game in-game advertisements often depict a real sport sponsorship that exists in the associated sport league. These advertisements add to the realism and authenticity of the game being played; therefore, they may be viewed as similar to a sponsorship, with the associated benefits that a sponsorship has over an advertisement (Cianfrone, Zhang, Trail & Lutz, 2008). Due to the inherent relationship between sport and sponsorship, sport video game sponsorships and in-game advertising may be viewed as beneficial. It is typical for a sport venue to have signage; therefore, a sport video game that includes venue signage in the virtual stadiums simply adds to the pragmatism. Parks Associates found male gamers aged 18-34 years old are in favor of in-game advertisements that enhance game play (Business Wire, 2006). Similarly, Nelson (2002) found gamers were open to in-game promotions provided that they fit the game. Meenaghan (2001a) explained that sport consumers view sponsorships as indirect and subtle. Sport video game sponsorships are subtle, because they are implemented similar to traditional sponsorships. Sport video game sponsorships are typically integrated into the game in a manner that does not affect game play, as logos may be incorporated through a sponsorship of a game segment (e.g., sponsor of the coin toss). A sport video game sponsorship in the form of a logo that is displayed when a team scores a touchdown in a football game, is not an obvious push for purchasing the product when compared to a television commercial that explains why a consumer should purchase.

An advantage of sport video game sponsorships or in-game advertising is the inability for gamers to skip them. Due to the integration of the sponsorships into the game, the gamers are exposed to the sponsorship when they play. The sponsorships are unavoidable. For example, a game sponsorship of the coin toss to start a football game would be integrated into the game in a manner where the commentators mention the sponsoring company and the gamer must select if they want to kick, receive, or defer to start the game. These sponsorships are different from a television commercial, which can be skipped by a viewer changing the channel. Further advantages of sport video game sponsorship include access to a highly coveted demographic, repetition of game use and therefore sponsorship exposure, and the interactive nature of the games (Lefton, 2004b).

Sport video game in-game advertising also differs from forms of internet game advertising or advergames. In those mediums, an online video game is created by a corporation purely to promote exposure of its brands. These games may be sport based in nature, but tend not to be as complex in terms of gaming levels and game play activity, nor likely have any relationships with the associated sport leagues or organizing bodies. These features in game simplicity, lack of authenticity, and direct commercial purpose make the games less attractive to sport consumers for repeated plays. While these games are often free to the consumer to play and provide interaction on the website, they are not likely to have as much repeated exposure when compared with a sport video console game. Console sport video games with ads imbedded in them offer a more subtle way to advertise a brand and, due to the complexity of the game, allow for more repetition than perhaps a one-shot game on a company's website, where the exposure to the brand is more limited because the gamer would not play as often.

## Impact of Stakeholder Relationships

The relationships among sport leagues, league sponsors, video game producers, and video game sponsoring companies are a critical part of the gaming industry. League licensing deals, such as MLB/2K Sports and NFL/EA Sports, influence in-game advertising. The current league sponsors have the first rights to utilize the sport video game as a form of advertising. If the current league sponsors defer this right, alternate companies may be selected; yet, leagues/sport organizations are treading a thin line and often select a different product category for an in-game advertiser. The licensing also allows brands, which are not league/sport organization sponsors, to get involved in the sport related games. For example, with no NFL license agreement, an alternate game, such as *Blitz: The League* by Midway, may have sponsorships with teams who cannot afford to sponsor the NFL, but may be reaching the same football fan base with the video game sponsorships.

## Measuring Effectiveness of In-Game Advertising and Sponsorships

The growth of sport video game sponsorships is certainly appealing to sport marketers; however, the effectiveness of the sponsorships has rarely been documented. The sponsoring corporations seek to verify some return on investment (ROI; Lefton, 2004b). Sport sponsorship studies have traditionally focused on ROI and examined effectiveness through intermediate measures (Sandage, 1983). Consumers' awareness of advertising or sponsoring brands (cognitive), attitudes or feelings towards sponsoring brands (affective), and behavioral purchase intentions (conative) of the sponsoring brands are often studied when determining advertising and sponsorship effectiveness; whereas, actual behavioral consumption is difficult for researchers to investigate. Similar to traditional forms of advertising and sponsorships, the question often arises of how much is it too much? If sport video games become littered with in-game advertisements or sponsorships, are they still effective? Are they annoying to gamers?

A few studies have been conducted by various researchers recently and have generally found that sport video games enhance consumer awareness of sponsorship products, and are a promising outlet for corporations (e.g., Cianfrone et al., 2008; Yang, Roskos-Ewoldsen, Dinu & Arpan, 2006).

Nelson (2002) conducted an experimental study to assess consumers' awareness of, and attitudes towards, local and national brand advertisements in computer and console video games. Recognition of sponsors was examined via a longitudinal study with 13 console gamers. The console gamers played a car racing game and were surveyed on recall and recognition. Findings of the experiments revealed that advertisements were initially recalled and recognized at a rate of about 25-35%; however, a follow-up survey five months later resulted in high levels of decay. In this study, an experiment was also carried out to examine attitude-toward-advertising, which involved 10 computer gamers of the same race car game. Attitude toward the in-game advertising was found to be favorable in general, although the gamers' playing habits were not taken into consideration in this study and the racing game was not a sport video game. Regardless, this investigation provided a foundation for other sport video game sponsorship studies. Yang et al. (2006) studied in-game advertising's impact on memory in sport video games. They examined explicit (recall and recognition measures) and implicit memory (word fragmentation measures) of the brands in a racing game, EA Sports *Formula 1 2001*, and a soccer game, EA Sports *FIFA 2002*. Prior to controlling for guessing in the recall and recognition measure, response rates were in the 40% range. Further analysis revealed explicit memory of in-game advertising showed no effect, but implicit memory was favorable, suggesting future purchase decisions may be influenced.

Another area of effectiveness that has been examined by academicians and industry researchers is attitudes toward in-game advertising. Nelson, Keum and Yaros (2004) studied consumer attitudes and purchase intentions, and found gamers related positively to in-game advertising when it added realism to the game. Participants who were not favorable towards in-game advertising had negative feelings towards advertising in general. Moreover, some gamers reported consumptive behaviors of purchasing brands that they learned through game play. Hernandez et al. (2004) examined in-game advertising of online advergaming: online games created solely for advertising purposes. These researchers found that when electronic games were overwhelmed by advertisements not congruent with the authentic sport competition events, high intrusiveness negatively influenced gamers' attitudes toward advertising. Conversely, advertising that was consistent with sport games elicited less intrusive feelings and more positive attitudes.

Cianfrone et al. (2008) examined sport video game sponsorship effects on gamers' awareness, affect, and purchase intentions. This study was conducted via an experimental design comparing unaided recall, aided recall, and recognition rates of console sport video gamers playing EA Sports *NCAA Football* games. Gamers who were exposed to the sponsorships showed about 40% recognition rates. The researchers found that sponsorship awareness levels were moderate to high; however, the sponsorships did not seem to create positive attitudes toward sponsoring brands, nor did they increase intentions to purchase sponsoring brands. Also, gamers who were not exposed to the sponsorships still had relatively high levels of recognition rates, indicating they were so accustomed to playing the sport video game with sponsorships that they did not recognize when they were not present.

Practitioners have also been trying to determine ROI and effectiveness. In 2006, Nielsen (2006) developed Nielsen GamePlay Metrics, an electronic video game console and computer gaming tracking system. The system tracks games played and usage level, which is valuable information for gaming stakeholders, including game publishers, developers and advertisers. This system may identify target markets of specific games and gaming tendencies. In July 2007, Nielsen Media Research joined Sony Computer Entertainment America Inc. to create a measurement tool for tracking "reach, frequency, and effectiveness of game network advertising" (Nielsen, 2007). They utilize PS3s and Playstation Network to monitor advertising effectiveness. Industry related research such as Nielsen's provides more information for gaming stakeholders to make educated decisions regarding implementation of in-game advertising.

## Motivation of Gamers

Factors that motivate gamers to play sport video games have received attention from marketers, programmers and researchers. In an effort to better understand how people consume sport video games and related advertising, it is important to understand why they play. This would be beneficial information for stakeholders to better market games, and more accurately develop and implement specific types of in-game advertising. Preliminary studies (Cianfrone, 2007; Kim & Ross, 2006; Kim, Ross & Ko, 2007; Sherry, Lucas, Greenberg & Lachlan, 2006) have shown a number of motives that may influence game play. Sherry et al. (2006) examined general video game play and identified Competition, Challenge, Social Interaction, Diversion, Fantasy, and Arousal as motives for game play. Kim and Ross (2006) assessed sport video game play and developed the Sport Video Game Motivation Scale. This scale specified that Fantasy, Social Interaction, Sport Knowledge Application, Enjoyment, Diversion, and Identification with Sport were factors of motivation. Kim et al. (2007) adopted this scale to study online sport video game play. These studies commonly followed the Uses and Gratifications Theory, which identifies personal needs to consume media sources (Katz, Blumler & Gurevitch, 1974). Cianfrone (2007) re-examined the Sport Video Game Motivation scale following motivation theories and research findings in both sport spectatorship and sport participation areas, in addition to the Uses and Gratification theoretical framework. This was because that sport video game play is both an interactive participation-based activity and a spectator-based activity like a televised sport event. Further studies need to be conducted to confirm previous findings and better understand gamer motivations, and subsequently, the consumption of games and in-game advertising. Another area that needs future research attention is the examination of market demand variables that affect the playing of sport video games. Market demand variables are a set of pull factors associated with the attributes of the core product (Zhang, Lam & Connaughton, 2003). For sport video games, market demand factors may include attractiveness, speed, competition, realism, and autonomy; this is an area that remains to be examined.

# SOLUTIONS AND RECOMMENDATIONS

## Stakeholder Relationships

Sport organizations need to continually monitor their relationships with video game publishers to further capitalize on this revenue stream. EA Sports is the market leader in sport video games and their relationships with sport organizations have created a monopoly in some types of games. The sport organizations will need to assess if their ROI from being a partner with EA is maximizing their profit. Revenue generated from in-game advertising and sponsorship may lead to financial conflicts if sport organizations consider that this revenue source needs to be shared with them. When business consolidation occurs, the market environment may change. It remains to be seen how Microsoft's (Xbox series) purchase of in-game advertising firm Massive affects the landscape of sport video games. As publicized in the media recently, many new EA Sports games on Xbox 360 or computer platforms will have in-game advertisements. Sony (PlayStation series) has not been as aggressive with in-game advertising; nonetheless, it appears that this company will likely follow the trend based on its relationship with Nielsen Ratings.

## Measuring Effectiveness

Nelson (2002) asserted that gamers were open to in-game promotions provided they fit the game well. Gaming publishers and advertisers should feel sure that sport games are a good fit for in-game advertising and sponsorships in terms of adding realism and authenticity. However, they will have to assess how much in-game advertising is appropriate and identify at what point advertising becomes too cluttered and ineffective. A major issue with sport video game in-game advertising and sponsorships is that growth potential is directly related to the effectiveness of video games as a marketing medium. In an effort to illicit more sponsorships and in-game advertisements, gaming corporations and publishers are searching to show a ROI to potential advertisers. Both academicians and practitioners have begun research on this topic; however, as in-game advertising technology continues to advance, timely examination of the effectiveness of in-game advertising and sponsorship is very much needed. Cianfrone et al. (2008) recommended exploring contributing variables that led to the growth of sport video game sponsorships. While one type of sport video game awareness has been shown to be favorable, its ability to predict or influence brand attitude remains to be determined.

Many factors can contribute to a consumer's response to a sport sponsorship, and consumption level of sport is usually one of the most influential factors (e.g., Levin, Joiner & Cameron, 2001; Madrigal, 2000; Meenaghan, 2001b). A sport fan usually consumes sport through many avenues including, but not limited to, event attendance, television, the Internet, and print media. Consumption of sports and the associated exposure to sport sponsors is a positive relationship that benefits the sponsoring corporations. For example, Cianfrone and Zhang (2006) found that various sport consumption forms were significantly correlated when studying the effects of advertising and sponsorship during a televised action-sport event. Due to the inherent relationship of sport consumption level and sponsorship effectiveness, it is reasonable to assume that an individual's sport video game consumption level is likely to influence his or her responses to sport video game sponsorships. The recurring nature of sport video game play forces a gamer to view the sport video game sponsorship repetitively. This interactive nature of gaming may lend itself to a different type of consumption than traditional sport spectatorship. Future studies should take this into consideration.

## FUTURE TRENDS

As technology advances, in-game advertising will be an area of continued focus for both practitioners and researchers. If gamers are shown to react positively to in-game advertising, and this effectiveness can be measured to show ROI for the sponsoring/advertising companies, the surge of in-game advertising will likely continue. The inherent relationship between sports and advertising allows in-game advertising and sponsorship signage to be more organic than other types of advertising. As gamers become more accustomed to in-game advertising, the question of whether or not it will sustain effectiveness in creating awareness, attitude changes, or purchase intentions remains to be investigated. The types of in-game advertising that are most influential in creating an attitudinal or behavior change are yet to be determined. The ever-changing technological advances in sport video games and the relationships of game publishers with the sport organizations will need to be followed closely. More research on sport video game advertising and sponsorship effectiveness will shape future growth. Future research on motivations for playing sport video games will also benefit stakeholders in the creation of game features and the marketing of products.

The technological advantage of online gaming remains an impacting factor on the growth of sport video games as a promotional tool. Online gaming, such as Xbox Live capabilities, allow complete customization of in-game advertising. The percentage of Xbox and PS3 gamers who play online is very high, suggesting this tracking method of game play can be used to an advertiser's advantage. Online game play allows for updated advertisements, thus, there is not as much repetition for a gamer. Customized advertising for a gamer makes it convenient to reach market segments with certain demographic or psychographic profiles. All in all, technological advances will play a major role in this invasive advertising method.

## CONCLUSION

In summary, sport video game in-game advertising and sponsorship is a new marketing area that is becoming increasingly popular. The sport video game industry has many stakeholders, including gaming corporations (e.g., Sony, Microsoft), publishers (e.g., EA Sports, 2K Sports), sport organizations (e.g., NCAA, NBA), in-game advertising firms (e.g., IGA Worldwide, Double Fusion), and outside sponsoring/advertising corporations, which rely on this advertising source. Practitioners and researchers have explored those variables that may affect the consumption of sport video games and subsequently the effectiveness of in-game advertising and sponsorships. These preliminary examinations have revealed very promising indications. Nonetheless, more systematic investigations are necessary for this segment of the sport industry to remain successful.

## REFERENCES

About us- The sports industry. (2008). *Street & Smith's SportsBusiness Journal.* Retrieved March 1, 2008, from http://www.sportsbusinessjournal.com/index.cfm?fuseaction=page.feature&featureId=43

Adams, R. (2005, January 24). ESPN gets $850 million to stay in the game. *Street & Smith's SportsBusiness Journal,* p. 5.

Adams, R. (2004, August 30). In-game advertising a growing, but not easy, sell. *Street & Smith's SportsBusiness Journal,* p. 19.

Browne, K. F. (2006, June). *Toward a digital advertising ecosystem for games.* Paper presented at the Game Developers Conference (GDC) Focus on: Game Advertising Summit, San Francisco, CA.

Business Wire. (2006, June 13). *Parks Associates: PC in-game advertising revenue to top $400 million by 2009; Male gamers 18 - 34 more open to ads.* Retrieved via Lexis Nexus Database and www.businesswire.com

Cianfrone, B.A. (2007). *The influence of motives and consumption of sport video games on sponsorship effectiveness.* Doctoral dissertation, University of Florida, Florida.

Cianfrone, B.A., & Zhang, J.J. (2006). Differential effects of television commercials, athlete endorsements, and venue signage during a televised action sports event. *Journal of Sport Management, 20*(3), 322-344.

Cianfrone, B.A., Zhang, J.J., Trail, G.T., & Lutz, R.J. (2008). Effectiveness of in-game advertisements in sport video games: An experimental inquiry on current gamers. *International Journal of Sport Communication, 1,* 195-218.

Crompton, J.L. (2004). Conceptualization and alternate operationalizations of the measurement of sponsorship effectiveness in sport. *Leisure Studies, 23*(3), 267-281.

Della Maggiora, E. (2006). *Nielsen Interactive Entertainment.* The Research perspective Paper presented at the Game Developers Conference (GDC) Focus on: Game Advertising Summit, San Francisco, CA.

Entertainment Software Association (ESA). (2007). *Essential facts about the computer and video gaming industry-Sales, demographics and usage data.* Retrieved December 8, 2007, from www.theesa.com

Fernando, A. (2004). Creating buzz: New media tactics have changed the PR and advertising game. *Communication World, 21*(6), 10-12.

Fisher, E. (2006, May 15). ESPN hits 'restart' in gaming. *Street & Smith's SportsBusiness Journal, 31.*

Forster, W. (2005). *The Encyclopedia of Game Machines - Consoles, handheld & home computers 1972-2005. Vancouver: GAMEplan.*

First-week sales of EA Sports *Madden NFL 07* hit $100M. (2006, September 1). *Sports Business Daily, XII*(235). Retrieved September 1, 2006, from http://www.sportsbusinessdaily.com

Hernandez, M. D., Chapa, S., Minor, M. S., Maldonado, C., & Barranzuela, F. (2004). Hispanic attitudes toward advergames: A proposed model of their antecedents. *Journal of Interactive Advertising, 5*(1).

Howard, T. (2006, July 11). As more people play, advertisers devise game plan. *USA Today,* B3.

Katz, E., Blumler, J.G., & Gurevitch, M. (1974). Uses and gratifications research. *The Public Opinion Quarterly, 37*(4), 509-523.

Kim, Y., & Ross, S. (2006). An exploration of motives in sport video gaming. *International Journal of Sport Marketing and Sponsorship, 8*(1), 34-46.

Kim, Y., Ross, S., & Ko, Y. (2007). Online sport video game motivations. *International Journal of Human Movement Science, 1*(1), 41-60.

King, B. (2006, April 17). Reaching the 18-34 demo. *Street & Smith's SportsBusiness Journal,* p. 19-26.

Lefton, T. (20 December, 2004a). Exclusive EA deal jolts category. *Street & Smith's SportsBusiness Journal, 10.*

Lefton, T. (22 November, 2004b). Video game marketers study how to take in-game ads to the next level. *Street & Smith's SportsBusiness Journal, 12.*

Levin, A.M., Joiner, C., & Cameron, G. (2001). The impact of sports sponsorship on consumers' brand attitudes and recall: The case of NASCAR fans. *Journal of Current Issues and Research in Advertising, 23*(2), 23-31.

Madrigal, R. (2000). The influence of social alliances with sports teams on intentions to purchase corporate sponsors' products. *Journal of Advertising, XXIX*(4), 13-24.

McClellan, S. (2005, December 5). Study: In-game integration reaps consumer recall. *Adweek, 46*(47), 9.

Meenaghan, T. (2001a). Sponsorship and advertising: A comparison of consumer perceptions. *Psychology & Marketing, 18*(2), 191-215.

Meenaghan, T. (2001b). Understanding sponsorship effects. *Psychology & Marketing, 18*(2), 95-122.

Microsoft. (2007, July 25). Premier EA titles to go live in Massive Network. Microsoft.com. Retrieved March 1, 2008, from www.microsoft.com/presspass/press/2007/july07/07-25EAMassivePR.mspx

Mullin, B., Hardy, S., & Sutton, W. (2007). *Sport marketing* (3rd ed). Champaign, IL: Human Kinetics.

Nelson, M.R. (2002). Recall of brand placements in computer/video games. *Journal of Advertising Research, 42*(2), 80-93.

Nelson, M.R., Keum, H., & Yaros, R.A. (2004). Advertainment or adcreep? Game players' attitudes toward advertising and product placements in computer games. *Journal of Interactive Advertising, 5*(1). Retrieved November 10, 2006, from http://www.jiad.org/vol5/no1/nelson/

Nielsen. (2006, October 18). Nielsen to provide video game rating service. *Nielsen.com.* Retrieved May 11, 2008, from http://www.nielsenmedia.com/nc/portal/site/Public/menuitem.55dc65b4a7d5adff3f65936147a062a0/?vgnextoid=aea42f5dde75e010VgnVCM100000ac0a260aRCRD

Nielsen. (2007, July 2). Nielsen and Sony Computer Entertainment America to develop measurement system for game network advertising. *Nielsen.com.* Retrieved May 11, 2008 from http://www.nielsen.com/media/pr_070702.html

Radia, S., & Harris, T. (2006, June). *Reaching the 18-34 demographic through games.* Play, Denuo Group. Paper presented at the Game Developers Conference (GDC) Focus on: Game Advertising Summit, San Francisco, CA.

Sandage, C. (1983). *Advertising theory and practice.* Homewood, IL: Richard D. Irwin.

Shields, M. (2006, April 12). In-game ads could reach $2 billion. *Mediaweek; Adweek.* Retrieved August 2007, from http://www.adweek.com/aw/national/article_display.jsp?vnu_content_id=1002343563

Shields, M. (2007, August 20). Less than dynamic. *Brandweek, 30*(48), 36-37.

Take-Two (2005, January 31). *Take-Two awarded long-term, third-party exclusives with Major League Baseball Properties, Major League Baseball Players Association and Major League Baseball Advanced Media to publish interactive MLB video game.* Retrieved April 30, 2008, from http://ir.take2games.com/pring_release.cfm?releaseid=154141

Thorsen, T. (2008, February 12). *EA Sports extends NFL deal through 2012 season.* Retrieved April 28, 2008 from http://www.gamespot.com/news/6185880.html?sid=6185880&part=rss&subj=6185880

Top Ten Sports Video Game Titles. (2007, January 15). *Street & Smith's SportsBusiness Journal,* p. 18.

Yang, M., Roskos-Ewoldsen, D.R., Dinu, L., & Arpan, L.M. (2006). The effectiveness of "in-game" advertising. *Journal of Advertising, 35*(4), 143-152.

Zhang, J.J., Lam, E.T.C., & Connaughton, D.P. (2003). General market demand variables associated with professional sport consumption. *International Journal of Sport Marketing & Sponsorship, 5*(1), 33-55.

# Chapter XVIII
# In-Game Advertising: Effectiveness and Consumer Attitudes

**Mark Lee**
*RMIT University, Australia*

**Rajendra Mulye**
*RMIT University, Australia*

**Constantino Stavros**
*RMIT University, Australia*

## ABSTRACT

*This chapter reports a recent research study involving a sports video game which sought to provide an overview on the use of in-game advertising, consumer attitudes towards the practice, and an empirical test to assess its effectiveness in terms of brand recall and recognition. Intervening variables such as attitude towards advertising in general and in-game advertising in particular, brand familiarity, and experience with gaming was also considered. A sample of 32 participants was asked to engage in video game play of a relatively new sports game and complete a series of measures examining attitudes, recall and recognition of in-game advertising. Findings supported all hypotheses with the exception of the hypothesis predicting a positive relationship between attitude toward in-game advertising and advertising effectiveness in terms of recall and recognition. User factors such as age, game experience, game likeability, and item specific factors such as characteristics of the display panel and relevance of the product to the game were found to play an important role in improving advertising effectiveness. Contrary to earlier studies, attitude towards in-game advertising was lower than expected, especially amongst the experienced game players.*

## INTRODUCTION

Decreasing use of traditional media by consumers and increasing advertising noise requires contemporary marketers to constantly seek new communication channels in order to achieve their promotional objectives. While marketers are continually experimenting with a variety of mediums, video games are emerging as a potential breakthrough for advertising placement because of their exponential growth and distinctive advantage over traditional mediums (Alpert, 2007, Gwinn, 2004; Hyman, 2006; Molesworth, 2006; Nelson, 2002; Nelson, Keum, & Yaros, 2004; Nelson, Yaros, & Keum, 2006; Schneider & Cornwell, 2005; Yang, Roskos-Ewoldsen, Dinu, & Arpan, 2006)

Worldwide sales of video games are expected to surpass the sales of the recorded music industry (Alpert, 2007; Young, 2004) with some experts forecasting sales to reach $48.9 billion in 2011 (Szalai, 2007). Individual game companies are already grossing annual sums in excess of the entire movie industry box office for the same period. The reach of video games has also experienced exponential growth. Surveys in the United States estimate that 68% of men aged 18-34 and 80% of men aged 12-17 have a video games console in their home (Cuneo, 2004; Massive-Incorporated, 2007). These surveys report that the young male segment spends on average two hours per day playing video games, with American males reporting they played video games more frequently than they watched television. The increasing trend of video games as a recreational pastime is likely to come at the expense of television viewing, a medium traditionally used by advertisers to reach target audiences.

The revenue from the placement of brands and products within video games is expected to reach $850 million in 2011, and is predicted to eventually mature in to a $5 billion industry (Wolf, 2007). Game developers have also been quick to utilise the additional revenue stream that in-game advertising brings to reduce the burden of increasing development costs (Gaudiosi, 2007; Gwinn, 2004; Hyman, 2006). Gaudiosi (2007) notes opportunities for convergence were created with the rapid proliferation of internet enabled game consoles, which were selling at a rate of 500,000 units per month in 2007. These opportunities have not gone unnoticed. Major industry brands such as Microsoft and Google have entered the industry through acquisition of large scale in-game advertising brokerage agents Massive Inc and Adscape Media respectively (Gaudiosi, 2007). Equally impressive are the reported marketing successes of brands that have made use of this medium (Lindstrom, 2001). For example, Red Bull, a high energy drink, was able to gain invaluable exposure to its target market via the hugely popular futuristic racing game Wipeout, while the Mitsubishi Lancer Evo found a market in gamers after it was introduced to the United States following a petition to the car's manufacturer by players of the Grand Turismo driving game (Nelson, 2004).

Despite the high growth and demonstrated potential of the in-game advertising medium, comparatively little empirical research has been conducted in this area to evaluate the effectiveness of this approach. This may be due to, in part, the sharp growth the practice has experienced, leaving researchers little time to examine the phenomenon. As the following literature review indicates, only a handful of studies have examined game players' attitudes towards in-game advertising or its effectiveness. Many of these studies have used a qualitative methodology and did not systematically examine the effect of player attitudes toward in-game advertising and its effect on brand recall and recognition or have controlled for the complex effects of extraneous factors such as user familiarity/usage level of video games and the manner in which advertisements are placed within the game. Yang et al (2006, p 150) suggest that future research should consider "how frequent players of video game experience the game playing" ... and "test the effects of different types of in-game placements on memory." The purpose of this research was to contribute to answering this call.

The first area of confirmatory investigation concerns the effectiveness of in-game advertising on brand awareness, recall and recognition. The second area of examination is consumer attitudes towards in-game advertising. Further this study sought to examine the relationships, if any, between the attitudes of gamers towards in-game advertising and the effectiveness of those advertisements.

## LITERATURE REVIEW AND BACKGROUND

Research into in-game advertising has been limited (Alpert, 2007; Molesworth, 2006; Nelson, 2002; Nelson et al., 2004; Nelson et al., 2006; Yang et al., 2006). Alpert (2007, p 87) described the topic as "virtually vacant" of written literature. In spite of this, research conducted has been successful in identifying specific outcomes. These outcomes have focused primarily on the effectiveness of in-game advertisements in increasing brand awareness and, perhaps as importantly, the attitudes of consumers towards the practice.

Due to the brief history of in-game advertising, the majority of related theory is based on the practice of product and brand placement within film and television. Product and brand placement are commonplace in the film and television mediums, dating back at least as far as 1940 (Gupta & Lord, 1998). The popularity of product placement is in part due to its acceptance by audiences and also for the opportunities it offers marketers to expose their products in their natural environment to a captive audience (DeLorme & Reid, 1999; Shrum, 2004). Films, television and games share a number of important similarities in the brand and product placement opportunities they offer, however in contrast to other mediums, video games add the element of interactivity. Marketers are beginning to realise the importance of this interactivity and budget share that was once destined for other communication mediums is now being diverted towards in-game advertising.

## ATTITUDES TOWARDS IN-GAME ADVERTISING

In general, previous research has noted that audiences seem to have more positive attitudes towards brand and product placements than towards traditional advertisements (Gupta & Lord, 1998; Nebenzahl & Secunda, 1993; Sung & De Gregorio, 2005). Despite this, little is known about the short or long-term effectiveness of this brand contact technique employed by advertisers or the attitudes toward such product placements among game players. One of the first studies to investigate this issue was conducted by Nelson (2002). Using a qualitative methodology, she identified a number of key themes and opinions of game players in reference to their attitudes towards in-game advertisements. While the opinions offered by gamers in the study ranged from the very positive to the very negative, overall attitudes were evenly represented and ranged from amusement and appreciation to scepticism and blame. Negative sentiment was often related to growing trends in perceived commercialism or overuse of specific placements, whereas positive comments centred on the realism created by the use of real brands.

These findings were replicated in a study by Molesworth (2006), who also examined the attitudes of gamers towards in-game advertising, but used a cultural perspective framework to focus on non-material aspects of consumer culture and in-game internal processes. He found varied responses by gamers to in-game advertising, however most individuals were positive towards brand placement due to its contribution to realism. Consumers encouraged in-game advertising as they perceived that the added revenue from the commercial activity would eventually result in cheaper retail pricing for games. In contrast to this however, it was also found that some felt uncomfortable with the concept that they had "paid for an advertisement". In-game advertising was seen by some as sneaky and subliminal and that this had cultivated a feeling that the in-game advertising could be viewed as an "invasion of privacy". Another identified theme noted by Molesworth (2006) was the consumer concerns regarding the potential of product placements to influence the artistic direction of game developers.

Both Molesworth (2006) and Nelson (2002) noted that while early studies had investigated if consumers were favourable or unfavourable towards product placement (in movies and television programs), it was not until recently that consideration had been given to consumers underlying beliefs towards advertising in general. It is likely that the attitude towards in-game advertising expressed in these two studies may be a reflection of the respondent's general attitude toward advertising i.e., advertising in television, print and radio mediums. The qualitative nature of the two studies make it difficult to infer whether this is indeed the case, however research on product placement in movies and television have found a correlation between the two (Gupta & Lord, 1998; Nebenzahl & Secunda, 1993; Sung & De Gregorio, 2005).

**H1a:** Attitude toward in-game advertisements will be positively correlated to attitude toward general advertisements.

Another observation that can be made from these studies is the slight increase in negative sentiments towards in-game advertising in recent years. In-game advertising may have been a novelty in Nelson's 2002 study and players may have welcomed it for its contribution to game realism. This novelty factor may have either worn out by Molesworth's 2006 study, or advertisers may have over stretched their mark within the game environment causing advertisement wearout and irritation amongst game players. It is therefore likely that heavy game users would be more likely to be negative towards in-game advertising than novice players.

**H1b:** Attitude towards in-game adverting will be dependent on a player's usage level or experience in playing video games.

## EFFECTIVENESS OF IN-GAME ADVERTISING

Cumulative evidence on product placement indicates a general positive attitude towards the practice. However more research is required to assess whether this positive attitude will translate into advertising effectiveness measures in the context of a video game in order for marketers to view it as a preferential medium. Previous research has defined and measured effectiveness of in-game advertising through brand awareness (Nelson, 2002; Nelson et al., 2006; Schneider & Cornwell, 2005; Yang et al., 2006). This measure consists of (unaided) brand recall and (aided) brand recognition, both of which require the target customer to associate the brands identity with the appropriate category need.

The first study to examine effectiveness of in-game advertisements was by Nelson (2002). In her study, brand awareness was examined by assessing free-recall measures directly after game-play and at a five month delay. Nelson (2002) found that even upon playing a game for the first time, and for a limited amount of time, game players were able to recall about 25 to 30 percent of brands placed in the games in the short term and about 10 to 15 percent at a delay. The author does acknowledge the various limitations of her preliminary investigation of the topic and that the high recall may be due to the task condition where a participant's brand exposure was a major part of the game play, for example, car selection in a racing game. Nelson et al. (2006) addressed some of these limitations in their most current study, and introduced the role of telepresence – a sense of being present or transported inside a virtual game environment; and the social context, that is the effect of actively playing the game or watching play. The study found that individuals playing video games recalled significantly fewer brand placements than watchers did, however the role had no effect on perceived brand attitudes. Telepresence and game liking were positively related to perceived persuasion of the brands advertised, but telepresence also mediated the influence of game liking on perceived persuasion of the brands.

While the work by Nelson and her colleagues is noteworthy, the brand recall measures used in both studies were based on the traditional recall measures of explicit memory as apposed to implicit memory. Implicit memory involves memory effects that occur at the subconscious level without intentional recollection of the event or advertisement and includes conditioned emotional reactions, motor skills and habits. Yang et al (2006) used a word fragment test to assess the role of implicit memory in processing in-game advertising. Contrary to earlier research on the topic, the study found higher levels of recall for implicit memory (word-fragment test) for brands, and low levels of recall for explicit memory (recognition test). The anomalous result of a lack of clear effect of brand placement on explicit memory was likely due, in part, to complex factors that the study had failed to account for and which may have influenced memory. One such factor could be the player's attitude towards in-game advertising which formed the basis of hypothesis 1. It could be argued that a player with a positive attitude towards in-game advertising is less likely to block in-game advertising from their perceptual field during game play and are therefore more likely to recall or recognise the advertisements than someone who had negative attitude towards in-game advertising. On basis of this reasoning the following hypothesis can be offered:

**H2a:** Recall of in-game advertising will be positively related to attitude toward in-game advertising.

**H2b:** Recognition of in-game advertising will be positively related to attitude toward in-game advertising.

## GAME EXPERIENCE

Schneider and Cornwell (2005) examined the effect of a player's experience level on their ability to recall and recognise in-game advertisements. Experienced players, defined as those playing five hours or more in an average week, recalled and recognised significantly more advertisements from game play than novice participants. There is some support for this hypothesis in the developmental psychology literature. In a classic study Greenfield et al (1996) examined the effects of game expertise on divided visual attention. Divided attention was measured through a luminance detection test in which response time to targets of varying probabilities (from 10% to 80%) that appeared at two locations on a computer screen was measured. Although there was no difference between the response times (attentional cost) of experienced and novice respondents in the 80% condition, experienced respondents outperformed the novice respondents in the more difficult 10% condition. To test the external validity of this result, a follow up study asked respondents to play five hours of a video game and then undertake the luminance test. The game play resulted in a significant decrease in response time at the 10% location, which was the focus of difference in the laboratory experiment. Schneider and Cornwell (2005) use the analogy of a camera lens to explain this phenomenon. As one brings an object in to focus, the resolution of the surrounding environment blurs. However, they propose that experts have a greater depth of field than novices, i.e., they do not lose focus on peripheral objects as much as novices do and therefore have a greater awareness of outlying stimuli. Schneider and Cornwell (2005) found support for this proposition in the context of a car racing game. It would be beneficial to verify whether this would hold true in case of a different game genre, such as a sports game, where the level of interaction with in-game advertising and representation of the advertising stimuli are very different.

**H3a:** Experienced players will report higher recall of in-game advertising than novice players.

**H3b:** Experienced players will report higher recognition of in-game advertising than novice players.

## INTERACTION, FAMILIARITY AND APPEARANCE

Interaction, familiarity and appearance have emerged through in-game advertising research as potential determinants of game players remembering in-game advertisements. Researchers have noted that the process of interaction has an influence on a game player's experience and thus has the potential to influence associated memory (Nelson et al., 2006; Schneider & Cornwell, 2005). Video games provide the potential for advertisers to deliver interactive communication in a specific environment; an opportunity that is absent from other popular communication mediums. Video games offer a simple way of achieving dynamics, such as a product being shown in use for a lengthy period of time and in a positive reference, that have been established as necessary for success of product placements in other mediums, such as film (Nelson et al., 2006; Schneider & Cornwell, 2005).

Previous research identified that an individual's positive opinions of a particular brand may influence their experience with that brand during an experiment (Machleit, Allen, & Madden, 1993; Yang et al., 2006). Thus it is necessary to investigate the varying levels of brand familiarity and the subsequent affect this may have on memory. Schneider and Cornwell (2005) noted that familiar brands were better recalled than unknown ones. However, Nelson et al (2006) pointed out that a true test of the effect of familiarity could only be made with the use of a fictitious brand against which a known brand could be compared. In their experiment they embedded fictitious brands in conjunction with well recognised brands within the game environment and found that market leading brands achieve superior brand awareness and that when utilised with fictitious or lesser known brands in a video game environment, familiar brands are recalled significantly more often. Nelson et al. (2006) did not dismiss the potential for new brands however, noting that they may also produce less 'wear out' effects over prolonged periods of game play, a genuine scenario for real world game players. In addition to familiarity and interaction, Schneider and Cornwell (2005) identified an advertisement's visual prominence as a key factor effecting brand memory. Further Schneider and Cornwell (2005) found that participants were able to remember a brand that they were unfamiliar with due to cultural factors, suggesting that visual prominence may be a more significant factor than familiarity.

**H4:** Participants will recall in-game advertisements and related details based on familiarity and interaction with the brand, and its visual characteristics represented on the advertising panel.

## METHOD

A convenience sample of 32 respondents participated in the research. The study was described as 'evaluation of a video game' and the incentive to participate was the opportunity of playing a latest release video game on a state of the art video game console. Participants consisted of 24 males and 8 females ranging in age from 22 to 36 years (M= 28.47, SD= 3.95). As part of the pre-screening process, respondents were asked about the number of hours they spent in an average week playing video games and whether they had played the experimental game. The purpose was to include an equal proportion of experienced and inexperienced players in the sample and to rule out those who had already played the experimental game.

Nine of the 32 participants engaged in over 10 hours of game play per week with the highest report of average time played as 20 hours per week. Seven participants reported playing either a minimum of 3 hours per week or a maximum of 7 hours, with a number of female players indicating at least half an hour of game play per week. Of the 32 participants a total of 12 did not play any video games in an average week. Using Schneider and Cornwell's (2005) convention, players who played video games for over 5 hours per week were classified as experienced players.

The video game selected for the experiment was Electronic Arts' EURO 2008 on the Playstation 3 console. The Playstation 3 is the most technologically advanced console video game system on the market and is sold throughout the world. The EURO 2008 game's relatively recent release meant that while individuals would be more interested in participating, they would be less likely to have played the game and as a result would not be familiar with the panel advertisements in the specific environment. A football game was selected due to the high level of standardization as well as the controllable amount of static advertising. A participant played the game for twelve minutes as this was thought to represent a typical playing experience for most users.

Prior to the commencement of the game play, participants were provided with a basic and clearly represented outline of the controls for the game and were encouraged to familiarise themselves with these before they began play. To control for the effects of extraneous factors, exposure was controlled through the pre-selection and standardization of game options. For example, in all games, the Italian and French teams were selected for consistency and control level balance, the games skill level was set to "amateur" ensuring an undifferentiated game for all participants and the environment that contained the panels was controlled by preselecting the stadium to "Stade De Geneve". This stadium was chosen to be the sole venue for the experiment as the sideline panels used to represent the brands and logos, situated around the field's boundary, were clearly visible, consistent and proportionately represented. Three individuals familiar with research design of this type analysed the stadium, resulting in the exclusion of three panels that were obscured from view. Immediately following the conclusion of game play participants were administered a three part questionnaire.

The first part of the questionnaire consisted of unaided recall measures based on the Memory Characteristics Questionnaire (MCQ) developed by Johnson, Foley and Suengas (1988). The format asks participants to describe any advertisement that appeared during the experiment and then encourages elaboration by asking where the advertisement appeared physically and why they remembered the advertisement. Participants are also asked to indicate, on a five point Likert scale, the confidence in the recollection in order to further establish memory levels associated with exposure to the advertisements with 1 indicating "very doubtful" and 5 indicating "very sure". This provided a possible score range of 5-25, with lower scores indicating doubt in recall and higher scores indicating strength of the recollection. The MCQ also utilised a number of qualitative questions in relation to the location, description and reason for recall in order to further understand participants reasoning of recall and recognition.

The second part of the questionnaire tested unaided recall and aided recognition based on the Advertising Panel Presentation (APP) procedure recommended by Schneider and Cornwell (2005). In this procedure participants are presented with graphically identical representations of in-game advertisements seen on the field boundary. In addition to the ten panels that actually appeared in-game, two 'foil' panels were presented. Participants were asked if they recalled the advertisement and were then asked to indicate on a five point Likert scale how confident they were of their recollection. If participants did not recall the panel, the score for that panel was zero. If they recalled the panel their confidence of recollection was the corresponding score, for example if the participant recalled the panel and indicated their recollection was very doubtful they were given a score of 1. Ten panels represented a possible score range of 0 to 50.

The final section of the questionnaire measured attitude towards general advertising and in-game advertising. This measure was originally based on a series of questions developed by Gupta and Gould (1997) for determining attitudes towards product placements within films and further modified by Nelson et al. (2004) for appropriateness to video games. Participants were asked to respond to a series of six questions relating to their attitudes towards general advertising, followed by six questions related to in-game advertisements. Reliability for this measure has been reported in earlier studies in the 0.79 - 0.87 alpha range (Nelson, 2004), and was found to be 0.81 in the current study.

Participants were also asked a number of demographic related questions, including age and gender, as well as their level of enjoyment and previous experience with the game.

## RESULTS

Data were analysed using bivariate and correlational techniques as they provided greater parsimony, and are not as sensitive to problems relating to small sample sizes. The results for the hypotheses proposed in this study can be found in Table 1.

Contrary to earlier studies, attitudes towards in-game advertising were lower compared to that of advertising in general and this difference is statistically significant on a paired sample t-test. (t=4.727, p=0.00). However when the sample is split according to the experience levels of the players and an independent sample t-test is carried out to test for differences in attitudes towards the two types of advertising, statistical

*Table 1. Hypotheses and experimental results*

| Hypotheses | t | df | Pearson r | Significance 2 -tailed |
|---|---|---|---|---|
| H1a: Attitude toward in-game advertisements will be positively correlated with attitude toward general advertisements. | | | 0.55 | 0.001 |
| H1b: Attitude towards in-game advertisng will depend on a player's level of experience in playing video game. | | | 0.45 | 0.011 |
| H2a: Recall of in-game advertising will be positively related to attitude toward in-game advertising. | | | 0.18 | 0.306 (ns) |
| H2b: Recognition of in-game advertising will be positively related to attitude toward in-game advertising | | | -.252 | 0.164 (ns) |
| H3a: Experienced players will report higher recall of in-game advertising than novice players. | 3.534 | 30 | | 0.001 |
| H3b: Experienced players will report higher recognition of in-game advertising than novice players | 3.103 | 30 | | 0.004 |
| H4: Participants will recall in-game advertisements and related details based on familiarity and interaction with the brand, and its visual characteristics represented on the advertising panel. | Tested through qualitative comments | | | |

significant difference was found only for in-game advertising. In other words, although the experts and novice do not differ in their attitude towards general advertising, the novice seem to have a more positive attitude towards in-game advertising than the experts.

As the use of means masks individual differences, a measure based on Pearson's r was used to test hypothesis 1 about the association between attitude toward general advertising and in-game advertising. The first hypothesis seeks to explore if attitude towards in-game advertising is a function of attitude towards general advertising and the experience level of the game player. A Person's correlation of 0.55 and 0.45 indicated that there is a moderate but statistically significant correlation in support of hypotheses H1a and H1b. Likewise, to test hypothesis H2a and H2b a Pearson's Correlation was conducted to examine possible relationships between attitudes toward in-game advertising and its influence on in-game advertising effectiveness using measures of recall and recognition. Results indicated no significant relationship between these two variables as can be seen in Table 1.

Hypothesis three posited that the effectiveness of in-game advertising, as measured by a brand recall and recognition test, is related to a game player's skill level as measured by hours of game played per week. Experienced players were expected to show a higher level of recall and recognition than novice players. An independent samples t-test provided support for the hypotheses, suggesting experienced players were able to recall and recognise significantly more advertisements than novice players. This was further supported by a modest, but statistically significant correlation, found between hours of game play per week and recall ($r=0.45$, $p=0.01$) and recognition ($r=0.35$, $p=0.04$) scores. Other factors that were found to significantly affect advertising effectiveness (recall and recognition) were game player's age ($r=-0.56$ and $-0.53$ resp) and the game's likeability ($r=0.48$ and $0.50$ respectively). A respondent's gender had no effect on any of the other variables examined in the study.

In addition to quantitative measures, the questionnaire also asked participants a number of qualitative questions relating to their unaided recall of in-game advertisements. Three prominent themes emerged from the qualitative questions. Firstly, familiarity was consistently listed as a key reason for recalling a panel. Secondly, participants responded that a panel's colours or graphical representation aided recall and finally, that scenes in which the participant was not actively involved enhanced the prominence and subsequent recall of panels. These themes that emerged from the qualitative research are discussed in more detail to follow.

Firstly, familiarity was a key concept in recall. Of the ten featured brands, Coca-Cola and McDonald's were the most commonly reported advertisements recalled unaided. This was repeatedly indicated to be due to the participant's familiarity with these brands. For example, one participant reported that the McDonald's panel stood out due to her familiarity with the McDonald's brand. *"It [McDonald's advertisement] caught my eye because it was familiar" (F, 28).* While some participants listed the entire Coca-Cola Zero as the advertisement name a number only listed the parent Coca-Cola brand neglecting the product level modifier 'Zero' indicating a familiarity with the Coca-Cola brand but not necessarily with associated products. Adidas was also mentioned by a number of participants, noting their familiarity with its specific distinctive logo and its team marketing associations. *"I know it is associated with the French team" (M, 29); "Adidas – Three stripes" (M, 23).*

Secondly, participants responded that a panel's colours or graphical representation allowed it to appear prominent enough to recall unaided. In particular participants reported that the MasterCard and Canon panels stood out due to their colour and bold text. Numerous participants recalled specific colours that panels had used to represent brands. *"Canon logo – red text" (M, 23)*. While a number of participants reported the bold, colourful and bright panels as the memorable details that enabled them to freely recall a specific brand. *"Eye catching bright and bold" (F, 28); "Big brands, colourful signage" (F, 22)*.

Thirdly, cut scenes emerged as a theme for participants recalling a panel. A cut scene occurs when, for a short period, the individual playing the video game experiences low, or completely absent control. Cut scenes in the experiment included times when a participant would pass the ball out of bounds or when they scored a goal and took the form of close ups of animated figures and slow motion replays. A number of participants identified brands from cut scenes, reporting that they specifically were able to recall a panel due to its inclusion in a distinctive cut scene such as the slow motion replays seen after a player attempted to, or actually scored a goal. *"Saw [the panel] during replay" (M,35). "Inbounding the ball and after scoring"(M,29)*.

In addition to these three key themes, it was also noted that the Castrol brand panel was the only brand not identified by any participant. This may be due to the lack of congruence between the brand (motor lubricant) and the games specific sport genre (football). It was also noted that every participant identified the location of the panels as being the playing field perimeter, however only one participant mentioned interaction with an in-game advertisement as the reason for recall. A number of novice game players reported that they failed to recall any banner because they were more concerned with the controls and winning the game.

## DISCUSSION

The first hypothesis confirms that attitude toward in-game advertising is a function of attitude toward advertising in general. Although the average attitude scores on a 5 point likert scale for both measures was above average, attitude toward in-game advertising ($\mu$=2.9) was found to be significantly lower than attitude toward general advertising ($\mu$=3.5). This result contradicts findings of earlier studies on product placement in movies, and particularly that of Nelson's (2002) pioneering study on attitude towards in-game advertising. However, this result may not seem unusual if the qualitative findings of Molesworth's (2006) study are taken into account. Molesworth (2006) reported a range of responses to in-game advertising from participants, noting that some felt resentful at having to endure advertising despite paying the full price of the game. Others saw in-game advertising as devious and felt that the potentially subliminal nature of in-game advertising was an invasion of their privacy. Nelson in a later study (Nelson et al., 2006) also reported some negative sentiments expressed by game players in regard to the incessant inclusion or inappropriately utilisation of in-game advertising. This was particularly evident when advertisements were overt, saturated, or did not fit the game context. In a minority of cases the negative reaction towards in-game advertising was reported to have lead to measures as extreme as boycotts of advertised brands and even the video games themselves.

Another possible explanation for this result may be the time difference between Nelson's first study and the more recent studies, including the one reported in this chapter. Video game advertising was less prominent prior to 2002 and hence may have been more favourably viewed as a novelty or exception. However, with the rapid proliferation of enhanced game consoles in homes and the trend towards prolonged play habits, game players may have been overexposed to recurring advertisements, leading to message wear out and frustration.

The aforementioned result not withstanding, marketers should not be overly concerned with the use of this medium as the study did not support the second hypothesis which posited a positive relationship between in-game advertising and advertising effectiveness, as measured by a respondent's ability to recall and recognise in-game advertisements. However, we would not discount the necessity of positive attitudes, because this finding fails to account for the substantial impact that an individual's positive attitude towards in-game advertisements may have on their actual brand purchase intention. Further, a positive attitude may be necessary to avoid actions such as brand boycott when a game player repeatedly experiences a specific in-game advertisement or becomes increasingly exposed to the process. Clearly more research is required to monitor this trend.

One possible explanation for the lack of support for hypothesis 2 may be the vast amount of complex factors that could potentially affect recall and recognition. One such factor tested as part of hypothesis 3 was the effect of game experience. This hypothesis was supported with results showing that experienced players recalled and recognised significantly more in-game advertisements compared to novice participants. This finding is consistent with Schneider and Cornwell's (2005) study which also tested this hypothesis, but in the context of a different game genre (car racing). Experienced players, defined as those playing five hours or more in an average week, recalled and recognised significantly more advertisements from game play than novice participants. They tested an interesting hypothesis related to the phenomenon of 'flow' experienced during game play. Flow was defined as a pleasant emotional response game players experience when they become immersed in the activity of game play. It is characterised by total concentration as well as, to varying degrees, an absence of self-consciousness. Schneider and Cornwell (2005) hypothesised that the state of flow would be encountered more readily by experienced players than novice players, because experienced players have a basic understanding of the primary elements of a game, such as controls, and therefore have an increased ability to focus on the game itself. This hypothesis was however, not supported. The authors speculated that novice players may have also achieved a variation of the flow state that in fact detracted from their ability to recall in-game advertisements.

There are many marketing implications of this for marketers and indeed game designers. Firstly, marketers need to consider that advertisements seen by experienced players will have significantly higher benefits in terms of levels of recall and recognition. In addition to this, advertisements that are exposed to game players when they become experienced with the particular game, for example those seen only once a player reaches higher levels or later stages of a game, may also allow significantly high levels of processing and thus represent increased opportunity for an advertisement to be recalled or recognised. In contrast, advertisements seen by novice players or during initial game play may have significantly lower impact on awareness. Secondly, in a real world setting, the varying difficulty and level of experience required for different genres and game play type will significantly affect awareness. While some video games are designed to be played out in complexity over a lengthy period of time, others such as those in the casual games genre are intended to be very simple in control and thus have shorter learning curves and longevity.

The final hypothesis relates to the effect of advertisement specific factors such as brand familiarity, interaction of the advertisement with game play, and visual characteristics of the advertising. Participants in this study cited brand familiarity followed by the advertisement's prominent colours and graphical logos as the most common factors that aided their recall of the advertising. Although there is consensus on the contribution of these factors to recall, the results on the effect of interaction with the advertisement during game play on recall is mixed. Schneider and Cornwell (2005) reported that direct interaction with the advertisement during game play contributed to recall, contrary to Nelson (2002) who reported that the prominence of an in-game advertisement had no significant impact on a participant's ability to recall that specific advertisement. In a later study however Nelson et al. (2006) did find that game player's recall of advertisements could be aided through the use of animated sequences or cut scenes and suggested that the watching of animated sequences or cut scenes, can offer an equally entertaining proposition as playing the game itself. The reason for the contrast in findings may be due to game genre and construct. While a number of researchers have found varying results, all have failed to include a measure of participant's potential interaction throughout game play. Interaction will vary on an individual level, for example one participant playing a car racing simulation game during an experiment may experience low levels of interaction, i.e., they do not crash into advertising signage on barricades, while another may spend more of their time interacting with the advertisements due to their lower skill level. In addition to this, Nelson (2002) also made note of the increased short-term brand awareness when the brand exposure is a major part of game play, for example car selection in a racing game. It is possible that findings regarding interaction vary due to levels of interaction in individual video games, which may rely heavily on game genre.

Another factor affecting in-game advertisement effectiveness may be the level of realism and relevance in which they are presented, and the likeability of the game itself. Although not the main intent of the study reported here, it was noted that the only panel that was not recalled or recognised by any of the participants was that for Castrol – a brand that can be viewed as having little relevance to the specific game genre. Furthermore, there was a significant correlation observed between likeability of the game and advertising effectiveness measures as well as verbal statements to that effect. Although these findings are tentative, marketers and game developers should give careful and conscious consideration to the aspect of game genre and advertisement fit, and the likeability of the game itself. These qualitative findings add strength to Schneider and Cornwell's (2005) argument that a player's reason for recalling an advertisement is considerably more complex than previous researchers have argued. This in itself is an area for further investigation.

In conclusion, while in-game advertising offers a potential breakthrough to the clutter of traditional media to deliver messages to prospects in a direct and naturalistic environment, the study presented here suggests four key implications that require careful consideration. These are:

First, attitude towards in-game advertising have become polarised in recent years and may be in the process of decline. This may be a result of the decline in overall attitude towards general advertising or as a reaction to specific issues with the use of in-game advertising.

Second, the effectiveness of in-game advertising does not solely depend on attitude towards the practice, but on factors that are unique to the game player and to the brand itself. In-game advertisements may be more effective for brands with high levels of existing brand awareness, brands that are well established and well known, and are related with the game genre.

Third, marketers need to ensure that advertisements that are utilised for in-game advertisements must contain colourful, eye catching and bold visual elements, ensuring that game players are able to notice the advertisements as distinct from their surroundings, whilst remaining appropriate.

Fourth, marketers must consider the impact of genre on effectiveness. Genre specific interactive elements offer varying key points for enhancing recall. Schneider and Cornwell (2005) reported that participants used advertisements as markers for forthcoming difficult corners in a car racing simulation, an observation that is not possible for participants playing a soccer game. This example illustrates one of many varying opportunities that are offered by different game genres that marketers need to consider in assessing the individual value obtained by a specific in-game advertisement.

## REFERENCES

Alpert, F. (2007). Entertainment software: Suddenly huge, little understood. *Asia Pacific Journal of Marketing and Logistics, 19*(1), 87-100.

Cuneo, Z.A. (2004, 19 January). Marketers game for action; Videogames: With men, gaming is bigger than the Sopranos. *Advertising Age,* s8.

DeLorme, E.D., & Reid, N.L. (1999). Moviegoers' experiences and interpretations of brands in films revisited. *Journal of Advertising, 28*(2), 71-94.

Gaudiosi, J. (2007). In-game ads reach the next level. *Business 2.0, 8*(6), 36-37.

Greenfield, P.M., DeWinstanley, P., Kilpatrick, H., & Keys, D. (1996). Action video games and informal education on strategies for dividing visual attention. *Journal of applied developmental psychology, 15*(1), 105-123.

Gupta, B.P., & Gould, J.S. (1997). Consumers' perceptions of ethics and acceptability of product placements in movies: Product category and individual differences. *Journal of Current Issues and Research in Advertising, 19*(1), 37-50.

Gupta, B.P., & Lord, R.K. (1998). Product placement in movies: The effect of prominence and mode on audience recall. *Journal of Current Issues and Research in Advertising, 20*(1), 47-59.

Gwinn, E. (2004, 21 April). Space invaders: Ads are infiltrating video games. *Chicago Tribune.*

Hyman, P. (2006). In-Game Advertising. *Game Developer, 13,* 11-47.

Johnson, K.M., Foley, A.M., & Suengas, G.A. (1988). Phenomenal characteristics of memories for perceived and imagined autobiographical events. *Journal of Experimental Psychology, 117*(4), 371-376.

Lindstrom, M. (2001). *Brand games: Are you ready to play?* Retrieved 6 October, 2007, from http://www.internet.com

Machleit, A.K., Allen, T.C., & Madden, J.T. (1993). The mature brand and bran interest: An alternative consequence of ad-evoked affect. *Journal of Marketing, 57,* 72-82.

Massive-Incorporated. (2007). *Massive study reveals in-game advertising increases average brand familiarity by up to 64 percent.* Retrieved 6 October 2007, from http://www.massiveincorporated.com/site_network/pr/08.08.07.htm

Molesworth, M. (2006). Real brands in imaginary worlds: Investigating players' experiences of brand placement in digital games. *Journal of Consumer Behaviour, 5,* 355-366.

Nebenzahl, D.I., & Secunda, E. (1993). Consumers attitudes toward product placement in movies. *International Journal of Advertising, 12*(1), 1-11.

Nelson, R.M. (2002). Recall of brand placements in computer/video games. *Journal of Advertising Research, March/April,* (pp. 80-92).

Nelson, R.M., Keum, H., & Yaros, A.R. (2004). Advertainment or adcreep? Game players' attitudes toward advertising and product placements in computer games. *Journal of Interactive Advertising, 4*(3), 1.

Nelson, R.M., Yaros, A.R., & Keum, H. (2006). Examining the influence of telepresence on spectator and player processing of real and fictitious brands in a computer game. *Journal of Advertising, 35*(4), 87-99.

Schneider, L., & Cornwell, B. T. (2005). Cashing in on crashes via brand placement in computer games: The effects of experience and flow on memory. *International Journal of Advertising, 24*(3), 321-343.

Shrum, L.J. (Ed.). (2004). *The Psychology of Entertainment Media: Blurring the Lines Between Entertainment and Persuasion*: Lawrence Erlbaum Associates.

Sung, Y., & De Gregorio, F. (2005). *New brand worlds: A comparison of college student attitudes toward brand placements in four media.* Paper presented at the AMA Winter Educators' Conference Chicago.

Szalai, G. (2007). Video game industry growth still strong. Retrieved 6 October 2007, from http://www.hollywoodreporter.com/hr/content_display/business/news/e3if5f9e6af-1f789e8c28399b0253e7b78d

Wolf, M. (2007). In-game advertising on consoles to reach $850 mil by 2011. *Multimedia Publisher, 18*(10), 1.

Yang, M., Roskos-Ewoldsen, R.D., Dinu, L., & Arpan, M.L. (2006). The effectiveness of "in-game" advertising'. *Journal of Advertising, 35*(4), 143-152.

Young, E. (2004). *EA makes ad play.* Retrieved 6 October, 2007, from http://sanfrancisco.bizjournals.com/sanfrancisco/stories/2004/02/09/story2.htm

# Chapter XIX
# The Effect of Arousal on Adolescent's Short-Term Memory of Brand Placements in Sports Advergames

**Monica D. Hernandez**
*The University of Texas-Pan American, USA*

**Sindy Chapa**
*Texas State University, USA*

## ABSTRACT

*The authors' study examined factors affecting Mexican adolescent's memory of brand placements contained in advergames. Specifically, two concerns were investigated: (1) the effect of high/moderate arousal on adolescent's short-term recognition, and (2) the effect of high/moderate arousal on brand confusion. Analyses indicated that high arousal advergames corresponded to both higher hit scores (better recognition) and lower false alarms (less confusion) than moderate arousal advergames. The findings revealed more accurate short-term memory when subjects were exposed to a high arousal condition than to a moderate arousal condition. Advertisers wishing to target adolescents could strengthen the recognition of their products and brands by relying on fast pace or competitive game genres.*

## INTRODUCTION

Young adults are considered one of the most vulnerable audiences to advertising as well as a unique online segment. One form of online advertising that adolescents are often exposed to is advergaming (Stoughton, 2005). Advergaming is the delivery of advertising messages through electronic games. The extensive visual exposure to the brands is a distinctive feature offered by this technique. In spite of its increasing popularity, scarce research has been conducted addressing effects of emotional responses to advergame playing on young consumers.

The term "arousal" and its synonyms "alertness," "activation," and "excitation" describe a process that energizes behavior and affects non-exclusively cognitive performance (Ragazzoni, 1998). To date, few studies had addressed the effect of arousal on different online aspects. The effect of arousal on online behavior has been examined for the Internet shopping experience (Menon & Khan, 2002), finding that the more stimulation and information load, the less the consumer will engage in shopping behavior. Lee, Suh and Whang (2003) found that some dimensions of emotion –including arousal- significantly influenced positive consumer attitudes. Yoo and Kim (2005) concluded that a U-shaped relationship exists between the level of animation and both recognition and attitudes toward the ad. High levels of arousal were found in highly entertaining web sites, resulting in positive site evaluations and increased purchase intention (Raney et al., 2003). However, no previous study has examined the effect of different levels of arousal on memory of brand placements in advergames. Moreover, no research to date has examined such effects on adolescent's memory of brands.

Our study examined how emotional responses affect adolescent's short-term recognition of brand placements contained in advergames. Specifically, two concerns were examined: (1) the effect of high/moderate arousal on adolescent's memory, and (2) the effect of arousal on brand confusion. In sum, our purpose was to uncover the effects of interaction via advergames with well-known and well-liked brand placements, and their extensions into adolescent's memory.

Latin America has the fastest growing Internet user population in the world (eMarketer, 2007). In spite of this, there is very little information about the Latin American online gaming audience. In order to fill this gap, Mexico was selected for the study due to their large number of users. The number of Mexican Internet users has exceeded 3.6 million, ranking second in the number of users in Latin America (Instituto Nacional de Estadística, Geografía e Informática, 2007). In particular, later elementary and junior high school children's memory will be examined because as opposed to pre-school and early elementary school children, the first group's brand preferences are more consistent (Bahn, 1986). In addition, this group is able to read and write with enough confidence to answer a simple survey. Accordingly, the purpose of our study was to examine factors having an effect on memory of advergames by later elementary and junior high school students in Mexico.

## BACKGROUND

Recent scholarly studies have addressed the effectiveness of product and brand placements in electronic games via assessment of brand recall. Nelson (2002) addressed short- and long-term brand recall and attitudes toward brand placements in commercial games among a small group of American players. Findings revealed that players were able to recall 25 to 30 percent of brands in the short term and about 10 to 15 percent in the long run. Brands demonstrated recall superiority when they were a major part of game-play, when they were local, new or atypical brands, or relevant to the player.

Hernandez et al. (2005) assessed brand recall in online settings, extending previous analysis of scripts by providing a comparison of recall by natives of one Eastern language (Korean) versus three groups of natives of two Western languages (Spanish and English) simultaneously. Subjects with the ability to process a biscriptal language (Korean) demonstrated recall superiority in one out of two experimental stimuli over subjects habitually processing language based on alphabetic script (English and Spanish). Overall, brand recall and attitudes results were consistent with Nelson's (2002) findings. Results also indicated that neither the level of expertise of players nor the perceived goal difficulty has an effect on brand recall.

## HYPOTHESES

Studies by Du Plessis (1994) and Walker and Dubitsky (1994) have concluded that positive attitude toward ads positively relates to advertising recall. The rationale behind this presumes that likeable advertising can affect an individual's processing by creating positive arousal. Contradictory findings of the effect of arousal on memory are found in the literature. According to Cahill and McGaugh (1995), aroused subjects exhibited enhanced long-term memory. Conversely, Tavassoli (1995) indicated arousal might influence cognitive capacity as generated by emotional intensity. Accordingly, we expect that adolescents exposed to high arousal advergames will exhibit an enhanced recognition performance. Therefore, the following hypothesis is derived:

**$H_1$:** *Adolescents exposed to a high arousal advergame will exhibit higher recognition scores than adolescents exposed to a moderate arousal advergame.*

On the other hand, following the theoretical perspective that arousal has an effect independent of valence, Shapiro, MacInnis and Park (2002) found support to the claim that subjects exposed to a moderate arousal condition were better able to discriminate target attributes from distractors than were those in the high arousal condition. Accordingly, the following hypothesis is generated:

**$H_2$:** *Adolescents exposed to a moderate arousal advergame will exhibit lower false alarm scores than adolescents exposed to a high arousal advergame.*

## METHOD

### Experimental Stimuli

Following Neeley and Schumann's (2004) study choice of products by children, advergames promoting fruit snacks, cheese crackers, and peanut butter cookies were selected for the study. Sports games with different characteristics and pace were selected. The games selected were 3D Dune Derby (http://www.nabiscoworld.com/games/nw_shock_nwdn.htm) (high arousal) and X-treme Ping-Pong (http://www.nabiscoworld.com/Games/game_large.aspx?gameid=10098) (moderate arousal) from nabiscoworld.com. Another criterion considered in the selection was that Nabisco is a multinational company and offer the same brands worldwide.

*High arousal stimulus.* The game 3D Dune Derby (game A) is a buggy racing game exhibiting four brand products and one corporate brand. The brand products are Ritz Bits® Sandwiches, Chips Ahoy!®, Oreo®, and Fun Fruits. The corporate brand is Nabisco. The brand names are displayed on buggies, dashboard, start banner, banner checkpoints, and signal flags. One of these brands is always visible in the frame of the gameplay window. Background music and sound effects (driving, crashing) are heard during the gameplay. 3D Dune Derby is a rapid game requiring quick reflexes.

*Moderate arousal stimulus.* The game X-treme Ping-Pong (game B) promotes three brand products and one corporate brand. The brand products are X-treme Jello, Chips Ahoy!®, and Oreo®. The brand names are displayed on the table and posters on the wall. The Jello brand is always visible on the table. Sound effects (hitting, applause) are heard during the gameplay.

### Measures

Prior to the gameplay, an adolescent background information survey was collected including age, gender, number of siblings, parent occupation, console game ownership, videogame playing, computer ownership, computer use, television viewing and experience with the foods promoted in the advergames.

To assess brand recognition after advergame playing, a two-alternative forced-choice recognition task was used. Participants were tested with several random pairs of items (in each pair, one item was seen and one not seen) from which they had to pick out the presented stimulus. Specifically, three pairs of items were presented for game A and three pairs for game B. Each pair was randomly created, and the position (left or right) of both target and distractor was also randomly selected. The distractors included other Nabisco brands (LifeSavers®, Triscuit®) and other comparable global snack brands (Cheez-It®, ChipsDeluxe®). To correct for successful guessing in recognition, the brand recognition scores were determined by summing all the brands recognized in each game (hit score) subtracting the number of brands not seen (distractors or false alarms).

*Table 1. Description of study participants*

| **Age** | |
|---|---|
| 10 | 31 (24%) |
| 11 | 35 (27%) |
| 12 | 27 (21%) |
| 13 | 20 (16%) |
| 14 | 10 (8%) |
| 15 | 5 (4%) |
| **Gender** | |
| Male | 77 (60%) |
| Female | 51 (40%) |
| **Computer usage per week** | Mean=12.76<br>Median= 7<br>Mode= 5 |
| **Videogame playing per week** | Mean= 10.50<br>Median= 5<br>Mode= 1 |
| **TV hours per week** | Mean= 6.68<br>Median= 3<br>Mode= 1 |

## Data Collection and Sampling Procedure

Three groups of a private elementary school (fourth, fifth and sixth grade) and three groups of a private junior high school from Matamoros, Mexico were invited to participate in the study. The sample was conveniently selected. One hundred and thirty-two students agreed to participate, yet four of them did not conclude the experiment. The final sample included a total of 128 adolescents. A description of the participants is shown in Table 1.

Five adults, including professionals and undergraduate students, were recruited to assist the researchers during the experiment. Parents of all participants were given an informal explanatory letter to inform them about the study. Consent was obtained from each student prior to the experiment. The experimental locations included the students' classrooms for pre-test and the school computer lab for post-test application. The pre-test was conducted a week prior to the lab experiment. Pre-test objectives were to measure brand experience along with product liking and to obtain participants' demographics. A week later, the post-test (lab experiment) was conducted in the school computer lab. Subjects participated on computers assigned individually. To avoid demand artifacts, the participants were told this was a video games study. Specifically, as opposed to explicitly telling them to be aware of the logos, the incidental exposure method was used. They were provided with instructions to the games in Spanish. The participants were instructed to play each advergame for 10 minutes. Following the gameplay, participants completed Spanish translated paper-and-pencil questionnaires.

## RESULTS

$H_1$ posited that adolescents would exhibit higher recognition scores when they are exposed to high arousal condition (advergame A) than when they are exposed to moderate arousal condition (advergame B). A simple t-test on the brand recognition for game A (high arousal) versus game B (moderate arousal) showed that a mean difference existed between the two advergames in terms of brand recognition, indicating that adolescents' cognitive capacity was enhanced after they were

*Table 2. t-Test*

| | t | Sig. | MD |
|---|---|---|---|
| **Advergame A (High arousal stimulus)** | 5.42 | .000 | 2.67 |
| **Advergame B (Moderate arousal stimulus)** | 133.37 | .000 | 2.93 |

exposed to the high arousal stimulus. Results are displayed in Table 2.

$H_2$ proposed that subjects exposed to the moderate arousal advergame would exhibit lower false alarms than when exposed to the high arousal advergame. Analyses were performed to determine whether a significant difference existed between the high and moderate arousal conditions. Three levels of brand recognition categorized as "no brand recognition" (participants selected distractors only), "false alarms" (participants selected both brand and distractors) and "perfect brand recognition" (participants selected seen brands only) assisted for $H_2$ testing and also to confirm $H_1$. Thus, in order to determine whether a significant difference existed among these three levels of recognition, an ANOVA test was performed.

Overall, the results indicated that a significant difference existed among groups in terms of brand recognition (mean difference = 3.15, p-value = .000). Specifically, the Bonferroni and Scheffée multiple comparison tests showed that a significant difference existed between "perfect brand recognition" when compared to "false alarms" and "no brand recognition." For the high arousal condition, results indicated that the brand recognition score was significantly greater when comparing hit scores of "perfect brand recognition" versus "false alarms" and "no brand recognition." In addition, no significant difference was found between "false alarms" and "no brand recognition." Furthermore, the direction of the means difference indicated that, contrary to expectations, the moderate arousal condition produced more false alarms than the high arousal condition. Results are displayed in Table 3.

Therefore, while $H_2$ was rejected, our results supported $H_1$. The findings indicate that the high arousal advergame produced higher hit scores

*Table 3. ANOVA tests and multiple comparisons (Scheffée and Bonferroni)*

| | Sum of Squares | Df | MS | F | Sig. |
|---|---|---|---|---|---|
| Between Groups | 6.30 | 2 | 3.15 | 13.81 | .000 |
| Within Groups | 57.7 | 253 | 2.28 | | |
| TOTAL | 64.00 | 255 | | | |

| Dependent Variable ADVERGAMES | A (High arousal) | B (Moderate arousal) | Mean Difference (I-J) | Sig. |
|---|---|---|---|---|
| **Recognition** | No brand recognition | False alarms | -.0731 | .918 |
| | | Perfect brand recognition | -.4631 | .018 |
| | False alarms | No brand recognition | .0731 | .918 |
| | | Perfect brand recognition | -.3900 | .000 |
| | Perfect brand recognition | No brand recognition | .4631 | .018 |
| | | False alarms | .3900 | .000 |

and lower false alarms than the moderate arousal advergame.

## CONCLUSION AND FUTURE TRENDS

Our study intended to uncover the effect of different levels of arousal on memory. Two concerns were examined: (1) the effect of high/moderate arousal on hit scores and (2) the effect of high/moderate arousal on false alarm scores. The results followed one position and resolved contradictory findings in the literature addressing the effect of arousal on memory of brand placements in online settings. Consistent to Cahill and McGaugh's (1995) findings, aroused subjects exhibited enhanced recognition scores.

Additionally, the results exposed the effect of arousal on incorrect responses, or guessing, when performing brand recognition. Contrary to Shapiro, MacInnis and Park's (2002) findings, our study revealed that subjects exposed to a high arousal condition were better able to discriminate target attributes from distractors than were those in the moderate arousal condition. This unexpected finding indicates that rapid and exciting advergames reinforced adolescents' memory.

Implications for advertisers suggest the creation of stimulating advergames to generate more accurate memory of brand placements. Advertisers wishing to target adolescents could strengthen the recognition of their products and brands by relying on fast pace or competitive game genres. Slow pace games may deflect the attention toward the brand placements. The inclusion of different levels of difficulty could assist to avoid moderate emotional responses from advergame players.

As an international experiment, the procedure used in our study was susceptible to some limitations. Several points should be stressed regarding the experimental stimuli. First, selected games represented brands from the same type of products. Second, only recognition of mature global brands was tested in this study. It would be particularly interesting for future research to test memory of new global brands. Finally, the experiment was conducted with a sample from one Latin American country.

Further work could replicate the experiment in another country or compare results from different countries. In addition, future research could extend the current findings by exploring other issues such as whether or not advergame playing will continue as a trend. Similarly, whether advergaming will be replaced by other types of online advertising remains unanswered. Finally, future work may compare other advergaming to traditional online ads effectiveness.

In sum, more accurate short-term memory was exhibited when subjects were exposed to high arousal condition than to moderate arousal condition. Our study contributes by providing insights about emotional responses determining young consumers' memory of brand placements in advergames. Furthermore, empirical evidence by means of a Mexican sample contributes in providing support about determinants of brand memory among Latin American adolescents. Nevertheless, advergaming remains an interesting and under researched topic to be addressed by advertisers, marketers and sport managers.

## REFERENCES

Bahn, K.D. (1986). How and when do brand perceptions and preferences first form? A cognitive developmental investigation. *Journal of Consumer Research, 13*(3), 382-393.

Cahill, L., & McGaugh, J.L. (1995). A novel demonstration of enhanced memory associated with emotional arousal. *Consciousness and Cognition, 4*(4), 410-421.

Du Plessis, E. (1994). Recognition versus recall. *Journal of Advertising Research, 34*(3), 75-91.

eMarketer (2007). Latin America online: Demographics, usage & e-commerce. Retrieved October 01, 2007, from http:// www.emarketer.com/Reports/All/Latam_aug06.aspx

Hernandez, M.D., Minor, M.S., Suh, J., Chapa, S., & Salas, J.A. (2005). Brand recall in the advergaming environment: A cross-national comparison. In M.R. Stafford & R.J. Faber (Eds.), *Advertising, Promotion and the New Media* (pp. 298-319). Armonk, NY: M.E. Sharpe, Inc.

Instituto Nacional de Estadística, Geografía e Informática (2007). Usuarios de Internet por paises seleccionados 1998 a 2004. Retrieved October 01, 2007, from http://www.inegi.gob.mx/est/contenidos/espanol/rutinas/ept.asp?t=tinf142&c=4870

Lee, M.B., Suh, K.S., & Whang, J. (2003). The impact of situation awareness information on consumer attitudes in the Internet shopping mall. *Electronic Commerce Research and Applications, 2*(3), 254-265.

Menon, S., & Kahn, B. (2002). Cross-category effects of induced arousal and pleasure on the Internet shopping experience. *Journal of Retailing, 78*(1), 31-40.

Neeley, S.M., & Schumann, D.W. (2004). Using animated spokes-characters in advertising to young children. *Journal of Advertising, 33*(3), 7-23.

Nelson, M.R. (2002). Recall of brand placements in computer/video games. *Journal of Advertising Research, 42*(2), 80-92.

Ragazzoni, A. (1998). Arousal: A neurophysiological view. *Abstracts/International Journal of Psychophysiology, 30*(9).

Raney, A.A., Arpan, L.M., Pashupati, K., & Brill, D.A. (2003). At the movies, on the web: An investigation of the effects of entertaining and interactive web content on site and brand evaluations. *Journal of Interactive Marketing, 17*(4), 38-53.

Shapiro, S., MacInnis, D.J., & Whan Park, C. (2002). Understanding program-induced mood effects: Decoupling arousal from valence. *Journal of Advertising, 31*(4), 15-26.

Stoughton, S. (2005). Skittles taps advergaming for product ads. *Marketing News* (September 1), *39*(14), 34.

Tavassoli, N.T. (1995). New research on limited cognitive capacity: Effects of arousal, mood and modality. *Advances in Consumer Research, 22*, 524-525.

Walker, D., & Dubitsky, T.M. (1994). Why liking matters. *Journal of Advertising Research, 34*(3), 9-18.

Yoo, C.Y., & Kim, K. (2005). Processing of animation in online banner advertising: The roles of cognitive and emotional responses. *Journal of Interactive Marketing, 19*(4), 18-34.

# Chapter XX
# Schemas of Disrepute: Digital Damage to the Code

**Ellen L. Bloxsome**
*Queensland University of Technology, Australia*

**Nigel K. Ll. Pope**
*Griffith University, Australia*

## ABSTRACT

*This chapter presents marketers, sporting management and sports organizations with a technique for analyzing consumer schemas associated with athletes. Correspondence analysis is a frequently used tool for social network analysis that evaluates relationships between actors and events. Correspondence analysis allows examination of the effectiveness of positioning efforts, as well as the assessment of potential brand damage caused by the off-field activities of athletes and endorsers. This technique can be used for snap-shot analyses of events, or longitudinal evaluation of changes in consumer and media schemas over time. The digital emphasis of the paper incorporates use of both non-traditional and external media commentary made about athletes and brands. Where media commentary is not controlled by management groups or brands, evaluation of potential schemas developing out of external media sources becomes important. Web logs, fan sites, and other digital information sources outside the control of sport managers contribute to the development of consumer perceptions, potentially affecting consumer sport involvement and merchandising revenue. When external media sources focus primarily on negative on- and off-field behavior by athletes, we suggest that there is potential for digital damage to sporting codes and brands.*

## INTRODUCTION

Viewing of major sporting events is growing rapidly. It is also changing in nature. Stadiums are expensive to build and take years in planning and permissions yet game attendance is constrained by stadium size. Revenue increase based on fan viewing is therefore dependent on developing new audiences, or alternative distribution networks to existing and potential audiences. For example, in Australia, the Australian Football League (hereafter, AFL) is working to expand its traditional consumer market to regions that do not traditionally boast a high level of support for the code. They have also incorporated a model of corporate responsibility in an appeal to women and indigenous Australians (AFL, 2007). The National Football League (hereafter, NFL) in the United States, engages in similar market expansion strategies, appealing to women, ethnic and international markets (Bhargava, 2007). This is because there is always a limit to the number of club memberships that can be sold, and seats that can be filled. New media audiences, those online: consuming streaming video and audio commentary, cable or high definition television join with electronic gaming and other merchandising activities to provide access to a larger market. New media audiences are not constrained by the size of physical stadiums; in fact, they may not be constrained by much at all.

This lack of constraint in new media usage and service provision is a frequently cited problem among moralists. The common complaints associated with new media include age-appropriate content; access to unregulated content; violence and nudity. These issues are not often associated with sport, except in the case of electronic or video games. However, the lack of control of new media content provides an interesting problem for sporting bodies. Emerging from the new media field is the potential that consumer perceptions may be influenced by sources external to the clubs and traditional news sources that have up until now provided sporting news and information.

In this chapter, we present the results of a study into the interactions between an athlete's name as discussed in user-generated internet communications, and attitudes to the sporting body to which he/she belongs as well as the brand names of the sporting body's sponsors. We begin with a discussion of the background to this research.

## BACKGROUND

As the penultimate game of the NFL, the Super Bowl is routinely investigated because of its economic impact and advertising spectacle. However, Super Bowl crowds have not increased much since the inception of the competition. In 1967 attendance at the first game was 61,946; while in the last 20 years it has averaged 73,600 (NFL, n.d). Despite this, revenues are strong. Following the 2008 event Arizona reportedly gained additional income of $500 million from an average tourist spend of $600 per day, during a four day visit (Associated Press, 2008). This type of figure alone indicates the importance of research into this area. That said, game attendance and tourist spending provide only half the picture of commercial sport as a business.

Speculation about attendance figures for the Indianapolis 500 has occurred each year since Tony Hulman purchased the Indianapolis Motor Speedway in 1945. It appears that this will be a perennially unresolved issue (Cavin, 2004). The Speedway provides approximately 250,000 ticketed seats, a large crowd by any standard, although unsubstantiated guess-estimates of race day crowds have reached as high as 400,000 (Cavin, 2004). Spectators who do not hold tickets for seats stand on spectator mounds, and it is this group that is hardest to count. Yet it is still only a fraction of the real audience.

For the 2008 race, 50 hours of broadcast were scheduled for television and radio (Powell, 2008). Podcasts and other multimedia events were provided by the series organizers and various race sponsors during the month of May. Authorized and scheduled media exposure of this event is extensive. Unofficial commentary, including websites endorsing all manner of racing paraphernalia, blog sites, and commercial sites that are effectively engaged in ambush marketing, are also to be found. The resultant noise surrounding the event can make it difficult to distinguish between the official sites and myriad others.

Similarly, Nielsen Media Research reported that 97.5 million viewers watched Super Bowl 2008 in the USA (AFP, February 5, 2008), making a worldwide audience in excess of 100 million easily credible. The online reaction to the Super Bowl is also notable, with advertising effects generating an increase in web traffic of roughly 24% for game advertisers during the weekend (Nielsen, 2008). Myspace.com reported an increase in site hits of 104% as viewers searched for Super Bowl advertisements, making the number of unique visitors 900,000 for the game weekend alone (Nielsen, 2008). Despite these indications that Super Bowl advertising is varied and successful, RepriseMedia used their fourth annual Search Marketing Scorecard to critique the lack of integration of web and television advertising. Overwhelmingly, RepriseMedia finds room for improvement, with the majority of Super Bowl advertisers failing to own their profile on social network sites; having a presence on video-sharing sites; or providing subscription incentives (RepriseMedia, 2008). The consumer call to action, and commitment to integrated advertising among big name advertisers, thus far, is not regarded as fully utilized.

We suspect that this lack of integration reflects a lack of understanding in two areas. The first is the nature of brand/event interaction (as in sport sponsorship) and the second, poor understanding of how new digital media work. Our chapter continues with an examination of how sponsorship operates within the perception of the consumer.

## CONSUMER BRAND PERCEPTIONS

Brands have both physical and abstract properties (Aaker, 1996; Brown & Dacin, 1997; Dacin & Smith, 1994). Any form of communication can affect perceptions of these properties, by interacting with pre-existing perceptions, or by arguments about a brand's qualities. Given the size of previously referred sporting audiences, it is unsurprising that sponsorship is a popular vehicle for presenting consumers with information and argument about brands.

Sponsorship is the "...provision of assistance either financial or in kind to an activity by a commercial organization for the purpose of achieving commercial objectives" (Meenaghan, 1983, p. 5). It can involve associating a brand's name with an event, a team, an individual, or a cause. Commonly cited objectives for sponsorship are to: 1) increase brand awareness and, 2) improve brand or corporate image (Gardner & Shuman, 1988; Gwinner, 1997; Gwinner & Eaton, 1999). Research has investigated this form of marketing communication from the points of view of sponsorship managers (Javalgi, Traylor, Gross & Lampman, 1994), spectators (Nicholls, Roslow & Laskey, 1994), and share price (Cornwell, Pruitt & Clark, 2005; Cornwell, Pruitt & Van Ness, 2001). In this chapter, we are more interested in the associative networks and schemas that consumers form as a result of sponsorship, or other media exposure.

Learnt knowledge – as opposed to intuition – is a network of memory nodes connected by associations. These associations occur when a node is activated. A brand name can be a node that is activated whenever a consumer is exposed to it. This may be through direct experience or communication. A fundamental proposition of branding is that brands are used by consumers as retrieval cues for learned associations, often regarding product performance or the image of the brand's manufacturer. Additionally, learned associations may provide the basis for new network nodes where direct experience or communications suggest meaning to consumers or information users.

This is most easily understood in terms of action and reaction, as in adaptive network models. These can be used to explain consumer learning, in which the focus is on benefits associated with a brand (Van Osselaer & Janiszewski, 2001). When an associative link between two nodes has been established (i.e., a brand and a benefit), an affective transfer occurs from one node to the other (Judd, Drake, Downing, & Krosnick, 1991). The action of information on one therefore creates a reaction in the other, causing reinforcement of a pre-existing belief or emotion.

Cues may also block or unblock other cues (VanOsselaer & Janiszewski, 2001). In such a case, a previously positive association with a brand may be blocked or unblocked by an association with another node. For example, an individual might believe that a particular brand of car is of good quality but revise that opinion (blocking) on seeing a person they dislike driving one. In the case of sponsorship, this becomes critical. It would seem that an individual's involvement with a sport will enhance or amplify the effect of a sponsorship (Celsi & Olson, 1988; dYdewalle, Vanden Abeele, Van Rensberger, & Coucke, 1988; Pham, 1992). If this is the case, any blocking of positive associations by the behavior of athletes will be reinforced by the consumer's involvement with the sport.

The problem that we see for firms engaged in sponsorship relates to how they can measure any nodal effect of their sponsees' activities on brand associations. To make such a measurement at an individual level would, of course be impractical. The standard means of making such measurements is to make inquiries, often by survey, of representatives of a target market segment. Such a process is open to several problems, such as 1) yea-saying (the desire of the individual to satisfy the interviewer), 2) order effect (in terms of earlier questions influencing responses to later ones) and, 3) generalizability. The first two problems can be resolved by direct observation of groups of people in response to various cues. The third can be resolved by several actions, one of which is to form a model from an aggregated group that reflects schemas as they may occur in individuals. In other words, one seeks to describe a composite individual that possesses several characteristics common to the group. In the following section, we discuss a socially based method for establishing nodes and schemas in the sport sponsorship context. From this commonality, we will then go on to describe how it may be used in context through the use of a worked example.

## MEASUREMENT TOOLS: SOCIAL NETWORK ANALYSIS

Social network analysis is used to investigate relations between actors or entities from a structuralist perspective, using a largely positivist epistemology. This school of analysis views actors as parties to exchange in social and structural relationships with other actors. The ties between actors in a network may be founded upon "kinship, material transactions, flows of resources, or support, behavioral interaction, group co-memberships or the affective evaluation of one person by another" (Wasserman & Faust, 1994, p. 8). Alternatively, network analysts might study the means by which, "gossip, ideas, or resources [are] diffused through the network" (Scott, 2000, p. 32). This includes identifying the paths facilitating exchange, as structural artifacts.

Social network analysis is appealing to sociologists, anthropologist, ethnologists, and management for several reasons. We can talk about groups (who is a group member, who merely wants to be); the strength of ties between people (Granovetter, 1983); and how information flows through organizational networks (Breiger, 1976; Carley, Hummon & Harty 1993; Doreian & Fararo, 1985; McNamee & Willis, 1994; Noma, 1982; Yeung, Liu, & Ng, 2005). This is not the full extent of social network analysis, but here we associate social network analysis with research that tries to explain human behavior and social organization.

This form of analysis has the advantage of a long association with technology and innovation. Early work in the discipline includes the studies by Freeman and others on the Electronic Information Exchange System (EIES) (Freeman, 1984; Wasserman & Faust, 1994). The EIES research sought to establish the state of awareness and acquaintanceship of various academics who had attended a single conference together in 1978, over time, by means of (mostly) asynchronous computer-mediated communication (Freeman, 1984). Work in the field of computer-mediated communication continues. Various authors have studied participation in online discussion groups or email networks (Adamic & Adar, 2005; Koku, Nazer, & Wellman, 2001; Matzat, 2004). Others are studying linking and conversations on blogs (see, for example, de Moor & Efimova, 2004), and the use of social networking sites (Boyd & Ellison, 2007; Ellison, Steinfield & Lampe, 2007). All of these allow for observation of behaviors, as opposed to the self-report of survey-based research.

The body of knowledge that follows the intellectual heritage of the EIES research focuses on relations of acquaintanceship. Freeman's paper had, as its purpose, an explanation of the growth of a discipline of study. He argued that:

*...participants could use the computer to communicate on a regular basis and perhaps arrive at the sorts of common understandings that constitute the basis for a collective effort focused on a new scientific approach.* (Freeman, 1984, p. 205)

Freeman's argument highlights regularity of contact, contestation, and the development of common understanding. We could argue, equally, that this applies to the day-to-day organization of a family, or the management of a business, or the cultural understanding of the value of sport. Regular interaction between group members and the development of common understanding are the means by which collective decisions are made and consensus reached.

Freeman also discusses one of the larger problems with social network analysis: that the structure of relations is given a primacy over the type, or qualitative value of relationships. He writes that:

*Structural studies of social networks typically ignore the content of the relations under examination; we act as if we expect to find some universal structural laws that can be applied equally to friendship and to corporate interlocks.* (Freeman, 1984, p.205)

This is a problem that has been recognized in other social network research. Of Padgett's extensive body of work on Florentine business and marriage alliances (McLean & Padgett, 1997; Padgett & Ansell, 1993; Padgett & McLean, 2006), criticism has been made of the arbitrary quality of relationships established (Wasserman & Faust, 1994). The data collected represent the ties of marriage and business partnership or alliance in Florence during the 15th century (Wasserman & Faust, 1994). It is the position of Wasserman and Faust that although the data set is impressive, there is no means by which we can truly know, in the way of universal structural laws, whether marriage ties influenced business partnerships, or vice versa. In terms of a philosophy of knowledge, this criticism is essentially valid.

In terms of a pragmatic understanding of history and culture, although we might seek certainty on the question of whether business partnerships were cemented by ties of marriage, we will find no universal law. In social network analysis, acquaintanceship and affiliation are frequent objects of study. Freeman's position on universal social laws is cautious. He believes that there is "...a still more basic relation that always underlies acquaintanceship, and that is awareness" (Freeman, 1984, p. 206). Thus, in terms of a philosophy of knowledge, the standard for social networks becomes more broad, easier to formalize, and less deterministic. Although we cannot say that marriage determines business partnerships or vice versa, we can say that social and institutional relationships exist between various individuals and groups.

We argue in this chapter that social network analysis, with its structural relationships between nodes, reflects the relationship between nodes held at an individual level in adaptive learning networks. If this is the case, analysis of a social network should reflect a community effect in response to, for example, sponsorship activity and the actors within a sponsorship agreement. To further justify this approach, we now discuss the methods and background to data analysis in social network research.

## SOCIAL NETWORK ANALYSIS APPROACHES TO DATA

In defining social network analysis, most authors refer to the history of the discipline. In the 1930's, Moreno developed the sociogram in order to visualize a network for explanation (Scott, 2000; Wasserman & Faust, 1994). The sociogram is, effectively, a graph depicting relations between actors, where nodes represent actors, and lines, the relationships between actors. As the focus of Social Network Analysis is the explanation of structural ties between actors, the sociogram provides the means to visualize the structure of networks.

Moreno's sociogram was developed out of a desire to map social configurations (Wasserman & Faust, 1994). Early research consisted of sociometric study of social groups where students were asked to name other students with whom they would and would not like to interact (Forsyth & Katz, 1946; Moreno, 1946). This type of research yields findings on positive (attraction) and negative (avoidance) social interaction, and notions of centrality or isolation. The sociogram was regarded as being particularly useful because it enabled visualization of social networks, allowing "an interrelated whole of the group structure to the eye of the student" (Moreno, 1946, p. 348). Visualizations remains an important concern for a large number of social network researchers, including some who say that visualization is "the first goal of social network analysis" (Dekker, 2001, p. 1).

The use of sociograms prompted the development of the mathematical forms of analysis that now provide the basis of network measurements. Criticism of the sociogram suggests that "...the sociogram has obvious advantages over verbal descriptions, but it is apt to be confusing to the reader, especially if the number of subjects is large..." (Forsyth & Katz, 1946, p. 341). The sociogram allows for the visualization of large amounts of information, the description of all of the actors in the network, and their relations, and strength of relations with every other network member. However, if we use the example of a network of 250 people, where ties in the network are asymmetrical, a sociogram used to depict this network would potentially hold 62, 250 lines – making a very difficult picture to interpret. Forsyth and Katz (1946), frustrated by uninformative large network sociograms, suggested a matrix approach to network data. Their position was that while sociograms retain a measure of utility, there had to be a better way to "preserve all the data and present it in an orderly, meaningful manner", (Forsyth & Katz, 1946, p. 341). The matrix approach, not only allows for the storage of a large amount of variable information, it also enables more complex and involved analyses of networks.

Two principal types of matrix are used to store Social Network data, the adjacency and the incidence matrix (Scott, 2000). The adjacency matrix is used to depict relations of each actor to every other actor in the network. Hence in an asymmetrical network, every cell in the matrix – barring the diagonal cells – represents a directed link from one actor to another. In an undirected network, only half the matrix is informative, as links in one half of the matrix are duplicated in the other. An incidence, or two-mode sociomatrix, is used to display the co-occurrence of two variables.

An incidence matrix may record the events that a group of actors attended over a period of time, corporate interlocks of directorships, or grazing habits of various species through a number of ecosystems. An incidence matrix may be broken down into two adjacency matrices, recording the links between people, and the links between corporations; or, an incidence matrix may be analyzed in order to yield understanding of the co-occurrence of variables.

## CORRESPONDENCE ANALYSIS

Correspondence analysis is a form of social network analysis that is used to evaluate the co-occurrence of matrix variables as they are mapped in low-dimensional graph space. It is described by Greenacre and Hastie (1987) as:

*...an exploratory multivariate technique that converts a matrix of nonnegative data into a particular type of graphical display in which the rows and columns of the matrix are depicted as points.* (p. 437)

Correspondence analysis was developed in France, principally by Benzecri (Benzecri, 1969; Berthon, Pitt, Ewing, Ramaseshan, & Jayaratna, 2001; Calantone, di Benedetto, Hakam, & Bojanic, 1989; Hoffman & Franke, 1986). The principles underlying correspondence analysis hold that relationships between row and column variables can be determined by representing the two in joint graph space. Patterns emerging in the graphical representation of points represent factors that explain the relationships between variables. Mathematically, principle components analysis contributes substantially to most correspondence analysis.

Hoffman and Franke (1986) find correspondence analysis to be of use in marketing research because it is able to map categorical data. Where chi-square and descriptive statistics might otherwise have been the limit of analysis for a wide range of marketing and categorical data, correspondence analysis provides an alternative.

Chi-square distances contribute to the measures of variance in row and column scores that inform discussion on 'inertia' within correspondence analysis. Inertia itself is "calculated as the weighted sum of the squared chi-square distances between the row profiles and the average row profile, where the weights are marginal row proportions" (Faust, 2005, p. 130). The same application is also made to column and average column profile for assessing that inertia.

Greenacre and Hastie (1987) explain the geometric analysis of inertia and chi-squared distances in correspondence analysis thus:

*...* $\chi^2/n$ *can be defined geometrically as the weighted average of the squared (chi-squared) distances of the row profiles to their centroid. The quantity* $\chi^2/n$*... is called the total inertia of the data matrix.... The null hypotheses of row-column independence, ... is equivalent to the hypothesis of homogeneity of the rows [thus] a significant* $\chi^2$ *can be interpreted geometrically as a significant deviation of the row profiles from their centroid, that is, from the homogeneity hypothesis.* (p. 438)

The graphical representation of a correspondence analysis therefore provides an indication of row-column dependencies, correspondences, or 'factors' that occur within the data matrix. And the geometrical analysis should inform of the significance of the deviation from homogeneity, and the tendency toward a factor-based distribution of the data.

Within marketing literature, correspondence analysis has been recommended as an application by for "...evaluating competing cigarette brands and- ... selecting a name for a new brand of cigarettes" (Hoffman & Franke, 1986, p. 214). Hoffman and Franke (1986) support the use of correspondence analysis for dealing with categorical data; they also provide an example demonstrating how consumer perceptions of brand and attribute differences can be mapped. Within the Hoffman and Franke paper, the authors note the possible uses of correspondence analysis as including: developing market segments; recognizing consumer perceptions of brands; positioning; monitoring the efficiency of advertising campaigns; and concept testing (Hoffman & Franke, 1986). Research since the Hoffman and Franke paper has established the suitability of correspondence analysis for each of these purposes, (see for example, Calantone, di Benedetto, Hakam, & Bojanic, 1989; Choi, Lehto, & Morrison, 2007; Kakai, Maskarinec, Shumay, Tatsumura, & Tasaki, 2003; Stipp and Schiavone, 1996).

This chapter suggests that future use of correspondence analysis in marketing research will involve the evaluation of consumer schemas developed from marketing communications, sponsorship efforts, and media sources uncontrolled by sport management or organizations. Where consumer knowledge schemas contribute to brand cognition and perceptions, we suggest that correspondence analysis provides the means to visualize links and aid analysis of consumer perceptions and effects of marketing communications. We further suggest that evaluation and monitoring of athlete and brand profiles will become increasingly necessary for clubs and sporting organizations given the expansion of digital media audiences and confusion between official and other information sources.

We now move on to a worked example of a study into consumer perceptions of sponsored athletes and their sport, using social network and correspondence analyses.

## METHOD

The current study provides an example of the links between awareness, knowledge, and consumer schemas within a sporting context. The sport of choice is the AFL. In Australia, average attendance at AFL games is approximately 36,800 people (AFL, 2007). The AFL Finals Series had a total attendance of 574, 424 people in 2007, up 7.9% from the previous year. Broadcast audiences, on the other hand, reached 4.97 million people each week of the AFL season; and for each of the five mainland capital cities, an average of 2.572 million people tuned in for the AFL Grand Final (AFL, 2007). For a country with a relatively small population (about 22 million), a weekly audience of 5 million provides a lot of branding time, with many potential consumers watching advertisements and receiving product information.

The text source for data describing players in this study is Wikipedia. This source is contrasted with the text from the Annual Report for the AFL. Annual Report text we treated as an exemplar for the organization's desired image, as they provide information on the commercial activities of the organization in order to impress or influence potential investors. They are also documents that outline company culture, providing information on the activities within the company that relate to corporate social responsibility principles, including, equity; community involvement; and ethical behavior.

The sample of players was selected with consideration of positive/negative profile, arrests, or controversy. An original list of fifteen players was limited to eight, as Wikipedia biographies were either not available for players, or not of sufficient length to produce an informative content/text analysis. Wikipedia was accessed from 17th to 21st April, 2008. As noted, text for analysis of the organization (the AFL) was sourced from the Annual Report of the organization (AFL, 2007).

For each of the players, and the organization, a text file was produced from biography or organization report. Each text file was content-analyzed using CATPAC II ™ (Woelfel & Woelfel, 1997), following the method used by Choi, Lehto and Morrison (2007), to evaluate travel blog websites. CATPAC II™, is a content-analysis program that extracts frequently used words from text files, and allows the exclusion of commonly used words such as "the", "and", "is", etc. CATPAC™ generally extracts 25 words per text file, ranked according to the frequency of occurrence, or alpha-order. For each player text file, their name was excluded from the content-analysis of their biography. As the subject of the biography, the subjects' names occurred frequently, however, as a means to evaluate how they are perceived, a high proportion of text devoted to a name is regarded as uninformative. Where a name may develop into a colloquial indicator, symbol, or brand, exclusion from analysis would be unhelpful. However, in this instance, it is implicit that the player should be the subject of their own biography, thus exclusion is justified.

Following the extraction of word-lists using CATPAC II™, processes of condensation and categorization were used to filter text data prior to correspondence analysis. Of the word-lists generated, only the top 10 ranked words for each subject (player or organization) were used to generate categories, with the expectation that these categories would contain the majority of memes (cultural communications of ideas or usage) encompassed in the data. Twenty categories were formed from the 49 unique words that comprised the top-10 ranked words. Table 1 shows the word

*Table 1. Extracted word-lists by player/ organization*

| Akermanis | Carey | Cousins | Didak | Fevola | Hird | Kerr | Voss | AFL |
|---|---|---|---|---|---|---|---|---|
| Lions | Melbourne | Eagles | Season | Season | Season | Cousins | Brisbane | AFL |
| Brisbane | Football | West | Against | Against | Games | AFL | Career | Commission |
| Club | Against | Coast | Round | Carlton | Essendon | Ben | Football | Game |
| Football | Captain | Drug | Goals | Round | Game | Charged | Lions | Club |
| AFL | Goals | AFL | Game | Career | Final | Assault | Against | Football |
| Braun | Team | Club | Goal | Goals | Against | Brownlow | Coaching | Players |
| Career | Club | Football | Played | Football | Team | Coast | Early | Community |
| Bulldogs | Face | Police | Collingwood | Forward | Best | Fined | Team | Australian |
| Early | North | Perth | Club | Medal | Football | Football | Game | Clubs |
| Field | Police | Player | First | AFL | Year | Fremantle | AFL | People |
| Season | Against | Season | Forward | Australian | Career | Player | Australian | Will |
| Channel | Coach | Stated | Games | Club | Memorable | Season | Bears | Year |
| Game | Neilson | Team | Hudson | Coleman | Round | Team | Life | Program |
| Match | Nine | Australian | Kicked | Final | Club | West | Play | Season |
| Often | Officers | Captain | Final | Goal | Disposals | Broken | Suffered | Policy |
| Performance | Footy | Following | Kick | Match | First | Club | Age | Media |
| Side | Former | Played | Match | Two | Injury | Court | Captain | Million |
| Ability | Goals | Best | Matches | Year | Three | Eagles | Coach | Indigenous |
| Abuse | Kate | Day | Year | Adelaide | Time | Fans | Home | Australian |
| AKA | Kick | February | AFL | Alex | AFL | Father | Injury | Coast |
| Australian | Later | Field | Australian | FeV | Medal | Field | Kilda | Sydney |
| Award | Network | First | Best | Line | Australian | Foot | Later | Key |
| Bears | Over | Found | Car | Media | Captain | Judd | Leadership | Women |
| Brownlow | Player | Game | Essendon | Outside | Coast | Known | Medal | Attendance |
| Coach | Quarter | Geelong | Half | Over | Early | Later | Premierships | Foundation |

lists generated for each player/organization. For each player/organization, a score was allocated to each category according to the CATPAC II™ frequency from their extracted word list. For example, the category "Match-play description" includes the words: against; field; match/es; game/s; kick; quarter; final; disposals; and played, among others. Some extracted words did not fit within any of the categories derived from the top-10 word lists, hence there has been a loss of potential meaning in their exclusion from the correspondence analysis. Table 2 presents contribution to category by player/ organization, and chi-square significance of categories. As shown, the categories contain between 64 and 80% of the extracted word lists per player/organization.

## FINDINGS

The correspondence analysis was conducted using UCINET 6.0™ (Borgatti, Everett, & Freeman, 2002). The results produced by UCINET 6.0™ indicate the presence of eight factors in the data, three factors were used to map the correspondence analysis graphically. Table 3 provides the percentage contribution and cumulative percentage for each of the factors found. As aforementioned, one intention of correspondence analysis is to map data in low-dimensional space. Table 3 presents the contribution of each factor. These are not particularly high. However, the graphical representation of the data remains interesting within the rubric of exploratory research.

*Table 2. Categories, contribution by player and chi-square significance.*

| Row | Category name | Akermanis | Carey | Cousins | Didak | Fevola | Hird | Kerr | Voss | AFL | Chi-square significance |
|---|---|---|---|---|---|---|---|---|---|---|---|
| R1 | AFL | 5.00 | 0.00 | 7.00 | 2.20 | 3.30 | 2.70 | 7.10 | 3.30 | 22.20 | 0.0000 |
| R2 | Australia/n | 2.00 | 0.00 | 2.60 | 2.20 | 2.70 | 2.30 | 0.00 | 3.30 | 5.90 | 0.2206 |
| R3 | Awards | 4.00 | 0.00 | 0.00 | 0.00 | 6.50 | 2.70 | 4.00 | 5.00 | 0.00 | 0.0072 |
| R4 | Captain | 0.00 | 5.10 | 2.60 | 0.00 | 0.00 | 2.30 | 0.00 | 2.50 | 0.00 | 0.0129 |
| R5 | Career | 0.00 | 0.00 | 0.00 | 0.00 | 6.00 | 3.60 | 0.00 | 8.30 | 0.00 | 0.0000 |
| R6 | Club name | 17.00 | 11.80 | 26.70 | 6.50 | 6.60 | 6.30 | 11.00 | 19.0 | 0.00 | 0.0000 |
| R7 | Club/s | 8.00 | 4.40 | 0.00 | 3.60 | 2.70 | 3.20 | 3.00 | 0.00 | 5.40 | 0.0624 |
| R8 | Coach/ing | 2.00 | 3.70 | 0.00 | 0.00 | 0.00 | 0.00 | 0.00 | 7.50 | 0.00 | 0.0000 |
| R9 | Commission | 0.00 | 0.00 | 0.00 | 0.00 | 0.00 | 0.00 | 0.00 | 0.00 | 5.90 | 0.0000 |
| R10 | Community | 0.00 | 0.00 | 0.00 | 0.00 | 0.00 | 0.00 | 0.00 | 2.50 | 3.80 | 0.0030 |
| R11 | Football | 6.00 | 9.50 | 4.40 | 0.00 | 4.90 | 4.50 | 4.00 | 6.60 | 5.10 | 0.2589 |
| R12 | Injury description | 0.00 | 0.00 | 0.00 | 0.00 | 0.00 | 3.20 | 3.00 | 2.50 | 0.00 | 0.0244 |
| R13 | Match-skill description | 3.00 | 8.00 | 2.20 | 9.40 | 8.70 | 8.10 | 0.00 | 0.00 | 0.00 | 0.0002 |
| R14 | Match-play description | 7.00 | 8.10 | 4.40 | 36.70 | 15.80 | 21.30 | 3.00 | 7.40 | 0.00 | 0.0000 |
| R15 | Origin | 10.00 | 0.00 | 3.90 | 0.00 | 2.20 | 0.00 | 4.00 | 2.50 | 0.00 | 0.0000 |
| R16 | Other player reference | 5.00 | 0.00 | 0.00 | 0.00 | 0.00 | 0.00 | 17.20 | 0.00 | 0.00 | 0.0000 |
| R17 | People | 0.00 | 0.00 | 0.00 | 0.00 | 0.00 | 0.00 | 0.00 | 0.00 | 8.00 | 0.0000 |
| R18 | Player/s | 0.00 | 8.00 | 3.50 | 0.00 | 0.00 | 0.00 | 4.00 | 5.00 | 4.00 | 0.0019 |
| R19 | Police | 0.00 | 8.10 | 11.90 | 0.00 | 0.00 | 0.00 | 16.10 | 0.00 | 0.00 | 0.0000 |
| R20 | Season | 4.00 | 0.00 | 3.10 | 16.50 | 18.10 | 15.40 | 4.00 | 0.00 | 6.30 | 0.0000 |
| Total contribution to categories | | 73.00 | 64.40 | 72.30 | 77.10 | 77.50 | 75.60 | 80.4 | 75.40 | 66.60 | |

*Table 3. Factor contribution and cumulative percentage*

| Factor | Percentage contribution | Cumulative% |
|---|---|---|
| 1 | 23.2 | 23.2 |
| 2 | 17.7 | 40.9 |
| 3 | 15.6 | 56.5 |
| 4 | 13.3 | 69.8 |
| 5 | 10.6 | 80.4 |
| 6 | 7.8 | 88.3 |
| 7 | 6.1 | 94.3 |
| 8 | 5.7 | 100.0 |

The graphical representation of the correspondence analysis (shown at Figure 1) indicates a clear separation of the AFL (organization) from the cluster of players. In this research we note that the AFL is relatively close to row variables 1 (AFL); 4 (captain); 10 (community); and 17 (people). The closeness of the AFL to R4 (captain) is misleading; however the closeness to community, people and AFL is reasonable. The AFL Annual Report word list rates community emphasis, social responsibility and various groups of people as key image targets (AFL, 2007).

Three clusters of players with other row variables can be seen in Figure 1. The player, Kerr, is clustered with R16 (other player references) and R19 (police). Kerr's profile in Wikipedia significantly links him to Ben Cousins (another player) in terms of friendship; to Cousins' drug use and other activities; and to an incident at a party to which the police were called. A second cluster shows Didak, Fevola and Hird linked with R5 (career); R13 (match-skill); R14 (match-play description); and R20 (season). This cluster explains a group for whom their Wikipedia profile focuses on their on-field activities – they are significantly distant from any mention of police, or off-field activities for which they might bring their club or code into disrepute. A more central group, including Carey, Cousins, Akermanis and Voss with R3; R6; R7; R8; R9; R11; R12; R15; and R18, is more difficult to interpret. These four players are mapped at an axis within the graph. The players are clustered at the same latitude as the AFL, despite their distance from the organization; and they have an association with all other players. The position of

*Figure 1. Perceptual map of correspondence between players and organization*

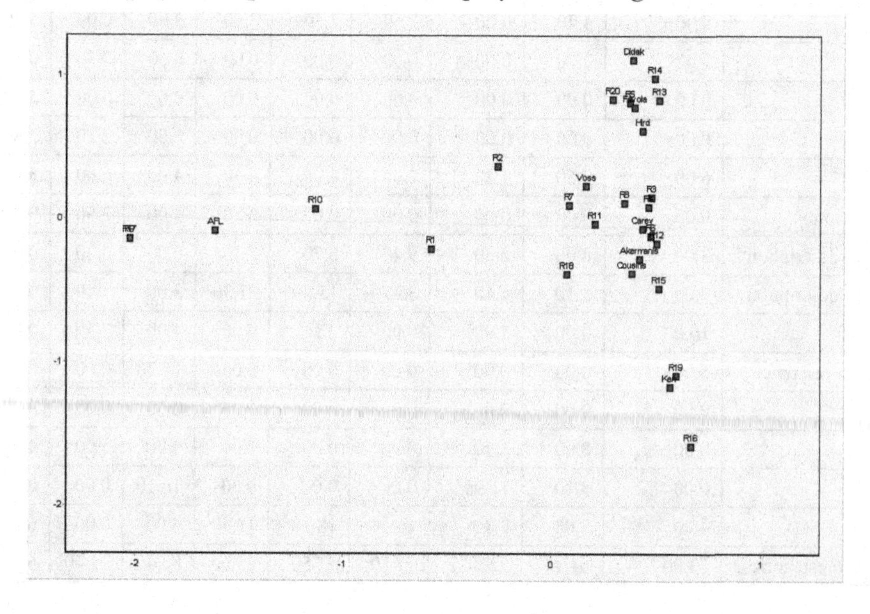

this cluster reflects the similarity of these players with other players, their proximity to numerous variables, and a relative closeness to the AFL.

## DISCUSSION

The focus of this chapter has been to recognize how consumer knowledge schemas can be identified in electronic communication. Further research should track how schemas transfer through the community in the form of memes. Wikipedia entries for various AFL players were analyzed to reveal the ways in which these players may be researched, or known, by the general public. Because Wikipedia is publicly authored, the possibility that incorrect or misleading information may be included is significant. This is analogous to the way information may pass through communities, where misunderstanding may lead to development of misinformation, and be passed on as fact. Wikipedia is not the only valid source for research into consumer schemas. In this example, one goal was to identify differences in player presentations, and players were selected on that basis. Players who were known to have been arrested or had negative publicity were sampled, along with some who had not. Wikipedia was the appropriate text source for selection on the basis that biographies existed for each of the players selected. It should be noted that not every AFL player has a Wikipedia profile, so implicit within this analysis is the expectation the each player sampled has a reasonably high level of consumer recognition, awareness or public profile.

Other text sources, such as blogs, discussion lists, and mainstream media would also be appropriate sources for analysis. In using any of these sources, the author, their purpose, and their readership should be considered. In the case of blogs, imagined readers, or the real comments of readers, have been known to influence how bloggers write, and the subjects they write on. Ewins (2005), a blogger and an academic, writes on the perils of blogging as he sees them. He notes:

> *Nobody wants to know what you had for breakfast unless you can turn it into a good story...The audience expects the politics-focused blogger to write about politics all of the time. Similarly, the humourous blogger is always expected to be funny, and the academic blogger always expected to be thoughtful.* (p. 373)

The implication is that when surveying blog posts (as opposed to discussion forums), we should expect a stylized account of events or presentations of people. As with mainstream media, some idea of a target audience exists for the blogger. Gossip or celebrity magazines produce articles on topics as varied as: which sex-symbol has cellulite, or which stars are having affairs with other peoples' partners. The same occurs in blogs. Authors are influenced by the sensational, the everyday is less extreme and less newsworthy. There may be no hard rules about blog postings, however, it should be considered that bloggers are producing a product, a text for consumption. If this is the case, while the language used may be more candid or colloquial, the overall impression provided for the reader is not unintentional or accidental, it has been scripted and produced.

The potential impact of blogs should not be underestimated. Already we have coaches and players' associations recommending that players do not read blog posts about themselves (Robinson, 2008). Highly critical or personal blog posts have led to the case where "AFL players have been banned from scouring fan websites, for fear that vicious player appraisals could lead to depression" (Robinson, 2008). The effect on players' state of mind is one thing, another is the question of authorship. Ambush marketing occurs where a company capitalizes upon association with an event, even though they are not official sponsors. Guerilla marketing, similarly, aims to capture consumer attention by stealth, where consumers may not be immediately aware that commercial messages are being passed. Pseudo/commercial blogs exist for commercial purposes – to exist within the blogosphere and exert a level of control over organization, brand or product image. The presentation of information may be as casual and informal as other blogs, the purpose of pseudo-blogs is to influence consumer word-of-mouth or purchasing. The source of information, intent of production of text, and authorship, should all be considered when conducting research using text produced by anonymous or pseudonymous authors.

## CONCLUSION

This research has compared profiles of athletes with their employing organization's desired profile in order to evaluate the degree of similarity or difference in these profiles. Where a significant and negative difference occurs between player profiles and organization image, the potential for damage to the organization image occurs. This can have obvious implications for both corporate image and the organization's ability to attract sponsorship and athletic staff.

The problem for a sporting bodies and sponsoring firms, is that few economically attractive means exist for establishing a picture of consumer perceptions of athletes and organizations. Surveys – a popular method – are prone to flaws in data collection and analysis. They are also expensive. With the availability of free data on the internet, a new method is therefore available. The use of social networking analysis and correspondence analysis, as we have demonstrated, goes a long way to addressing this problem.

## REFERENCES

Aaker, D.A. (1996). *Building Strong Brands*. New York: The Free Press.

Adamic, L., & Adar, E. (2005). How to search a social network. *Social Networks, 27*, 187-203.

AFL (2007). *111th Annual Report*. Retrieved May 18, 2008, from http://www.afl.com.au

AFP (2008). Super Bowl XLII pulls record audience. *ABC News*, 5 February. Retrieved March 22, 2008, from http://www.abc.net.au/news/stories/2008/02/05/2154417.htm

Associated Press (2008). Report: Arizona benefited over $500 million from Super Bowl. *National Football League, 16*. April. Retrieved 4 May, 2008 from http://www.nfl.com/superbowl/story?id=09000d5d807cc5e7

Benzecri, J.P. (1969). Statistical analysis as a tool to make patterns emerge from data. In S, Watanabe (Ed.), *Methodologies of Pattern Recognition* (pp. 35–74). New York: Academic Press, Inc.

Berthon, P., Pitt, L., Ewing, M., Ramaseshan, B., & Jayaratna, N. (2001). Positioning in cyberspace: Evaluating telecom web sites using correspondence analysis. *Information Resources Management Journal, 14*(1), 13-21.

Bhargava, R. (2007). 5 Reasons the NFL is the dominant sports brand in America. *Digital Media Wire, Inc., 5*. Retrieved 5 May, 2008, from http://www.dmwmedia.com/news/2006/09/20/5-reasons-the-nfl-is-the-dominant-sports-brand-in-america.htm

Borgatti, S.P., Everett, M.G. & Freeman, L.C. (2002). *Ucinet for Windows: Software for Social Network Analysis*. Harvard, MA: Analytic Technologies.

Boyd, D.M., & Ellison, N.B. (2007). Social network sites: Definition, history, and scholarship. *Journal of Computer-Mediated Communication, 13*(1). Retrieved April 23, 2008, from http://jcmc.indiana.edu/vol13/issue1/boyd.ellison.html

Breiger, R.L. (1976). Career attributes and network structure: A blockmodel study of a biomedical research specialty. *American Sociological Review, 41*(1), 117-135.

Brown, T.J. & Dacin, P.A. (1997). The Company and the product: Corporate associations and consumer product responses. *Journal of Marketing, 61*(January), 68-84.

Calantone, R.J., Di Benedetto, C.A., Hakam, A., & Bojanic, D.C. (1989). Multiple multinational tourism positioning using correspondence analysis. *Journal of Travel Research, 28*(2), 25-32.

Carley, K.M., Hummon, N.P., & Harty, M. (1993). Scientific influence: An analysis of the main path structure in the Journal of Conflict Resolution. *Science Communication, 14*, 417-447.

Cavin, C. (2004). Take a seat: Study puts Indy's capacity at 257,324. *USA Today*, 27 May. Retrieved March 18, 2008, from http:www.usatoday.com/sports/motor/irl/indy500/2004-05-27-attendance-count_x.htm

Celsi, R.L. & Olson, J.C. (1988). The role of involvement in attention and comprehension processes. *Journal of Consumer Research, 15*(September), 210-224.

Choi, S., Lehto, X.Y., & Morrison, A.M. (2007). Destination image representation on the web: Content analysis of Macau travel related websites. *Tourism Management, 28*, 118-129.

Cornwell, T.B., Pruitt, S., & Clark, J.M. (2005). The relationship between major-league sports' official sponsorship announcements and the stock prices of sponsoring firms. *Journal of the Academy of Marketing Science, 33*(4), 401-412.

Cornwell, T.B., Pruitt, S., & Van Ness, R. (2001). The value of winning in motor sports: sponsorship-linked marketing. *Journal of Advertising Research, 41*(1), 17-31.

Dacin, P. A. & Smith, D. C. (1994). The effect of brand portfolio characteristics on consumer evaluations of brand extensions. *Journal of Marketing Research*, 31(May), 229-242.

Dekker, A. (2001). Visualisation of social networks using CAVALIER™ . Paper presented at *Australian Symposium on Information Visualisation*, Sydney.

Doreian, P., & Fararo, T.J. (1985). Structural equivalence in a journal network. *Journal of the American Society for Information Science*, 36(1), 28-37.

dYdewalle, G., Vanden Abeele, P., Van Rensberger, J. & Coucke, P. (1988). Incidental processing of advertisements while watching soccer-games and broadcasts. In M. Gruneberg, P. Morris & R. Sykes (Eds.), *Practical Aspects of Memory: Current Research and Issues* (pp. 478-483). New York: John Wiley.

Ellison, N.B., Steinfield, C., & Lampe, C. (2007). The benefits of Facebook "friends": Social capital and college students' use of online social network sites. *Journal of Computer-Mediated Communication, 12*(4). Retrieved May 23, 2008, from http://jcmc.indiana.edu/vol12/issue4/ellison.html

Ewins, R. (2005). Who are you? weblogs and academic identity. *E-Learning, 2*(4), 368-377.

Faust, K. (2005). Using correspondence analysis for joint displays of affiliation networks. In P.J. Carrington, J. Scott, & S. Wasserman (Eds.), *Models and Methods in Social Network Analysis* (pp. 117-147). New York: Cambridge University Press.

Forsyth, E., & Katz, L. (1946). A matrix approach to the analyses of sociometric data: Preliminary report. *Sociometry, 9*(4), 340-347.

Freeman, L.C. (1984). The impact of computer based communication on the social structure of an emerging scientific specialty. *Social Networks, 6*, 201-221.

Gardner, M.P., & Shuman, P. (1988). Sponsorships and small businesses. *Journal of Small Business Management, 26*(4), 44-52.

Granovetter, M. (1983). The strength of weak ties: a network theory revisited. *Sociological Theory, 1*, 201-223.

Greenacre, M., & Hastie, T. (1987). The geometric interpretation of correspondence analysis. *Journal of the American Statistical Association, 82*(398), 437-447.

Gwinner, K.P. (1997). A model of image creation and image transfer in event sponsorship. *International Marketing Review, 14*(3), 170-182.

Gwinner, K.P., & Eaton, J. (1999). Building brand image through event sponsorship: The role of image transfer. *Journal of Advertising, 28*(Winter), 47-57.

Hoffman, D.L., & Franke, G.R. (1986). Correspondence analysis: graphical representation of categorical data in marketing research. *Journal of Marketing Research, 23*, 213-227.

Javalgi, R.G., Traylor, M.B., Gross, A.C., & Lampman, E. (1994). Awareness of sponsorship and corporate image: An empirical investigation. *Journal of Advertising, 23*(4), 47-58.

Judd, C.M., Drake, R.A., Downing, J.W., & Krosnick, J.A. (1991). Some dynamic properties of attitude structures: Context-induced response facilitation and polarization. *Journal of Personality and Social Psychology, 60*(2), 193-202.

Koku, E., Nazer, N., & Wellman, B. (2001). Netting scholars: Online and offline. *The American Behavioural Scientist, 44*(10), 1752-1774.

Matzat, U. (2004). Academic communication and Internet Discussion Groups: Transfer of information or creation of social contacts? *Social Networks, 26*, 221-255.

McLean, P.D., & Padgett, J.F. (1997). Was Florence a perfectly competitive market? Transactional evidence from the Renaissance. *Theory and Society, 26*(2/3), 209-244.

McNamee, S.J., & Willis, C.L. (1994). Stratification in science: A comparison of publication patterns in four disciplines. *Science Communication, 15*, 396-416.

Meenaghan, T. (1983). Commercial sponsorship. *European Journal of Marketing, 17*(7), 5-73.

de Moor, A., & Efimova, L. (2004). An argumentation analysis of weblog conversations. In M. Aakhus & M. Lind (Eds.), *Proceedings of the 9th International Working Conference on the Language-Action Perspective on Communication Modeling.* Retrieved May 17, 2008, from http://www.scils.rutgers.edu/lap04/lap04.htm

Moreno, J.L. (1946). Sociogram and sociomatrix: A note to the paper by Forsyth and Katz. *Sociometry, 9*(4), 348-349.

NFL (n.d.) *Super Bowl tickets and rings.* Retrieved June 22, 2008, from http://www.nfl.com

Noma, E. (1982). Untangling citation networks. *Information Processing & Management, 18*(2), 43-53.

Padgett, J.F., & Ansell, C.K. (1993). Robust action and the rise of the Medici, 1400-1434. *American Journal of Sociology, 98*(6), 1259-1319.

Padgett, J.F., & McLean, P.D. (2006). Organizational invention and elite transformation: The birth of partnership systems in Renaissance Florence. *American Journal of Sociology, 111*(5), 1463-1568.

Pham, M.T. (1992). Effect of involvement, arousal and pleasure on the recognition of sponsorship stimuli. *Advances in Consumer Research, 19*, 1-9.

Powell, E. (2008). ABC, ESPN, IMS radio network providing in-depth Indy 500 coverage. *Indy500.com*, 2 May. Retrieved May 29 from http://www.indy500.com/news/11179/ABC_ESPN_IMS_Radio_Network_Providing

RepriseMedia (2008). *Super Bowl advertisers fumble in search and social media.* Retrieved March 19, 2008, from http://www.reprisemedia.com/pressreleases

Robinson, M. (2008). Players cop a blogging. *Herald Sun*. Retrieved June 8, 2008, from http://www.news.com.au/heraldsun/sport/afl/story/0,26576,23732964-19742,00.html

Scott, J. (2000). *Social Network Analysis: A Handbook* (2nd. Ed.). Newbury Park: Sage.

Van Osselaer, S.M.J., & Janiszewski, C. (2001). Two ways of learning brand associations. *Journal of Consumer Research, 28*(September), 202-223.

Wasserman, S. & Faust, K. (1994). *Social Network Analysis: Methods and Applications*. New York: Cambridge University Press.

Woelfel, J.K., & Woelfel, J.D. (1997). *CATAPC 2.0: Software for Content Analysis*. Buffalo: Galileo Company.

Yeung, Y., Liu, T.C., & Ng, P. (2005). A social network analysis of research collaboration in physics education. *American Journal of Physics, 73*(2), 145-150.

# Compilation of References


(TC)². (nd). Retrieved May 12, 2008, from http://www.tc2.com

Aaker, D.A. (1996). *Building Strong Brands*. New York: The Free Press.

Abdel-Aziz, Y.I., & Karara, H.M. (1971). Direct linear transformation from computer coordinates into object coordinates in close-range photogrammetry. In *Proceedings of the ASP Symposium on Close-Range Photogrammetry* (pp.1-18). Falls Church, VA: American Society of Photogrammetry.

About us- The sports industry. (2008). *Street & Smith's SportsBusiness Journal*. Retrieved March 1, 2008, from http://www.sportsbusinessjournal.com/index.cfm?fuseaction=page.feature&featureId=43

ABS, Australian Bureau of Statistics. (2007). *Sports attendance, Australia, 2005-06,* (Catalog No. 4174.0). Canberra: ABS. Retrieved February 21, 2008, from http://abs.gov.au/AUSSTATS/abs@.nsf/Lookup/4174.0Main+Features12005-06?OpenDocument.

Adamic, L., & Adar, E. (2005). How to search a social network. *Social Networks, 27*, 187-203.

Adams, E. (2005). The Designer's Notebook: Bad game designer, no Twinkie! VI. *Gamasutra*. Retrieved August 8, 2007, from http://www.gamasutra.com/features/20050603/adams_01.shtml.

Adams, E. (2006). Cheer up! Video games are in great shape. *Gamasutra*. Retrieved August 8, 2007, from http://www.gamasutra.com/features/20060421/adams_01.shtml.

Adams, R. (2004, August 30). In-game advertising a growing, but not easy, sell. *Street & Smith's SportsBusiness Journal, 19*.

Adams, R. (2005, January 24). ESPN gets $850 million to stay in the game. *Street & Smith's SportsBusiness Journal, 5*.

Adcock, M. (2008). *Are we there yet?* Retrieved April 30, 2008, from http://web.media.mit.edu/~matta/projects.html.

Adler, M. (1985). Stardom and talent. *American Economic Review, 75*(1), 208-212.

AFL (2007). *111th Annual Report*. Retrieved May 18, 2008, from http://www.afl.com.au

AFP (2008). Super Bowl XLII pulls record audience. *ABC News*, 5 February. Retrieved March 22, 2008, from http://www.abc.net.au/news/stories/2008/02/05/2154417.htm

Aggarwal, J.K., & Cai, Q. (1994). Human Motion Analysis: A Review. In J.K. Aggarwal (Ed.) *Proceedings of IEEE Workshop on Motion of Non-Rigid and Articulated Objects* (pp. 2-14), Austin, TX:IEEE.

Ahmadi, A., Rowlands, D.D., & James, D. (2006). Investigating the translational and rotational motion of the swing using accelerometers for athlete skill assessment. In *Proceedings of the 5th International IEEE Senors Conference* (pp. 980-983). Daegu Exhibition & Convention Centre (EXPO), Daegu, Korea.

Alciatore, D.G. (2004). *The Illustrated Principles of Pool and Billiards*. Sterling Publishing.

Alian, M., & Shouraki, S. (2004). A fuzzy pool player robot with learning ability. In *WSEAS Trans. on Electronic, 1*, 422–425.

Alian, M., Lucas, C., & Shouraki, S. (2004). Evolving game strategies for pool player robot. In *4th WSEAS Intl. Conf. on Sim., Mod. and Opt.*

Alian, M.E., Shouraki, S., Shalmani, M., Karimian, P., & Sabzmeydani, P. (2004b). Roboshark: A gantry pool player robot. In *ISR 2004: 35th Intl. Sym. Rob.*

Alpert, F. (2007). Entertainment software: Suddenly huge, little understood. *Asia Pacific Journal of Marketing and Logistics, 19*(1), 87-100.

Alreck, P.L., & Settle, R.B. (2004). *The Survey Research Handbook* (Third ed.). Boston, MA: McGraw-Hill Irwin.

Al-Zubaidi, K., & Stevens, G. (2004). *CSCP at Work.* Paper presented at the Proceedings of the Conference Mensch und Computer 2004 (MC 2004).

Anderson, C.A., & Bushman, B.J. (2001). Effects of violent video games on aggressive behavior, aggressive cognition, aggressive affect, physiological arousal, and prosocial behavior: A meta-analytic review of the scientific literature. *Psychological Science, 12*(5), 353-359.

Anderson, J., & Done, S. (2003). *The Official Liverpool FC Illustrated Encyclopedia.* London: Carlton Publishing Group.

Andrisani, J. (2002). *Think like Tiger: An analysis of Tiger Woods' mental game.* New York: Penguin Putnam.

Anshel, M. A. (Ed.). (1991). *Dictionary of the Sports and Exercise Sciences.* Illinois.

Apple. (2008). *Apple – Nike + iPod.* Retrieved April 30, 2008, from http://www.apple.com/ipod/nike/.

Ariely, D. (2000). Controlling the information flow: Effects on consumers' decision making and preferences. *Journal of Consumer Research, 27*(2), 233-249.

Assael, Pope, N., Brennan, L., & Voges, K. (2007). *Consumer Behaviour.* Brisbane: John Wiley and Sons.

Associated Press (2008). Report: Arizona benefited over $500 million from Super Bowl. *National Football League, 16.* April. Retrieved 4 May, 2008 from http://www.nfl.com/superbowl/story?id=09000d5d807cc5e7

Australian Bureau of Statistics. (2001). *Voluntary work Australia,* (Catalog No. 4441.0). Canberra: ABS.

Australian Bureau of Statistics. (2003). *General social survey. Summary results Australia,* (Catalog No. 4159.0). Canberra: ABS.

Australian Bureau of Statistics. (2003). *Sport and recreation: A statistical overview,* (Catalog No. 4156.0). Canberra: ABS.

Australian Bureau of Statistics. (2006). *Sport and recreation: A statistical overview, Australia, 2006 Edition 2,* (Catalog No. 4156.0). Canberra: ABS.

Australian Government. (2005). *Giving Australia: Research on philanthropy in Australia.* Canberra: ACOSS.

Australian Institute of Sport (Sport Science and Medicine) (2008). Retrieved 12 March, from http://www.ausport.gov.au/ais/sssm/quality_assurance

Babiel, S., Hartmann, U., Spitzenpfeil, P., & Mester, J. (1997). Ground reaction forces in alpine skiing, cross-country skiing and ski jumping. In E. Muller, H. Schwameder, E. Kornexl & C. Raschner (Eds.), *Science and skiing* (pp. 200-207). USA: Taylor & Francis.

Badenoch, D. (2005). *Interdisciplinary Curriculum Design and Teaching for Integrative Learning and Development of Graduate Qualities: A case study of reconceptualising Physical Education: University of South Australia.* Paper presented to the Division of Education, Arts & Social Sciences' Teaching & Learning Colloquium, University of South Australia.

Bahamonde, R. (2000). Changes in angular momentum during the tennis serve. *Journal of Sports Sciences, 18,* 579-592.

Bahamonde, R.E. (1994). *Biomechanical analysis of serving during the performance of flat and slice ten-*

*nis serves.* Unpublished doctoral dissertation, Indiana University, USA.

Bahn, K.D. (1986). How and when do brand perceptions and preferences first form? A cognitive developmental investigation. *Journal of Consumer Research, 13*(3), 382-393.

Bailey, K.D. (1994). *Methods of Social Research* (Fourth ed.). New York, NY: The Free Press.

Bailey, R. (2005). Evaluating the relationship between physical education, sport and social inclusion. *Educational Review, 57*(1), 71-90.

Balabanis, G., & Reynolds, N.L. (2001). Consumer attitudes towards multi-channel retailers' web sites: The role of involvement, brand attitude, internet knowledge and visit duration. *Journal of Business Strategies, 18*(2), 105-132.

Balague, G. (2005, September 12). *Why Liverpool FC Reign in Spain.* Retrieved September 11, 2007, from http://www.liverpoolfc.tv/news/drilldown/N149947050912-1150.htm

Baldaro, B., Tuozzi, G., Codispoti, M., Montebarocci, O., Barbagli, F., Trombini, E., & Rossi, N. (2004). Aggressive and non-violent videogames: Short-term psychological and cardiovascular effects on habitual players. *Stress and Health, 20*(4), 203-208.

Bambauer, S. (2006). Effects of brand placement in PC/video games on the change of the attitude toward the advertised brand. In D. Grewal (Ed.), *Proceedings of the 2006 Summer Marketing Educators' Conference* (pp. 231-240). Chicago: American Marketing Association.

Baranowski, T., Baranowski, J., Cullen, K.W., Marsh, T., Islam, N., Zakeri, I., Honess-Morreale, L., & deMoor, C. (2003). Squire's Quest!: Dietary outcome evaluation of a multimedia game. *American Journal of Preventive Medicine, 24*(1), 52-61.

Barkhuus, L., Maitland, J., Anderson, I., Sherwood, S., Hall, M., & Chalmers, M. (2006). Shakra: Sharing and motivating awareness of everyday activity. In *Ubicomp 2006.* California, USA: ACM Press.

Bartle, R., & Trubshaw, R. (1979). *Multi-User Dungeon.* Colchester, UK: Essex University.

Bassanini, A., & Scarpetta, S. (2002). Growth, technological change, and ICT diffusion:

Baumann, M. (2007). Slimming down with Wii™ Sports. *Information Today, 24*(4), 47.

Bavelier, D., & Green, C. (2003). Action video game modifies visual selective attention. *Nature, 423*(6939), 534-537.

BCA (2008). Billiard congress of america. Retrieved from http://www.bca-pool.com.

Beattie, A.E., & Mitchell, A.A. (1985). The relationship between advertising recall and persuasion: An experimental investigation. In L.F. Alwitt & A.A. Mitchell (Ed.), *Psychological processes and advertising effects* (pp. 129-155). Hillsdale, NJ: Lawrence Erlbaum Associates.

Bebie, T., & Bieri, H. (1998). *SoccerMan - reconstructing soccer games from video sequences.* Paper presented at the International Conference on Image Processing, Chicago, USA.

Beebe, L.H. (2007). What Can We Learn From Pilot Studies? *Perspectives in Psychiatric Care, 43*(4), 213-218.

Behrenshausen, B. G. (2007). Toward a (Kin) Aesthetic of Video Gaming: The Case of Dance Dance Revolution. *Games and Culture, 2*(4).

Behrenshausen, B.G. (2007). Toward a (kin)aesthetic of video gaming. *Games and Culture, 2*(4), 335-354.

Bellah, R., Madsen, R., Sullivan, W., Swidler, A., & Tipton, S. (1985). *Habits of heart: Individualism and commitment in American life.* New York: Harper and Row.

Ben-Porat, A. (2000). Overseas Sweetheart: Israeli Fans of English Football. *Journal of Sport & Social Issues, 24*(4), 344-350.

Benz, M-A. (2007). *Strategies in markets for experience and credence goods.* Wiesbaden: DUV Gabler Edition.

Benzecri, J.P. (1969). Statistical analysis as a tool to make patterns emerge from data. In S, Watanabe (Ed.), *Methodologies of Pattern Recognition* (pp. 35–74). New York: Academic Press, Inc.

Bernhard, B.J., & Eade, V.H. (2005). Gambling in a fantasy world: An exploratory study of rotisserie baseball games. *UNLV Gaming Research & Review Journal, 9*(1), 29-42.

Bernstein, N. (1967). *The Co-ordination and Regulation of Movements.* Oxford, England, Pergamon Press.

Berthon, P., Pitt, L., Ewing, M., Ramaseshan, B., & Jayaratna, N. (2001). Positioning in cyberspace: Evaluating telecom web sites using correspondence analysis. *Information Resources Management Journal, 14*(1), 13-21.

Bezjian-Avery, A., Calder, B., & Iacobucci, D. (1998). New media interactive advertising vs traditional advertising. *Journal of Advertising Research, 38*(4), 23-32.

Bhargava, R. (2007). 5 Reasons the NFL is the dominant sports brand in America. *Digital Media Wire, Inc., 5.* Retrieved 5 May, 2008, from http://www.dmwmedia.com/news/2006/09/20/5-reasons-the-nfl-is-the-dominant-sports-brand-in-america.htm

Bianchi-Berthouze, N., Kim, W., & Patel, D. (2007). Does body movement engage you more in digital game play? And why? In *Affective computing and intelligent interaction* (pp. 102-113). (vol. 4738). Berlin: Springer.

Big Dawg Baseball. (2007). 2007 fantasy projections now available! Retrieved February 5, 2007, from http://www.bigdawgbaseball.com/?gclid=CIOQpcP4l4oCFRsZVAod_3S3nA

Birch, D. (2004, August 25). Fantasyland. *Modesto Bee*, F1.

Björk, S., Holopainen, J., Ljungstrand, P., & Mandryk, R. (2002). Special Issue on Ubiquitous Games. *Personal Ubiquitous Comput., 6*(5-6), 358-361.

Black, J., & Ellis, T. (2006). Multi camera image tracking. *Image and Vision Computing, 24*(11), 1256-1267.

Blajer, W., & Czaplicki, A. (2001). Modeling and Inverse Simulation of Somersaults on the Trampoline. *Journal of Biomechanics, 34*, 1619-1629.

Blajer, W., & Czaplicki, A. (2003). Contact Modeling and Identification of Planar Somersaults on the Trampoline. *Multibody System Dynamics, 10*, 289-312.

Bocij, P., Chaffey, D., Greasley, A., & Hickie, S. (2006). *Business information systems.* (3rd ed.). Harlow: Pearson Education Limited.

Bodypad. (2008). *Home.* Retrieved April 30, 2008, from http://www.bodypad.com.

Boehret, K. (2008). The Mossberg Solution: For Wii™ lovers, something worth sweating over. *Wall Street Journal, May 14*, D1.

Bogost, I. (2004). *Asynchronous multiplay: Futures for casual multiplayer experience.* Paper presented at the Other Players Conference on Multiplayer Phenomena, Copenhagen, Denmark.

Boivin, S., & Gagalowicz, A. (2001). Image-based Rendering of Diffuse, Specular and Glossy Surface from a Single Image. *SIGGRAPH*, (pp. 107-116).

Bolter, J.D., & Grushin, R. (2000). *Remediation-Understanding new mMedia.* USA: MIT Press Paperback.

Bonis, J. (2007). Acute Wiiitis. *New England Journal of Medicine, 356*(23), 2431.

Borgatti, S.P., Everett, M.G. & Freeman, L.C. (2002). *Ucinet for Windows: Software for Social Network Analysis.* Harvard, MA: Analytic Technologies.

Bottino, A., & Laurentini, A. (2001). A Silhouette Based Technique for the Reconstruction of Human Movement. *Computer Vision and Image Understanding, 83*, 75-95.

Bouguet, J-Y. (nd). *Camera calibration toolkit for Matlab and in OpenCV.* Retrieved May 12, 2008, from http://www.vision.caltech.edu/bouguetj/calib_doc/

Boyd, D.M., & Ellison, N.B. (2007). Social network sites: Definition, history, and scholarship. *Journal of Computer-Mediated Communication, 13*(1). Retrieved

April 23, 2008, from http://jcmc.indiana.edu/vol13/issue1/boyd.ellison.html

Boyle, A., Macleod, M., Slevin, A., Sobecka, N., & Burton, P. (1993). The use of information technology in the voluntary sector. *International Journal of Information Management,* (pp. 94-112).

Branscombe, N. R., & Wann, D. L. (1991). The Positive Social and Self Concept Consequences of Sports Team Identification. *Journal of Sport and Social Issues, 15*(2), 115-127.

Brasington, R. (1990). Nintendinitis. *New England Journal of Medicine, 322*(20), 1473-1474.

Bregler, C., Malik, J., & Pullen, K. (2004). Twist Based Acquisition and Tracking of Animal and Human kinematics. *International Journal of Computer Vision, 56*, 179-194.

Breiger, R.L. (1976). Career attributes and network structure: A blockmodel study of a biomedical research specialty. *American Sociological Review, 41*(1), 117-135.

Brightman, J. (2007). WMS: PS3 to 'win' console war because of Blu-ray. *GameDaily.* Retrieved July 2, 2008, from http://www.gamedaily.com/articles/features/wms-ps3-to-win-console-war-because-of-blu-ray/70379/?biz=1.

Brookfield, R. (1983). *Adult learners, adult education and the community.* New York: Teachers Collage Press.

Brown, D. (2006). Playing to win: Video games and the fight against obesity. *Journal of the American Dietetic Association, 106*(2), 188-189.

Brown, D.L., & Wheatley, G.H. (1989).Relationship between spatial Knowledge. In C. Maher, G. Goldin, & R. Davis (Eds.), *Proceedings of the 11th Annual Meeting, North America Chapter of the International Group for the Psychology of Mathematic Education* (pp. 143-148). New Brunswick, NJ: International Group for the Psychology of Mathematics Education.

Brown, E., & Cairns, P. (2004). A grounded investigation of game immersion. In *Extended Abstracts of the 2004 Conference on Human Factors in Computing Systems* (pp. 1297-1300). New York: ACM Press.

Brown, J.B., Collins, A., & Duguid, P. (1989). *Situated cognition and the culture of learning.* USA: Educational Researcher.

Brown, T.J. & Dacin, P.A. (1997). The Company and the product: Corporate associations and consumer product responses. *Journal of Marketing, 61*(January), 68-84.

Browne, K. (1999). The branding of soccer. *Finance Week*(January 29), 26.

Browne, K. F. (2006, June). *Toward a digital advertising ecosystem for games.* Paper presented at the Game Developers Conference (GDC) Focus on: Game Advertising Summit, San Francisco, CA.

Bueng R.S., Bying-Ji, Y., & Sang-Rok, O., & Yound Soo, K. (2004). Landing Motion Analysis of Human-Body Model Considering Impact and ZMP condition. *In Proceedings of 2004 IEEE/RSJ Intl. Conf. Intelligent Robots and Systems*, Sendeai, Japan, (pp. 1972-1978).

Bull, S.J., Albinson, J.G., & Shambrook, C.J. (1996). *The mental game plan: Getting psyched for sport.* Eastbourne, UK: Sports Dynamics.

Burgess, D. (1992). Techniques for low-cost spatial audio. In *Proceedings of UIST'92.* California, USA: ACM.

Busch, A., & James, D.A. (2007). Analysis of cricket shots using inertial sensors. In F. Fuss, A. Subic & S. Ujihashi (Eds.), *The impact of technology on sport II* (pp. 317-322). Netherlands: Taylor & Francis.

Business Wire. (2006, June 13). *Parks Associates: PC in-game advertising revenue to top $400 million by 2009; Male gamers 18 - 34 more open to ads.* Retrieved via Lexis Nexus Database and www.businesswire.com

Cahill, L., & McGaugh, J.L. (1995). A novel demonstration of enhanced memory associated with emotional arousal. *Consciousness and Cognition, 4*(4), 410-421.

Calantone, R.J., Di Benedetto, C.A., Hakam, A., & Bojanic, D.C. (1989). Multiple multinational tourism positioning using correspondence analysis. *Journal of Travel Research, 28*(2), 25-32.

Calvert, S.L., & Tan, S. (1994). Impact of virtual reality on young adults' physiological arousal and aggressive thoughts: Interaction versus observation. *Journal of Applied Developmental Psychology, 15*(1), 125-139.

Carley, K.M., Hummon, N.P., & Harty, M. (1993). Scientific influence: An analysis of the main path structure in the Journal of Conflict Resolution. *Science Communication, 14*, 417-447.

Carnagey, N.L., Anderson, C.A., & Bushman, B.J. (2007). The effect of video game violence on physiological desensitization to real-life violence. *Journal of Experimental Social Psychology, 43*(3), 489-496.

Caron, F. (2008). Gaming expected to be a $68 billion business by 2012. *Ars Technica*. Retrieved July 2, 2008, from http://arstechnica.com/news.ars/post/20080618-gaming-expected-to-be-a-68-billion-business-by-2012.html.

Carr, N. (2001). *The digital enterprise*. Boston: Harvard Business School Press.

Carranza, J., Theobalt, C., Magnor, M., & Seidel, H-P. (2003). Free-Viewpoint Video of Human Actors. *SIGGRAPH*, (pp. 569-577).

Castronova, E. (2005). *Synthetic worlds: The business and culture of online games*. Chicago: University of Chicago Press.

Cavin, C. (2004). Take a seat: Study puts Indy's capacity at 257, 324. *USA Today*, 27 May. Retrieved March 18, 2008, from http:www.usatoday.com/sports/motor/irl/indy500/2004-05-27-attendance-count_x.htm

CBSSportsLine.com. (2006). CBSSportsLine.com fantasy rules. Retrieved October 21, 2006, from http://football.sportsline.com/splash/football/spln/single/rules

Celsi, R.L. & Olson, J.C. (1988). The role of involvement in attention and comprehension processes. *Journal of Consumer Research, 15*(September), 210-224.

Chambers, J.H., & Ascione, F.R. (1987). The effects of prosocial and aggressive videogames on children's donating and helping. *Journal of Genetic Psychology, 148*(4), 499-505.

Chan, E., & Vorderer, P. (2006). Massively multiplayer online games. In P. Vorderer & J. Bryant (Ed.), *Playing video games: Motives, responses and consequences* (pp. 77-90). USA: Routledge.

Chang, S.W.S. (1994). *Automating Skills Using a Robot Snooker Player*. PhD thesis, Bristol University.

Charity Commission for England and Wales. (2002). *Giving confidence in charities: Annual report 2001–2002*. London.

Cheng, B., Li, J., & Yang, J. (2004). Design of the neural-fuzzy compensator for a billiard robot. In *IEEE Intl. Conf. Networking, Sensing & Control*, (pp. 909–913).

Cheung, G., Baker, S., & Kanade, T. (2003a). Shape-from-silhouette for articulated objects and its use for human body kinematics estimation and motion capture. In Y. Matsushita (Ed.), *Proceedings of the IEEE Conference on Computer Vision and Pattern Recognition*, (pp. 77-84), Madison, WI: IEEE.

Cheung, G., Baker, S., & Kanade, T. (2003b). Visual Hull Alignment and Refinement Across Time: A 3D Reconstruction Algorithm Combining Silhouette with Stereo. In Y. Matsushita (Ed.), *Proceedings of the IEEE Conference on Computer Vision and Pattern Recognition*, (pp. 375-382), Madison, WI: IEEE.

Cho, C.H., & Leckenby, J.D. (1999). Interactivity as a measure of advertising effectiveness: Antecedents and consequences of interactivity in web advertising. In M.S. Roberts (Ed.), *Proceedings of the 1999 Conference of the American Academy of Advertising* (pp. 162-179). Gainesville, FL: American Academy of Advertising.

Choi, K., & Seo, Y. (2004, May). *Probabilistic tracking of the soccer ball*. Paper presented at the International Workshop on Statistical Methods in Video Processing, in conjunction with ECCV, Prague, Czech Republic.

Choi, K., Park, B., Lee, S., & Seo, Y. (2006). Tracking the ball and players from multiple football videos. *International Journal of Information Acquisition, 3*(2), 121-129.

Choi, S., Lehto, X.Y., & Morrison, A.M. (2007). Destination image representation on the web: Content analysis of Macau travel related websites. *Tourism Management, 28*, 118-129.

Chou, T.J., & Ting, C.C. (2003). The role of flow experience in cyber-game addiction. *CyberPsychology & Behavior, 6*(6), 663-675.

Chow, J., Carlton, L., Lim, Y., Chae, W., Shim, J., Kuenster, A.F., & Kokuban, K. (2003). Comparing the pre- and post-impact ball and racquet kinematics of elite tennis players' first and second serves: A preliminary study. *Journal of Sports Sciences, 21*, 529-537.

Chua, S., Wong, E., Tan, A.W., & Koo, V. (2002). Decision algorithm for pool using fuzzy system. In *iCAiET 2002: Intl. Conf. AI in Eng. & Tech.*, (pp. 370–375).

Chua, S.C., Wong, E.K., & Koo, V.C. (2005). *Intelligent Pool Decision System Using Zero-Order Sugeno Fuzzy System, 44*. Springer Netherlands.

Chua, S.C., Wong, E.K., & Koo, V.C. (2007). Performance evaluation of fuzzy-based decision system for pool. *Appl. Soft Comput.*, *7*(1), 411–424.

Cialdini, R.B., Borden, R.J., Thorne, A., Walker, M.R., Freeman, S., & Sloan, L.R. (1976). Basking in Reflected Glory: Three (Football) Field Studies. *Journal of Personality and Social Psychology, 34*(3), 366-375.

Cianfrone, B.A. (2007). *The influence of motives and consumption of sport video games on sponsorship effectiveness.* Doctoral dissertation, University of Florida, Florida.

Cianfrone, B.A., & Zhang, J.J. (2006). Differential effects of television commercials, athlete endorsements, and venue signage during a televised action sports event. *Journal of Sport Management, 20*(3), 322-344.

Cianfrone, B.A., Zhang, J.J., Trail, G.T., & Lutz, R.J. (2008). Effectiveness of in-game advertisements in sport video games: An experimental inquiry on current gamers. *International Journal of Sport Communication, 1*, 195-218.

Ciarli, R., & Rabellotti, R. (2007). ICT in industrial districts: An empirical analysis on adoption, use and impact. *Industry and Innovation,* (pp. 277-303).

Clary, E., & Snyder, M. (1991). *A functional analysis and prosocial behaviour: The case of volunteerism.* Newbury Park: Sage.

Cohen, J.B. (1983). Involvement and you: 1000 great ideas. In R.P. Bagozzi & A.M. Tybout (Eds.), *Advances in consumer research* (vol. 10, pp. 325-328). Ann Arbor, MI: Association for Consumer Research.

Coleman, R. (2002). Characteristics of volunteering in UK sport: Lessons from cricket. *Managing Leisure,* (pp. 220-238).

Comer, D.R. (2001). Not just a Mickey Mouse exercise: Using Disney's The Lion King to teach leadership. *Journal of Management Education, 25*(4), 430-436.

Committee on Sporting Singapore (CoSS) (2000). Retrieved May 12, 2008, from http://app.mcys.gov.sg/web/sprt_towards_committeesporting.asp

comScore. (2007). Worldwide online gaming community reaches 217 million people. Retrieved July 20, 2008, from http://www.comscore.com/press/release.asp?press=1521.

Consolvo, S., Everitt, K., Smith, I., & Landay, J.A. (2006). Design requirements for technologies that encourage physical activity. In *Proceedings of CHI 2006* (pp. 457-466). Quebec, Canada: ACM Press.

Cootes, T., & Taylor, C. (1992). *Active shape models - 'smart snakes'.* Paper presented at the British Machine Vision Conference, Leeds, UK.

Corazza, S., Mundermann, L., & Andriacchi, T. (2006). The Evolution of Methods for the Capture of Human Movement Leading to Markerless Motion Capture for Biomechanical Applications. *Journal of Neuroengineering and Rehabilitation, 3*, 6.

Coriolis, G.-G. (1835). *Théorie mathématique des effets du jeu de billard.* J. Gabay, Paris, France.

Cornwell, T.B., Pruitt, S., & Clark, J.M. (2005). The relationship between major-league sports' official sponsorship announcements and the stock prices of sponsoring firms. *Journal of the Academy of Marketing Science, 33*(4), 401-412.

Cornwell, T.B., Pruitt, S., & Van Ness, R. (2001). The value of winning in motor sports: sponsorship-linked marketing. *Journal of Advertising Research, 41*(1), 17-31.

Coutinho, C., Pezarat, P., & Veloso, A. (2004). EMG patterns of the upper limb muscles in the first (flat) and second (topspin) serve. *Medicine and Science in Tennis, 9*(3), 14-15.

Coyle, J.R., & Thorson, E. (2001). The effects of progressive levels of interactivity and vividness in Web marketing sites. *Journal of Advertising, 30*(3), 65-77.

Crichton, S., & Kinash, S. (2003). Virtual Ethnography: Interactive Interviewing Online as Method [Electronic Version]. *Canadian Journal of Learning and Technology, 29.* Retrieved July 13, 2006 from http://www.cjlt.ca/content/vol29.2/cjlt29-2_art-5.html.

Cricket Australia. (2006). *Annual report 2005-2006.* Melbourne: Cricket Australia.

Crompton, J.L. (2004). Conceptualization and alternate operationalizations of the measurement of sponsorship effectiveness in sport. *Leisure Studies, 23*(3), 267-281.

Csikszentmihalyi, M. (1977). *Beyond boredom and anxiety.* San Francisco: Jossey-Bass.

Csikszentmihalyi, M., & LeFevre, J. (1989). Optimal experience in work and leisure. *Journal of Personality and Social Psychology, 56*(5), 815-822.

Cui, C., & Shi, J. (2004). Analysis of feature extraction in 3D model retrieval. *Journal of Computer-aided Design & Computer Graphics, 16*(7), 882-889.

Cuneo, Z.A. (2004, 19 January). Marketers game for action; Videogames: With men, gaming is bigger than the Sopranos. *Advertising Age,* s8.

Cuskelly, G. (1995). The influence of committee functioning on the organisational commitment of volunteer administrators in sport. *Journal of Sport Behaviour,* (pp. 254-270).

Cutmore, T.R.H., & James, D.A. (1999). Identifying and reducing noise in psychophysiological recordings. *International Journal of Psychophysiology, 32*(2), 129-150.

Cyberathlete Professional League. (2008). About us. Retrieved July 17, 2008, from http://www.thecpl.com/index.php?page_id=1090.

Cyberware (n.d.). Retrieved May 12, 2008, from http://www.cyberware.com

Czyz, J., Ristic, B., & Macq, B. (2007). A particle filter for joint detection and tracking of color objects. *Image and Vision Computing, 25*(8), 1271-1281.

D'Orazio, T., Leo, M., Spagnolo, P., Mazzeo, P.L., Mosca, N., & Nitti, M. (2007). *A Visual Tracking Algorithm for Real Time People Detection.* Paper presented at the Eighth International Workshop on Image Analysis for Multimedia Interactive Services. WIAMIS'07, Santorini, Greece.

Dacin, P. A. & Smith, D. C. (1994). The effect of brand portfolio characteristics on consumer evaluations of brand extensions. *Journal of Marketing Research,* 31(May), 229-242.

Damasio, A. (2000). *The feeling of what happens: Body, emotion and the making of consciousness.* Washington: Harvest Books.

Dambreville, S., Rathi, Y., & Tannenbaum, A. (2006). *Shape-based approach to robust image segmentation using kernel pca.* Paper presented at the IEEE Conference on Computer Vision and Pattern Recognition, New York, USA.

Daniel, E., Wilson, H., & Myers, A. (2002). Adoption of e-commerce by SMEs in the UK:

Daveri, R. (2002). *The new economy in Europe (1992–2001).* Working Paper Series, WP No. 213 IGIER, Universita Bocconi, Milan.

Davey, N.P., James, D.A., & Anderson, M.E. (2004). Signal analysis of accelerometry data using gravity based modeling. In *Proceedings of SPIE: Vol. 5274* (pp. 362-370). USA: SPIE.

Davis, N.W., & Duncan, M.C. (2006). Sports knowledge is power: Reinforcing masculine privilege through fantasy sport league participation. *Journal of Sport & Social Issues, 30*(3), 244-264.

Dawley, D., Stephens, R., & Stephens, B. (2005). Dimenstionality of organisational commitment in volunteer workers: Chamber of Commerce board members and role fulfilment. *Journal of Vocational Behaviour,* (pp. 511-525).

de Castell, S., & Jenson, J. (2004). Paying attention to attention: New economies for learning. *Educational Theory, 54*(4), 381-398.

De Lisi, R., & Wolford, J.L. (2002). Improving children's mental rotation accuracy with computer game playing. *Journal of Genetic Psychology, 163*(3), 272-282.

de Moor, A., & Efimova, L. (2004). An argumentation analysis of weblog conversations. In M. Aakhus & M. Lind (Eds.), *Proceedings of the 9th International Working Conference on the Language-Action Perspective on Communication Modeling.* Retrieved May 17, 2008, from http://www.scils.rutgers.edu/lap04/lap04.htm

De Rivera, J. (1984). Development and the full range of emotional experience. In C. Malatesta & C. Izard (Eds.), *Emotion in adult development* (pp. 45-63). Beverly Hills, CA: Sage.

DeBell, M., & Chapman, C. (2003). *Computer and internet use by children and adolescents in 2001.* (NCES 2004-2014). Washington, DC: U.S. Department of Education, National Center for Education Statistics.

Dekker, A. (2001). Visualisation of social networks using CAVALIER™. Paper presented at *Australian Symposium on Information Visualisation*, Sydney.

Delaney, K.J. (2004). Space Invaders: Ads in videogames pose a new threat to media industry; Marketers pay for placement at the expense of TV; A small but growing area; The terrorist on the cellphone. *Wall Street Journal, Jul 28*, A.1.

Della Maggiora, E. (2006). *Nielsen Interactive Entertainment.* Paper presented at the Game Developers Conference (GDC) Focus on: Game Advertising Summit, San Francisco, CA.

Dellaert, B.G.C., & Kahn, B.E. (1999). How tolerable is delay?: Consumers' evaluation of internet web sites after waiting. *Journal of Interactive Marketing, 13*(1), 41-54.

DeLorme, E.D., & Reid, N.L. (1999). Moviegoers' experiences and interpretations of brands in films revisited. *Journal of Advertising, 28*(2), 71-94.

Dementhon, D.F., & Davis, L. (1995). Model-based Object Pose in 25 Lines of Code. *Intl. Journal of Computer Vision, 15*, 123-141.

Denman, H., Rea, N., & Kokaram, A. (2003). Content-based analysis for video from snooker broadcasts. *Computer Vision and Image Understanding, 92*(2/3), 176–195.

Derbaix, C., Decrop, A., & Cabossart, O. (2002). Colors and Scarves: The Symbolic Consumption of Material Possessions by Soccer Fans. *Advances in Consumer Research, 29*, 511-518.

Desurvire, H., Caplan, M., & Toth, J.A. (2004). Using heuristics to evaluate the playability of games. In *Extended Abstracts of the 2004 Conference on Human Factors in Computing Systems* (pp. 1509-1512). New York: ACM Press.

Devaney, P. (2005). America: Advertisers find a new playmate in the computer games industry. *Marketing Week, 28*(33), 28-29.

DFC Intelligence. (2004). New DFC Intelligence reports look at growing convergence of video game, music and movie industries. Retrieved September 23, 2004, from http://www.dfcint.com/news/prmarch192004.html.

DFC Intelligence. (2007). DFC Intelligence estimates worldwide portable game market to exceed $10 billion in 2007. Retrieved August 5, 2008, from http://www.dfcint.com/wp/?p=74.

DFC Intelligence. (2008). DFC Intelligence forecasts video game market to reach $57 billion in 2009. Retrieved August 5, 2008, from http://www.dfcint.com/wp/?p=222.

Diamond, D. (2004). Rotisserie baseball—what is it, and why should I play? Retrieved August 4, 2004, from http://www.kcmets.com/RotoWorld

Diaz, J. (1977). *Communication and rural development.* Paris: United Nations Educational, Scientific and Cultural Organisation (UNESCO).

Dipietro, M., Ferdig, R.E., Boyer, J., & Black, E.W. (2007). Towards a framework for understanding electronic educational gaming. *Journal of Educational Multimedia and Hypermedia, 16*(3), 225-248.

Dishman, R.K., Motl, R.W., Saunders, R., Felton, G., Ward, D.S., Dowda, M., & Pate, R.R. (2005). Enjoyment mediates effects of a school-based physical-activity intervention. *Medicine and Science in Sports and Exercise, 37*(3), 478-487.

Doherty, A., & Carron, A. (2003). Cohesion in volunteer sport executive committees. *Journal of Sport Management,* (pp. 116-141).

Doherty, A., Patterson, M., & Van Bussel, M. (2004). What do we expect? An examination of perceived committee norms in non-profit sport organisations. *Sports Management Review,* (pp. 109-132).

Dolance, S. (2005). "A Whole Stadium Full": Lesbian Community at Women's National Basketball Association Games. *The Journal of Sex Research, 42*(1), 74-83.

Dominick, J.R. (1984). Media effects on the young videogames, television violence, and aggression in teenagers. *Journal of Communication, 34*(2), 136-148.

Dorai, C., & Venkatesh, S. (2002). Media Computing: Computational Media Aesthetics (The International Series in Video Computing). *Springer,* Edition 1.

Doreian, P., & Fararo, T.J. (1985). Structural equivalence in a journal network. *Journal of the American Society for Information Science*, 36(1), 28-37.

Dorman, S. (1997). Video and computer games: Effect on children and implications for health education. *Journal of School Health, 67*(4), 133-138.

Dorval, M., & Pepin, M. (1986). Effect of playing a video game on a measure of spatial visualization. *Perceptual and Motor Skills, 62*(1), 159-162.

Dower, T. (2007). A parallel universe. *Marketing, Feb,* 16-19.

Du Plessis, E. (1994). Recognition versus recall. *Journal of Advertising Research, 34*(3), 75-91.

Duckworth, F.C., & Lewis, A.J. (1998). A fair method of resetting the target in interrupted one-day cricket matches. *Journal of the Operational Research Society, 49*(3), 220-227.

Duncan, M.C., & Brummett, B. (1989). Types and sources of spectating pleasure in televised sports. *Journal of Sport & Social Issues, 6,* 195-211.

Durkin, K., & Barber, B. (2002). Not so doomed: Computer game play and positive adolescent development. *Applied Developmental Psychology, 23*(4), 373-392.

Durlach, N., Allen, G., Darken, R., Garnett, R.L., Loomis, J., Templeman, J., & von Wiegand, T. E. (2000). Virtual environments and the enhancement of spatial behavior: Towards a comprehensive research agenda. *PRESENCE-Teleoperators and Virtual Environments, 9,* 593-615.

Durrant-Whyte, H.F. (1988). Sensor models and multisensor integration. *International Journal of Robotic Research, 7,* 97-113.

Dussault, J.-P., & Landry, J.-F. (2006a). Optimization of a billiard player—position play. In H.J. van den Herik, S. chin Hsu, T. sheng Hsu, and H.H.L.M., Donkers (Eds.), *ACG, 4250 of Lecture Notes in Computer Science,* (pp. 263–272). Springer.

Dussault, J.-P., & Landry, J.-F. (2006b). Optimization of a billiard player—tactical play. In *Proceedings of the*

*Computer and Games Conference, Italy, May.* Springer Verlag Heidelberg Germany, lncs series. CG06.

dYdewalle, G., Vanden Abeele, P., Van Rensberger, J. & Coucke, P. (1988). Incidental processing of advertisements while watching soccer-games and broadcasts. In M. Gruneberg, P. Morris & R. Sykes (Eds.), *Practical Aspects of Memory: Current Research and Issues* (pp. 478-483). New York: John Wiley.

Ed, H., Chi, E.H., Song J., & Corbin, G. (2004). *Killer App of wearable computing: wireless force sensing body protectors for martial arts.* Paper presented at the 17th Annual ACM Symposium On User Interface Software And Technology, Santa Fe, NM.

Edwards, C., & Greene, J. (2005). Who's got game now? *Business Week, May 16*(3933), 40.

Edwards, J. (2007, July 4). Liverpool play tough to land Torres for a record £20m. Retrieved September 11, 2007, from http://www.dailymail.co.uk/pages/live/articles/sport/football.html?in_article_id=465972&in_page_id=1779&ito=1490

Edwards, P., & Watts, A. (1983). Volunteerism and human service organizations: Trends and prospects. *Journal of Applied Social Sciences,* (pp. 225-245).

Eisenberg, D. (2003, March 9). *The NBA's Global Game Plan.* Retrieved March 16, 2007, from http://www.time.com/time/magazine/article/0,9171,430855,00.html

Ekin, A., Teklap, M., & Mehrotra, R. (2003). Automatic soccer video analysis and summarization. *IEEE Transactions on Image Processing, 12*(7), 796-807.

Eley, D., & Kirk, D. (2002). Developing citizenship through sport: The impact of a sport based volunteer programme on young sport leaders. In D. Eley & D. Kirk (Eds.), *Sport, education and society* (pp. 151-166). London: Routledge.

Elliott, B. (2006). Biomechanics and tennis. *British Journal of Sports Medicine, 40,* 392-396.

Elliott, B., & Wood, G. (1983). The biomechanics of the foot-up and foot-back tennis serves techniques. *Australian Journal of Sports Sciences, 3*(2), 3-6.

Elliott, B., Fleisig, G., Nicholls, R., & Escamilia, R. (2003). Technique effects on upper limb loading in the tennis serve. *Journal of Science and Medicine in Sport, 6*(1), 76-87.

Elliott, B., Marsh, A., & Blanksby, B. (1986). A three-dimensional cinematographic analysis of the tennis serve. *Journal of Applied Biomechanics, 2*(4), 260-271.

Elliott, B., Marshall, R., & Noffal, G. (1995). Contributions of upper limb segment rotations during the power serve in tennis. *Journal of Applied Biomechanics, 11*(4), 433-442.

Ellis, D. (1984). Video arcades, youth, and trouble. *Youth and Society, 16*(1), 47-65.

Ellison, N.B., Steinfield, C., & Lampe, C. (2007).The benefits of Facebook "friends": Social capital and college students' use of online social network sites. *Journal of Computer-Mediated Communication, 12*(4). Retrieved May 23, 2008, from http://jcmc.indiana.edu/vol12/issue4/ellison.html

eMarketer (2007). Latin America online: Demographics, usage & e-commerce. Retrieved October 01, 2007, from http:// www.emarketer.com/Reports/All/Latam_aug06.aspx

Empirical evidence. *European Sport Management Quarterly, 8*(2), 145-164.

End, C.M. (2001). An Examination of NFL Fans' Computer Mediated BIRGing. *Journal of Sport Behavior, 24*(2), 162-181.

End, C.M., Dietz-Uhler, B., Harrick, E.A., & Jacquemotte, L. (2002). Identifying With Winners: A Reexamination of Sport Fans' Tendency to BIRG. *Journal of Applied Social Psychology, 32*(5), 1017-1030.

Entertainment Software Association (ESA). (2007). *Essential facts about the computer and video gaming industry-Sales, demographics and usage data.* Retrieved December 8, 2007, from www.theesa.com

Erickson, E.H. (1950). *Childhood and society.* New York: Norton.

ESA. (2004). Americans playing more games, watching less movies and television. Retrieved September 23, 2004, from http://www.theesa.com/5_12_2004.html.

ESA. (2004). Games, parents and ratings. Retrieved September 23, 2004, from http://www.theesa.com/gamesandratings.html.

ESA. (2006). Essential facts about the computer and video game industry. Retrieved July 4, 2006, from http://www.theesa.com/archives/files/Essential%20Facts%20 2006.pdf.

ESA. (2006). Facts and research: Game player data. Retrieved January 12, 2008, from http://www.theesa.com/facts/gamer_data.php.

ESA. (2008). Essential facts about the computer and video game industry. Retrieved September 4, 2008, from http://www.theesa.com/facts/pdfs/ESA_EF_2008.pdf.

ESA, Entertainment Software Association. (2003). Game players are a more diverse gender, age and socio-economic group than ever, according to new poll. Retrieved September 23, 2004, from http://www.theesa.com/8_26_2003.html.

ESPN.com. (2006). *Rules – legal restrictions.* Retrieved October 21, 2006, from http://sports.espn.go.com/fantasy/football/ffl/story?page=fflruleslegal

Estridge, H.L. (2007, May 11). Rangers owner talks sports business. Retrieved September 11, 2007, from http://dallas.bizjournals.com/dallas/stories/2007/05/07/daily47.html

ESWC, Electronic Sports World Cup™. (2008). ESWC overview. Retrieved July 17, 2008, from http://www.eswc.com/info/overview_eswc.

Etter, R., & Specht, M. (2005). Melodious Walkabout – Implicit navigation with contextualized personal audio contents. In *Adjunct Proceedings of the Third International Conference on Pervasive Computing* (pp. 43-49). (vol. 3468). Germany: Springer.

Euchner, C.C. (1993). *Playing the Field: Why Sports Teams Move and Cities Fight to Keep Them.* Baltimore, MD: Johns Hopkins University Press.

Ewins, R. (2005). Who are you? weblogs and academic identity. *E-Learning, 2*(4), 368-377.

EyeToy Kinetic. (2008). *Home.* Retrieved April 30, 2008, from http://www.eyetoykinetic.com

Fantasy Baseball. (n.d.). Retrieved October 2, 2006, from http://www.answers.com/main

Fantasy Sports Trade Association. (n.d.). *Welcome to the official site of the FSTA.* Retrieved October 24, 2006, from http://www.fsta.org

Farred, G. (2004). Anfield envy: How the spirit of Liverpool hangs over Manchester United. In D.L. Andrews (Ed.), *Manchester United: A Thematic Study* (pp. 222-238). London: Routledge.

Faust, K. (2005). Using correspondence analysis for joint displays of affiliation networks. In P.J. Carrington, J. Scott, & S. Wasserman (Eds.), *Models and Methods in Social Network Analysis* (pp. 117-147). New York: Cambridge University Press.

Federoff, M. (2002). *Heuristics and usability guidelines for the creation and evaluation of fun in video games.* Unpublished thesis, Indiana University, Bloomington.

Fennema, E., & Sherman, J. A. (1977). Sex-related differences in mathematics achievement, spatial visualization, and affective factors. *American Educational Research Journal, 23*(1), 51-71.

Ferguson, C.J. (2007). Evidence for publication bias in video game violence effects literature: A meta-analytic review. *Aggression and Violent Behavior, 12*(4), 470-482.

Fernando, A. (2004). Creating buzz: New media tactics have changed the PR and advertising game. *Communication World, 21*(6), 10-12.

Fifield, A. (2007). Limbering up for an e-sport revolution Korean-style. *Financial Times, Sep 15*, 10.

Fink, J.S., Trail, G.T., & Anderson, D.F. (2002). An Examination of Team Identification: Which Motives are Most Salient to its Existence? *International Sports Journal, Summer*, (pp. 195-207).

First-week sales of EA Sports™ *Madden NFL 07* hit $100M. (2006, September 1). *Sports Business Daily, XII*(235). Retrieved September 1, 2006, from http://www.sportsbusinessdaily.com

Fisher, E. (2006, May 15). ESPN hits 'restart' in gaming. *Street & Smith's SportsBusiness Journal, 31.*

Fisher, E. (2006, May 8). Lawsuit over fantasy stats set for trial. *Sports Business Journal, 6.*

Fisher, E. (2006, August 14). Fantasy ruling could force MLBAM to revisit deal with union. *Sports Business Journal, 7.*

Fisher, E. (2006, May 8). New group aims to be voice of fantasy sports. *Sports Business Journal*, 6.

Fisher, E. (2006, June 12). MySpace race: Fan participation drives change on Web. *Sports Business Journal, 1.*

Fisher, E. (2006, July 31). Fantasy football gets early jump on season. *Sports Business Journal*, 7.

Fisher, E. (2006, February 20). New stat for fantasy: More than 6M players. *Sports Business Journal, 5.*

Fisher, E. (2006, September 4). NBC Sports will roll out redesigned site on Tuesday. *Sports Business Journal, 44.*

Fisher, E. (2007, April 30). Women fuel NASCAR's fantasy growth. *Sports Business Journal, 1.*

Fisher, R.J. (1998). Group-Derived Consumption: The Role of Similarity and Attractiveness in Identification with a Favorite Sports Team. *Advances in Consumer Research, 25*, 283-288.

Fisher, R.J., & Wakefield, K. (1998). Factors Leading to Group Identification: A Field Study of Winners and Losers. *Psychology & Marketing, 15*(1), 23-40.

Fitsync. (2008). *Products.* Retrieved April 30, 2008, from http://fitsync.com/website/pages/productsPersonal.html

Fitzgerald, S.G., Cooper, R.A., Thorman, T., Cooper, R., Guo, S., & Boninger, M.L. (2004). The GAME(Cycle) exercise system: Comparison with standard ergometry. *Journal of Spinal Cord Medicine, 27(5),* 453–459.

Foege, A. (2005). All the young dudes. *Mediaweek, 15*(31), 16-20.

Fogelgren-Pedersen, A. (2005). The Mobile Internet: The pioneering users' adoption decisions. In *Proceedings of the 38th Hawaii International Conference on Systems Sciences.* Hawaii.

Fontana, A., & Frey, J.H. (2000). The Interview: From Structured Questions to Negotiated Text. In N.K. Denzin & Y.S. Lincoln (Eds.), *Handbook of Qualitative Research* (pp. 645-672). Thousand Oaks, CA: SAGE Publications.

Football Stadia After Taylor. (2002, March). Retrieved September 6, 2007, from http://www.le.ac.uk/football-research/resources/factsheets/fs2.pdf

*Forster, W. (2005). The Encyclopedia ofGame Machines - Consoles, handheld & home computers 1972-2005. Vancouver: GAMEplan.*

Forsyth, E., & Katz, L. (1946). A matrix approach to the analyses of sociometric data: Preliminary report. *Sociometry, 9*(4), 340-347.

*Fortin, D.R., & Dholakia, R.R. (2005). Interactivity and vividness effects on social presence and involvement with a web-based advertisement. Journal of Business Research, 58(3), 387-396.*

Franck, E., & Neusch, S. (2008). Mechanisms of superstar formation in German soccer:

Frasca, G. (2003). Simulation versus narrative: Introduction to ludology. In M.J.P. Wolf & B. Perron (Eds.), *The video game theory reader* (pp. 221-236). London: Routledge.

Freeman, L.C. (1984). The impact of computer based communication on the social structure of an emerging scientific specialty. *Social Networks, 6*, 201-221.

Friedl, M. (2003). *Online Game Interactivity Theory.* Hingham: Charles River Media.

Fritz, J., & Fehr, W. (Eds.). (1997). *Handbuch Medien und Computer-Spiele*. Bonn.

Fukuda, T., Michelini, R., Potkonjak, V., Tzafestas, S., Valavanis, K., & Vukobratovic, M. (2001). How Far Away is "Artificial Man". *IEEE Robotics and Automation Magazine*, *March issue*, (pp. 66-73).

Funk, J.B., & Buchman, D.D. (1996). Playing violent video and computer games and adolescent self-concept. *Journal of Communication, 46*(2), 19-32.

Gagalowicz, A. (1990). Collaboration between Computer Graphics and Computer Vision.In S. Nayar (Ed.), *Proceedings of the IEEE Conference on Computer Vision* (pp. 733-737). London, UK: IEEE.

Gagalowicz, A., & Gerard, P. (2000). Three Dimensional Object Tracking using Analysis/Synthesis Techniques. In A. Leonardis, F. Solina, R. Bajcsy, & F. Solina (Eds.), *Confluence of Computer Vision and Computer Graphics* (pp. 307-330), Dordrecht, NL: Kluwer.

Gage, J. (2006, April 17). Winner With Losses. Retrieved May 16, 2007, from http://www.forbes.com/free_forbes/2006/0417/084.html

Gaither, C. (2003). Battle for the sexes. *Boston Globe, Jan 13*. Retrieved May 20, 2008, from http://www.there.com/pressBostonGlobe_011303.html.

Gallagher, S., & Park, S.H. (2002). Innovation and competition in standard-based industries: A historical analysis of the U.S. home video game market. *IEEE Transactions on Engineering Management, 49*(1), 67-82.

Game Guru. (2006). Electronic Sports World Cup™ India Lan gaming tournament kicks off. Retrieved August 25, 2008, from http://www.gameguru.in/general/2006/19/electronic-sports-world-cup-india-lan-gaming-tournament-kicks-off/.

Gamebike. (2008). *Home*. Retrieved April 30, 2008, from http://www.cateyefitness.com/GameBike/index.html

GamersCircle. (2008). PC gaming a $10.7 billion industry (Games Convention 2008). Retrieved October 5, 2008, from http://www.gamerscircle.net/2008/08/19/pc-gaming-a-107-billion-industry-games-convention-2008/.

Gamersize. (2008). *Home*. Retrieved April 30, 2008, from http://www.gamercize.net/

Games Universe. (2004). V8 Supercars Race Driver. Retrieved September 24, 2004, from http://gamesuniverse.com.au/ProductPage.asp?ProductID=3131.

Gantz, W., & Wenner, L.A. (1995). Fanship and the television sports viewing experience. *Sociology of Sport Journal, 12*, 56-74.

Gardner, D.G., Dukes, R.L., & Discenza, R. (1993). Computer use, self-confidence and attitudes: A causal analysis. *Computers in Human Behavior, 9*(4), 427-440.

Gardner, M.P., & Shuman, P. (1988). Sponsorships and small businesses. *Journal of Small Business Management, 26*(4), 44-52.

Gartner. (2008). Gartner says worldwide mobile gaming revenue to surpass $4.5 billion in 2008. Retrieved August 5, 2008, from http://www.gartner.com/it/page.jsp?id=706407.

Gaudiosi, J. (2007). In-game ads reach the next level. *Business 2.0, 8*(6), 36-37.

Gavrila, D.M. (1999). The Visual Analysis of Human Movement: A Survey. *Computer Vision and Image Understanding, 73*, 82-98.

Gee, J.P. (2003). *What video games have to teach us about learning and literacy*. New York: Palgrave/Macmillan.

Gee, J.P. (2003). *What video games have to teach us about learning and literacy*. New York: Palgrave Macmillan.

Gee, J.P. (2004). Learning by design: Games as learning machines. *Gamasutra*. Retrieved August 8, 2007, from http://www.gamasutra.com/gdc2004/features/20040324/gee_01.shtml.

Gentile, D.A. (2005). Examining the effects of video games from a psychological perspective: Focus on violent games and a new synthesis. *National Institute on Media and the Family*. Retrieved August 13, 2007, from http://www.psychology.iastate.edu/faculty/dgentile/pdfs/Gentile_NIMF_Review%20_2005.pdf.

Gentile, D.A., & Stone, W. (2005). Violent video game effects on children and adolescents: A review of the literature. *Minerva Pediatrica, 57*(6), 337-358.

Gerard, P., & Gagalowicz, A. (2003). Human Body Tracking using a 3D Generic Model Applied to Golf Swing Analysis. Paper presented at *MIRAGE 2003Conference,* Rocquanecort, France.

Gibb, G.D., Bailey, J.R., Lambirth, T.T., & Wilson, W.P. (1983). Personality differences between high and low electronic video game users. *Journal of Psychology, 114*(2), 159-165.

Gieske, C., & Forato, M. (2004). *The most valuable football brands in Europe*. London: FutureBrand.

Girard, O., Micallef, J., & Millet, G.P. (2005). Lower-limb activity during the power serve in tennis: Effects of performance level. *Medicine and Science in Sports and Exercise, 37,* 1021-1029.

Giulianotti, R. (2002). Soccer Goes Glocal. *Foreign Policy, 131*(July-August), 82-83.

Goh, W.B., & Chan, K.Y. (2007). The Multiresolution Gradient Vector Field Skeleton. *Pattern Recognition, 40*, 1255-1269.

Goldhaber, M. (1997). The attention economy and the Net. *First Monday.* Retrieved 15 July, 2008, from www.firstmonday.org/issues/issue2_4/goldhaber.

Golftek (2008). Retrieved 13 May, 2008 from http://www.golftek.com/

Goncalves, L., Di Bernardom, E., Ursella, E., & Perona, P. (1995). Monocular Tracking of the Human Arm in 3D. In X. Yalin (Ed.), *Proceedings of the International Conference on Computer Vision,* (pp. 764-770). Cambridge, Mass.: IEEE.

Gordon, B.J., & Dapena, J. (2006). Contributions of joint rotations to racquet speed in the tennis serve. *Journal of Sports Sciences, 24*(1), 31-49.

Gordon-Larsen, P., Adair, L.S., & Popkin, B.M. (2002). Ethnic differences in physical activity and inactivity patterns and overweight status. *Obesity Research, 10*(3), 141–149.

Gould, D., & Petlichkoff, L. (1988). Participation motivation and attrition in young athletes. In F.L. Smoll, R.A. Magill, & M.J. Ash (Eds.), *Children in sport* (3rd ed., pp. 161-178). Champaign, IL: Human Kinetics Books.

Granovetter, M. (1983). The strength of weak ties: a network theory revisited. *Sociological Theory, 1*, 201-223.

Gratton, C., & Henry, I. (Eds.). (2001). *Sport in the city: The role of sport in economic and social regeneration.* UK: Routledge.

Graves, L., Stratton, G., Ridgers, N. D., & Cable, N.T. (2007). Comparison of energy expenditure in adolescents when playing new generation and sedentary computer games: cross sectional study. *BMJ, 335*(7633), 1282-1284.

Gray, G. (1999). Discussion on future directions of global digital museum: Museum and GUI. In S. Sugita, J.K. Hong, J. Reeve & G. Gay (Ed.), *Senri Ethnological Report, Global Digital Museum for Museum Education on the Internet.* Osaka, Japan: National Museum of Ethnology.

Gray, G. (2001). Co-construction of digital museums. *Spectra, 28*(1), 12-14.

Graybill, D., Strawniak, M., Hunter, T., & O'Leary, M. (1987). Effects of playing versus observing violent versus non-violent video games on children's aggression. *Psychology, 24*(3), 1-8.

Green, C.S., & Bavelier, D. (2003). Action v*ideo* game modifies visual selective attention. *Nature, 423*(6939), 534-537.

Greenacre, M., & Hastie, T. (1987). The geometric interpretation of correspondence analysis. *Journal of the American Statistical Association, 82*(398), 437-447.

Greenfield, P.M., DeWinstanley, P., Kilpatrick, H., & Keys, D. (1996). Action video games and informal education on strategies for dividing visual attention. *Journal of applied developmental psychology, 15*(1), 105-123.

Greenspan, M. (2005). Uofa wins the pool tournament. *Intl. Comp. Gaming Ass. Journal, 28*(3), 191–193.

Greenspan, M. (2006). Pickpocket wins pool tournament. *Intl. Comp. Gaming Ass. Journal, 29*(3), 153–156.

Greenspan, M., Lam, J., Leckie, W., Godard, M., Zaidi, I., Anderson, K., Dupuis, D., & Jordan, S. (2008). Toward a competitive pool playing robot. *IEEE Computer Magazine, 41*(1), 46–53.

Greenwood, P.B. (2001). *Sport Fan Team Identification in a Professional Expansion Setting.* Unpublished Masters, North Carolina State University, Raleigh, NC.

Greenwood, P.B., Kanters, M. A., & Casper, J. M. (2006). Sport Fan Team Identification Formation in Mid-Level Professional Sport. *European Sport Management Quarterly, 6*(3), 253-265.

Grest, D., Woetzel, J., & Koch, R. (2005). Nonlinear Body Pose Estimation from Depth Images. In W. Kropatsch (Ed.), *Proceedings of the DAGM - German Association for Pattern Recognition Conference*, (pp. 285-292). Vienna, Austria: DAGM.

Griffith, J.L., Voloschin, P., Gibb, G.D., & Bailey, J.R. (1983). Differences in eye-hand motor coordination of video-game users and non-users. *Perceptual and Motor Skills, 57*(1), 155-158.

Griffiths, M.D., Davies, M.N.O., & Chappell, D. (2003). Breaking the stereotype: The case of online gaming. *CyberPsychology and Behavior, 6*(1), 81-91.

Grigorovici, D. (2003). *Persuasive effects of presence in immersive virtual environments.* In G. Riva, F. Davide & W.A. Ijsselsteijn (Eds.), *Being there: Concepts, effects and measurements of user presence in synthetic environments* (pp. 191-207). Amsterdam: IOS Press.

Grigorovici, D., & Constantin, C. (2004). Experiencing interactive advertising beyond rich media: Impacts of ad type and presence on brand effectiveness in 3D gaming immersive virtual environments. *Journal of Interactive Advertising, 5*(1). Retrieved August 24, 2005, from http://www.jiad.org/vol5/no1/grigorovici/index.htm.

Grodal, T. (2000). Video games and the pleasures of control. In D. Zillmann & P. Vorderer (Eds.), *Media entertainment: The psychology of its appeal* (pp. 197-214). Mahwah, NJ: Lawrence Erlbaum.

Gros, B. (2003). The impact of digital games in education. *First Monday Journal.* Retrieved September 24, 2004, from http://www.firstmonday.org/issues/issue8_7/xyzgros/index.html.

Grossman, L. (2005). Out of the X Box. *Time, 165*(21), 44-52.

Guay, R.B., & McDaniel, E. (1977). The relationship between mathematics achievement and spatial abilities among elementary school children. *Journal for Research in Mathematics Education, 8*(2), 211-215.

Gupta, B.P., & Gould, J.S. (1997). Consumers' perceptions of ethics and acceptability of product placements in movies: Product category and individual differences. *Journal of Current Issues and Research in Advertising, 19*(1), 37-50.

Gupta, B.P., & Lord, R.K. (1998). Product placement in movies: The effect of prominence and mode on audience recall. *Journal of Current Issues and Research in Advertising, 20*(1), 47-59.

Gwinn, E. (2004, 21 April). Space invaders: Ads are infiltrating video games. *Chicago Tribune.*

Gwinner, K., & Swanson, S.R. (2003). A model of fan identification: antecedents and sponsorship outcomes. *Journal of Services Marketing, 17*(3), 275-294.

Gwinner, K.P. (1997). A model of image creation and image transfer in event sponsorship. *International Marketing Review, 14*(3), 170-182.

Gwinner, K.P., & Eaton, J. (1999). Building brand image through event sponsorship: The role of image transfer. *Journal of Advertising, 28*(Winter), 47-57.

Haag, H. (Ed.). (1996). *Sportphilosophie.* Schorndorf: Verlag Karl Hofmann.

Hall, M., & Banting, K. (2002). *The nonprofit sector in Canada: An introduction.* Working Paper, School of Policy Studies, Queen's University.

Hamamatsu (n.d.). Retrieved May 12, 2008, from http://usa.hamamatsu.com

Hamilton, E., & Cunningham, P. (1989). Community-based adult education. In S. Merriam & P. Cunningham (Eds.), *Handbook of adult and continuing education* (pp. 439-450). San Francisco: Jossey-Bass.

Hamman, R. (2000). Forget the web ads: Online promos capture attention. *Marketing News, 34*(23), 28-29.

Hammond, B. (2007). *A computer vision tangible user interface for mixed reality billiards*. Master's thesis, Pace University.

Harrison, L.K., Denning, S., Easton, H.L., Hall, J.C., Burns, V.E., Ring, C., & Carroll, D. (2001). The effects of competition and competitiveness on cardiovascular activity. *Psychophysiology, 38*(4), 601-606.

Hartig, K. (2006). Next-generation consoles drive global game revenues to more than $45 billion by 2010. *PricewaterhouseCoopers*. Retrieved September 5, 2007, from http://www.pwc.com/extweb/pwcpublications.nsf/docid/C987CEB2D179131F852572090083B4B3/$FILE/VideoGames_KH_ls.pdf.

Harvard, Å., & Løvind, S. (2002). "Psst"-ipatory Design. Involving artists, technologists, students and children in the design of narrative toys. In T. Binder, J. Gregory & I. Wagner (Eds.), *Proceedings of the PDC 2002 Participatory Design Conference*. Malmö.

Hawkeye (2008). Retrieved 13 May, 2008 from http://www.hawkeyeinnovations.co.uk/

Hazlett, T. (1986). Private monopoly and public interest: An economic analysis of the cable television franchise. *University of Pennsylvania Law Review, 134*(6), 1335-1409.

Heckhausen, H. (1967). *The anatomy of achievement motivation*. New York: Academic Press.

Hein, K. (2004). Getting in the game. *Brandweek, 45*(7), 26-27.

Hein, K. (2008). EA opts not to keep it real with new 'Freestyle' brand. *Brandweek, 49*(18), 7.

Heinemann, K. (1998). *Einführung in die Soziologie des Sports*. Schorndorf: Verlag Karl Hofmann.

Heinonen, H. (2002). Finnish Soccer Supporters Away from Home: A Case Study of Finnish National Team Fans at a World Cup Qualifying Match in Liverpool, England. *Soccer and Society, 3*(3), 26-50.

Henriksen, F. (2004, September 13). Lure of United dims as Real reel in fans. Retrieved August 7, 2006, from http://www.realmadrid.dk/news/article/default.asp?newsid=5194

Hernandez, M. D., Chapa, S., Minor, M. S., Maldonado, C., & Barranzuela, F. (2004).

Hernandez, M.D., Minor, M.S., Suh, J., Chapa, S., & Salas, J.A. (2005). Brand recall in the advergaming environment: A cross-national comparison. In M.R. Stafford & R.J. Faber (Eds.), *Advertising, Promotion and the New Media* (pp. 298-319). Armonk, NY: M.E. Sharpe, Inc.

Herren, R., Sparti, A., Aminian, K., & Schutz, Y. (1999). The prediction of speed and incline in outdoor running in humans using accelerometry. *Medicine & Science in Sports & Exercise, 31*(7), 1053-1059.

Hickok (2008). *Billiard—history*. Retrieved from http://www.hickoksports.com/history/billiard.shtml.

Hiemstra, R. (1993). *The educative community*. (3rd ed.). Syracuse: Syracuse University Adult Education Publications.

Hispanic attitudes toward advergames: A proposed model of their antecedents. *Journal of Interactive Advertising, 5*(1).

Hockey Australia. (2006). *2006 Hockey Australia Annual Report*. Melbourne: Hockey Australia.

Hoffman, D.L., & Franke, G.R. (1986). Correspondence analysis: graphical representation of categorical data in marketing research. *Journal of Marketing Research, 23*, 213-227.

Hoffman, D.L., & Novak, T.P. (1996). Marketing in hypermedia computer-mediated environments: Conceptual foundations. *Journal of Marketing, 60*(3), 50-69.

Holbrook, M.B., & Hirschman, E.C. (1982). The experiential aspects of consumption: Consumer fantasies, feelings, and fun. *Journal of Consumer Research, 9*(2), 132-140.

Hollan, J., & Stornetta, S. (1992). Beyond being there. In *CHI 1992: Proceedings of the SIGCHI conference on Human factors in computing systems* (pp. 119-125). New York: ACM Press.

Holt, D.B. (1995). How consumers consume: A typology of consumption practices. *Journal of Consumer Research, 22*(1), 1-16.

Hoppe, H., DeRose, T., Duchamp, T., McDonald, J., & Stuetzle, W. (1992). Surface reconstruction from unorganized points. *ACM SIGGRAPH Computer Graphics, 26*(2), 71-78.

Howard, T. (2006, July 11). As more people play, advertisers devise game plan. *USA Today,* B3.

Howe, T., & Sharkey, P.M. (1998). Identifying likely successful users of virtual reality systems. *Presence, 7*(3), 308-316.

Hoyle, R.H., Harris, M.J., & Judd, C.M. (2002). *Research Methods in Social Relations.* Fort Worth, TX: Wadsworth.

Hrehocik, M. (2008). ICAA defines 'active aging.' *Long-Term Living, 57*(4), 12-14.

Hsu, C.L., & Lu, H.P. (2004). Why do people play online games? An extended TAM with social influences and flow experience. *Information and Management, 41*(7), 853-868.

Hu, J. (2003). *Sites see big season for fantasy sports.* Retrieved October 5, 2006, from http://news.com/2102-1026_3-5061351.html

Huggins, T. (2005, May 2). Chelsea seek to paint world blue. Retrieved November 18, 2005, from http://www.rediff.com/sports/2005/may/02foot1.htm

Hurley, O. (2008). Game on again for coin-operated arcade titles. *The Guardian, Feb 7.* Retrieved September 30, 2008, from http://www.guardian.co.uk/technology/2008/feb/07/games.it.

Hutchins, B. (2006). Computer gaming, media and e-sport. In V. Colic-Peisker, F. Tilbury & B. McNamara (Eds.), *Proceedings of the 2006 Australian Sociological Association Annual Conference* (pp. 1-9). Australia: University of Western Australia and Murdoch University.

Huysman, M., & Wulf, V. (Eds.). (2004). *Social capital and information technology.* London: MIT Press.

Hyman, P. (2005). Sponsors go ape over advergames. *Hollywood Reporter.* Retrieved July 16, 2007, from http://www.hollywoodreporter.com/hr/search/article_display.jsp?vnu_content_id=1001523697.

Hyman, P. (2006). In-Game Advertising. *Game Developer, 13,* 11-47.

IDO – MachineDance European Championships. (2006). *Machine Dance.* Retrieved April 30, 2008, from http://www.machinedance.nl/arrowdance/uploadedmedia/ec_machinedance_information_260506.pdf

IDSA. Interactive Digital Software Association. Retrieved at http://www.idsa.com

Inoue, Y. (1997). A study on dynamics of golf swing. In *Proceedings of Dynamics and Design Conference* (pp. 99-103). Japan: Japanese Society of Mechanical Engineering.

Institute Discussion Paper No. TI 2001-022/1, Rotterdam.

Instituto Nacional de Estadística, Geografía e Informática (2007). Usuarios de Internet por paises seleccionados 1998 a 2004. Retrieved October 01, 2007, from http://www.inegi.gob.mx/est/contenidos/espanol/rutinas/ept.asp?t=tinf142&c=4870

Interaction Laboratories. (2008). *Powergrid.* Retrieved April 30, 2008, from http://www.ia-labs.com/ViewVideoGameMarket.aspx?ID=7

Intille, S., & Bobick, A. (1999). *A framework for recognizing multi-agent action from visual evidence.* Paper presented at the National Conference on Artificial Intelligence, Cambridge, UK.

Isard, M., & Blake, A. (1996). *Contour tracking by shochastic propagation of conditional density.* Paper presented at the 4th European Conference Computer Vision, Cambridge, UK.

Ishii, H., Wisneski, C., Orbanes, J., Chun, B., & Paradiso, J. (1999). PingPongPlus: Design of an Athletic-Tangible Interface for Computer-Supported Cooperative Play. In *Proceedings of CHI'99* (pp. 394-401). New York: ACM-Press.

Izard, C. (1977). *Human emotions*. New York: Plenum Press.

Jacobson, B.P. (2003). *Rooting for Laundry: An Examination of the Creation and Maintenance of a Sport Fan Identity.* Unpublished Doctoral dissertation, University of Connecticut.

James, D., Gibson, T., & Uroda, W. (2005). Dynamics of a swing: A study of classical Japanese swordsmanship using accelerometers. In A. Subic & S. Ujihashi (Ed.), *The impact of technology on sport* (pp. 355-360). Melbourne, Australia: Australasian Sports Technology Alliance.

James, D.A., Davey, N., & Rice, T. (2004). An accelerometer based sensor platform for insitu elite athlete performance analysis. In *IEEE Sensor Proceedings*. Vienna: IEEE.

James, J.D., & Trail, G.T. (2005). The Relationship Between Team Identification and Sport Consumption Intentions. *International Sports Journal, Winter*, (pp. 1-10).

Janoff, B. (2005, April 11). Marketers of the next generation. *Brandweek*, *46*(15), 38.

Janoff, B. (2005, April 11). Marketers of the next generation. *Brandweek*, *46*(15), 38.

Jansz, J., & Martens, L. (2005). Gaming at a LAN event: The social context of playing video games. *New Media and Society*, 7(3), 333-355

Javalgi, R.G., Traylor, M.B., Gross, A.C., & Lampman, E. (1994). Awareness of sponsorship and corporate image: An empirical investigation. *Journal of Advertising*, *23*(4), 47-58.

Ji, Q., Pan, Z., et al. (2004). Virtual arrangement and rehearsal of group calisthenics. *Journal of Computer-aided Design & Computer Graphics, 16*(9), 1185-1190.

Jih, H., & Reeves, T.C. (1992). Mental models: A research focus for interactive learning systems. *Educational Technology Research and Development, 40*(3), 39-53.

Johansen, R. (1988). Current User Approaches to Groupware. In R. Johansen (Ed.), *Groupware* (pp. 12-44). New York: Freepress.

Johanson, G. (2002). Historical research. In K. Williamson (Ed.), *Research methods for students, academics and professionals – Information management and systems.* Wagga Wagga: Centre for Information Studies, Charles Sturt University.

Johnson, D., & Wiles, J. (2003). Effective affective user interface design in games. *Ergonomics, 46*(13/14), 1332-1345.

Johnson, G.J., Bruner II, G.C., & Kumar, A. (2006). Interactivity and its facets revisited. *Journal of Advertising, 35*(4), 35-52.

Johnson, K.M., Foley, A.M., & Suengas, G.A. (1988). Phenomenal characteristics of memories for perceived and imagined autobiographical events. *Journal of Experimental Psychology, 117*(4), 371-376.

Jones, D., Parkes, R., & Houlihan, A. (2006). *Football Money League: Changing of the Guard.* Manchester: Sports Business Group at Deloitte.

Jones, I. (1997). A Further Examination of the Factors Influencing Current Identification with a Sports Team, A Response to Wann, et al. (1996). *Perceptual and Motor Skills, 85*, 257-258.

Jones, I. (1997, December). Mixing Qualitative and Quantitative Methods in Sports Fan Research. Retrieved June 20, 2007, from http://www.nova.edu/ssss/QR/QR3-4/jones.html

Jones, I. (1998). *Football Fandom: Football Fan Identity and Identification at Luton Town Football Club.* Unpublished Doctoral dissertation, University of Luton, Luton.

Jones, M.B., Kennedy, R.S., & Bittner, A.C. (1981). A video game for performance testing. *American Journal of Psychology, 94*(1), 143-152.

Jorgensen, T.P. (1999). *The physics of golf.* USA: Springer.

Judd, C.M., Drake, R.A., Downing, J.W., & Krosnick, J.A. (1991). Some dynamic properties of attitude structures: Context-induced response facilitation and polarization. *Journal of Personality and Social Psychology, 60*(2), 193-202.

Judge, T.A., Bono, J.E., & Locke, E.A. (2000). Personality and job satisfaction: The mediating role of job characteristics. *Journal of Applied Psychology, 85*(2), 237-249.

Juul, J. (2004). Working with the player's repertoire. *International Journal of Intelligent Games and Simulation, 3*(1), 54-61.

JWT Intelligence. (2006). Gaming. Retrieved May 20, 2008, from https://016fd0d.netsolstores.com/index.asp?pageaction=viewprod&prodid=4.

Kakadiaris, I.A., & Metaxas, D. (1998). 3D Human Body Acquisition from Multiple Views. *International Journal of Computer Vision, 30*, 191-218.

Kap, J.T. (2004). But is it art? *PBS.* Retrieved September 23, 2004, from http://www.pbs.org/kcts/videogamerevolution/impact/art.html.

Kaps, P., Schwameder, H., & Engstler, G. (1997). Inverse dynamic analysis of take-off in ski jumping. In E. Muller, H. Schwameder, E. Kornexl, & C. Raschner (Eds.), *Science and skiing* (pp. 72-83). USA: Taylor & Francis.

Karanasios, S., Sellitto, C., Burgess, S., Johanson, G., Schauder, D., & Denison, T. (2006). The role of the Internet in building capacity: Small businesses and community based organisations in Australia. In *Proceedings of the 7th Working for E-Business Conference.*

Karim, S.A. (2005). Playstation thumb – a new epidemic in children. *South African Medical Journal, 95*(6), 412.

Katz, E., Blumler, J.G., & Gurevitch, M. (1974). Uses and gratifications research. *The Public Opinion Quarterly, 37*(4), 509-523.

Kehl, R., & van Gool, L. (2006). Markerless Tracking of Complex Human Motions from Multiple Views. *Computer Vision and Image Understanding, 104*, 190-209.

Kellogg, W. (1999). Community-based organisations and neighbourhood environmental problem solving: A framework for adoption of information technologies. *Journal of Environmental Planning,* (pp. 445-469).

Ken'ichiro, N., et al. (2004). Integrated Motion Control for Walking, Jumping and Running of a Small Bipedal Entertainment Robot. In *Proceedings of the IEEE International Conference on Robotics, & Automation,* (pp. 3189-3194).

Kendall, J., Tung, L., Chua, K.H., Ng, C.H.D., & Tan, S.M. (2001). Receptivity of Singapore's SMEs to electronic commerce adoption. *Journal of Strategic Information Systems, 10*(3), 223-242.

Kennedy, E. (2000). You talk a good game. *Men and Masculinities, 3*(1), 57-84.

Kennedy, R.S., Bittner, A.C., & Jones, M.B. (1981). Video game and conventional tracking. *Perceptual and Motor Skills, 53*(1), 310.

Kerr, A.K. (2008, July 16-19). *Team Identification and Satellite Supporters: The Potential Value of Brand Equity Frameworks.* Paper presented at the Sixth Annual Conference of the Sport Marketing Association, Gold Coast, Australia.

Kerr, A.K. (in press). Australian Football Goes For Goal: The Team Identification of American A.F.L. Sports Fans. *Football Studies, 11*(1).

Kerr, A.K., & Gladden, J.M. (2008). Extending the understanding of professional team brand equity to the global marketplace. *International Journal of Sport Management and Marketing, 3*(1/2), 58-77.

Kestenbaum, G.I., & Weinstein, L. (1985). Personality, psychopathology, and developmental issues in male adolescent video game use. *Journal of the American Academy of Child Psychiatry, 24*(3), 329-333.

Kieras, D., & Polson, P.G. (1985). An approach to the formal analysis of user complexity. *International Journal of Man-Machine Studies, 22*(4), 365-394.

Kim, K.H., Park, J.Y., Kim, D.Y., Hak, M., & Chun, H.C. (2002). E-Lifestyle and motives to use online games. *Irish Marketing Review, 15*(2), 71-77.

Kim, K.H., Park, J.Y., Kim, D.Y., Moon, H.I., & Chun, H.C. (2002). E-lifestyle and motives to use online games. *Irish Marketing Review, 15*(2), 71-77.

Kim, S.W. (1995). *The motive of a video game use and the types of enjoying of the youths.* Unpublished master's thesis, Han Yang University, Korea.

Kim, T., Seo, Y., & Hong, K. (1998). *Physics-based 3D position analysis of a soccer ball from monocular image sequences.* Paper presented at the Sixth IEEE International Conference on Computer Vision, Bombay, India.

Kim, Y., & Ross, S. (2006). An exploration of motives in sport video gaming. *International Journal of Sport Marketing and Sponsorship, 8*(1), 34-46.

Kim, Y., & Ross, S. (2006). An exploration of motives in sport video gaming. *International Journal of Sport Marketing and Sponsorship, 8*(1), 34-46.

Kim, Y., Ross, S., & Ko, Y. (2007). Online sport video game motivations. *International Journal of Human Movement Science, 1*(1), 41-60.

King, B. (2005, November 14). Magazines cater to growing industry. *Sports Business Journal*, 20.

King, B. (2006, April 17). Reaching the 18-34 demo. *Street & Smith's SportsBusiness Journal*, (pp. 19-26).

King, B. (2006, April 17). Reaching the 18-34 demo. *Street & Smith's SportsBusiness Journal*, (pp. 19-26).

Klaassen, A. (2006, August 7). That's real money—$1.5B—pouring into made-up leagues. *Advertising Age*, *77*(32), 4.

Klamer, A., & van Dalen, H. (2001). *Attention and the art of scientific publishing*. Tinbergen

Kloeck, B., & de Rooij, N. (1994). Mechanical sensors. In S.M. Sze (Ed.), *Semiconductor sensors* (pp. 153-204). New York: Wiley.

Klosterman, C. (2004). *Sex, drugs, and cocoa puffs: A low culture manifesto.* New York: Scribner.

Klug, G.C., & Schell, J. (2006). Why people play games: An industry perspective. In P. Vorderer & J. Bryant (Ed.), *Playing video games: Motives, responses and consequences* (pp. 91-100). USA: Routledge.

Knudson, D. (2003). *Fundamentals of biobmechanics.* New York: Kluwer Academic Plenum Publishers.

Knudson, D. (2006). *Biomechanical principles of tennis technique using science to improve your strokes.* California: Racquet Tech Publishing.

Knuesel, H., Geyer, H., & Seyfarth, A. (2005). Influence of Swing Leg Movement on running stability. *Human Movement Science*, *24*(4), 532-543

Koepp, M.J., Gunn, R.N., Lawrence, A.D., Cunningham, V.J., Dagher, A., Jones, T., Brooks, D.J., Bench, C.J., & Grasby, P.M. (1998). Evidence for striatal dopamine release during a video game. *Nature, 393*(6682), 266-268.

Koh, M., & Anwari, K. (2004). Integrating video and computer technology in teaching – an example in gymnastics initial PE teacher-training in Singapore. *British Journal of Teaching Physical Education*, 35(3), 43 -46.

Koh, T.H. (2000). Ulcerative "nintendinitis": A new kind of repetitive strain injury. *Medical Journal of Australia, 173*(11), 671.

Koku, E., Nazer, N., & Wellman, B. (2001). Netting scholars: Online and offline. *The American Behavioural Scientist, 44*(10), 1752-1774.

Kolbe, R.H., & James, J.D. (2000). An Identification and Examination of Influences That Shape the Creation of a Professional Team Fan. *International Journal of Sports Marketing & Sponsorship, February/March*, (pp. 23-37).

Komi, P.V., & Virmavirta, M. (1997). Ski-jumping takeoff performance: Determining factors and methodological advances. In E. Muller, H. Schwameder, E. Kornexl & C. Raschner (Eds.), *Science and skiing* (pp. 3-26). USA: Taylor & Francis.

Kretchmer, S.B. (2004). Advertainment: The evolution of product placement as a mass media marketing strategy. *Journal of Promotion Management, 10*(1/2), 37-54.

Lafortune, M.A., Cavanagh, P.R., Sommer, H.J., & Kalenak, A. (1992). Three-Dimensional Kinematics of the Human Knee during Walking. *Journal of Biomechanics, 25*, 347-357.

Lai, A., James, D.A., Hayes, J.P., & Harvey, E.C. (2003). Application of triaxial accelerometers in rowing kinematics measurement. In *Proceedings of SPIE: Vol. 5274* (pp. 531-542). Bellingham, WA: SPIE.

Lam, J., & Greenspan, M. (2007). An iterative algebraic approach to tcf matrix estimation. In *IEEE/RSJ 2007 Intl. Conf. Intell. Rob. Sys.*, (pp. 3848–3853).

Lam, J., & Greenspan, M. (2008). Eye-in-hand visual servoing for accurate shooting in pool robotics. In *5th Can. Conf. Comp. Rob. Vis.*

Lam, J., Long, F., Roth, G., & Greenspan, M. (2006). Determining shot accuracy of a robotic pool system. In *CRV 2006: 3rd Can. Conf. Comp. Rob. Vis.*

Lancaster, G.A., Dodd, S., & Williamson, P. R. (2004). Design and analysis of pilot studies: recommendations for good practice. *Journal of Evaluation in Clinical Practice, 10*(2), 307-312.

Landry, J.-F., & Dussault, J.-P. (2007). AI optimization of a billiard player. *Journal of Intelligent and Robotic Systems, 50*(4), 399–417.

Lanningham-Foster, L., Jensen, T.B., Foster, R.C., Redmond, A.B., Walker, B.A., Heinz, D., & Levine, J.A. (2006). Energy expenditure of sedentary screen time compared with active screen time for children. *Pediatrics, 118*(6), 2535.

Lanningham-Foster, L., Jensen, T.B., Foster, R.C., Redmond, A.B., Walker, B.A., Heinz, D., & Levine, J.A. (2006). Energy expenditure of sedentary screen time compared with active screen time for children. *Pediatrics, 118*(6), 1831-1835.

Larsen, L., Jensen, M., & Vodzi, W. (2002). Multi modal user interaction in an automatic pool trainer. In *ICMI 2002: 4th IEEE Intl. Conf. Multimodal Interfaces*, (pp. 361–366).

Larson, P., & Van Rooyen, A. (1998). The Virtual Reality Mental Rotation Spatial Skills Project. *CyberPsychology and Behavior, 1*, 113-120.

Lascu, D., Giese, T.D., Toolan, C., Guehring, B., & Mercer, J. (1995). Sport Involvement: A Relevant Individual Difference Factor in Spectator Sports. *Sport Marketing Quarterly, 4*(4), 41-46.

Lavrakas, P.J. (2004). Questionnaire. In M. S. Lewis-Beck, A. Bryman & T. Futing Liao (Eds.), *The SAGE Encyclopedia of Social Science Research Methods* (pp. 902-903). Thousand Oaks, CA: SAGE Publications.

Lazear, E., & Rosen, S. (1981). Rank-order tournaments as optimum labor contracts. *Journal of Political Economy, 89*(3), 841-864.

Lazzaro, N. (2004). Why we play games: Four keys to more emotion without story. *XEODesign*. Retrieved August 8, 2007, from http://www.xeodesign.com/whyweplaygames/xeodesign_whyweplaygames.pdf.

Lazzaro, N., & Keeker, K. (2004). What's my method? A game show on games. In *Extended Abstracts of the 2004 Conference on Human Factors in Computing Systems* (pp. 1093-1094). New York: ACM Press.

Leckie, W. and Greenspan, M. (2005). Pool physics simulation by event prediction 1: Motion transitions. *Intl. Comp. Gaming Ass. Journal*, 28(4):214–222.

Leckie, W. and Greenspan, M. (2006a). Monte carlo methods in pool strategy game. In *Advances in Computer Games 11*. To appear.

Leckie, W., & Greenspan, M. (2006b). Pool physics simulation by event prediction 2: Collisions. *Intl. Comp. Gaming Ass. Journal, 29*(1), 24–31.

Lee, K., Lev, B., & Yeo, G. (2008). Executive pay dispersion, corporate governance, and firm performance. *Review of Quantitative Finance and Accounting, 30*(3), 315-328.

Lee, M.B., Suh, K.S., & Whang, J. (2003). The impact of situation awareness information on consumer attitudes in the Internet shopping mall. *Electronic Commerce Research and Applications*, *2*(3), 254-265.

Lefton, T. (20 December, 2004a). Exclusive EA™ deal jolts category. *Street & Smith's SportsBusiness Journal, 10.*

Lefton, T. (22 November, 2004b). Video game marketers study how to take in-game ads to the next level. *Street & Smith's SportsBusiness Journal, 12.*

Legnani, G., & Marshall, R.N. (1993). Evaluation of the joint torques during the tennis serve: Analysis of the experimental data and simulations. In *Proceedings of the VIth International Symposium on Computer simulation in Biomechanics* (pp. 8-11). Paris: International Society of the Biomechanics, Technical Group on Computer Simulation.

Lehrer, J. (2006). How the Nintendo Wii will get you emotionally invested in video games. *Seedmagazine.com. Brain & Behavior,* Nov 16. Retrieved April 30, 2008, from http://www.seedmagazine.com/news/2006/11/a_console_to_make_you_wiip.php

Lerasle, F., Rives, G., & Dhome, M. (1999). Tracking of Human Limbs by Multicular Vision. *Computer Vision and Image Understanding, 75*, 229-246.

Lessiter, J., Freeman, J., Keogh, E., & Davidoff, J.A. (2001). A cross-media presence questionnaire: The ITC-sense of presence inventory. *Presence, 10*(3), 282-297.

Levin, A.M., Joiner, C., & Cameron, G. (2001). The impact of sports sponsorship on consumers' brand attitudes and recall: The case of NASCAR® fans. *Journal of Current Issues and Research in Advertising, 23*(2), 23-31.

Levy, D. (2005). *Sports fanship habitus: An investigation of the active consumption of sport, its effects and social implications through the lives of fantasy sports enthusiasts.* Unpublished doctoral dissertation, University of Connecticut.

Lewis, M. (2001). Franchise Relocation and Fan Allegiance. *Journal of Sport & Social Issues, 25*(1), 6-19.

Li, H., Daugherty, T., & Biocca, F. (2001). Characteristics of virtual experience in electronic commerce: A protocol analysis. *Journal of Interactive Marketing, 15*(3), 13-30.

Li, H., Daugherty, T., & Biocca, F. (2002). Impact of 3-D advertising on product knowledge, brand attitude and purchase intention: The mediating role of presence. *Journal of Advertising*, 31(3), 43-57.

Li, Y., & Lindner, J.R. (2005). Faculty adoption behaviour about web-based distance education: A case study from China Agricultural University. *British Journal of Educational Technology, 38*(1), 83-94.

Libaw, O.Y. (2000). Dance machine: Quirky arcade game draws legions of fans. *ABC News*. Retrieved August 5, 2008, from http://www.ddrfreak.com/newpress/ABC%20News.htm.

Lienert, A. (2004). Video games open new path to market cars. *Detroit News*. Retrieved September 24, 2004, from http://www.detnews.com/2004/autosinsider/0402/15/b01-64356.htm.

Lim, W.Y., & Koh, M. (2006). Effectiveness of Learning Technologies in the Teaching and Learning of Gymnastics. *Pacific Asian Education, 18*(2), 69–77.

Lin, Z. M., Yang, J., & Yang, C. (2004). Grey decision-making for a billiard robot. In *IEEE Intl. Conf. Systems, Man and Cybernetics.*

Lindstrom, M. (2001). *Brand games: Are you ready to play?* Retrieved 6 October, 2007, from http://www.internet.com

Liu, Y., & Shrum, L.J. (2002). What is interactivity and is it always such a good thing? Implications of definition, person and situation for the influence of interactivity on advertising effectiveness. *Journal of Advertising, 31*(4), 53-64.

Liverpool F.C. (2007, September 2). Retrieved September 3, 2007, from http://en.wikipedia.org/wiki/Liverpool_F.C.

Liverpool wins Champions League. (2005, May 30). Retrieved May 16, 2007, from http://www.cbc.ca/sports/story/2005/05/25/uefa050525.html

Lloyd, D.G., Alderson, J., & Elliott, B.C. (2000). An upper limb kinematic model for the examination of cricket bowling: A case study of Muttiah Muralitharan. *Journal of Sports Sciences, 18*(12), 975-982.

Lo, K., Wang, L., Wu, C., & Su, F. (2004). Biomechanical analysis of trunk and lower extremity in the tennis serve. In *Proceedings of the XXII International Symposium of Biomechanics in Sport* (pp. 261-264). Ottawa: University of Ottawa.

Lombard, M., & Ditton, T. (1997). At the heart of it all: The concept of presence. *Journal of Computer-Mediated Communication, 3*(2). Retrieved February 11, 2005, from http://jcmc.indiana.edu/vol3/issue2/lombard.html.

Long, F., Herland, J., Tessier, M.-C., Naulls, D., Roth, A., Roth, G., & Greenspan, M. (2004). Robotic pool: An experiment in automatic potting. In *IROS 2004: IEEE/RSJ Intl. Conf. Intell. Rob. Sys.*, (pp. 361–366).

Long, J., & Sanderson, I. (2001). The social benefits of sport. Where is the proof? In C. Gratton & I. Henry (Ed.), *Sport in the city: The role of sport in economic and social regeneration* (pp. 187-203). London, UK: Routledge.

Lu, W. L., Okuma, K., & Little, J. J. (2008). Tracking and recognizing actions of multiple hockey players using the boosted particle filter. *Image and Vision Computing*, in press.

Ma, C. M., & Sonka, M. (1996). A Fully Parallel 3D Thinning Algorithm and Its Applications. *Computer Vision and Image Understanding*, 64, 420-433.

MacCormick, J., & Blake, A. (1999). *A probabilistic exclusion principle for tracking multiple objects.* Paper presented at the Int. Conf. on Computer Vision.

MacDonald, G. (1988). The economics of rising stars. *American Economic Review, 78*(1), 155-167.

Macgregor, D.M. (2000). Nintendonitis? A case report of repetitive strain injury in a child as a result of playing computer games. *Scottish Medical Journal, 45*(5), 150.

Machleit, A.K., Allen, T.C., & Madden, J.T. (1993). The mature brand and bran interest: An alternative consequence of ad-evoked affect. *Journal of Marketing, 57*, 72-82.

MacKay, N., Parent, M., & Gemino, A. (2004). A model of electronic commerce adoption by small voluntary organisations. *European Journal of Information Systems*, (pp. 147–159).

Maddison, R., Mhurchu, C.N., Jull, A., Jiang, Y., Prapavessis, H., & Rodgers, A. (2007). Energy expended playing video console games: An opportunity to increase children's physical activity? *Pediatric Exercise Science, 19*(3), 334-343.

Madon, S. (1999). International NGOs: Networking, information flows and learning. *Journal of Strategic Information Systems*, (pp. 251-261).

Madrigal, R. (2000). The Influence of Social Alliances with Sports Teams on Intentions to Purchase Corporate Sponsors' Products. *Journal of Advertising, 29*(4), 13-24.

Madrigal, R. (2000). The influence of social alliances with sports teams on intentions to purchase corporate sponsors' products. *Journal of Advertising, XXIX*(4), 13-24.

Madrigal, R. (2004). A Review of Team Identification and Its Influence on Consumers' Responses Toward Corporate Sponsors. In L.R. Kahle & C. Riley (Eds.), *Sports Marketing and the psychology of marketing communication* (pp. 241-255). Mahwah, NJ: Lawrence Erlbaum Associates.

Magerkurth, C., & Stenzel, R. (2003). Computerunterstutztes Kooperartives Spielen -- Die Zukunft des Spieltisches. In J. Ziegler & G. Szwillus (Eds.), *Proceedings of Mensch & Computer 2003 (MC ,03), Stuttgart*: Teubner.

Mahony, D.F., Howard, D.R., & Madrigal, R. (2000). BIRGing and CORFing Behaviors by Sport Spectators: High Self-Monitors Versus Low Self-Monitors. *International Sports Journal, Winter*, (pp. 87-106).

Mahony, D.F., Nakazawa, M., Funk, D.C., James, J.D., & Gladden, J.M. (2002). Motivational Factors Influencing the Behaviour of J. League Spectators. *Sport Management Review, 5*, 1-24.

Major League Gaming. (2006). About Major League Gaming. Retrieved July 17, 2008, from http://www.mlgpro.com/about/index.

Manchester, W. (1993). *A world lit only by fire: The medieval mind and the renaissance: Portrait of an age*, USA: Back Bay Books.

Marcia, J.E. (1994). Ego identity and object relations. In J.M. Masling & R.F. Bornstein (Eds.), *Empirical perspectives on object relations theory* (pp.59-93). Washington, DC: American Psychological Association.

Marentakis, G., & Brewster, S. (2006). Effects of feedback, mobility and index of difficulty on deictic spatial audio target acquisition in the horizontal plane. In *Proceedings of CHI 2006* (pp. 359-368). Quebec, Canada: ACM Press.

Marlow, W.C. (1995). *The physics of pocket billiards.* MAST, Palm Beach Gardens, Florida.

Marshall, R.N., & Elliott, B.C. (2000). Long-axis rotation: The missing link in proximal-to- distal segmental sequencing. *Journal of Sports Sciences, 18*(4), 247-254.

Martin, B.A. (2004). Using the imagination: Consumer evoking and thematizing of the fantastic imaginary. *Journal of Consumer Research, 31*(1), 136-149.

Massive-Incorporated. (2007). *Massive study reveals in-game advertising increases average brand familiarity by up to 64 percent.* Retrieved 6 October 2007, from http://www.massiveincorporated.com/site_network/pr/08.08.07.htm

Matsuyama, T., Wu, X., Takai, T., & Nobuhara, S. (2004). Real-time 3D Shape Reconstruction, Dynamic 3D Mesh Deformation, and High Fidelity Visualization for 3D Video. *Computer Vision and Image Understanding, 96*, 393-434.

Matzat, U. (2004). Academic communication and Internet Discussion Groups: Transfer of information or creation of social contacts? *Social Networks, 26*, 221-255.

Mayagoitia, R., Nene, A., & Veltink, P. (2002). Accelerometer and rate gyroscope measurement of kinematics: An inexpensive alternative to optical motion analysis systems. *Journal of Biomechanics, 35*(4), 537-542.

Mayra, F. (2007). The contextual game experience: On the socio-cultural contexts for meaning in digital play. In *Situated play: Proceedings of DiGRA 2007 Conference* (pp. 810-814). Tokyo: Digital Games Research Association.

McCandless, D. (2003). Just one more go.... *The Guardian, Apr 3.* Retrieved August 11, 2007, from http://www.guardian.co.uk/technology/2003/apr/03/onlinesupplement3.

McClellan, S. (2005, December 5). Study: In-game integration reaps consumer recall. *Adweek, 46*(47), 9.

McClure, R.F., & Mears, F.G. (1984). Video game players: Personality characteristics and demographic variables. *Psychological Reports, 55*(1), 271-276.

McKay, P. (2004). Polites takes aim to protect V8 Supercars. *Sydney Morning Herald.* Retrieved September 24, 2004, from http://www.smh.com.au/cgi-bin/common/popupPrintArticle.pl?path=/articles/2004/02/23/1077497513603.html.

McKenna, S. J., Jabri, S., Duric, Z., Rosenfeld, A., & Wechsler, H. (2000). Tracking Groups of People. *Computer Vision and Image Understanding, 80*, 42-56.

McKenna, S., Jabri, S., Duric, Z., & Wechsler, H. (2000). *Tracking interacting people.* Paper presented at the Fourth IEEE International Conference Automatic Face and Gesture Recognition, Washington, USA.

McLean, P.D., & Padgett, J.F. (1997). Was Florence a perfectly competitive market? Transactional evidence from the Renaissance. *Theory and Society, 26*(2/3), 209-244.

McMillan, D., & Chavis, D. (1986). Sense of community: A definition and theory. *American Journal of Community Psychology,* (pp. 6-23).

McMillan, S.J. (2002). Exploring models of interactivity from multiple research traditions: Users, documents and systems. In L.A. Lievrouw & S. Livingstone (Eds.), *Hand-*

*book of new media: Social shaping and consequences of ICTs* (pp. 163-182). London: Sage.

McNamee, S.J., & Willis, C.L. (1994). Stratification in science: A comparison of publication patterns in four disciplines. *Science Communication, 15*, 396-416.

Meenaghan, T. (1983). Commercial sponsorship. *European Journal of Marketing, 17*(7), 5-73.

Meenaghan, T. (2001). Sponsorship and advertising: A comparison of consumer perceptions. *Psychology & Marketing, 18*(2), 191-215.

Meenaghan, T. (2001). Understanding sponsorship effects. *Psychology & Marketing, 18*(2), 95-122.

Meier, R., & Cahill, V. (2005). Taxonomy of Distributed Event-Based Programming Systems. *Computing Journal, 48*, 602-626.

Mellecker, R.R., & McManus, A.M. (2008). Energy expenditure and cardiovascular responses to seated and active gaming in children. *Archives of Pediatrics & Adolescent Medicine, 162*(9), 886-891.

Menon, S., & Kahn, B. (2002). Cross-category effects of induced arousal and pleasure on the Internet shopping experience. *Journal of Retailing, 78*(1), 31-40.

Merleau-Ponty, M. (2002). *Phenomenology of perception: An introduction.* (2nd Ed.). USA: Routledge.

Messner, M.A., Dunbar, M., & Hunt, D. (2000). The televised sports manhood formula. *Journal of Sport & Social Issues, 24*(4), 380-394.

Meyer, M. (1992). *Zur Entwicklung der Sportbedürfnisse.* Unpublished Dissertation, German Sports University, Cologne.

Mhurchu, C.N., Maddison, R., Jiang, Y., Jull, A., & Prapavessis, H., (2008). Couch potatoes to jumping beans: A pilot study of the effect of active video games on physical activity in children. *International Journal of Behavioral Nutrition and Physical Activity, 5*(8). Retrieved August 30, 2008, from http://www.ijbnpa.org/content/5/1/8.

Microsoft. (2007, July 25). Premier EA™ titles to go live in Massive Network. Microsoft.com. Retrieved March 1, 2008, from www.microsoft.com/presspass/press/2007/july07/07-25EAMassivePR.mspx

Mikic, I., Trivedi, M., Hunter, E., & Cosman, P. (2003). Human Body Model Acquisition and Tracking using Voxel Data. *International Journal of Computer Vision, 53*, 199-223.

Minichiello, V., Aroni, R., Timewell, E., & Alexander, L. (1995). *In-depth Interviewing: Principles, Techniques, Analysis*. South Melbourne: Longman.

Mitchell, A., & Savill-Smith, C. (2004). *The use of computer and video games for learning: A review of the literature*. London: Learning and Skills Development Agency.

Mitrano, J.R. (1999). The "Sudden Death" of Hockey in Hartford: Sports Fans and Franchise Relocation. *Sociology of Sport Journal, 16*, 134-154.

Moe-Nilssen, R., & Helbostad, J. (2004). Estimation of gait cycle characteristics by trunk accelerometry. *Journal of Biomechanics, 37*(1), 121-126.

Moeslund, T.B., & Granum, E. (2001).A Survey of Computer Vision-based Human Motion Capture. *Computer Vision and Image Understanding, 81*, 231-268.

Moeslund, T.B., Hilton, A., & Krüger, V. (2006). A survey of advances in vision-based human motion capture and analysis. *Computer Vision and Image Understanding, 104*(2-3), 90-126.

Mokka S., Väätänen A., & Välkkynen, P. (2003). Fitness computer games with a bodily user interface. In *Proceedings of ICEC 2003* (pp. 1-3). Pittsburgh: ACM Press.

Mokka, S., Väätänen, A., & Välkkynen, P. (2003). Fitness Computer Games with a Bodily User Interface, *Proceedings of the Second International Conference on Entertainment Computing* (pp. 1-3). Pittsburgh, Pennsylvania: ACM International Conference Proceeding Series.

Molesworth, M. (2006). Real brands in imaginary worlds: Investigating players' experiences of brand placement in digital games. *Journal of Consumer Behavior, 5*(4), 355-366.

Montazemi, A. (1988). Factors affecting information satisfaction in the context of the small business environment. *MIS Quarterly,* (pp. 239-256).

Montoye, H.J., Washburn, R., Servais, S., Ertl, A., Webster, J.G., & Nagle, F.J. (1983). Estimation of energy expenditure by a portable accelerometer. *Medicine & Science in Sports & Exercise, 15*(5), 403-407.

Moore, E.S. (2006). *It's child's play: Advergaming and the online marketing of food to children.* Menlo Park, CA: Kaiser Family Foundation.

Moran, A.P. (2004). *Sport and exercise psychology: A critical introduction.* Hove, UK: Routledge.

Moreno, J.L. (1946). Sociogram and sociomatrix: A note to the paper by Forsyth and Katz. *Sociometry, 9*(4), 348-349.

Morgon, G. (1995). ITEM: A strategic approach to information systems in voluntary organisations. *Journal of Strategic Information Systems,* (pp. 225-237).

Moritz, E. F. (2004). *Systematic Innovation in Popular Sports.* Paper presented at the 5th Conference of the International Sports Engineering Association.

Moritz, E. F., & Steffen, J. (2003). Test For Fun – ein Konzept für einen nutzerorientierten Sportgerätetest. In K.v. Roemer, J. Edelmann-Nusser, K. Witte & E.F. Moritz (Eds.), *Sporttechnologie zwischen Theorie und Praxis* (pp. 43-63). Aachen.

Morlier, J., & Mesnard, M. (2007). Influence of the Moment Exerted by the Athlete on the Pole in Pole-Vaulting Performance. *Journal of Biomechanics, 40*(10), 2261-2267.

Morris, L., Sallybanks, J., & Willis, K. (2003). Sport, physical activity and antisocial behaviour in youth. *Australian Institute of Criminology Research and Public Policy Series, 249,* 1-6.

Moselund, T.B., Hilton, A., & Kruger, V. (2006). A Survey of Advances in Vision-based Human Motion Capture and Analysis. *Computer Vision and Image Understanding, 104,* 90-126.

Moser, C., & Kalton, G. (1971). *Survey Methods in Social Investigation* (Second ed.). London: Heinemann Educational Books.

Motamedi, M.E., & White, R.M. (1994). Acoustic sensors. In S.M. Sze (Ed.), *Semiconductor sensors* (pp. 97-151). New York: Wiley.

Motion Analysis (nd). Retrieved May 12, 2008, from http://www.motionanalysis.com

Mueller, F., & Gibbs, M. (2007). *A Physical Three-Way Interactive Game Based on Table Tennis.* Paper presented at the IE '07: Proceedings of the 4th Australasian conference on Interactive entertainment (2007), RMIT University, Melbourne, Australia.

Mueller, F., & Gibbs, M.R. (2007). *Evaluating a distributed physical leisure game for three players.* Paper presented at the Proc. of Australia on Computer-Human interaction OZCHI '07.

Mueller, F., Agamanolis, S., & Picard, R. (2003). Exertion Interfaces: Sports over a distance for social bonding and fun. In *Proceedings of CHI 2003.* Fort Lauderdale, USA: ACM Press.

Mueller, F., Cole, L., O'Brien, S., & Walmink, W. (2006). *Airhockey Over a Distance – A Networked Physical Game to Support Social Interactions.* Paper presented at the Advances in Computer Entertainment Technology ACE 2006.

Mueller, F., O'Brien, S., & Thorogood, A. (2007). Jogging over a Distance. In *Extended Abstracts CHI 2007.* California, USA: ACM Press.

Mueller, F., Stevens, G., Thorogood, A., O'Brian, S., & Wulf, V. (2007). Sports over a Distance. *Personal and Ubiquitous Computing (PUC), 11*(8), 633–645.

Mullin, B., Hardy, S., & Sutton, W. (2007). *Sport marketing* (3rd ed). Champaign, IL: Human Kinetics.

Murphy, S.M. (2007). A social meaning framework for research on participation in social online games. *Journal of Media Psychology, 12*(3). Retrieved May 20, 2008, from http://www.calstatela.edu/faculty/sfischo/A_Social_Meaning_Framework_for_Online_Games.html.

Nagao, N., & Sawada, Y. (1977). A kinematic analysis of the golf swing by means of fast motion picture in connection with wrist action. Journal of Sports Medicine & Physical Fitness, *17*(4), 413–419.

Nash, R. (2000). Globalised Football Fandom: Scandinavian Liverpool FC Supporters. *Football Studies, 3*(2), 5-23.

Nebenzahl, D.I., & Secunda, E. (1993). Consumers attitudes toward product placement in movies. *International Journal of Advertising, 12*(1), 1-11.

Neeley, S.M., & Schumann, D.W. (2004). Using animated spokes-characters in advertising to young children. *Journal of Advertising, 33*(3), 7-23.

Nelson, M.R. (2002). Recall of brand placements in computer/video games. *Journal of Advertising Research, 42*(2), 80-93.

Nelson, M.R., Keum, H., & Yaros, R.A. (2004). Advertainment or adcreep? Game players' attitudes toward advertising and product placements in computer games. *Journal of Interactive Advertising, 5*(1). Retrieved November 10, 2006, from http://www.jiad.org/vol5/no1/nelson/

Nelson, R.M. (2002). Recall of brand placements in computer/video games. *Journal of Advertising Research, March/April*, (pp. 80-92).

Nelson, R.M., Keum, H., & Yaros, A.R. (2004). Advertainment or adcreep? Game players' attitudes toward advertising and product placements in computer games. *Journal of Interactive Advertising, 4*(3), 1.

Nelson, R.M., Yaros, A.R., & Keum, H. (2006). Examining the influence of telepresence on spectator and player processing of real and fictitious brands in a computer game. *Journal of Advertising, 35*(4), 87-99.

Neuborne, E. (2001). For kids on the Web, it's an ad, ad, ad, ad, world; How to help yours see the sales pitches behind online games. *Business Week, Aug 13*(3725), 108.

NFL (n.d.) *Super Bowl tickets and rings.* Retrieved June 22, 2008, from http://www.nfl.com

Nicovich, S.G. (2005). The effect of involvement on ad judgment in a video game environment: The mediating role of presence. *Journal of Interactive Advertising, 6*(1). Retrieved August 9, 2007, from http://www.jiad.org/vol6/no1/nicovich/index.htm.

Nielsen. (2006, October 18). Nielsen to provide video game rating service. *Nielsen.com.* Retrieved May 11, 2008, from http://www.nielsenmedia.com/nc/portal/site/Public/menuitem.55dc65b4a7d5adff3f65936147a062a0/?vgnextoid=aea42f5dde75e010VgnVCM100000ac0a260aRCRD

Nielsen. (2007). The state of the console. Retrieved August 6, 2008, from *www.nielsenmedia.com/nc/nmr_static/docs/Nielsen_Report_State_Console_03507.pdf.*

Nielsen. (2007, July 2). Nielsen and Sony Computer Entertainment America to develop measurement system for game network advertising. *Nielsen.com.* Retrieved May 11, 2008 from http://www.nielsen.com/media/pr_070702.html

Nintendo. (2008a). *Wii Sports.* Retrieved April 30, 2008, from http://www.nintendo.com/games/detail/1OTtO06SP7M52gi5m8pD6CnahbW8CzxE.

Nintendo. (2008b). *Wii Fit.* Retrieved April 30, 2008, from http://www.nintendo.com/wiifit

Nobuhara, S., & Matsuyama, T. (2006). Deformable Mesh Model for Complex Multi-Object 3D Motion Estimation from Multi-Viewpoint Video. In M. Pollefeys (Ed.), *Proceedings of the 3rd Intl. Symposium on 3D Processing, Visualization and Transmission (3DPVT 2006)*, (pp. 264-271). Chapel Hill: University of North Carolina.

Noma, E. (1982). Untangling citation networks. *Information Processing & Management, 18*(2), 43-53.

Novak, T.P., Hoffman, D.L., & Yung, Y.F. (2000). Modeling the flow construct in online environments: A structural modeling approach. *Marketing Science, 19*(1), 22-42.

NPD Group. (2004). The NPD Group reports on sales of licensed video game titles. Retrieved September 23, 2004, from http://www.npd.com/press/releases/press_040608.htm.

NPD Group. (2007a). Playing video games viewed as family/ group activity and stress reducer. Retrieved July 15, 2008, from http://www.npd.com/press/releases/press_071212.html.

NPD Group. (2007b). Amount of time kids spend playing video games is on the rise. Retrieved July 15, 2008, from http://www.npd.com/press/releases/press_071016a.html.

Nuesch, S. (2007). *The economics of superstars and celebrities*. Wiesbaden: DUV Gabler Edition.

O'Brien, S., Mueller, F., & Thorogood, A. (2007). Jogging the Distance. In *Proceedings of CHI 2007*. California, USA: ACM Press.

O'Guinn, T.C., Allen, C.T., & Semenik, R.J. (2006). *Advertising and integrated brand promotion*. (4th ed.). Mason: Thomson/South-Western.

Ohgi, Y., & Baba, T. (2005). Uncock timing in driver swing motion. In A. Subic & S. Ujihashi (Ed.), *The impact of technology on sport* (pp. 349-354). Melbourne, Australia: Australasian Sports Technology Alliance.

Ohgi, Y., Seo, K., Hirai, N., & Murakami, M. (2006). Measurement of jumper's body motion in ski jumping. In E. Morriz & S. Haake (Eds.), *The Engineering of Sports, 6*, 275-280. USA: Springer.

Ohgi, Y., Seo, K., Hirai, N., & Murakami, M. (2007). Aerodynamic forces acting in ski jumping. *Journal of Biomechanics, 40*, S402.

Ohgi, Y., Yasumura, M., Ichikawa, H., & Miyaji, C. (2000). Analysis of stroke technique using acceleration sensor IC in freestyle swimming. *The Engineering of Sport*, (pp. 503-511).

Ohno, Y., Miurs, J., & Sharai, Y. (1999). *Tracking players and a ball in soccer games*. Paper presented at the IEEE International Conference on Multisensor Fusion and Integration for Intelligent Systems, Taipei, Taiwan.

Ohshima, T., Satoh, K., Yamamoto, H., & Tamura, H. (1998). AR2 Hockey, *Conference Abstracts and Applications of SIGGRAPH'98* (pp. 268). New York: ACM-Press.

Ohta, K., Ohgi, Y., Kimura, H., & Hirotsu, N. (2005). *Sports data*. Japan: Kyoritsu Publishing.

Ok, H., Seo, Y., & Hong, K. (2002). *Multiple soccer players tracking by condensation with occlusion alarm probability*. Paper presented at the Statistical Methods in Video Processing Workshop, Copenhagen, Denmark.

Oliver, J. (2005, November 21). Sports media & technology conference. *Sports Business Journal*, 10.

Oliver, N., & Flores-Mangas, F. (2006). MPTrain: A mobile, music and physiology-based personal trainer. In *Proceedings of HCI-Mobile 2006* (pp. 21-28). USA: ACM Press.

Orum, A.M., Feagin, J.R., & Sjoberg, G. (1991). The Nature of the Case Study. In J.R. Feagin, A.M. Orum & G. Sjoberg (Eds.), *A Case for the Case Study* (pp. 1-26). Chapel Hill, NC: The University of North Carolina Press.

Padgett, J.F., & Ansell, C.K. (1993). Robust action and the rise of the Medici, 1400-1434. *American Journal of Sociology, 98*(6), 1259-1319.

Padgett, J.F., & McLean, P.D. (2006). Organizational invention and elite transformation: The birth of partnership systems in Renaissance Florence. *American Journal of Sociology, 111*(5), 1463-1568.

Paffenbarger, R.S., Patrick, K., Pollock, M.L., Rippe, J.M., Sallis, J., & Wilmore, J.H. (1995). Physical activity and public health: A recommendation from the Centers for Disease Control and Prevention and the American College of Sports Medicine. *Journal of the American Medical Association, 273*(5), 402–407.

Pagulayan, R., Keeker, K., Wixon, D., Romero, R., & Fuller, T. (2003). User-centered design in games. In J.A. Jacko & A. Sears (Eds.), *The human-computer interaction handbook: Fundamentals, evolving techniques and emerging applications* (pp. 883-905). Mahwah, NJ: Lawrence Erlbaum Associates.

Pan, Z.G., & Xu, W.W. (2003). Easybowling: A small bowling machine based on virtual simulation. *Computers & Graphics, 27*(2), 231-238.

Past Manager Profile - Bill Shankly (1959-1974). Retrieved September 4, 2007, from http://www.liverpoolfc.tv/team/past_players/managers/shankly/

Pate, R.R., Pratt, M., Blair, S.N., Haskell, W.L., Macera, C.A., Bouchard, C., Buchner, D., Ettinger, W., Heath, G.W., King, A.C., Kriska, A., Leon, A.S., Marcus, B.H., Morris, J.,

Paterson, P.A. (2003). Synergy and expanding technology drive booming video game industry. *TD Monthly*. Retrieved September 23, 2004, from http://www.toydirectory.com/monthly/Aug2003/Games_Booming.asp.

Payne, A.H. (1978). Comparison of the ground reaction forces in golf drive and tennis service. *Aggressologie, 19*, 53-54.

Peak Performance. (nd). Retrieved May 12, 2008, from http://www.peakperform.com

Peak Training System Sensor Kit. (2008). *Sensors*. Retrieved April 30, 2008, from http://www.riderunrow.com/products_sensors.htm

Pearce, J. (1993). *Volunteers: The organisational behaviour of unpaid workers*. London: Routledge.

Pearson, E., & Bailey, C. (2007). Evaluating the potential of the Nintendo Wii™ to support disabled students in education. In *ICT: Providing choices for learners and learning. Proceedings ascilite Singapore 2007* (pp. 833-836). Singapore: Center for Educational Development, Nanyang Technological University.

Pedersen, J.B. (2002). Are professional gamers different? - Survey on online gaming. *Game Research*. Retrieved April 30, 2008, from www.gameresearch.com/art_pro_gamers.asp

Pei-Yan, Z., & Tian-Sheng, L. (2007). Real-Time Motion Planning for a Volleyball Robot Task Based on a Multi-Agent Technique. *J. of Intelligent Robot Systems, 49*, 355-366.

Pennington, R. (2001). Signs of marketing in virtual reality. *Journal of Interactive Advertising, 2*(1). Retrieved August 10, 2007, from http://www.jiad.org/vol2/no1/pennington/index.htm.

Pereira, J. (2004). Junk-food games; online arcades draw fire for immersing kids in ads; Ritz Bits Wrestling, anyone? *Wall Street Journal, May 3*, B1.

Pers, J., & Kovacic, S. (2000, 06/14/2000 - 06/15/2000). *Computer vision system for tracking players in sports games*. Paper presented at the Image and Signal Processing and Analysis, IWISPA 2000. First International Workshop on Image and Signal Processing and Analysis. in conjunction with 22nd International Conference on Information Technology Interfaces., Pula, Croatia.

Petit, R. (2004). *Billard: Théorie du jeu*. Chiron Éditeur.

Pham, M.T. (1992). Effect of involvement, arousal and pleasure on the recognition of sponsorship stimuli. *Advances in Consumer Research, 19*, 1-9.

Pingali, G., Opalach, A., & Jean, Y. (2000). *Ball tracking and virtual replays for innovative tennis broadcasts*. Paper presented at the International Conference on Pattern Recognition, Barcelona, Spain.

Pitie, F., Berrani, S. A., Kokaram, A., & Dahyot, R. (2005). *Off-Line Multiple Object Tracking Using Candidate Selection and the Viterbi Algorithm*. Paper presented at the International Conference on Image Processing, ICIP 2005, Genova, Italy.

Planker, R., & Fua, P. (2001). Tracking and Modeling People in Video Sequences. *Computer Vision and Image Understanding, 81*, 285-302.

Porcari, P., Foster, C., Dehart-Beverly, M., Shafer, N., Recalde, P., & Voelker, S. (2002). *Prescribing exercise using the talk test*. Retrieved April 30, 2008, from http://www.fitnessmanagement.com/FM/tmpl/genPage.asp?p=/information/articles/library/cardio/talk0801.html

Potkonjak, V., & Vukobratovic, M (1979). Two New Methods for Computer Forming of Dynamic Equations of Active mechanisms. *J. Mechanism and Machine Theory, 14*(3).

Potkonjak, V., & Vukobratovic, M. (2005). A Generalized Approach to Modeling Dynamics of Human and Humanoid Motion. *Intl. Journal of Humanoid Robotics, 2*(1), 1-24.

Potkonjak, V., Vukobratovic, M., Babkovic, K., & Borovac, B. (2006). General Model of Dynamics of Human and Humanoid Motion: Feasibility, Potentials and Verification. *Intl. Journal of Humanoid Robotics, 3*(1), 21-48.

Pourcelot, P., Audigie, F., Degueurce, C., Geiger, D., & Denoix, J.M. (2000). A method to synchronise cameras using the direct linear transformation technique. *Journal of Biomechanics, 33*(12), 1751-1754.

Powell, E. (2008). ABC, ESPN, IMS radio network providing in-depth Indy 500 coverage. *Indy500.com*, 2 May. Retrieved May 29 from http://www.indy500.com/news/11179/ABC_ESPN_IMS_Radio_Network_Providing

Powell, W. (2008). *Virtually Walking? Developing Exertion Interfaces for Locomotor Rehabilitation.* Paper presented at the Paper presented at the CHI 2008. Workshop submission to "Exertion Interfaces". Retrieved from http://workshopchi.pbwiki.com/f/PowellW_Exertion08V1.2.pdf

PQ Media. (2006). Alternative media research series II: Alternative advertising and marketing outlook 2006. Retrieved July 5, 2006, from http://www.pqmedia.com/execsummary/AlternativeAdvertisingMarketingOutlook2006-ExecutiveSummary.pdf.

Pretty, G., Andrewes, L., & Collett, C. (1994). Exploring adolescent's sense of community and its relationship to loneliness. *Journal of Community Psychology,* (pp. 346-357).

PricewaterhouseCoopers. (2007). Industry previews: Video games. Retrieved September 5, 2007, from http://www.pwc.com/extweb/industry.nsf/docid/8CF0A9E084894A5A85256CE8006E19ED?opendocument&vendor=#video.

PricewaterhouseCoopers. (2008). Entertainment and media companies face a collaboration imperative for next five years, says PricewaterhouseCoopers annual outlook report. Retrieved August 5, 2008, from http://www.pwc.com/extweb/ncpressrelease.nsf/docid/6DD913426F4A05108525746B004C3C42.

Privette, G. (1983). Peak experience, peak performance and flow: A comparative analysis of positive human experience. *Journal of Personality and Social Psychology, 45*(6), 1361-1368.

Product Review. (2004). V8 Supercars Australia racing video game details. Retrieved September 23, 2004, from http://www.productreview.com.au/showitem.php?item_id=1075.

Provenzo, E.F. (1991). *Video Kids: Making Sense of Nintendo.* MA: Harvard University Press Cambridge.

ProZone (2008). ProZone Sports Retrieved 12th May 2008, from http://www.pzfootball.co.uk/index.htm

Purvis, G. (2002, October 18). *The Heysel Tragedy.* Retrieved September 6, 2007, from http://www.lfconline.com/news/loadsngl.asp?cid=EDB3

Purvis, G. (2002, October 18). *The Hillsborough Tragedy.* Retrieved September 6, 2007, from http://www.lfconline.com/news/loadfeat.asp?cid=EDB2&id=69043

Purvis, G. (2002, October 18). The History of Liverpool Football Club. Retrieved September 3, 2007, from http://www.lfconline.com/news/loadsngl.asp?cid=EDB1

Putnam, R. (2000). *Bowling alone.* New York, USA: Touchstone, Simon & Schuster.

Quah, C.K., Gagalowicz, A., & Seah, H.S. Marker-less 3D Video Motion Capture in Cluttered Environments. In F.K. Fuss (Ed.), *Proceedings of the Asia-Pacific Congress on Sports Technology,* (pp. 121-126). Singapore: Nanyang Technological University.

Quah, C.K., Gagalowicz, A., Roussel, R., & Seah, H.S. (2005). 3D modeling of humans with skeletons from uncalibrated wide baseline views. In A. Gagalowicz (Ed.), *Proceedings of the International Conference on Computer Analysis of Images and Patterns,* (pp. 379-389). Rocquencourt, France: INRIA.

Qualysis. (nd). Retrieved May 12, 2008, from http://www.qualysis.com

Raabe, P.B. (2001). *Philosophical counseling: Theory and practice.* Westport, CT: Praeger.

Radia, S., & Harris, T. (2006, June). *Reaching the 18-34 demographic through games.* Play, Denuo Group. Paper presented at the Game Developers Conference (GDC) Focus on: Game Advertising Summit, San Francisco, CA.

Ragazzoni, A. (1998). Arousal: A neurophysiological view. *Abstracts/International Journal of Psychophysiology, 30*(9).

Randel, J.M., Morris, B.A., Wetzel, C.D., & Whitehill, B.V. (1992). The effectiveness of games for educational purposes: A review of recent research. *Simulation and Gaming, 23*(3), 261-276.

Raney, A.A., Arpan, L.M., Pashupati, K., & Brill, D.A. (2003). At the movies, on the web: An investigation of the effects of entertaining and interactive web content on site and brand evaluations. *Journal of Interactive Marketing, 17*(4), 38-53.

Rauterberg, M. (2003). *Emotional Aspects of Shooting Activities: 'Real' versus 'Virtual' Actions and Targets.* Paper presented at the 2nd International Conference on Entertainment Computing (ICEC 2003), Pittsburgh, PA.

Rauterberg, M. (2004). Positive effects of entertainment technology on human behavior. In R. Jacquart (Ed.), *Building the information society* (pp. 51-58). Toulouse, France: Kluwer Academic Publishers.

Raymond, L. (1985). An empirical study of management information systems sophistication in small business. *MIS Quarterly,* 37-52.

Recent evidence from OECD countries. *Oxford Review of Economic Policy,* (pp. 324-344).

Redman, R.W. (2006). The challenge of interdisciplinary teams. *Research and Theory for Nursing Practice, 20*(2), 105-107.

Reimer, B. (2004). For the love of England. Scandinavian football supporters, Manchester United and British popular culture. In D.L. Andrews (Ed.), *Manchester United. A Thematic Study* (pp. 265-277). Abingdon, England: Routledge.

Reisinger, D. (2008). Why the Xbox 360™ will win the console war. *cnet.* Retrieved July 31, 2008, from http://news.cnet.com/8301-13506_3-9939276-17.html.

Reportlinker. (2007). Western world MMOG market: 2006 review and forecasts to 2011. Retrieved July 31, 2008, from http://www.reportlinker.com/p046468/online-games.html.

RepriseMedia (2008). *Super Bowl advertisers fumble in search and social media.* Retrieved March 19, 2008, from http://www.reprisemedia.com/pressreleases

Resnick, M. (1998). Learning in school and out. *Educational Researcher, 16*(9), 13-20.

Rice-Oxley, M. (2007, June 8). English fans pool case to buy their own soccer team. Retrieved June 11, 2007, from http://www.csmonitor.com/2007/0608/p01s03-woeu.html

Richardson, B., & O'Dwyer, E. (2003). Football Supporters and Football Team Brands: A Study in Consumer Brand Loyalty. *Irish Marketing Review, 16*(1), 43-53.

Richelieu, A., & Pons, F. (2006). Toronto Maple Leafs vs Football Club Barcelona: how two legendary sports teams built their brand equity. *International Journal of Sports Marketing & Sponsorship, May,* (pp. 231-250).

Roberts, H. (1979). *Community development: Learning and action.* Toronto: University of Toronto Press.

Robertson, J., & Good, J. (2005). Story creation in virtual game worlds. *Communications of the ACM, 48*(1), 61-65.

Robinson, M. (2008). Players cop a blogging. *Herald Sun.* Retrieved June 8, 2008, from http://www.news.com.au/heraldsun/sport/afl/story/0,26576,23732964-19742,00.html

Robinson, R.J., Barron, D.A., Grainger, A.J., & Venkatesh, R. (2008). Wii knee. *Emergency Radiology, 15*(4), 255-257.

RocSearch. (2004). Video game industry. Retrieved September 15, 2004, from http://www.rocsearch.com/pdf/Video%20Game%20Industry.pdf.

Roehm, H.A., & Haugtvedt, C.P. (1999). Understanding interactivity of cyberspace advertising. In D.W. Schumann & E. Thorson (Eds.), *Advertising and the World Wide Web* (pp. 27-39). Mahwah, NJ: Lawrence Erlbaum.

Rogers, E. (2005). *Diffusion of innovations.* (5th ed.). New York: The Free Press.

Rosales, R., & Sclaroff, S. (2006). Combining Generative and Discriminative Models in a Framework for Articulated Pose Estimation. *International Journal of Computer Vision, 63*, 251-276.

Rosas, R., Nussbaum, M., Cumsille, P., Marianov, V., Correa, M., Flores, P., Grau, V., Lagos, F., Lopez, X., Lopez, V., Rodriguez, P., & Salinas, M. (2003). Beyond Nintendo: Design and assessment of educational video games for first and second grade students. *Computers and Education, 40*(1), 71-94.

Rosen, S. (1981). The economics of superstars. *American Economic Review, 71*(1), 845-848.

Rosen, S. (1986). Prizes and incentives in elimination tournaments. *American Economic Review, 76*(4), 701-716.

Rosenhahn, B., He, L., & Klette, R. (2005). Automatic Human Model Generation In A. Gagalowicz (Ed.), *Proceedings of the International Conference on Computer Analysis of Images and Patterns*, (pp. 41-48). Rocquencourt, France: INRIA.

Rosewater, A. (2004). Hey, I'm practicing here, not playing: Video games so precise they help many drivers. *USA Today, Aug 8*, 6f.

Rosewater, A. (2004). Hey, I'm practicing here, not playing: Video games so precise they help many drivers. *USA Today, Aug 8*, 6f.

Rosser, J.C., Lynch, P.J., Cuddihy, L., Gentile, D.A., Klonsky, J., & Merrell, R. (2007). The impact of video games on training surgeons in the 21st century. *Archives of Surgery, 142*(2), 181-186.

Russo, C., & Walker, C. (2006, May 8). Fantasy sports growth hinges on marketing, offline efforts. *Sports Business Journal, 21.*

Sakamoto, A. (1994). Video game use and the development of sociocognitive abilities in children: Three surveys of elementary school students. *Journal of Applied Social Psychology, 24*(1), 21-42.

Salen, K., & Zimmerman, E. (2003). This is not a game: Play in cultural environments. Retrieved September 15, 2008, from http://www.gamesconference.org/digra2003/2003/index.php?Abstracts/Salen%2C+et+al.

Sall, A., & Grinter, R.E. (2007). Let's get physical! In, out and around the gaming circle of physical gaming at home. *Computer Supported Cooperative Work, 16*(1-2), 199-229.

Sandage, C. (1983). *Advertising theory and practice.* Homewood, IL: Richard D. Irwin.

Sanford, K., & Madill, L. (2007). Understanding the power of new literacies through videogame play and design. *Canadian Journal of Education, 30*(2), 432-455.

Sanneblad, J., & Holmquist, L.E. (2003). *Designing collaborative games on handheld computers.* Paper presented at the SIGGRAPH 2003 Conference on Sketches & Applications.

Sargent, S. L., Zillman, D., & Weaver, J.B. (1998). The gender gap in the enjoyment of televised sports. *Journal of Sport & Social Issues*, 22(1), 46-64.

Saxton, J., & Game, S. (2001). *Virtual promise: Are charities making the most of the Internet revolution.* London: Third Sector.

Schiesel, S. (2006). The land of the video geek. *New York Times, Oct 8*, 2.1.

Schilling, M.A. (2003). Technological leapfrogging: Lessons from the U.S. video game console industry. *California Management Review, 45*(3), 6-32.

Schneider, L., & Cornwell, B. T. (2005). Cashing in on crashes via brand placement in computer games: The effects of experience and flow on memory. *International Journal of Advertising, 24*(3), 321-343.

Schoffeleers, M. (1991). *Twins and unilateral figures in Central and Southern Africa: Symmetry and asymmetry in the symbolization of the sacred.* Journal of Religion in Africa.

Schubert, T., Friedmann, F., & Regenbrecht, H. (2001). The experience of presence: Factor analytic insights. *Presence, 10*(3), 266-281.

Scott, J. (2000). *Social Network Analysis: A Handbook* (2nd. Ed.). Newbury Park: Sage.

Seay, A., Jerome, W., Sang Lee, K., & Kraut, R. (2003). *Project Massive 1.0: Organizational Commitment, Sociability and Extraversion in Massively Multiplayer Online Games.* Paper presented at the LEVEL UP Digital Games Research Conference.

Segal, N.L. (2006). *Twinsburg, Ohio; Twinsburg Research Institute Twin Study Summaries; The Outside World.*

Seidman, I.E. (1998). *Interviewing as Qualitative Research: A Guide for Researchers in Education and Social Sciences.* New York, NY: Teachers College Press.

Sénéchal, D. (1999). *Mouvement d'une boule de billard entre les collisions.* Unpublished manuscript.

Seo, K., Murakami, M., & Yoshida, K. (2004). Optimal flight technique for V-style ski jumping. *Sports Engineering, 7*(2), 97-103.

Seo, K., Watanabe, I., & Murakami, M. (2004). Aerodynamic force data for a V-style ski jumping flight. *Sports Engineering, 7*(1), 31-39.

Shamos, M. (1995). *A brief history of the noble game of billiards.* http://www.bca-pool.com/aboutus/history/start.shtml.

Shapiro, M.A., & McDonald, D.G. (1992). I'm not a real doctor, but I play one in virtual reality: Implications of virtual reality for judgments about reality. *Journal of Communication, 42*(4), 94-114.

Shapiro, S., MacInnis, D.J., & Whan Park, C. (2002). Understanding program-induced mood effects: Decoupling arousal from valence. *Journal of Advertising, 31*(4), 15-26.

Shepard, R. (1997). *Amateur physics for the amateur pool player.* self published.

Shibli, S., Taylor, P., Nichols, T., Gratton, C., & Kokolakakis, T. (1999). The characteristics of volunteers in UK sports clubs. *European Journal for Sport Management,* (pp.10-27).

Shields, M. (2005). Overload of game ads could defeat purpose. *Adweek, 46*(46), 9.

Shields, M. (2006, April 12). In-game ads could reach $2 billion. *Mediaweek; Adweek.* Retrieved August 2007, from http://www.adweek.com/aw/national/article_display.jsp?vnu_content_id=1002343563

Shields, M. (2007, August 20). Less than dynamic. *Brandweek, 30*(48), 36-37.

Shimai, S., Masuda, K., & Kishimoto, Y. (1990). Influences of TV games on physical and psychological development of Japanese kindergarten children. *Perceptual and Motor Skills, 70*(3), 771-776.

Shipman, F.M. (2001). Blending the real and virtual: Activity and spectatorship in fantasy sports. *Proceedings of the Conference on Digital Arts and Culture.* Retrieved October 3, 2006, from http://www.csdl.tamu.edu/~shipman/papers/dac01.pdf

Shrum, L.J. (Ed.). (2004). *The Psychology of Entertainment Media: Blurring the Lines Between Entertainment and Persuasion*: Lawrence Erlbaum Associates.

Siebel, N., & Maybank, S. (2004, May). *The ADVISOR Visual Surveillance System.* Paper presented at the Applications of Computer Vision (ACV'04), Prague, Czech Republic.

Siekpe, J.S., & Hernandez, M.D. (2006). The effect of system and individual characteristics on flow, and attitude formation toward advergames. In P. Rutsohn (Ed.), *Proceedings of the Annual Meeting of the Association of Collegiate Marketing Educators* (pp. 131-137). Oklahoma City, OK: Association of Collegiate Marketing Educators.

Silberman, L.B. (2005). *Athletes' use of video games to mediate their play: College students' use of sport video games.* Paper delivered at the 2005 Seminar Series, Caladonian University School of Computing and Mathematical Sciences, Glasgow, Scotland.

Silk, M., & Chumley, E. (2004). Memphis United? Diaspora, s(t)imulated spaces and global consumption economies. In D.L. Andrews (Ed.), *Manchester United: A Thematic Study* (pp. 249-264). Abingdon: Routledge.

Sinclair, J., Hingston, P., & Masek, M. (2007). Considerations for the design of exergames. In *Proceedings of the 5th International Conference on Computer Graphics and Interactive Techniques in Australia and Southeast Asia* (pp. 289-295). Perth, Australia: ACM.

Smith, A.W.B., & Lovell, B.C. (2003). Autonomous Sports Training from Visual Cues. In B.C. Lovell, D.A. Campbell, C. E. Fookes, & A. Maeder (Eds.), *Proceedings of the Eighth Australian and New Zealand Intelligent Information Systems Conference* (pp. 171-175), Brisbane: The Australian Pattern Recognition Society.

Smith, D., & Holmes, P. (2004). The effect of imagery modality on golf putting performance. *Journal of Sport and Exercise Psychology, 26*, 385-395.

Smith, M. (2006). Running the table: An ai for computer billiards. In *AAAI 2006: The 21st Nat. Conf. on AI.*

Smith, S.L. (1979). *Comparison of selected kinematic and kinetic parameters associated with the flat and slice serves of male intercollegiate tennis players.* Unpublished doctoral dissertation, Indiana University, USA.

Smoll, F.L., & Smith, R.E. (2002). *Children and youth in sport: A biosychosocial perspective.* Madison, WI: Kendall Hunt.

Snider, M. (2003). Study surprise: Video games enhance college social life. *USA Today.* Retrieved September 23, 2004, from http://www.usatoday.com/tech/news/2003-07-06-games_x.htm.

Snider, M. (2008). Designer Miyamoto makes video games pulse with life. *USA Today.* Retrieved April 30, 2008, from http://news.yahoo.com/s/usatoday/20080515/tc_usatoday/designermiyamotomakesvideogamespulsewithlife

So., B.R., Yi, B.J., & Kim, W.K. (2002). Impulse Analysis and Its Applications to Dynamic Environment. *In Proc of ASME Biennial Mechanisms Conf.*, Montreal, Canada.

SoftSport (2008). SoftSport Inc. Retrieved 12th May, 2008, from http://www.softsport.com/wc/

Soloway, E. (1991). How the Nintendo Generation learns. *Association for Computing Machinery. Communications of the ACM, 34*(9), 23-27.

Sony. (2007). PlayStation®2 celebrates its seventh anniversary, more than 120 million consoles sold worldwide. Retrieved July 31, 2008, from http://www.us.playstation.com/News/PressReleases/431.

Souvignier, E. (2001). Training räumlicher Fähigkeiten. [Training spatial abilities.]. In K.J. Klauer (Ed.), *Handbuch Kognitives Training* (pp. 293-319). Göttingen: Hogrefe.

SportsAnalytica (2008). Retrieved 12th May, 2008, from http://www.briggspalmer.com/new/index.html

Sprigings, E., Marshall, R., Elliott, B., & Jennings, L. (1994). A three-dimensional kinematic method for determining the effectiveness of arm segment rotations in producing racquet-head speed. *Journal of Biomechanics, 27*(3), 245-254.

Squire, K. (2007). *Open-ended video games: A model for developing learning for the interactive age.* USA: MIT.

Stake, R.E. (1995). *The Art of Case Study Research.* Thousand Oaks, CA: Sage Publications.

Stake, R.E. (2001). Case Studies. In N.K. Denzin & Y.S. Lincoln (Eds.), *Handbook of Qualitative Research* (pp. 435-454). Thousand Oaks, CA: Sage Publications.

Starbuck, W.J., & Webster, J. (1991). When is play productive? *Accounting, Management and Information Technology, 1*(1), 71-90.

Starck, J., & Hilton, A. (2003). Model-based Multiple View Reconstruction of People In Y. Matsushita (Ed.), *Proceedings of the IEEE Conference on Computer Vision and Pattern Recognition*, (pp. 915-922), Madison, WI: IEEE.

Stepanenko, Yu., & Vukobratovic, M. (1976). Dynamics of Articulated Open-Chain Active mechanisms. *Mathematical Biosciences, 28*(½).

Stettler, N., Signer, T.M., & Suter, P.M. (2004). Electronic games and environmental factors associated with childhood obesity in Switzerland. *Obesity, 12*(6), 896-903.

Steuer, J. (1992). Defining virtual reality: Dimensions determining telepresence. *Journal of Communication, 42*(4), 73-93.

Stoll, C. (1999). *High tech heretic – Reflections of a Computer Contrarian.* New York: First Anchor Books.

Stoughton, S. (2005). Skittles taps advergaming for product ads. *Marketing News* (September 1), *39*(14), 34.

Subic, A.J., & Haake, S.J. (Eds.). (2000). *The Engineering of Sport 3.* Cambridge.

Sullivan, D.B. (1991). Commentary and viewer perception of player hostility: Adding punch to televised sports. *Journal of Broadcasting & Electronic Media, 35*(4), 487-504.

Sung, Y., & De Gregorio, F. (2005). *New brand worlds: A comparison of college student attitudes toward brand placements in four media.* Paper presented at the AMA Winter Educators' Conference Chicago.

Suraya, R. (2005). Internet diffusion and e-business opportunities amongst Malaysian travel agencies. In *Proceedings of the Hawaii International Conference on Business* (pp. 1-13). Honolulu.

Sutton, W.A., McDonald, M.A., Milne, G.R., & Cimperman, J. (1997). Creating and Fostering Fan Identification in Professional Sports. *Sport Marketing Quarterly, 6*(1), 15-22.

Sweetser, P., & Johnson, D. (2004). Player-centered game environments: Assessing player opinions, experiences and issues. In M. Rauterberg (Ed.), *Entertainment computing – ICEC 2004: Third International Conference, LNCS 3166* (pp. 321-332). New York: Springer Verlag.

Sweetser, P., & Wyeth, P. (2005). GameFlow: A model for evaluating player enjoyment in games. *ACM Computers in Entertainment, 3*(3), 1-24.

Szalai, G. (2007). Video game industry growth still strong. Retrieved 6 October 2007, from http://www.hollywoodreporter.com/hr/content_display/business/news/e3if5f9e6af1f789e8c28399b0253e7b78d

Take-Two (2005, January 31). *Take-Two awarded long-term, third-party exclusives with Major League Baseball® Properties, Major League Baseball® Players Association and Major League Baseball® Advanced Media to publish interactive MLB® video game.* Retrieved April 30, 2008, from http://ir.take2games.com/pring_release.cfm?releaseid=154141

Tanner, L. (2008). New form of physical therapy: Wii™ games. *LiveScience.* Retrieved August 19, 2008, from http://www.livescience.com/health/080209-ap-wii-therapy.html.

Tapp, A., & Clowes, J. (2000). From "carefree casuals" to "professional wanderers". Segmentation possibilities for football supporters. *European Journal of Marketing, 36*(11/12), 1248-1269.

Taub, E.A. (2004). Video game makers play it safe. *International Herald Tribune.* Retrieved September 23, 2004, from http://www.iht.com/articles/540148.html.

Tavassoli, N.T. (1995). New research on limited cognitive capacity: Effects of arousal, mood and modality. *Advances in Consumer Research, 22*, 524-525.

Taylor, C. (2005). Who is playing games – and why. *Time, 165*(21), 52.

Tellis, W. (1997, September). Application of a Case Study Methodology. Retrieved June 2, 2007, from http://www.nova.edu.ssss/QR/QR3-3/tellis2.html

Terzopolus, D. (1999). Visual Modeling for Computer Animation: Graphics with a Vision. *Computer Graphics, 33*, 42-45.

The Age. (2006). The next gaming wave. Retrieved August 19, 2008, from http://www.theage.com.au/news/games/the-next-gaming-wave/2006/11/28/1164476212099.html?page=fullpage.

*The Brand Champions League: Europe's Most Valuable Football Clubs.* (2005). London: Brand Finance.

Theobalt, C., Aguiar, E., Magnor, M., Theisel, H., & Seidel H-P. (2004). Marker-free Kinematic Skeleton

Estimation from Sequence of Volume Data. In R. Lau & D. Baciu (Eds.), *Proceedings of the ACM Virtual Reality Software and Technology Conference* (pp. 57-64). New York, USA: ACM.

Thompson, A. (2007, March 10). In fantasy land, sports judges hear imaginary cases. *The Wall Street Journal*, A1.

Thorsen, T. (2008, February 12). *EA Sports™ extends NFL® deal through 2012 season*. Retrieved April 28, 2008 from http://www.gamespot.com/news/6185880.html?sid=6185880&part=rss&subj=6185880

Top Ten Sports Video Game Titles. (2007, January 15). *Street & Smith's SportsBusiness Journal, 18*.

Toscos, T., Faber, A., Shunying, A., & Mona Praful, G. (2006). Chick Clique: Persuasive technology to motivate teenage girls to exercise. In *Extended Abstracts CHI 2006* (pp. 1873-1878). Quebec, Canada: ACM Press.

Towards a stage model. *International Small Business Journal,* (pp. 253-270).

Tsoulouhas, T., Knoeber, C., & Agrawal, A. (2007). Contests to become CEO: Incentives, selection and handicaps. *Economic Theory, 30*(2), 195-221.

Turban, E., Leidner, D., McLean, E., & Wetherbe, J. (2006). *Information technology and management*. Hoboken: John Wiley & Sons.

Turner, J.R., & Carroll, D. (1985). Heart rate and oxygen consumption during mental arithmetic, a video game, and graded exercise: Further evidence of metabolically-exaggerated cardiac adjustments? *Psychophysiology, 22*(3), 261-267.

Ujihashi, S., & Haake, S.J. (Eds.). (2002). *The Engineering of Sport 4*. Cambridge.

Umegaki, K. et al., (1998). Influences of magnitude of vertical rotational torques and timing of grip torque generation on golf swing motion. In *Proceedings of the Symposium on Sports Engineering* (pp. 111-115). Japan: Japanese Society of Mechanical Engineering.

Unnithan, V.B., Houser, W., & Fernhall, B. (2006). Evaluation of the energy cost of playing a dance simulation video game in overweight and non-overweight children and adolescents. *International Journal of Sports Medicine, 27*(10), 804-809.

Usaka, T. (1996). *A large scaled multi-user virtual environment system for digital museums*. Master's thesis, Division of Science, Graduate School of University of Tokyo, Japan.

V8 Supercars. (2007). A success story - V8 Supercars Australia. Retrieved July 6, 2007, from http://www.v8supercar.com.au/content/about_avesco/the_v8_supercars_australia_success_story/?ind=M.

Van Gheluwe, B., & Hebbelinck, M. (1985). The kinematics of the service movement in tennis: A three-dimensional cinematographical approach. In B. Johnsson (Ed.), *Biomechanics IX-B* (pp. 521-525). Champaign, IL: Human Kinetics.

Van Gheluwe, B., De Ruysscher, I., & Craenhals, J. (1987). Pronation and endorotation of the racket arm in a tennis serve. In B. Johnsson (Ed.), *Biomechanics X-B* (pp. 667-672). Champaign, IL: Human Kinetics.

Van Osselaer, S.M.J., & Janiszewski, C. (2001). Two ways of learning brand associations. *Journal of Consumer Research, 28*(September), 202-223.

van Teijlingen, E., & Hundley, V. (2001). The importance of pilot studies. *Social Research Update*(35), 1-4.

Vandewater, E.A., Shim, M.S., & Caplovitz, A.G. (2004). Linking obesity and activity level with children's television and video game use. *Journal of Adolescence, 27*(1), 71-85.

Vanrenterghem, M., Lees, A., Lenor, M., Aerts, P., & De Clercq, D. (2004). Performing the Vertical Jump: Movement Adaptations for Sub Maximal Jumping. *Human Movement Science, 22*(6) (2004), 1713-1727.

Veal, A.J. (2005). *Business Research Methods: A Managerial Approach* (Second ed.). Frenchs Forest, Sydney: Pearson Addison Wesley.

Vicon. (nd). Retrieved May 12, 2008, from http://www.vicon.com

Vincente, K.J., Hayes, B.C., & Williges, R.C. (1987). Assaying and isolating individual differences in searching a hierarchical file system. *Human Factors, 29*(3), 349-359.

Vision IQ (2008). Retrieved 13 May, 2008 from http://poseidon-tech.com/us/index.html

Volunteering Australia. (2005). *Definitions and principles of volunteering.* Retrieved May 28,2007, from http://www.volunteeringaustralia.org.

Vorderer, P. (2000). Interactive entertainment and beyond. In D. Zillmann & P. Vorderer (Eds.), *Media entertainment: The psychology of its appeal* (pp. 21-36). Mahwah, NJ: Lawrence Erlbaum.

Vukobratovic, M., & Borovac, B. (2004). Zero-Moment Point, Thirty-Five Years of Its Life. *Intl. J. Humanoid Robotics, 1*(1), 157-173.

Vukobratovic, M., & Juricic D. (1969). Contribution to the Synthesis of the Biped Gait. *IEEE Trans. on Bio-Medical Engineering, 16*(1), 1-6.

Vukobratovic, M., & Potkonjak, V. (1982). *Dynamics of Manipulation Robots.* Berlin, Heidelberg, Germany: Springer-Verlag.

Vukobratovic, M., & Stemanenko, Yu. (1973). Mathematical Models of General Anthropomorphic Systems. *Mathematical Biosciences, 17,* 191-222.

Vukobratovic, M., Herr, H., Borovac, B., Rakovic, M., Popovic, M., Hofmann, A., & Potkonjak, V. (in press). Biological Principles of Control Selection for a Humanoid Robot's Dynamic Balance Preservation. *Intl. J. Humanoid Robotics.*

Vukobratovic, M., Hristic, D., & Stojiljkovic, Z. (1974). Development of Active Anthropomorphic Exoskeletons. *Medical and Biological Engineering, 12*(1).

Vukobratovic, M., Potkonjak, V., & Matijevic, V. (2003). *Dynamics of Robots with Contact Tasks.* Dordrecht, The Netherlands: Kluwer Academic Publishers.

Vukobratovic, M., Potkonjak, V., Babkovic, K., & Borovac, B. (2007). Simulation Model of General Human and Humanoid Motion. *Intl. J. Multibody System Dynamics, 17*(1), 71-96.

Wakai, M., & Linthome, N. P. (2005). Optimum take-off angle in the standing long jump. *Human Movement Science, 24*(1), 81-96.

Wakefield, K.L., & Blodgett, J.G. (1994). The importance of servicescapes in leisure service settings. *Journal of Services Marketing, 8*(3), 66-76.

Wakefield, K.L., & Blodgett, J.G. (1996). The effect of the servicescape on customers' behavioral intentions in leisure service settings. *Journal of Services Marketing, 10*(6), 45-61.

Wakefield, K.L., Blodgett, J.G., & Sloan, H.J. (1996). Measurement and management of the sportscape. *Journal of Sport Management, 10*(1), 15-31.

Walker, D., & Dubitsky, T.M. (1994). Why liking matters. *Journal of Advertising Research, 34*(3), 9-18.

Walker, S. (2007). *Fantasyland: A sportswriter's obsessive bid to win the world's most ruthless fantasy baseball league.* New York: Penguin Group.

Walton, J.S. (1981). *Close-range cine-photogrammetry: A generalized technique for quantifying gross human motion.* Unpublished doctoral dissertation, Pennysylvania State University, University Park, PA.

Wan, C.S., & Chiou, W.B. (2006). Psychological motives and online games addiction: A test of flow theory and humanistic needs theory for Taiwanese adolescents. *CyberPsychology & Behavior, 9*(3), 317-324.

Wang, X., & Perry, A.C. (2006). Metabolic and physiologic responses to video game play in 7- to 10-year-old boys. *Archives of Pediatrics and Adolescent Medicine, 160*(4), 411-415.

Wankel, L., & Bonnie, G. (1990). The psychological and social benefits of sport and physical activity. *Journal of Leisure Research, 22*(2), 167-182.

Wann, D.L., & Branscombe, N.R. (1993). Sports Fans: Measuring Degree of Identification with Their Team. *International Journal of Sport Psychology, 24,* 1-17.

Wann, D.L., Melnick, M.J., Russell, G.W., & Pease, D.G. (2001). *Sports fans: The psychology and social impact of spectators*. New York: Routledge.

Wann, D.L., Tucker, K.B., & Schrader, M.P. (1996). An Exploratory Examination of the Factors Influencing the Origination, Continuation, and Cessation of Identification with Sports Teams. *Perceptual and Motor Skills, 82*, 995-1001.

Wasserman, S. & Faust, K. (1994). *Social Network Analysis: Methods and Applications*. New York: Cambridge University Press.

WCG, World Cyber Games. (2008). About WCG: WCG concept. Retrieved July 17, 2008, from http://www.worldcybergames.com/6th/inside/WCGC/WCGC_structure.asp.

Webster, J., & Martocchio, J.J. (1992). Microcomputer playfulness: Development of a measure with workplace implications. *MIS Quarterly, 16*(2), 201-226.

Weekley, D. (2004, September 7). Fantasy football numbers on the rise. *Charleston Gazette*, 3B.

Weinberg, H. (1999). Dual axis, low g, fully integrated accelerometers. *Analog Dialogue, 33*(1), 1-2.

Weiser, M. (1991). The Computer for the 21st Century. *Scientific American*, 265, 94-104.

White, M., Lehmann, H., & Trent, M. (2007). 31: Disco dance video game-based interventional study on childhood obesity. *Journal of Adolescent Health, 40*(2), S32.

Whiting, R. (2003, April 21). Hideki Matsui. Godzilla vs. the Americans. Retrieved November 18, 2005, from http://www.time.com/time/asia/2003/heroes/hideki_matsui.html

Widman, L.M., McDonald, C.M., & Abresch, R.T. (2006). Effectiveness of an upper extremity exercise device integrated with computer gaming for aerobic training in adolescents with spinal cord dysfunction. *Journal of Spinal Cord Medicine, 29*(4), 363-370.

Williams, D. (2002). Structure and competition in the U.S. home video game industry. *International Journal on Media Management, 4*(1), 41-54.

Williams, D. (2006). Why game studies now? Gamers don't bowl alone. *Games and Culture, 1*(1), 13-16.

Williams, J., & Hopkins, S. (2005). *The Miracle of Istanbul: Liverpool FC from Paisley to Benitez*. Edinburgh: Mainstream Publishing.

Williams, J., & Llopis, R. (2006). *Groove Armada: Rafa Benitez, Anfield and the New Spanish Fury*. Edinburgh: Mainstream Publishing.

Williams, K. (2005). Manufacturers finally learn how to sell game networking to operators. *Vending Times, 45*(12). Retrieved August 5, 2008, from http://vendingtimes.com/ME2/dirmod.asp?sid=&nm=&type=Publishing&mod=Publications%3A%3AArticle&mid=8F3A7027421841978F18BE895F87F791&tier=4&id=EB40CAAFDA0643598C31564B24E0206A.

Williams, W. (2008). In electronic age, sports stores compete for youth. *The State Journal*. Retrieved September 8, 2008, from http://www.statejournal.com/story.cfm?func=viewstory&storyid=43197.

Williamson, R., & Andrews, B.J. (2001). Detecting absolute human knee angle and angular velocity using accelerometers and rate gyroscopes. *Medical and Biological Engineering and Computing, 39*(3), 294-302.

Wilson, J. (2000). Volunteering. *Annual Review of Sociology*, (pp. 215-240).

Wingbermuhle, J., Liedtke, C-E., & Solodenko, J. (2001). Automated Acquistion of Lifelike 3D Human Models from Multiple Posture Data. In A. Gagalowicz & W. Philips (Eds.) *Proceedings of the International Conference on Computer Analysis of Images and Patterns* (pp. 400-409). New York, USA: Springer.

Winkel, M., Novak, D.M., & Hopson, H. (1987). Personality factors, subject gender, and the effects of aggressive video games on aggression in adolescents. *Journal of Research in Personality, 21*(2), 211-223.

Winograd, T., & Flores, F. (1987). Understanding computers and cognition: A new foundation for design. Norwood, NJ: Addison-Wesley Professional.

Witmer, B.G., & Singer, M.J. (1998). Measuring presence in virtual environments: A presence questionnaire. *Presence, 7*(3), 225-240.

Woelfel, J.K., & Woelfel, J.D. (1997). *CATAPC 2.0: Software for Content Analysis*. Buffalo: Galileo Company.

Wolf, M. (2007). In-game advertising on consoles to reach $850 mil by 2011. *Multimedia Publisher, 18*(10), 1.

Wolmar, C. (2007). *Fire and steam: How the railways transformed Britain*. London: Academic Books.

Woltring, H.J. (1986). A Fortran package for generalized, cross-validation spline smoothing and differentiation. *Advanced Engineering Software, 8*, 104-113.

Wood, C. (2007, February 7). Americans Kop Liverpool deal. Retrieved September 3, 2007, from http://www.dailymail.co.uk/pages/live/articles/sport/football.html?in_article_id=434446&in_page_id=1779&in_a_source=&ct=5

Wreden, N. (1997). Business boosting technologies. *Beyond Computing, 27*.

Wren, C., Azarbayejani, A., Darrell, T., & Pentland, A. (1997). Pfinder: real-time tracking of the human body. *IEEE Transactions on Pattern Analysis and Machine Intelligence, 19*, 780-785.

Wu, G. (1999). Perceived interactivity and attitude toward web sites. In M.S. Roberts (Ed.), *Proceedings of the 1999 Conference of the American Academy of Advertising* (pp. 254-262). Gainesville, FL: American Academy of Advertising.

Wulf, V., Moritz, E.F., Henneke, C., Al-Zubaidi, K., & Stevens, G. (2004). Computer supported collaborative sports: Creating social spaces filled with sports activities. In *Proceedings of the Third International Conference on Entertainment Computing ICEC 2004* (pp. 80-89). Heidelberg: Springer LNCS.

Xu, M., Orwell, J., & Jones, G. (2004). *Tracking football players with multiple cameras*. Paper presented at the International Conference on Image Processing (ICIP'04), Singapore, Singapore.

Xu, W., Pan, Z., & Zhang, M. (2003). Footprint sampling based motion editing. *International Journal of Image and Graphics, 3*(2), 311-324.

Yamanobe, K., & Watanabe, K. (1999). Measurement of take-off forces in ski jumping com-petition. *Japanese Journal of Biomechanics in Sports and Exercise, 3*(4), 277-286.

Yang, M., Roskos-Ewoldsen, D.R., Dinu, L., & Arpan, L.M. (2006). The effectiveness of "in-game" advertising. *Journal of Advertising, 35*(4), 143-152.

Yang, S., Smith, B., & Graham, G. (2008). Healthy video gaming: Oxymoron or possibility? *Journal of Online Education, 4*(4). Retrieved September 30, 2008, from http://innovateonline.info/index.php?view=article&id=186&action=synopsis.

Yeung, Y., Liu, T.C., & Ng, P. (2005). A social network analysis of research collaboration in physics education. *American Journal of Physics, 73*(2), 145-150.

Yin, R.K. (2003). *Case Study Research: Design and Methods* (Third ed.). Thousand Oaks, CA: SAGE Publications.

Yoo, C.Y., & Kim, K. (2005). Processing of animation in online banner advertising: The roles of cognitive and emotional responses. *Journal of Interactive Marketing, 19*(4), 18-34.

Youn, S., & Lee, M. (2003). Antecedents and consequences of attitude toward the advergame in commercial web sites. In L. Carlson (Ed.), *Proceedings of the 2003 Conference of the American Academy of Advertising* (p. 128). Pullman, WA: American Academy of Advertising.

Young, E. (2004). *EA makes ad play*. Retrieved 6 October, 2007, from http://sanfrancisco.bizjournals.com/sanfrancisco/stories/2004/02/09/story2.htm

Yu, J., & Rajkumar, B. (2005). A taxonomy of scientific workflow systems for grid computing. *ACM SIGMOD Record, 34*, 44-49.

Zhang, J.J., Lam, E.T.C., & Connaughton, D.P. (2003). General market demand variables associated with professional sport consumption. *International Journal of Sport Marketing & Sponsorship, 5*(1), 33-55.

Zhang, Z. (2000). A Flexible New Technique for Camera Calibration. *IEEE Trans. Pattern Analysis and Machine Intelligence, 22*, 1330-1334.

Zhang, Z., & Yang, B. (2006). Visual projection similarity for matching similarity among 3D models. *Journal of Computer-Aided Design & Computer Graphics, 18*(7), 1049-1053.

Zhang, Z., Pan, Z., et al. (2005). Multiscale generic fourier descriptor for gray image retrieval. *Journal of Image and Graphics, 10*(5), 611-615.

Zimmerman, B.J. (2000). Self-efficacy: An essential motive to learn. *Contemporary Educational Psychology, 25*(1), 82-91.

Zivkovic, Z. (2004). *Improved adaptive Gaussian mixture model for background subtraction.* Paper presented at the 17th International Conference on Pattern Recognition (ICPR'04), Cambridge, UK.

# About the Contributors

**Nigel K. Ll. Pope** took his undergraduate degree in politics and Anglo-Saxon literature at the University of Queensland. He later took his MBA from the University of Central Queensland and his doctorate from Griffith University, specializing in sport sponsorship. His research has appeared in the *Journal of Advertising*, *European Journal of Marketing* and *Sport Marketing Quarterly* amongst other publications. His current interests are in digitization of entertainment and the structural foundations of the entertainment industry. He is currently an associate professor with Griffith Business School.

**Kerri-Ann L. Kuhn** is a Post-Doctoral Research Fellow and Lecturer at the Queensland University of Technology (QUT), where she teaches e-Marketing Strategies. She holds a PhD from QUT (Australia) and a Masters of Marketing Management with Honours from Griffith University (Australia), where she also earned her Bachelor degrees in International Business and Commerce, and a Graduate Diploma in Japanese. She is also the recipient of an Australian Postgraduate Award. Her research interests are in the areas of marketing communications and interactive technologies, particularly video gaming, and the effects on consumer behavior. She has published and presented papers in this area, which includes appearances in national print media and radio interviews with the Australian Broadcasting Corporation (ABC).

**John J.H. Forster** is currently on secondment from Griffith University, Australia (where he teaches economics) to The American University of Sharjah, UAE (teaching management). In one way or another, John has been associated with Canberra CAE, Queensland University of Technology and Sydney University (all Australia); University of the South Pacific (Fiji); University of Waikato (NZ); North East London Polytechnic, Keele University and Queen Mary College (all UK); and McMaster University (Canada). His early interest was in urban networks and urban labour markets. John has also been an Australian public servant, as well as a member of the crew of the historic tug, *SS Forceful*, on the Brisbane River and Moreton Bay. He has published eight books on topics such as strategic management, sports governance and public management. His interest in digital technologies include both sport and the operation of electronic markets.

* * *

**Amin Ahmadi** is currently a PhD scholar at Griffith University in conjunction with the Queensland Academy of Sport. He is currently receiving scholarship from the Centre of Excellence in sports science. He completed his MSc program in computer and communication engineering at Griffith, Australia in 2005. He is a member of IEEE, Sports Medicine Australia and the Centre of Wireless Monitoring and

Application at Griffith University. His main research area of interest is in monitoring and enhancing the performance of Tennis Players using inertial sensors. He also won the Minister Excellence Award in "Innovation and Creativity" in 2006.

**Kalman Babković** was born in Kikinda, Serbia, 1975. He received his BSc and MSc degrees in electrical engineering from the University of Novi Sad in 2000 and 2005 respectively. He is currently a teaching assistant at the Faculty of Technical Sciences, University of Novi Sad. He is coauthor of 3 scientific papers in the field of robotics published in international journals.

**Peter Barron** received a PhD in Computer Science from Trinity College in 2005. He was formerly a member of the Distributed Systems Group in Trinity College where he was a Research Fellow in the areas of middleware and programming models for ubiquitous and mobile computing. He is now working for NewBay Software Ltd.

**Scott Bingley** completed an honours degree in Information Systems in 2005. After this he became a Software Test Analyst before being awarded an Australian postgraduate scholarship to commence his PhD, using an innovation diffusion approach to examine the adoption of Internet applications in local sporting bodies. As part of his degree, Scott has undertaken study as part of a university exchange program at Slippery Rock University in the USA. Scott's research interests include the use of information systems in community based organisations, an area where he has been involved with a number of research projects.

**Ellen L. Bloxsome** spent more than one and a half decades in the hospitality industry. She took an Honours degree in Sociology and Marketing at Murdoch University and is currently completing her doctoral dissertation in the electronic communication of cultural memes. She is now a freelance scholar in South East Queensland, Australia.

**Branislav Borovac** was born in Leskovac, Serbia, 1951. He received the MSc and PhD degrees in Mechanical Engineering from the University of Novi Sad in 1982 and 1986 respectively. He became assistant professor of engineering design 1987, assistant professor of robotics 1988, associate professor of robotics 1993 and since 1998 he has been full professor of robotics, all at the Faculty of Technical Sciences, University of Novi Sad. He is coauthor of two research monographs published by Springer - Verlag, 1990 and CRC Press, 2001. He is author/coauthor of 30 scientific papers in the field of robotics published in international journals, as well as author/coauthor of about 80 papers in proceedings of international conferences and congresses.

**Stephen Burgess** completed his PhD in the School of Information Management and Systems at Monash University. His thesis was in the area of small business interactions with customers via the Internet. He has research and teaching interests that include the use of ICTs in small businesses (particularly in the tourism field), the strategic use of ICTs, and B2C electronic commerce. He has received a number of competitive research grants in these areas. More recently his small business research has extended to the use of the Internet in local community based organisations. He has recently edited two books and special editions of journals in topics related to the use of ICTs in small business, and been track chair at the ISOneWorld, IRMA, Conf-IRM and ACIS conferences in these areas.

**Brendan Burkett**, associate professor, joined the University in 1998 following an international sporting career and as a professional engineer employed in Australia and in Europe. As a past international sports person Brendan was fortunate enough to represent Australia as is the paralympic champion, world champion, world record holder, Commonwealth Games and Australian multiple medallist, and is an inductee in the Swimming Queensland Hall of Fame. In 2000 Brendan was the Institution of Engineers, Australia, professional engineer of the year, and is a fellow of IEAust. The combination of sporting and professional achievements has been recognised as Brendan is the recipient of several awards such as the Australia Day Sporting Award, the Order of Australia Medal (OAM). The impact of Brendan's research in sports technology is demonstrated with research grants and publications, as an invited professor for the European masters in physical activity, and as part of the Prime Ministers 2020 Summit.

**Andrew Busch** received BEng and BIT degrees from the Queensland University of Technology, Australia, in 1998, and completed his PhD in 2004 at the same institution. He is currently a lecturer in the School of Engineering at Griffith University. Dr. Busch is an active researcher in the area of sports and biomedical engineering, with emphasis on the use of inertial sensors for the study of the biomechanics of cricket. Other areas of expertise are signal and image processing, including the use of speech, audio and image analysis for biometric authentication.

**Vinny Cahill**, professor, holds a personal chair in computer science at Trinity College Dublin where he also serves as Head of the Department of Computer Science and director of research for computer science and statistics. His research addresses many aspects of distributed systems, in particular, middleware and programming models for ubiquitous and mobile computing with application to intelligent transportation systems, global business systems and personal healthcare/independent living. He has published over 100 peer-reviewed publications in international conferences and journals.

**Sindy Chapa**, Ph.D, is an assistant professor and associate director of the Center for the Study of Latino Media and Markets at the School of Journalism and Mass Communication at Texas State University. Professor Chapa's research design involves both quantitative and ethnographic elements. Her research concentrates on political advertising, advergames, and a series of topics in the consumer behavior area, including cross-cultural preferences for counterfeit products and family decision-making. Her publications have appeared in the *Journal of Customer Behaviour*, *Journal of International Consumer Marketing*, *Consumption, Markets and Culture: Resonant Representations*, *International Journal of Business Disciplines* and *Journal of Interactive Advertising*. She has also co-authored book chapters in several textbooks, including the *Global Consumer Behavio*r, edited by Chantal Ammi; and *Advertising, Promotion and New Media*, edited by Marla Stafford and Ronald Faber.

**Beth A. Cianfrone** is an assistant professor of sport administration in the Department of Kinesiology and Health at Georgia State University in Atlanta, Georgia. She has two main areas of research interest: (a) sport sponsorship and advertising effectiveness, and (b) sport consumer behavior. Her research aims to quantify the impacts of various sponsorship mediums and examine theoretical models that explain consumer behavior related to the impact of sport sponsorship. Dr. Cianfrone has been a prolific scholar with manuscripts published in major sport management journals such as the Journal of Sport Management and International Journal of Sport Communication. She has delivered over 40 presentations at international or national conferences on sport management or sport marketing.

**Jean-Pierre Dussault** is a professor in the "Département d'Informatique" at the University of Sherbrooke, since 1982. He received his BSc in mathematics in 1977, a MSc in 1979 and his PhD in operations research in 1983, all from the University of Montréal. His main research interests concern the design and convergence analysis of optimization algorithms. He is also interested by various applications of optimization techniques, among which the optimization of a billiard player, image synthesis and reconstruction, network optimization, etc.

**Andre Gagalowicz** is a research director at INRIA, FRANCE. He was the creator of the first laboratory involved in image analysis/synthesis collaboration techniques. He graduated from Ecole Superieure d'Electricite in 1971 (engineer in Electrical Engineering), obtained his PhD in automatic control from the University of Paris XI, Orsay, in 1973, and his state doctorate in mathematics (doctoctorat d'Etat es Sciences) from the University of Paris VI (1983). He is fluent in English, German, Russian and Polish and got a bachelor degree in Chinese from the University of Paris IX, INALOCO in 1983. His research interests are in 3D approaches for computer vision, computer graphics, and their cooperation and also in digital image processing and pattern recognition. He received the prices of the best scientific communication and the best technical slide at the Eurographics'85 conference. He was awarded the second price of the Seymour Cray competition in 1991 and one of his papers was selected by *Computers and Graphics* journal as one of the three best publications of this journal from the last ten years. He took part to the redaction of eight books and wrote more than one hundred publications.

**Marc Godard** is a senior undergraduate student studying cognitive science at Queen's University, Kingston. Marc's academic interests include computer gaming systems, computer vision, robotics, artificial intelligence and neural networks. Marc can consistently run a pool table.

**Benjamin D. Goss** received his bachelor's and master's degrees at Louisiana Tech University and his doctorate at The University of Southern Mississippi. He serves as an associate professor in the entertainment management program within the Department of Management and the College of Business Administration at Missouri State University in Springfield, Missouri, USA. Goss has taught in Missouri State's joint business degree program with Liaoning Normal University in Dalian, China. He served as a visiting instructor at The University of Southern Mississippi and held academic appointments as an assistant professor and sport management program coordinator at Winthrop University and Clemson University. He has authored or co-authored 15 peer-reviewed refereed journal articles, two book chapters, and several trade journal articles. Goss' vita includes 56 professional conference presentations, including 35 at the international and national levels. In 2007, he co-founded the Journal of Sport Administration & Supervision and now serves as its editor.

**Michael Greenspan** is with the Department of Electrical and Computer Engineering at Queen's University in Kingston, Ontario, Canada, where he has been as associate professor since 2001. From 1991 to 2001, he was employed by the Institute for Information Technology of the National Research Council of Canada, as a researcher and ultimately as the group leader of the Computational Video Group. He was awarded a PhD from Carleton University, Ottawa, in 1999. Dr. Greenspan holds membership with the Professional Engineers of Ontario, the IEEE Computer Society, the Canadian Image Processing and Pattern Recognition Society (CIPPRS), and he serves on the Expert Advisory Panel of Precarn

Associates. He was the recipient of the CIPPRS 2003 Young Investigators Award, and was awarded the Best Paper Award with his co-author at the 5th Canadian Conference on Computer and Robot Vision in 2008. Dr. Greenspan's research interests include computer vision, especially object recognition and tracking, entertainment robotics, and biomedical applications of computer vision.

**Mads Haahr**, PhD, works as lecturer in the School of Computer Science and Statistics. He is a true multidisciplinarian whose research interests include technical subjects, such as mobile and ubiquitous computing, self-organisation in peer-to-peer networks and artificial intelligence for games, as well as softer subjects such as locative media, computer game studies and the art/technology interface. He is also co-founder and editor-in-chief of *Crossings: Electronic Journal of Art and Technology*, a peer-reviewed academic journal that explores the areas where art, science and technology intersect.

**Gaoqi He** received his bachelor's degree and master's degree from Department of Mathematics in 1997 and 2000 from East China Normal University respectively, and PhD Degree in 2007 from State Key Laboratory of CAD&CG, Zhejiang University. Now he is working as a lecturer at Department of Computer Science, East China University of Science & Technology. His main research interest focuses on computer graphics, virtual reality and mobile computing.

**Monica D. Hernandez** is a visiting assistant professor at The University of Texas-Pan American. She holds a PhD in international business from the same university. She has published in *Journal of Consumer Behaviour*, *Young Consumers*, *Journal of Interactive Advertising*, and *Journal of Business & Entrepreneurship* among others. Her work also has appeared in various book chapters, including *Global Consumer Behavior* and *Advertising, Promotion and New Media*, and numerous conference proceedings. Areas of research interest include Internet advertising, international marketing and consumer psychology.

**Daniel A. James** received a PhD from Griffith University, Brisbane, Australia in 1998 after completing his BSc and MPhil there. He is currently a senior research fellow with the Centre of Wireless Monitoring and Applications at Griffith University and was recently elevated to senior member of the IEEE by the Engineering in Medicine and Biology Society. His principle areas of research interest are in applying sensors to sporting and biomedical applications. He has around 50 publications in this area including several international patents. His work in sports monitoring technologies includes the sports of swimming, rowing, snowboard, combative sports, running, cricket and martial arts. His current partners include the Australian Institute of Sport, Queensland Academy of Sport, Cricket Australia, International Cricket Council, and Hearts First Cardiology Clinic.

**Anthony K. Kerr** is a doctoral student at the University of Technology, Sydney (UTS). He received his MBA from the University of Oregon and MS in Sport Management from the University of Massachusetts and has worked extensively in professional and collegiate sports worldwide. His research interests focus largely on the antecedents of fandom in foreign markets and the contribution these fans can make to professional sports teams. He has published articles on the brand equity of sports teams and the team identification of foreign team supporters in a variety of journals including the *International Journal of Sports Marketing and Sponsorship* and *Football Studies*.

**Michael Koh** is the director of the School of Sports, Health and Leisure at the Republic Polytechnic, Singapore. He remains active in research and is currently involved in nationally funded and institutionally funded research grants. He also publishes regularly in the areas of sports biomechanics, motor learning and development and physical activity. As School Director, he has put together several research teams that have won research grants in nationally funded projects from agencies such as the National Research Fund - Interactive Digital Media and the A*STAR Science and Engineering Research Council Fund of Singapore.

**Joseph Lam** received a BSc (Eng) in Computer Engineering and a BSc in Life Sciences in 2004, and an MSc (Eng) in 2006 from Queen's University, Kingston, Ontario. He was awarded both the Best Paper and Best Student Paper at the 5th Canadian Conference on Computer and Robotic Vision (CRV' 08). In 2007, Joseph worked as a research engineer with the robotics and vision group at MDA Space Missions in Brampton, Ontario. He is currently a PhD candidate in the Department of Electrical and Computer Engineering at Queen's University. His research interests include robot calibration, eye-in-hand systems, object recognition in range data, and urban scene analysis.

**Jean-François Landry** received his bachelor's degree in computer science at Université de Moncton (Moncton, Canada), to then pursue his studies at Université de Sherbrooke (Sherbrooke, Canada) to acquire his Master's degree in the same field in 2008. The main focus of his research was about using optimization and artificial intelligence methods to solve problematics related to the creation of an automatic pool player. During his studies, he worked as an intern for 6 months at SAP Research Labs France (Nice, France) in 2007 in the field of artificial intelligence in the context of public security. Since April 2008 he's been holding a position as Development Engineer at Amadeus France in Sophia Antipolis and will start his PhD studies at Université de Sherbrooke in January 2009.

**Will Leckie** is an electro-optics hardware designer at Nortel in Ottawa, Canada. He received the MSc in Electrical Engineering from Queen's University at Kingston and the BSc in Applied Physics from the University of Waterloo, Canada. His research interests include optical communications, control systems and intelligent systems.

**Mark Lee** completed a master's of business (marketing) at RMIT University, during which time he developed a research interest in video game advertising. He is now a marketing consultant specialising in emerging marketing opportunities and entrepreneurial ventures.

**Jianfeng Liu** received his bachelor's degree from East China Jiaotong University in 2007. Now he is a master's student in the State Key Laboratory of CAD&CG, Zhejiang University. His main research interests are Virtual Reality, Augmented Reality and Human-Computer Interaction.

**Kieran Moran**, PhD, is currently a senior lecturer and director of the Sports Biomechanics Research Group in the School of Health and Human Performance, Dublin City University, Ireland. He has overseen the development of a number of technologies for assessing and enhancing neuromuscular capacity and sports performance.

**Eckehard Fozzy Moritz** has studied mechanical engineering at the Technical University of Munich, and received his doctorate degree at the Technical University of Tokyo. He is currently director of the SportKreativWerkstatt, a company masterminding, moderating, and facilitating innovation projects for companies like BMW and BASF. He holds a honorary professorship from Qufu University, China, and is adjunct professor at the Universidad de las Américas, Puebla, Mexico. Professor Moritz has been chairman of the 6th World Conference of the International Sports Engineering Association, and has been teaching and consulting at universities world-wide.

**Florian 'Floyd' Mueller** is a researcher in novel interfaces that gap the bridge between human bodies and technology. His latest research interest is the concept of *Exertion Interfaces*: interfaces that deliberately require intense physical effort to facilitate social connectedness, to enable what he calls *Sports over a Distance*. Floyd has three degrees from three continents, including a Masters degree in Media Arts and Sciences from the MIT Media Lab in Cambridge, USA. He received two scholarships to study in the USA and Australia, and was also offered a Fulbright scholarship and a Media Lab Europe Fellowship. He also has extensive work experience from the USA, Australia, Ireland and Germany, where he worked for industry and academic research organizations such as Springer Verlag, Xerox PARC, FX Palo Alto Laboratory, MIT, Media Lab Europe, CSIRO and the University of Melbourne.

**Rajendra Mulye**, PhD, is a senior lecturer and program leader in the School of Economics, Finance and Marketing at RMIT University. His area of research includes the country-of-origin effect, choice modelling and more recently education. He has recently taken up a secondment as protagonist, Petty Officer John–117 or Master Chief, in the Halo of the universe.

**Yuji Ohgi** received a PhD from Keio University after completing a masters and undergraduate degree at the university of Tsukuba in sports and health sciences. Currently an associate professor he is the editor in chief of the Japanese Society of Science in Swimming and Water Exercise. A former member of the JOC and coaching staff for the Atlanta, games he pioneered the use of inertial sensors in swimming and other sporting applications. He currently heads a spin out company in Japan as well as leading a collaborative effort between Japanese hi-tech firms to create a single chip monitoring solution. He holds a number of national (Japan) and international patents in sports technology and has published extensively in the Japanese and international literature.

**Alex Ong** is the assistant director of the School of Sports, Health and Leisure at the Republic Polytechnic, Singapore. Alex has a BSc Joint Honours in Sports Science / Physical Education / Mathematics as well as MSc in biomechanics (medicine) and a MA (Education). He also holds a PhD in education. He has been in the education industry for the past 15 years and is involved in e-learning, classroom and outdoor teaching in schools, as well as education policy formulation. Whilst teaching, he was actively involved in coaching, relying primarily on biomechanical principles to enhance the technique of his developmental athletes. His passion in applying theory to practice has benefited many student athletes. In the area of education, he is currently involved in sports science experiential interdisciplinary education using interactive digital media.

**Weimin Pan** got his bachelor's degree from Department of Computer Science, Central South University in 2006. He received his master's degree from State Key Laboratory of CAD&CG, Zhejiang University in 2008. His main research interests are virtual reality, information security and data encryption.

**Zhigeng Pan** received his bachelor's degree and master's degree from the Computer Science Department in 1987 and 1990 from Nanjing University respectively, and PhD Degree in 1993 from Zhejiang University. Since 1993, he has been working at the State Key Lab of CAD&CG on a number of academic and industrial projects related with distributed graphics, virtual reality and multimedia. He has published more than 70 papers in international journals, national journals and international conferences. Currently, he is the editor-in-chief of *The International Journal of Virtual Reality.*

**Veljko Potkonjak** was born in Belgrade in 1951. He received his B.Sc., M.Sc., and Ph.D. degrees from Faculty of Electrical Engineering, University of Belgrade in 1974, 1977, and 1981 respectively. He became an assistant professor at the Faculty of Electrical Engineering in 1985, associate professor in 1990, and the full professor in 1995. During his educational career, Professor Potkonjak has been teaching mechanics, robotics and biomechanics. He was also a teacher or a visiting researcher at the Faculty of Electronics, University of Nish, Technical Faculty in Cacak, National Technical University of Athens, and American University of Athens. Research interests of Prof. Veljko Potkonjak primarily concern robotics, focusing on the dynamic modeling of robotic system and the implementation of these models to design and control. He is the author/coauthor of three international research monographs (in English, some of them translated into Japanese and Chinese), two book chapters, several textbooks, 60 international journal papers, and a number conference papers. Within the field of robotics and automation he was engaged in a large number of projects (research and commercial).

**Chee Kwang Quah** is a project manager at the School of Sports, Health and Leisure at the Republic Polytechnic, Singapore. He obtained his BEng(Hons) in Electronic and Electrical Engineering at University of Leeds in England. He then obtained his MSc in Multimedia Technology and Systems from University of Surrey (England), and PhD in computer engineering from Nanyang Technological University of Singapore. His research interests are in the areas of computer vision, computer graphics and image analysis. His current focus is on capturing 3D human motion by using video cameras.

**Sean Reilly** is a PhD candidate in the Distributed Systems Group at the School of Computer Science and Statistics in Trinity College Dublin. His research interests include computer-augmented sports, embedded systems, middleware and event-based programming models for ubiquitous computing.

**David D. Rowlands** received his BSc (physics) in 1986 and PhD in 2000 from Griffith University. He is a member of Centre of Wireless Monitoring and Application at Griffith University. He is also a member of IEEE. His main area of research is in the area of sports monitoring and technology. His other current research is in the area of biological monitoring. This covers the visualization of biological data, the analysis of biological signals, and medical information systems. He has also been researching in the areas of mobile computer security. His PhD research was in the area of deep sub-micron semiconductor device simulation and fabrication.

**Donald P. Roy** (PhD, University of Memphis) is an associate professor of marketing in the Jennings A. Jones College of Business at Middle Tennessee State University in Murfreesboro, Tennessee, USA. His research and teaching focus on marketing communications and sports marketing. He has published articles in the *Journal of Advertising, Psychology & Marketing, International Journal of Sports Marketing & Sponsorship, Marketing Management Journal, Services Marketing Quarterly, Journal of Marketing* *for Higher Education,* and *Sport Marketing Quarterly.* He has also authored several cases that have been published in marketing textbooks and is a member of the Marketing Management Association.

**Hock Soon Seah** is a professor and director of the gameLAB at the School of Computer Engineering at Nanyang Technological University (NTU), Singapore. Concurrently, he is also a Co-Director of the NTU Institute for Media Innovation and a Visiting Research Fellow of the University of New South Wales, Australia. He is the founding president of the Singapore Chapter of ACM SIGGRAPH since 1998. He serves on the editorial board of *Computers & Graphics Journal, International Journal of Virtual Reality,* and *International Journal of Computer Games Technology.* He has more than 20 years experience in computer graphics and animation research. His current research areas are in geometric modeling, image sequence analysis with applications to digital film effects, automatic in-between frame generation from hand-drawn sketches, augmented reality, and advanced medical visualization.

**Lauren Silberman** is a graduate student in MIT's Comparative Media Studies Program and a researcher at the Education Arcade. She graduated with a BA in English from the University of Wisconsin-Madison, where she also spent four years as a research assistant in the Games, Learning, and Society program. Using commercial sport video games as a model, her core research investigates how sport video games mediate athletes' physical play. She has observed and interviewed numerous professional and college level athletes about their virtual game-play. Her research has been published in *The Journal of Physical Education* and she has presented her research in various forums throughout the United States and abroad. She has worked for NBC and other leading media companies as a researcher and project assistant.

**Constantino Stavros**, PhD, is a senior lecturer and program leader with the School of Economics, Finance & Marketing at RMIT University. He has research interests in strategic marketing communications and, much to the disappointment of his wife, is an avid video gamer.

**Gunnar Stevens** is a senior researcher on the Fraunhofer FIT in Germany. He has received his diploma degree in computer science at the University of Bonn. For several years he has worked in private industry as well as in academia. Currently, he leads a public funded project where he develops software systems in cooperation with several software companies. The software systems are easy to adapt and are based on the metaphor of software components. His research interests include the interdisciplinary areas of human computer interaction and computer supported cooperative work. Primarily he studies phenomena of appropriating technology and related to this, he is interested on ethnographical methods based on a design perspective. Additionally, he is co-author of papers about the methodological concept of the *Business Ethnography* as a methodology for reflexive technology development.

**Miomir Vukobratović** was born in Botos, Serbia, 26. 12. 1931. He received the BSc (1957) and PhD degrees (1964) in mechanical engineering, University of Belgrade, and his second DSc degree from the Institute Mashinovedeniya, Russian Academy of Sciences, Moscow, 1972. He is director of Robotics Center, at Mihailo Pupin Institute, Belgrade. He is a holder of "Joseph Engelberger" award in robotics (Robotic Industries Association in USA, 1996.), and holder of highest national scientific awards and recognitions. Professor Vukobratović is a member of Serbian Academy of Sciences and Arts, foreign member of Russian Academy of Sciences, president of Serbian Academy of Engineering Sciences, a foreign member of Chinese, Japanese, Hungarian and several other national engineering academies.

He is doctor honoris causa of Moscow State University (Lomonosov) and several other universities in Europe. His major interest is in the development of efficient computer aided modeling of robotic system dynamics as well as dynamic modeling, stability and control in legged locomotion, especially humanoid robotics. Main pioneering results are: zero-moment point (ZMP) concept and semi-inverse method; centralized feed-forward control and robot dynamic control; Force feedback in dynamic control and application of practical stability tests in robotics; Unified approach to control laws synthesis for robot interacting with dynamic environment; pioneering achievement in active exoskeletons and humanoid robotics. He is cited more than 2000 times according to SCOPYS citation index.

**Volker Wulf** is a professor in Information Systems and the director of the Media Research Institute at the University of Siegen. At Fraunhofer FIT, he heads the research group user-centred software-engineering (USE). He has published more than 190 papers. He has edited 10 books, among which, *Expertise Sharing: Beyond Knowledge Management* and *Social Capital and Information Technology* both with MIT Press Cambridge MA, and *End User Development* with Springer Dordrecht, are probably best known. His research interests lie primarily in the area of computer supported cooperative work, computer supported cooperative learning, human computer interaction, participatory design, ubiquitous computing, and entertainment computing. Recently he is interested in new ways to interface physical activities with virtual representations.

**James J. Zhang** is a professor of sport management in the Department of Tourism, Recreation, and Sport Management at the University of Florida (UF). His primary research interests are applied measurement and/or applied studies examining sport consumer and organizational behaviors. Adopting an integrated approach, the following perspectives have been investigated for the purpose of predicting sport consumption behaviors and formulating effective marketing strategies: (a) market demand and competition, (b) socio-motivation, (c) program and service quality, (d) lifestyle, and (e) sociodemographic backgrounds. He has also studied sport leadership with a focus on formal and informal leadership associated with athletic program management. Classical and contemporary leadership theories, such as trait, behavioral, contingency, and transformation, are often used as research frameworks. Dr. Zhang has published extensively in all related sport management journals that currently exist in North America and made presentations at all major conferences in the field.

# Index